SERIES ON SEMICONDUCTOR SCIENCE AND TECHNOLOGY

Series Editors

R. J. Nicholas   University of Oxford
H. Kamimura   University of Tokyo

SERIES ON SEMICONDUCTOR SCIENCE AND TECHNOLOGY

1. M. Jaros: *Physics and applications of semiconductor microstructures*
2. V. N. Dobrovolsky and V. G. Litovchenko: *Surface electronic transport phenomena in semiconductors*
3. M. J. Kelly: *Low-dimensional semiconductors*
4. P. K. Basu: *Theory of optical processes in semiconductors*
5. N. Balkan: *Hot electrons in semiconductors*
6. B. Gil: *Group III nitride semiconductor compounds: physics and applications*
7. M. Sugawara: *Plasma etching*
8. M. Balkanski and R. F. Wallis: *Semiconductor physics and applications*
9. B. Gil: *Low-dimensional nitride semiconductors*
10. L. Challis: *Electron-phonon interaction in low-dimensional structures*
11. V. Ustinov, A. Zhukov, A. Egorov, N. Maleev: *Quantum dot lasers*
12. H. Spieler: *Semiconductor detector systems*
13. S. Maekawa: *Concepts in spin electronics*
14. S. D. Ganichev, W. Prettl: *Intense terahertz excitation of semiconductors*
15. N. Miura: *Physics of semiconductors in high magnetic fields*
16. A. V. Kavokin, J. J. Baumberg, G. Malpuech, F. P. Laussy: *Microcavities*
17. S. Maekawa, S. O. Valenzuela, E. Saitoh, T. Kimura: *Spin current*

Spin Current

Edited by

Sadamichi Maekawa
Sergio O. Valenzuela
Eiji Saitoh
Takashi Kimura

OXFORD

UNIVERSITY PRESS

Great Clarendon Street, Oxford, OX2 6DP,
United Kingdom

Oxford University Press is a department of the University of Oxford.
If furthers the University's objective of excellence in research, scholarship,
and education by publishing worldwide. Oxford is a registered trade mark of
Oxford University Press in the UK and in certain other countries

First Edition published in 2012

Impression: 1

British Library Cataloguing in Publication Data

Data available

Library of Congress Cataloging in Publication Data

Library of Congress Control Number: 2012936272

ISBN 978–0–19–960038–0

Printed and bound by
CPI Group (UK) Ltd, Croydon, CR0 4YY

Preface

Since the discovery of the giant magnetoresistance (GMR) effect in magnetic multilayers in 1988, a new branch of physics and technology, called spin-electronics or spintronics, has emerged, where the flow of electrical charge as well as the flow of electron spin, the so-called "spin current," are manipulated and controlled together. Whereas charge current flows without decay (owing to fundamental charge conservation), spin current decays on a length-scale of less than a few micrometers. In other words, it exists only at nanometer scales. Recent progress in the physics of magnetism and the application of spin current has progressed in tandem with the nanofabrication technology of magnets and the engineering of interfaces and thin films.

This book is intended to provide an introduction and guide to the new physics and application of spin current. The emphasis is place on the interaction between spin and charge currents in magnetic nanostructures.

The International Conference on Magnetism (ICM), the largest conference in the physics of magnetism, has been held triennially since the first one organized by Louis Néel at Grenoble, France in 1958. The Eighteenth Conference in the ICM series took place in Karlsruhe, Germany in July 2009, where a paradigm in physics was epitomized by "a flood of spin current," which introduces a new front in the evolution of traditional research in magnetism.

I am glad to note that the achievements of the research in spin current by Sergio O. Valenzuela, Eiji Saitoh, and Takashi Kimura were recognized through the Young Scientist in Magnetism Awards at the Eighteenth ICM sponsored by the International Union of Pure and Applied Physics (IUPAP), an agency under the auspices of UNESCO (United Nations Educational, Scientific, and Cultural Organization).

In this book, three of them give introductions to spin current, the spin Hall effect, spin torques, and the spin Seebeck effect based on their achievements. Although the chapters make up a coherent whole, each chapter is self-contained and may be read independently. The physics based on spin current is growing rapidly. Therefore, we have tried to introduce the most recent results up to August 2011. I hope this book is a sound guide to the new physics and technology.

August 2011

Sadamichi Maekawa
(On behalf of the Editors)

Contents

Contributors   xv

Part I Spin current

1 Introduction   3
   E. Saitoh

   1.1   Spin of electrons   3
         1.1.1   Spin angular momentum   3
         1.1.2   Dirac equation and spin   5
         1.1.3   Nonrelativistic approximation   7
   1.2   Spin current   9
         1.2.1   Concept of spin current   9
         1.2.2   An exact definition of spin current   10
   References   14

2 Incoherent spin current   15
   K. Ando and E. Saitoh

   2.1   Fermi–Dirac distribution   15
   2.2   Diffusion equation   18
   2.3   Spin diffusion equation   19
   References   24

3 Exchange spin current   25
   E. Saitoh and K. Ando

   3.1   Magnetic order and exchange interaction   25
   3.2   Exchange spin current   25
         3.2.1   Landau–Lifshitz–Gilbert equation   25
         3.2.2   Rewriting the Landau–Lifshitz–Gilbert
                 equation   27
   3.3   Spin-wave spin current   28
         3.3.1   Spin-wave formulation   28
         3.3.2   Spin current carried by a spin wave   31
   References   32

4 Topological spin current   33
E. Saitoh

4.1 Bulk topological spin current   33
4.2 Surface topological spin current   35
References   35

5 Spin polarization in magnets   36
K. Takanashi and Y. Sakuraba

5.1 Spin polarization in ferromagnets   36
5.2 Half-metallic ferromagnets   37
5.3 Experimental techniques for spin-polarization
    measurement   39
    5.3.1 Point-contact Andreev reflection (PCAR) 39
    5.3.2 Superconducting tunneling spectroscopy (STS) 41
    5.3.3 Spin-resolved photoemission spectroscopy (SP-PES) 43
5.4 Magnetoresistive devices with half-metals   43
    5.4.1 Magnetic tunnel junctions with half-metals 43
    5.4.2 Current-perpendicular-to-plane magnetoresistive
    device with half-metals   45
References   46

6 Optically induced and detected spin current   49
A. Hirohata

6.1 Introduction   49
    6.1.1 Optical generation of spins   49
    6.1.2 Spin polarization in GaAs   49
    6.1.3 Photoexcitation model   50
6.2 Optical spin injection   55
    6.2.1 Photoexcitation   55
    6.2.2 Schottky diodes   55
    6.2.3 Spin-polarized scanning tunneling microscopy
    (spin STM)   57
6.3 Optical spin detection   58
    6.3.1 Spin-polarized light-emitting diodes (spin LED) 58
    6.3.2 Schottky diodes   60
    6.3.3 Spin injection into Si   62
References   62

7 Theory of spinmotive forces in ferromagnetic structures   65
S. E. Barnes

7.1 Introduction   65
7.2 Spin Faraday's law   65

7.2.1   Magnetic momentum, spin electric, and magnetic
        fields   66
7.2.2   Calculation of magnetic momentum for adiabatic
        electrons   66
7.2.3   The spin fields $\vec{E}_s$ and $\vec{B}_s$ and Faraday's law 68
7.2.4   Landau–Lifshitz equations   69
7.2.5   Spin-torque-transfer (STT), spin valves, and magnetic
        tunnel junctions (MTJs)   69
7.2.6   Spinmotive force (smf), spin valves and MTJs   71
7.2.7   The spin electric field $\vec{E}_s$ 73
7.3   Examples of spinmotive forces   74
7.3.1   A plain Néel domain wall   74
7.3.2   Reciprocal relations   75
7.3.3   The spinmotive force and magnons   76
7.3.4   Spin forces and Doppler shifts for phonons   77
7.3.5   Voltage steps and magnetoresistance   77
7.3.6   Static magnetic vortices   78
7.3.7   Field-driven magnetic vortices   78
7.4   Ferromagnetic resonance (FMR) and spinmotive forces   80
7.4.1   Field-driven FMR   80
7.4.2   STT-FMR and a-FMR resonance   80
7.4.3   The spin Seebeck effect and the smf   80
7.4.4   FMR with thermal gradients   81
7.5   Spinmotive forces, magnons, and phonons   82
7.5.1   The $s = 1$ nature of magnons and phonons   82
7.5.2   Realization of smf effects with magnons and
        phonons   83
References   84

8   Spin pumping and spin transfer   87

**A. Brataas, Y. Tserkovnyak, G. E. W. Bauer,
and P. J. Kelly**

8.1   Introduction   87
8.1.1   Technology pull and physics push   87
8.1.2   Discrete versus homogeneous   87
8.1.3   This chapter   88
8.2   Phenomenology   89
8.2.1   Mechanics   89
8.2.2   Spin-transfer torque and spin pumping   89
8.2.3   Onsager reciprocity relations   99
8.3   Microscopic derivations   103
8.3.1   Spin-transfer torque   103
8.3.2   Spin pumping   109

8.4 First-principles calculations 114
    8.4.1 Alpha 115
    8.4.2 Beta 120
8.5 Theory versus experiments 123
8.6 Conclusions 124
References 125

9 Spin caloritronics 136
G. E. W. Bauer

9.1 Introduction 136
9.2 Basic physics 137
9.3 Spin-dependent thermoelectric phenomena in metallic structures 138
    9.3.1 Magneto-Peltier and Seebeck effects 138
    9.3.2 Thermal Hall effects 139
9.4 Thermal spin-transfer torques 140
    9.4.1 Spin valves 140
    9.4.2 Magnetic tunnel junctions 140
    9.4.3 Textures 140
9.5 Magneto-heat resistance 141
9.6 Spin caloritronic heat engines and motors 143
9.7 Spin Seebeck effect 144
9.8 Conclusions 145
References 145

10 Multiferroics 149
N. Nagaosa

10.1 Introduction 149
10.2 Multiferroics—a generic consideration 152
10.3 Spin-current model of ferroelectricity 154
10.4 Spin Hamiltonian for $R\text{MnO}_3$ 157
10.5 Electromagnons in multiferroics 161
10.6 Ultrafast switching of spin chirality by optical excitation 165
10.7 Quasi-one-dimensional quantum multiferroics 167
10.8 Summary and conclusions 168
References 170

Part II Spin Hall effect

11 Introduction 177
S. O. Valenzuela

11.1 Historical background 177
11.2 Spin–orbit interaction 181

11.3 The family of spin Hall effects   184
11.4 Experimental observation   185
References   190

12 Spin Hall effect   194

S. Maekawa and S. Takahashi

12.1 Introduction   194
12.2 Spin Hall effect due to side jump and skew scattering
      in diffusive metals   195
12.3 Spin and charge currents   199
12.4 Spin–orbit coupling   200
12.5 Nonlocal spin Hall effect   201
12.6 Anomalous Hall effect (AHE) in a ferromagnet   203
12.7 Summary   204
References   205

13 Spin generation and manipulation based on
        spin–orbit interaction in semiconductors   209

J. Nitta

13.1 Origin of spin–orbit interaction (SOI) in semiconductors   209
13.2 Gate controlled Rashba SOI   212
13.3 Spin relaxation and its suppression   214
13.4 Spin Hall effect based on Rashba and Dresselhaus spin–orbit
      interaction   216
13.5 Aharonov–Casher spin interference; theory   217
13.6 Aharonov–Casher spin interference; experiment   220
References   224

14 Experimental observation of the spin Hall effect
        using electronic nonlocal detection   227

S. O. Valenzuela and T. Kimura

14.1 Observation of the spin Hall effect   227
14.2 Nonlocal spin injection and detection   227
14.3 The electronic spin Hall experiments   230
References   239

15 Experimental observation of the spin Hall effect
        using spin dynamics   244

E. Saitoh and K. Ando

15.1 Inverse spin Hall effect induced by spin pumping   244
15.2 Spin-Hall-effect induced modulation of magnetization
      dynamics   248
References   250

16 Spin-injection Hall effect   252

**J. Wunderlich, L. P. Zârbo, J. Sinova,
and T. Jungwirth**

16.1   Spin-dependent Hall effects   252
16.2   The spin-injection Hall effect experiment   253
16.3   Theory discussion   258
       16.3.1   Nonequilibrium polarization dynamics along
                the $[1\bar{1}0]$ channel   259
       16.3.2   Hall effect   261
       16.3.3   Spin diffusion and spin precession in narrow
                2DEG bars   262
16.4   Spin Hall effect transistor   265
16.5   Prospectives of spin-injection Hall effect   269
References   269

**17 Quantum spin Hall effect and topological
    insulators**   272

S. Murakami and T. Yokoyama

17.1   Quantum spin Hall systems   272
       17.1.1   Introduction   272
       17.1.2   Topology and topological insulators   275
       17.1.3   Topological numbers   277
17.2   Two-dimensional (2D) topological insulators   278
       17.2.1   Edge states of 2D topological insulators   278
       17.2.2   Experiments on edge states of 2D topological
                insulators   280
17.3   Three-dimensional (3D) topological insulators   281
       17.3.1   Surface states of three-dimensional topological
                insulators   281
       17.3.2   Properties of surface states of 3D topological
                insulators   281
       17.3.3   Materials for 3D topological insulators   284
       17.3.4   3D topological insulators and Majorana fermions   288
17.4   Summary   290
References   291

18 Spin Seebeck effect   296

E. Saitoh and K. Uchida

18.1   Introduction   296
18.2   Sample configuration and measurement mechanism   297
18.3   Detection of spin Seebeck effect using inverse spin
       Hall effect   299
18.4   Theoretical concept of spin Seebeck effect   301

18.5 Thermal spin injection by spin-dependent Seebeck effect 303
18.6 Conclusion 305
References 305

Part III Spin-transfer torque

19 Introduction
T. Kimura

19.1 Theoretical description of spin-transfer torque 312
19.2 Perpendicular spin torque 315
19.3 Diffusive picture for injecting spin current 317
19.4 Experimental study of magnetization reversal due to spin torque 319
19.5 Magnetization dynamics due to spin-current injection 322
19.6 Domain wall displacement due to spin-current injection 327
19.7 Theoretical description of the spin-current-induced domain wall displacement 329
19.8 Dynamics of magnetic domain wall under spin-current injection 330
19.9 Vortex motion due to spin-current injection 332
19.10 Other new phenomena 334
References 336

20 Spin-transfer torque in uniform magnetization
Y. Suzuki

20.1 Torque and torquance in magnetic junctions 343
20.2 Voltage dependence and field-like torque 347
20.3 Landau–Lifshitz–Gilbert (LLG) equation in Hamiltonian form 349
20.4 Small-amplitude dynamics and anti-damping 352
     20.4.1 Linearized LLG equation 352
     20.4.2 Spin-torque diode effect 353
20.5 Spin-transfer magnetization switching 357
20.6 Large-amplitude dynamics and auto-oscillation 362
References 367

21 Magnetization switching due to nonlocal spin injection
T. Kimura and Y. Otani

21.1 Generation and absorption of pure spin current 372
21.2 Efficient absorption of pure spin current 375
21.3 Efficient injection of spin current 377
21.4 Magnetization switching due to injection of pure spin current 378
References 381

22 Magnetic domains and magnetic vortices   382

Y. Otani and R. Antos

22.1 Micromagnetic equations   383
22.2 Analytical approaches   386
22.3 Experimental techniques   388
22.4 Steady state motion phenomena   389
22.5 Dynamic switching   391
22.6 Magnetostatically coupled vortices   393
22.7 Conclusions and perspectives   395
References   396

23 Spin-transfer torque in nonuniform magnetic structures   402

T. Ono

23.1 Magnetic domain wall   402
    23.1.1 Magnetic vortex   402
23.2 Current-driven domain wall motion   405
    23.2.1 Basic idea of current-driven domain wall motion   405
    23.2.2 Direct observation of current-driven domain wall
        motion by magnetic force microscopy   407
    23.2.3 Beyond the adiabatic approximation: Non-adiabatic
        torque   408
    23.2.4 Domain wall motion by adiabatic torque and intrinsic
        pinning   411
    23.2.5 Toward applications of current-driven domain wall
        motion   412
23.3 Current-driven excitation of magnetic vortices   413
    23.3.1 Current-driven resonant excitation of magnetic
        vortices   413
    23.3.2 Switching a vortex core by electric current   417
References   420

24 Spin torques due to large Rashba fields   424

S. Zhang

24.1 Introduction   424
24.2 Rashba coupling at metallic interfaces   425
24.3 Current-driven spin torque with Rashba coupling   427
24.4 Manipulating magnetization by the current   429
24.5 Discussions and conclusions   430
References   430

Index   433

Contributors

K. Ando
Institute for Materials Research
Tohoku University
2-1-1 Katahira
Aoba-ku
Sendai 980-8577
Japan
ando@imr.tohoku.ac.jp

R. Antos
Institute of Physics
Faculty of Mathematics and Physics
Charles University in Prague
Ke Karlovu 5
CZ-12116 Prague
Czech Republic
antos@karlov.mff.cuni.cz

S. E. Barnes
Physics Department
University of Miami
Coral Gables, Florida 33146
USA
barnes@physics.miami.edu

G. E. W. Bauer
Institute for Materials Research
Tohoku University
2-1-1 Katahira, Aoba-ku
Sendai 980-8577
Japan
g.e.w.bauer@imr.tohoku.ac.jp

A. Brataas
Department of Physics
Norwegian University of
Science and Technology
7491 Trondheim
Norway
Arne.Brataas@ntnu.no

A. Hirohata
Department of Electronics
The University of York
Heslington, York, North Yorkshire
YO10 5DD
UK
atsufumi.hirohata@york.ac.uk

T. Jungwirth
Department of Physics
Texas A&M University
College Station, TX 77843–4242
USA
jungw@fzu.cz

P. J. Kelly
Faculty of Science and Technology
and MESA+ Institute for
Nanotechnology
University of Twente
The Netherlands
P.J.Kelly@utwente.nl

T. Kimura
Advanced Electronics Research
Division, INAMORI Frontier
Research Center, Kyushu University
Motooka 744, Nishi-ku
Fukuoka 819-0395
Japan
kimura@ifrc.kyushu-u.ac.jp

S. Maekawa
Advanced Science Research Center
Japan Atomic Energy Agency
Tokai, Ibaraki, 319-1195

Japan
maekawa.sadamichi@jaea.go.jp

S. Murakami
Department of Physics
Tokyo Institute of Technology
2-12-1 H44 Ookayama, Meguro-ku
Tokyo 152-8551
Japan
murakami@stat.phys.titech.ac.jp

N. Nagaosa
University of Tokyo
7-3-1-6-212 Hongo, Bunkyo-ku
Tokyo 113-8656
Japan
nagaosa@ap.t.u-tokyo.ac.jp

J. Nitta
Department of Materials Science
Tohoku University
6-6-02 Aramaki-Aza Aoba
Aoba-ku, Sendai 980-8579
Japan
nitta@material.tohoku.ac.jp

T. Ono
Institute for Chemical Research
Kyoto University
Gokasho, Uji, Kyoto 611-0011
Japan
ono@scl.kyoto-u.ac.jp

Y. Otani
Institute for Solid State Physics
University of Tokyo
5-1-5 Kashiwanoha
Kashiwa 277-8581 Japan
yotani@issp.u-tokyo.ac.jp

E. Saitoh
Institute for Materials Research
Tohoku University
2-1-1 Katahira, Aoba-ku
Sendai 980-8577
Japan
eizi@imr.tohoku.ac.jp

Y. Sakuraba
Institute for Materials Research
Tohoku University
2-1-1 Katahira, Aoba-ku
Sendai 980-8577
Japan
y.sakuraba@imr.tohoku.ac.jp

J. Sinova
Department of Physics
Texas A&M University
College Station
TX 77843–4242
USA
sinova@physics.tamu.edu

Y. Suzuki
Graduate School of
Engineering Science
Osaka University
1-3 Machikaneyamacho
Toyonaka, Osaka, 560-8531
Japan
suzuki-y@mp.es.osaka-u.ac.jp

S. Takahashi
Institute for Materials Research
Tohoku University
2-1-1 Katahira, Aoba-ku
Sendai 980-8577
Japan
takahasi@imr.tohoku.ac.jp

K. Takanashi
Institute for Materials Research
Tohoku University
2-1-1 Katahira, Aoba-ku
Sendai 980-8577
Japan
koki@imr.tohoku.ac.jp

Y. Tserkovnyak
Department of Physics and
Astronomy
University of California
Los Angeles

California 90095
USA
yaroslav@physics.ucla.edu

K. Uchida
Institute for Materials Research
Tohoku University
2-1-1 Katahira, Aoba-ku
Sendai 980-8577
Japan
kuchida@imr.tohoku.ac.jp

S. O. Valenzuela
Institució Catalana de Recerca
i Estudis Avançats (ICREA)
Catalan Institute of Nanotechnology
(ICN) and Universitat Autònoma
de Barcelona, Campus UAB
Bellaterra, Barcelona 08193
Spain
SOV@icrea.cat

J. Wunderlich
Hitachi Cambridge Laboratory
J.J. Thomson Avenue

Cambridge CB3 OHE
UK
jw526@cam.ac.uk

T. Yokoyama
Department of Physics
Tokyo Institute of Technology
2-12-1 Ookayama
Meguro-ku
Tokyo 152-8551 Japan
yokoyama@stat.phys.titech.ac.jp

L. P. Zârbo
Institute of Physics ASCR, v.v.i.
Cukrovarnická 10
162 53 Praha 6
Czech Republic
zarbo@fzu.cz

S. Zhang
University of Missouri Columbia
Columbia
Missouri 65211
USA
zhangs@physics.arizona.edu

Part I   Spin current

1 Introduction

E. Saitoh

An electron is an elementary particle that carries negative electric charge and governs various properties of condensed matter. Besides charge, an electron has internal angular momentum. This internal angular momentum, similar to the rotation of a classical particle, is named spin. Spin is the dominant origin of magnetism, thus, when spins of electrons in a solid are aligned to some extent in the same direction, the solid becomes a magnet.

In condensed matter, there are some types of flow carried by electrons. A flow of electron charge is a charge current, or an electric current. The physics of charge current has been developed in the previous century and is now an essential contributor to our understanding of electronics. Since an electron carries both charge and spin, the existence of a charge current naturally implies the existence of a flow of spin. This flow is called a spin current.

Experiments carried out in the previous century did not focus on spin current because of its relatively short decay time-scale τ. However, the rapid progress in nanofabrication technology in this century has allowed researchers to access spin currents. From the theoretical point of view, the detailed formulation of spin currents is not simple and is still a challenging undertaking. Nevertheless, spin current is a very useful and versatile concept; it has given birth to many phenomena in condensed matter science and spintronics.

In this chapter, we introduce the concept of spin current. We begin with an introduction to the general concept of spin and spin current, which is followed by a discussion of particular spin currents.

1.1 Spin of electrons

1.1.1 *Spin angular momentum*

An electron has a spin angular momentum besides orbital angular momenta. This concept was first introduced by Uhlenbeck and Goudsmit for the interpretation of atomic spectra. Later, Dirac provided a theoretical foundation for spin in terms of relativistic quantum mechanics.

The spin angular momentum is expressed by a vector operator $\boldsymbol{S}$. Since $\boldsymbol{S}$ represents angular momentum, it satisfies the commutation relation

$$[S_i, S_j] = i\hbar\varepsilon_{ijk}S_k. \tag{1.1}$$

The z-component of spin has only two values in the spin space: $1/2$ in units of $\hbar$. The spin variable is written as s, and the z-component of spin is written as s. An eigenfunction of spin for the state $s = 1/2$ is written as

$$\chi_{\frac{1}{2}}(s) = \begin{pmatrix} 1 \\ 0 \end{pmatrix}. \tag{1.2}$$

The right-hand side is a two-component vector with bases $s = 1/2$ (up spin) and $-1/2$ (down spin) describing the state. Similarly, the state for $s = -1/2$ is written as

$$\chi_{-\frac{1}{2}}(s) = \begin{pmatrix} 0 \\ 1 \end{pmatrix}. \tag{1.3}$$

The spin angular momentum is expressed by a matrix $\boldsymbol{\sigma}$ with $\chi_{\frac{1}{2}\frac{1}{2}}(s)$ and $\chi_{\frac{1}{2}-\frac{1}{2}}(s)$ as bases:

$$\boldsymbol{S} = \frac{\hbar}{2}\boldsymbol{\sigma}. \tag{1.4}$$

Here, $\boldsymbol{\sigma}$ are the Pauli spin matrices

$$\sigma_x = \begin{pmatrix} 0 & 1 \\ 1 & 0 \end{pmatrix}, \quad \sigma_y = \begin{pmatrix} 0 & -i \\ i & 0 \end{pmatrix}, \quad \sigma_z = \begin{pmatrix} 1 & 0 \\ 0 & -1 \end{pmatrix}. \tag{1.5}$$

Equations (1.2) and (1.3) form a complete basis set of spin wavefunctions. The orthonormal relationship

$$\sum_{s=\pm\frac{1}{2}} = \chi_s^*(s)\chi_{s'}(s) = \delta_{ss'} \tag{1.6}$$

holds for the wavefunctions corresponding to different eigenvalues of S_z. A state of a particle is defined with a spatial wavefunction describing the probability amplitude at points in space and a spin wavefunction giving the direction of the spin.

A charge under rotational motion has a magnetic moment. A particle with charge $-e$ and mass m moving with orbital angular momentum $\boldsymbol{L}$ has a magnetic moment

$$\boldsymbol{m}_{\text{orb}} = \frac{-e}{2mc}\boldsymbol{L}. \tag{1.7}$$

Similarly, spin in quantum mechanics also has a corresponding magnetic moment. The magnetic moment of spin is given by

$$\boldsymbol{m}_{\text{spin}} = -g_0 \frac{e}{2mc}\boldsymbol{S}, \tag{1.8}$$

where g_0 is the g-factor, which is about 2 for electrons, as discussed in the next subsections.

1.1.2 Dirac equation and spin

Dirac showed that electron spin is, in fact, naturally derived from quantum mechanics combined with special relativity [1].

Here, for simplicity, a free electron is examined. Remember that the Schrödinger equation for a nonrelativistic free electron is obtained from the nonrelativistic energy dispersion relation

$$\epsilon = \frac{p^2}{2m} \qquad (1.9)$$

by substituting $p = -i\hbar\nabla$ and $\epsilon = i\hbar(d/dt)$. p is the momentum of the particle.
According to special relativity, the energy dispersion relation becomes

$$\epsilon^2 = (cp)^2 + (mc^2)^2, \qquad (1.10)$$

where m and c are the electron's rest mass and the speed of light, respectively. By substituting $p = -i\hbar\nabla$ and $\epsilon = i\hbar(d/dt)$, we obtain

$$\left[\nabla^2 - \frac{1}{c^2}\frac{\partial^2}{\partial t^2} - \left(\frac{mc}{\hbar}\right)^2\right]\psi = 0. \qquad (1.11)$$

This equation is called the Klein–Gordon equation. However, this equation cannot be directly applicable to electrons since it contains a second-order time differential, which allows us to choose two initial condition parameters for ψ and $d\psi/dt$, respectively. This situation is inconsistent with the probability interpretation of the wavefunction $\psi(\boldsymbol{r})$. In this interpretation, $\psi(\boldsymbol{r})\psi^*(\boldsymbol{r})$ is equal to the chance of finding an electron at the position $\boldsymbol{r}$ and $\int \psi(\boldsymbol{r})\psi^*(\boldsymbol{r})d\boldsymbol{r}$ should be a constant with t, namely,

$$\frac{d}{dt}\int \psi(\boldsymbol{r})\psi^*(\boldsymbol{r})d\boldsymbol{r} \qquad (1.12)$$

should be zero. This means that ψ and $d\psi/dt$ cannot be chosen independently.
Let us try to find a first-order time-differential equation by factorizing the relativistic energy dispersion

$$\epsilon^2 = (cp)^2 + (mc^2)^2.$$

This dispersion relation is factorized to obtain

$$(\epsilon + c\boldsymbol{\alpha}\cdot\boldsymbol{p} + \beta mc^2)(\epsilon - c\boldsymbol{\alpha}\cdot\boldsymbol{p} - \beta mc^2) = 0. \qquad (1.13)$$

Here, the coefficients α and β should satisfy the following relations

$$\alpha_i^2 = \beta^2 = 1 \ (i = x, y, z),$$
$$\alpha_i \alpha_j + \alpha_j \alpha_i = 0 \ (i \neq j), \tag{1.14}$$
$$\alpha_i \beta + \beta \alpha_i = 0.$$

Dirac showed that these relations are not satisfied if α and β are simple scalars, and at least α and β must be 4×4 matrices. The following is one possible combination, called the Dirac representation, of α and β that satisfies Eq. (1.14):

$$\alpha_x = \begin{pmatrix} 0 & 0 & 0 & 1 \\ 0 & 0 & 1 & 0 \\ 0 & 1 & 0 & 0 \\ 1 & 0 & 0 & 0 \end{pmatrix}, \quad \alpha_y = \begin{pmatrix} 0 & 0 & 0 & -i \\ 0 & 0 & i & 0 \\ 0 & -i & 0 & 0 \\ i & 0 & 0 & 0 \end{pmatrix}, \quad \alpha_z = \begin{pmatrix} 0 & 0 & 1 & 0 \\ 0 & 0 & 0 & -1 \\ 1 & 0 & 0 & 0 \\ 0 & -1 & 0 & 0 \end{pmatrix},$$

$$\beta = \begin{pmatrix} 1 & 0 & 0 & 0 \\ 0 & 1 & 0 & 0 \\ 0 & 0 & -1 & 0 \\ 0 & 0 & 0 & -1 \end{pmatrix}. \tag{1.15}$$

Equation (1.1.2) is always satisfied when

$$(\epsilon - c\boldsymbol{\alpha} \cdot \boldsymbol{p} - \beta mc^2)\psi = 0. \tag{1.16}$$

By substituting $p = -i\hbar\nabla$ and $\epsilon = i\hbar(d/dt)$, we obtain

$$i\hbar\frac{\partial\psi}{\partial t} + i\hbar c\boldsymbol{\alpha} \cdot \nabla\psi - \beta mc^2\psi = 0. \tag{1.17}$$

This is a relativistic expansion of the Schrödinger equation for a free electron, called the Dirac equation. Since α and β are 4×4 matrices, a solution ψ has four components.

Recall that the system has rotational symmetry and, therefore, the Dirac Hamiltonian

$$\mathcal{H} = c\boldsymbol{\alpha} \cdot \boldsymbol{p} + \beta mc^2 \tag{1.18}$$

should conserve angular momentum in general. However we can see that the orbital angular momentum of electrons $\boldsymbol{L} = \boldsymbol{r} \times \boldsymbol{p}$ is not a constant of motion. In other words, $\boldsymbol{L}$ does not commute with the Hamiltonian in Eq. (1.18). As an example, for $L_x = yp_z - zp_y$,

$$[L_x, \mathcal{H}] = c\,[(yp_z - zp_y), (\boldsymbol{\alpha} \cdot \boldsymbol{p} + \beta mc)]$$

$$= c \sum_j \alpha_j \,[(yp_z - zp_y), p_j]$$

$$= -\,i\hbar c\,(\alpha_z p_y - \alpha_y p_z), \qquad (1.19)$$

which means that L_x is not conserved. Since the total angular momentum must still be conserved, we hypothesize that electrons have an internal spin angular momentum $\boldsymbol{S}$ in addition to the orbital angular momentum so that the sum of these angular momenta $\boldsymbol{J} = \boldsymbol{L} + \boldsymbol{S}$ commutes with the Hamiltonian:

$$[\boldsymbol{J}, \mathcal{H}] = [\boldsymbol{L}, \mathcal{H}] + [\boldsymbol{S}, \mathcal{H}] = 0. \qquad (1.20)$$

In other words, the commutation relation

$$[\boldsymbol{S}, \mathcal{H}] = -\,[\boldsymbol{L}, \mathcal{H}] \qquad (1.21)$$

is assumed. One can see that $\boldsymbol{S} = (\hbar/2)\tilde{\boldsymbol{\sigma}}$, where the components of $\tilde{\boldsymbol{\sigma}}$ are

$$\tilde{\sigma}_x = \begin{pmatrix} 0 & 1 & 0 & 0 \\ 1 & 0 & 0 & 0 \\ 0 & 0 & 0 & 1 \\ 0 & 0 & 1 & 0 \end{pmatrix}, \quad \tilde{\sigma}_y = \begin{pmatrix} 0 & -i & 0 & 0 \\ i & 0 & 0 & 0 \\ 0 & 0 & 0 & -i \\ 0 & 0 & i & 0 \end{pmatrix}, \quad \tilde{\sigma}_z = \begin{pmatrix} 1 & 0 & 0 & 0 \\ 0 & -1 & 0 & 0 \\ 0 & 0 & 1 & 0 \\ 0 & 0 & 0 & -1 \end{pmatrix},$$

$$(1.22)$$

satisfies (1.21), which means that $\boldsymbol{S}$ can be regarded as a spin operatior in the Dirac equation.

Next, consider an electron with charge $-e$ placed in an electromagnetic field. The operators in Eq. (1.18) are changed by replacing $\boldsymbol{p}$ with $\boldsymbol{p} + e\boldsymbol{A}$ and adding the electrostatic potential $-e\phi$, resulting in

$$\left[i\hbar\frac{\partial}{\partial t} + e\phi - c\boldsymbol{\alpha} \cdot (\boldsymbol{p} + e\boldsymbol{A}) - \beta mc^2 \right] \psi = 0. \qquad (1.23)$$

ϕ represents the electric potential.

1.1.3   Nonrelativistic approximation

Multiplying Eq. (1.23) by the operator of $\{ i\hbar\frac{\partial}{\partial t} + e\phi + c\boldsymbol{\alpha} \cdot (\boldsymbol{p} + e\boldsymbol{A}) + \beta mc^2 \}$ from the left, and using (1.14) and

$$\alpha_x \alpha_y = i\tilde{\sigma}_z, \quad \alpha_y \alpha_z = i\tilde{\sigma}_x, \quad \alpha_z \alpha_x = i\tilde{\sigma}_y \qquad (1.24)$$

give

$$\left[\left(i\hbar\frac{\partial}{\partial t} + e\phi \right)^2 - c^2(\boldsymbol{p} + e\boldsymbol{A})^2 - m^2 c^4 + ic\hbar\boldsymbol{\alpha} \cdot \boldsymbol{E} + c^2 e\hbar\tilde{\boldsymbol{\sigma}} \cdot \boldsymbol{B} \right] \psi = 0. \quad (1.25)$$

Actually, the four components of ψ represent the four degrees of freedom of an electron with up spin, one with down spin, an antiparticle of the electron with up spin, and one with down spin. To extract the pure electron degree of freedom, we take a nonrelativistic approximation. We assume a solution for (1.25)

$$\psi(\boldsymbol{r}, t) = \begin{pmatrix} \psi_1(\boldsymbol{r}) \\ \psi_2(\boldsymbol{r}) \\ \psi_3(\boldsymbol{r}) \\ \psi_4(\boldsymbol{r}) \end{pmatrix} \exp\left(-i\frac{\epsilon}{\hbar}t\right), \tag{1.26}$$

where

$$\epsilon = mc^2 + \epsilon', \tag{1.27}$$

where ϵ' corresponds to the nonrelativistic energy. Substituting (1.26), (1.27), and (1.22) into (1.25) gives

$$\left[(\epsilon' + e\phi)^2 + 2mc^2(\epsilon' + e\phi) - c^2(\boldsymbol{p} + e\boldsymbol{A})^2 - c^2 e\hbar\boldsymbol{\sigma} \cdot \boldsymbol{B}\right]$$

$$\begin{pmatrix} \psi_1 \\ \psi_2 \end{pmatrix} + ic\hbar e\boldsymbol{\sigma} \cdot \boldsymbol{E} \begin{pmatrix} \psi_3 \\ \psi_4 \end{pmatrix} = 0. \tag{1.28}$$

Deviding (1.28) by $2mc^2$ and neglecting terms containing $1/c^2$ or $1/c$ gives

$$\left[\frac{1}{2m}(\boldsymbol{p} + e\boldsymbol{A})^2 + \frac{e\hbar}{2m}\boldsymbol{\sigma} \cdot \boldsymbol{B} - e\phi\right] \begin{pmatrix} \psi_1 \\ \psi_2 \end{pmatrix} = \epsilon' \begin{pmatrix} \psi_1 \\ \psi_2 \end{pmatrix}, \tag{1.29}$$

where $\boldsymbol{\sigma}$ consists of the Pauli matrices

$$\sigma_x = \begin{pmatrix} 0 & 1 \\ 1 & 0 \end{pmatrix}, \quad \sigma_y = \begin{pmatrix} 0 & -i \\ i & 0 \end{pmatrix}, \quad \sigma_z = \begin{pmatrix} 1 & 0 \\ 0 & -1 \end{pmatrix}. \tag{1.30}$$

This equation is a nonrelativistic Schrödinger equation in which the interaction between the spin degree of freedom $\boldsymbol{\sigma}$ and the magnetic field $\boldsymbol{B}$ is embedded naturally. The terms in the bracket in the left-hand side constitute Schrödinger Hamiltonian. We introduce the electron spin operator $\boldsymbol{s} = 1/2\boldsymbol{\sigma}$ and $\mu_{\mathrm{B}} = \frac{e\hbar}{2mc}$. Then, the second term in the bracket in the left hand side of (1.29) becomes

$$\frac{e\hbar}{2m}\boldsymbol{\sigma} \cdot \boldsymbol{B} = 2\mu_{\mathrm{B}}\boldsymbol{s} \cdot \boldsymbol{B}. \tag{1.31}$$

This term represents the Zeeman interaction between the spin $\boldsymbol{s}$ and the magnetic field $\boldsymbol{B}$. In this way, the existence of spin and the Zeeman interaction of an electron is naturally derived from relativistic quantum mechanics.

1.2 Spin current

1.2.1 *Concept of spin current*

In this chapter, we will introduce the concept of spin current. Before dealing with spin current, we give a quick review of the definition of charge current for comparison.

Charge current is defined in terms of the charge conservation law. Consider a charge Q within a region enclosed by a closed surface Ω. When the total charge Q within this region is increasing, the increase is due to the charge flowing into this region across the surface Ω, owing to the charge conservation law. This flow of charge is described by the equation

$$\iiint_V \dot{\rho}\,d\boldsymbol{r} = -\iint_\Omega \boldsymbol{j}_c \cdot d\boldsymbol{\Omega}, \tag{1.32}$$

where ρ is the charge density and $\boldsymbol{j}_c$ is the charge current density. The left-hand side of this equation is the increase in charge in the volume surrounded by the surface Ω. By applying Gauss's theorem to this equation, we obtain

$$\dot{\rho} = -\mathrm{div}\boldsymbol{j}_c. \tag{1.33}$$

This equation, called the continuity equation of charge, which is a representation of the charge conservation law, defines a charge current density [2].

We are now in a position to consider spin current. The spin current density $\boldsymbol{j}_s$ is introduced similarly in terms of spin angular momentum conservation. If spin angular momentum is fully conserved, the continuity equation for spin angular momentum can be written as

$$\frac{d\boldsymbol{M}}{dt} = -\mathrm{div}\boldsymbol{j}_s, \tag{1.34}$$

and the spin current density is defined via this equation. $\boldsymbol{M}$ denotes the local magnetization (magnetic-moment density). Since a spin current has two orientations—the spatial flowing orientation and the spin orientation, the expectation value of the spin current density is not a vector but a second-rank tensor.

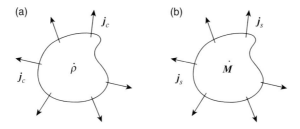

FIG. 1.1. The sum of the charge variations over the volume surrounded by a surface is equal to the sum of the currents penetrating the surface.

In practice, in nanoscales in solids, this angular momentum conservation is often a good approximation. However, in general, spin angular momentum is not conserved completely, due to spin relaxation, and it thus obeys the following modified equation of continuity:

$$\frac{d\boldsymbol{M}}{dt} = -\mathrm{div}\boldsymbol{j}_s + \boldsymbol{T}. \tag{1.35}$$

The last term $\boldsymbol{T}$ represents the nonconservation of spin angular momentum, namely, the relaxation and generation of spin angular momentum. As shown later, in fact, $\boldsymbol{T}$ can be calculated if Lagrangian of the system is fully given. However, if not, this term should be treated phenomenologically. The simplest phenomenological model for the relaxation is

$$\boldsymbol{T} = -(\boldsymbol{M} - \boldsymbol{M}_0)/\tau, \tag{1.36}$$

which is called single pole relaxation. τ is a decay time constant and $(\boldsymbol{M} - \boldsymbol{M}_0)$ is the nonequilibrium magnetization measured from its equilibrium value $\boldsymbol{M}_0$.

1.2.2 An exact definition of spin current

In some case, when a proper Lagrangian is given concretely, a spin current can be exactly defined in terms of the conservation law. In this section, we show an example of such a definition. We start with reviewing a quantum mechanical definition of charge current.[1]

Microscopic description of conduction electrons   For free electrons, the Hamiltonian described using field operators becomes

$$\mathcal{H} = \int \sum_\sigma \left[\frac{\hbar^2}{2m} |\nabla c_\sigma\left(\boldsymbol{r}\right)|^2 - \mu c_\sigma^\dagger\left(\boldsymbol{r}\right) c_\sigma\left(\boldsymbol{r}\right) \right] d\boldsymbol{r}, \tag{1.37}$$

using the electron mass m. Electrons are fermions, therefore c and $c^\dagger$ anticommute. σ are indexes to show the two spin states. μ is the chemical potential. The Fourier transformation of this Hamiltonian yields

$$\mathcal{H} = \sum_{\boldsymbol{k},\sigma} \left(\frac{\hbar^2 k^2}{2m} - \mu \right) c_{\boldsymbol{k},\sigma}^\dagger\left(\boldsymbol{r}\right) c_{\boldsymbol{k},\sigma}\left(\boldsymbol{r}\right). \tag{1.38}$$

For the following discussion, the Lagrangian formalism is more convenient. The Lagrangian of the system of electrons defined with the operators $c^\dagger$ and c can be written using the Hamiltonian $\mathcal{H}$ as

$$L = i\hbar \int \left[\sum_\sigma c_\sigma^\dagger \frac{\partial c_\sigma}{\partial t} \right] d\boldsymbol{r} - \mathcal{H}. \tag{1.39}$$

---

[1] This section draws heavily from *Basics of Spintronics* by G. Tatara (Baifukan, 2009).

Conservation of charge First, we show that the charge conservation law is related to the rotational symmetry in the phase factor of the wavefunction: the U(1) gauge symmetry. Important physical quantities in electron systems are the charge current density $\boldsymbol{j}_c$ and the charge density ρ. As shown above, these quantities satisfy the charge conservation rule, or the continuity equation of charge. Let us confirm this using the U(1) symmetry. For the Lagrangian of free electrons

$$L = \int \left[i\hbar c^\dagger \frac{\partial c}{\partial t} - \frac{\hbar^2}{2m} \nabla c^\dagger \nabla c \right] d\boldsymbol{r}, \tag{1.40}$$

we consider the phase transformation of the electron field

$$
\begin{aligned}
c\,(\boldsymbol{r},\,t) &\to e^{i\varphi(\boldsymbol{r},\,t)} c\,(\boldsymbol{r},\,t), \\
c^\dagger\,(\boldsymbol{r},\,t) &\to e^{-i\varphi(\boldsymbol{r},\,t)} c^\dagger\,(\boldsymbol{r},\,t).
\end{aligned}
\tag{1.41}
$$

The scalar quantity $\varphi(\boldsymbol{r},t)$ is the phase which is dependent on spacetime. The derivative of the electron field is converted into

$$\frac{\partial c}{\partial x_\mu} \to e^{i\varphi} \left(\frac{\partial}{\partial x_\mu} + i \frac{\partial \varphi}{\partial x_\mu} \right) \tag{1.42}$$

by the phase transformation. In turn, the Lagrangian is converted into

$$L = \int \left[i\hbar c^\dagger \left(\frac{\partial}{\partial t} + i \frac{\partial}{\partial t}\varphi \right) c - \frac{\hbar^2}{2m} \left(\nabla - i\nabla\varphi \right) c^\dagger \left(\nabla + i\nabla\varphi \right) c \right] d\boldsymbol{r}. \tag{1.43}$$

If the phase φ is a single-valued function of spacetime and is differentiable, physical phenomena must be invariant with respect to phase transformation. Therefore, the Lagrangian must be invariant, so the first-order term of φ in Eq. (1.43) must be zero. Using the variation of the action, or time integral of the Lagrangian $I \equiv \int_{-\infty}^{\infty} L dt$, integration by parts results in

$$\delta I = \int_{-\infty}^{\infty} dt \int \hbar \left[\frac{\partial \left(c^\dagger c \right)}{\partial t} - \frac{i\hbar}{2m} \mathrm{div} \left(c^\dagger \overset{\leftrightarrow}{\nabla} c \right) \right] \varphi \, d\boldsymbol{r} = 0. \tag{1.44}$$

In other words, the charge conservation law

$$\frac{\partial \rho}{\partial t} + \mathrm{div} \boldsymbol{j}^{(0)} = 0 \tag{1.45}$$

is obtained. Here,

$$\rho \equiv e\langle c^\dagger c \rangle, \quad j_i^{(0)} \equiv e \left\langle -i\frac{\hbar}{2m} c^\dagger \overset{\leftrightarrow}{\nabla}_i c \right\rangle \tag{1.46}$$

are the charge density and current density, respectively. Equation (1.42) is called a covariant differential. A U(1) gauge field is defined as $A_\mu \equiv \partial_\mu \varphi$ and Eq. (1.43) is the Lagrangian of electrons interacting with the gauge field. However, the quantity $F_{\mu\nu} \equiv \partial_\mu A_\nu - \partial_\nu A_\mu$ corresponding to the physical field satisfies $F_{\mu\nu} = 0$ if φ is single valued and is differentiable.

Clearly this reflects the fact that a continuous transform of phase does not change the phenomena. In contrast, if there is an electromagnetic field, φ is multivalued or is not differentiable. In this case $F_{\mu\nu}$ becomes a finite nonzero value because the partial differential of φ depends upon the order of the differential. The following discussion will consider situations where magnetic or electric fields exist, and the U(1) gauge field $A_{\rm em}$ corresponding to the electromagnetic field is separated from the differentiable part of the phase degree of freedom φ. The Lagrangian of free electrons including the electromagnetic field is

$$L_{\rm em} = \int \left[i\hbar c^\dagger \frac{\partial c}{\partial t} - e\phi c^\dagger c - \frac{\hbar^2}{2m} \left(\nabla + i\frac{e}{\hbar} \boldsymbol{A}_{\rm em} \right) c^\dagger \left(\nabla - i\frac{e}{\hbar} \boldsymbol{A}_{\rm em} \right) c \right] d\boldsymbol{r}, \quad (1.47)$$

where $\phi \equiv \hbar A_{{\rm em}\,t}$ and $A_{{\rm em}\,i}$ are the scalar potential and vector potential, respectively.

The continuity equation of charge current is derived also in the case that there are spin–orbit interactions and an external electromagnetic field. The Lagrangian becomes

$$L_{\rm em,so} = \int \left[i\hbar c^\dagger \frac{\partial c}{\partial t} - e\phi c^\dagger c - \frac{\hbar^2}{2m} \left(\nabla + i\frac{e}{\hbar} \boldsymbol{A}_{\rm em} \right) c^\dagger \left(\nabla - i\frac{e}{\hbar} \boldsymbol{A}_{\rm em} \right) c \right] d\boldsymbol{r} - \mathcal{H}_{\rm so}.$$
$$(1.48)$$

$$\mathcal{H}_{\rm so} = - i\frac{e\hbar^2}{4m^2 c^2} \int c^\dagger \left\{ \nabla\phi_{\rm so} \cdot \left[\left(\nabla - \frac{e}{\hbar} \boldsymbol{A}_{\rm em} \right) \times \boldsymbol{\sigma} \right] \right\} c\, d\boldsymbol{r}$$
$$= - i\lambda_{\rm so} \int c^\dagger \left\{ \nabla\phi_{\rm so} \cdot \left[\left(\nabla - \frac{e}{\hbar} \boldsymbol{A}_{\rm em} \right) \times \boldsymbol{\sigma} \right] \right\} c\, d\boldsymbol{r}. \quad (1.49)$$

The requirement of invariance under phase transformation of the electron field (1.41) analogous to the derivation of Eq. (1.45) for free electrons yields

$$\frac{\partial \rho}{\partial t} + {\rm div}\boldsymbol{j}_c = 0. \quad (1.50)$$

Here, ρ is given by Eq. (1.46) as in the case of free electrons. The current density is

$$j_i \equiv e \left\langle -i\frac{\hbar}{2m} c^\dagger \overleftrightarrow{\nabla}_i c - \frac{e}{m} A_{{\rm em}\,i} c^\dagger c - \frac{\lambda_{\rm so}}{\hbar} \sum_{jk} \epsilon_{ijk} (\nabla_j \phi_{\rm so})(c^\dagger \sigma_k c) \right\rangle. \quad (1.51)$$

Conservation of spin and spin current   The law of spin conservation is derived by looking at the change in $L_{\rm em,so}$ under the rotation in the spin space (SU(2) rotation)

$$c\left(\boldsymbol{r},\,t\right)\to e^{i\varphi(\boldsymbol{r},\,t)\cdot\boldsymbol{\sigma}}c\left(\boldsymbol{r},\,t\right),$$
$$c^{\dagger}\left(\boldsymbol{r},\,t\right)\to c^{\dagger}\left(\boldsymbol{r},\,t\right)e^{-i\varphi(\boldsymbol{r},\,t)\cdot\boldsymbol{\sigma}},\tag{1.52}$$

where φ is a three-component vector. In the absence of spin–orbit interaction, the free electron part L_{em} has a similar form as Eq. (1.50):

$$\frac{\partial\rho_s^{(0)\alpha}}{\partial t}+\mathrm{div}\boldsymbol{j}_s^{(0)\alpha}=0,\tag{1.53}$$

$$\hat{j}_{si}^{(0)\alpha}\equiv e\left(-i\frac{\hbar}{2m}c^{\dagger}\overset{\leftrightarrow}{\nabla}_i\sigma_{\alpha}c-\frac{e}{m}A_{\mathrm{em}i}c^{\dagger}\sigma_{\alpha}c\right),\tag{1.54}$$

$$\rho_s^{(0)\alpha}\equiv e\langle c^{\dagger}\sigma_{\alpha}c\rangle.\tag{1.55}$$

This is a continuity equation of spin. The spin current is defined as $\boldsymbol{j}_s^{\alpha(0)}$. This equation represents spin angular momentum conservation under the Lagrangian (1.47), which is due to the action, the integral of the Lagrangian, $I_{\mathrm{em}}=\int_{-\infty}^{\infty}dtL_{\mathrm{em}}$, being unchanged by the rotational transformation in the spin sector (1.52): the spin rotational symmetry.

However, in the presence of the spin–orbit interaction term, total spin is not conserved because the interaction breaks the spin rotational symmetry. In fact, the variation of the spin–orbit interaction is

$$\delta\mathcal{H}_{\mathrm{so}}=\frac{\hbar}{e}\int\sum_{\alpha}\varphi_{\alpha}\left(\mathrm{div}\delta\boldsymbol{j}_s^{\alpha}-T^{\alpha}\right)d\boldsymbol{r}.\tag{1.56}$$

There is a term $\boldsymbol{T}$ that cannot be written as a divergence. Here,

$$\delta\hat{j}_{si}^{\alpha}\equiv-\frac{e}{\hbar}\lambda_{\mathrm{so}}\sum_j\epsilon_{ij\alpha}\left(\nabla_j\phi_{\mathrm{so}}\right)\left(c^{\dagger}c\right)\tag{1.57}$$

and

$$T^{\alpha}\equiv\frac{2m}{\hbar^2}\lambda_{\mathrm{so}}\sum_{ijk\beta}\epsilon_{ijk}\epsilon_{\alpha\beta k}\langle\left(\nabla_i\phi_{\mathrm{so}}\right)\hat{j}_{sj}^{(0)\beta}\rangle.\tag{1.58}$$

Then, the requirement of invariance under spin rotation of the electron field analogous to the derivation of Eq. (1.50) yields

$$\frac{\partial\rho_s^{\alpha}}{\partial t}+\mathrm{div}\boldsymbol{j}_s^{\alpha}=T^{\alpha}.\tag{1.59}$$

Here, the spin density and the density of the total spin current is

$$j_{si}^{\alpha}\equiv e\langle\hat{j}_{si}^{(0)\alpha}+\delta\hat{j}_{si}^{\alpha}\rangle.\tag{1.60}$$

The spin–orbit interaction adds a correction δj_{si}^{α} to the spin current density, resulting in a term $\boldsymbol{T}$ corresponding to nonconservation of spin. This term corresponds to a source or sink of spin. In the presence of the spin–orbit interaction, spin conservation is only an approximation because of this effect. This nonconservation of spin is due to the angular momentum transfer of spin into the orbit and eventually into macroscopic degrees of freedom such as the lattice system.

References

[1] Dirac, P. A. M. (1982). *The Principles of Quantum Mechanics*; 4th edition. Oxford University Press, New York.

[2] Jackson, J. D. (1998). *Classical Electrodynamics*; 3rd edition. John Wiley, New York.

2 Incoherent spin current

K. Ando and E. Saitoh

2.1 Fermi–Dirac distribution [1–3]

The thermal equilibrium properties of a free N-electron system at temperature T are calculated using the distribution function or occupation probability $f(E,T)$. We consider an N-electron system with one-electron energy levels E_i. The degeneracy of the levels g_i and their occupation number n_i must satisfy $n_i \leq g_i$ because of the Pauli exclusion principle. In the equilibrium condition, the free energy F of the total system must be minimized with respect to a variation in the relative occupation numbers of the levels as

$$\delta F = \sum_i \frac{\partial F}{\partial n_i} \delta n_i = 0, \tag{2.1}$$

with the conservation of the electron number:

$$\sum_i \delta n_i = 0. \tag{2.2}$$

For the simple case of exchange of electrons between two levels, k and l, Eqs. (2.1) and (2.2) become $(\partial F/\partial n_k)\delta n_k + (\partial F/\partial n_l)\delta n_l = 0$ and $\delta n_k + \delta n_l = 0$, respectively. These conditions are satisfied only when $\partial F/\partial n_k = \partial F/\partial n_l$. As shown in this simple example, in the equilibrium condition, all $\partial F/\partial n_i$ must be equal. Thus, we define the chemical potential μ^c as

$$\frac{\partial F}{\partial n_i} = \mu^c. \tag{2.3}$$

The free energy discussed above is expressed as $F = U - TS$ with the internal energy $U = \sum_i n_i E_i$ and the entropy $S = k_B \ln W$. W is the number of ways of distributing the N electrons among the states. Since electrons cannot be distinguished from each other, the number of ways of distributing n_i electrons in the energy level E_i is $g_i!/[n_i!(g_i - n_i)!]$. Thus, the number of ways W of the total system is

$$W = \prod_i \frac{g_i!}{n_i!(g_i - n_i)!}, \tag{2.4}$$

which yields the entropy as

$$S = k_B \sum_i [\ln g_i! - \ln n_i! - \ln(g_i - n_i)!]. \tag{2.5}$$

For large n, the factorials can be replaced by using Stirling's approximate formula, $\ln n! \approx n \ln n - n$, and one finds the chemical potential as

$$\mu^c = \frac{\partial F}{\partial n_i} = E_i + k_B T \ln \frac{n_i}{g_i - n_i}, \tag{2.6}$$

or the occupation number n_i:

$$n_i = g_i \frac{1}{e^{(E_i - \mu^c)/k_B T} + 1}. \tag{2.7}$$

Equation (2.7) shows that the probability that a quantum mechanical state is occupied is given by

$$f(E, T) = \frac{1}{e^{(E - \mu^c)/k_B T} + 1}, \tag{2.8}$$

which is known as the Fermi–Dirac distribution function.

The behavior of the Fermi–Dirac function $f(E, T)$ is shown in Fig. 2.1. At $T = 0$, $f(E, T = 0)$ becomes the step function; at $T = 0$, all the states with energy lower than μ^c are occupied and all the states with energy higher than μ^c are empty. Thus at $T = 0$, the chemical potential is equal to the Fermi energy E_F:

$$\mu^c(T = 0) = E_F. \tag{2.9}$$

Here, E_F is defined as the energy of the highest occupied quantum state in a system at $T = 0$. At finite temperature T, some of the electrons are excited to the states above E_F and $f(E, T)$ deviates from the step function only in the thermal energy range of order $k_B T$ around $\mu^c(T)$.

The chemical potential μ^c is temperature dependent; μ^c is determined by the constraint that the total number of electrons must remain constant. Using the Fermi-Dirac function in Eq. (2.8), one can calculate the particle number density as

$$n = \int_{-\infty}^{\infty} N(E) f(E) dE, \tag{2.10}$$

where $N(E)$ is the density of states and $N(E) dE$ is the number of electrons per unit volume of r-space with energies between E and $E + dE$. At almost all temperatures of interest in metals, T is much less than the Fermi temperature $T_F = E_F/k_B$: $T \ll T_F$. Since $f(E, T)$ at temperature T differs from $f(E, T = 0)$ only in a region of the order of $k_B T$ around μ^c, the difference between $\int_{-\infty}^{\infty} N(E) f(E) dE$ and their zero-temperature values, $\int_{-\infty}^{E_F} N(E) dE$,

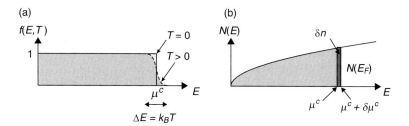

FIG. 2.1. (a) The Fermi–Dirac function $f(E,T)$ at $T = 0$ (solid line) and $T > 0$ (dotted curve). The two curves differ only in a region of order $k_B T$ around μ^c. At $T > 0$, some electrons just below E_F have been excited to levels just above E_F. (b) The density of states $N(E)$ for free electrons.

is entirely determined by the form of $N(E)$ near $E = \mu^c$. If $N(E)$ does not vary rapidly in the energy range of the order of $k_B T$ around $E = \mu^c$, Eq. (2.10) can be approximated using the Taylor expansion of $N(E)$ around $E = \mu^c$ and then we can use the Sommerfeld expansion:

$$\int_{-\infty}^{\infty} H(E) f(E) dE = \int_{-\infty}^{\mu^c} H(E) dE + \frac{\pi^2}{6} (k_B T)^2 \frac{dH(E)}{dE} \bigg|_{E=\mu_c} + O(T^4). \quad (2.11)$$

By expanding the upper limit of the integral about E_F, and keeping terms up to second order in T, we have

$$n = \int_0^{E_F} N(E) dE + (\mu^c - E_F) N(E_F) + \frac{\pi^2}{6} (k_B T)^2 \frac{dN(E)}{dE} \bigg|_{E=E_F}. \quad (2.12)$$

Since the first term is equal to n, we obtain

$$\mu^c = E_F - \frac{\pi^2}{6} (k_B T)^2 \frac{d}{dE} \ln N(E) \bigg|_{E=E_F}. \quad (2.13)$$

For free electrons, $N(E) \propto E^{1/2}$, this becomes

$$\mu^c = E_F \left[1 - \frac{1}{3} \left(\frac{\pi k_B T}{2 E_F} \right)^2 \right] = E_F \left[1 - \frac{\pi^2}{12} \left(\frac{T}{T_F} \right)^2 \right]. \quad (2.14)$$

This relation gives μ^c as a function of T and shows that the temperature dependence of μ^c is small for the free-electron gas because of $T \ll T_F$.

As described in Eq. (2.3), the chemical potential is the energy necessary to add a particle to the system. When the particle density n becomes $n + \delta n$, the chemical potential changes from μ^c to $\mu^c + \delta \mu^c$. Assuming $N(E) \approx N(E_F)$ around $E = \mu^c$ and using Eq. (2.12), we have

$$\delta\mu^c = \frac{\delta n}{N(E_F)}. \tag{2.15}$$

This relation can be seen in Fig. 2.1(b).

2.2 Diffusion equation

When the electron density is nonuniform, the gradient of the electron density drives a current called a diffusive current. For simplicity, we consider one-dimensional electron diffusion as shown in Fig. 2.2(a). Let the electron density at position x at time t be $n(t, x)$. Here, we assume that at each step Δt, the each electron at position x: (1) moves to the right by a step a with probability p, (2) moves to the left by a step a with probability p, and (3) stay on x with probability $1 - 2p$. At time t, there are $a \cdot n(t, x)$ electrons in the region from $x - a/2$ to $x + a/2$. After the time evolution of Δt, $pa \cdot n(t, x)$ electrons move to the right and $pa \cdot n(t, x)$ electrons move to the left, respectively. Similarly, $pa \cdot n(t, x + a)$ electrons move to x from the right and $pa \cdot n(t, x - a)$ electrons move to x from the left. Thus we have

$$a \cdot n(t + \Delta t, x) = a \cdot n(t, x) - 2pa \cdot n(x, t) + pa \cdot n(t, x - a) + pa \cdot n(t, x + a). \tag{2.16}$$

Now, we assume that the step a is very small compared to the scales on which n varies. We then have a Taylor expansion

$$n(t, x \pm a) \simeq n(t, x) \pm a\frac{\partial n(t, x)}{\partial x} + \frac{a^2}{2}\frac{\partial^2 n(t, x)}{\partial x^2}. \tag{2.17}$$

If n changes only slightly during the time step Δt, we can approximate

$$n(t + \Delta t, x) \simeq n(t, x) + \Delta t\frac{\partial n(t, x)}{\partial t}. \tag{2.18}$$

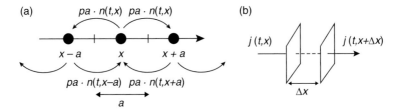

FIG. 2.2. (a) Electron diffusion in a one-dimensional system. Each electron at position x moves to neighboring sites by a step a with probability p or stays with probability $1 - 2p$. (b) The flow of flux j of n.

Substituting Eqs. (2.17) and (2.18) into Eq. (2.16), in the limit of $a \to 0$ and $\Delta t \to 0$ with keeping $a^2/\Delta t$ constant, we find the diffusion equation:

$$\frac{\partial}{\partial t} n(t, x) = D \frac{\partial^2}{\partial x^2} n(t, x), \tag{2.19}$$

where $D = pa^2/\Delta t$ is the diffusion constant.

Let $j(t, x)$ denote the flux of n, the net rate at which n is passing from the left of x to the right of x at time t (see Fig. 2.2b). Since the electron density n is a conserved quantity, for arbitrary Δx,

$$\frac{d}{dt} \int_x^{x+\Delta x} n(t, x) dx = j(t, x) - j(t, x + \Delta x). \tag{2.20}$$

In the limit of $\Delta x \to 0$, Eq. (2.20) yields

$$\frac{\partial n}{\partial t} = -\frac{\partial j}{\partial x}, \tag{2.21}$$

or in a three-dimensional system

$$\frac{\partial}{\partial t} n(t, \boldsymbol{x}) + \mathrm{div} \boldsymbol{j}(t, \boldsymbol{x}) = 0, \tag{2.22}$$

which is known as the continuity equation. The diffusion equation (2.19) and the continuity equation (2.21) give a current that is proportional to the local gradient in the density:

$$j_{\mathrm{diffusion}} = -D \frac{\partial n}{\partial x}. \tag{2.23}$$

This is known as Fick's law. Equation (2.23) says that electrons diffuse on average from regions of high density toward regions of low density.

2.3   Spin diffusion equation [4]

Here, we will discuss a diffusion spin current due to a spatial inhomogeneous spin density and a drift spin current in the absence of coherent dynamics of spin. Conduction electrons in a semiconductor or in a metal can be regarded as an electron gas. First, we consider spinless electrons. In the presence of an electric field $\mathbf{E}$, the drift current density is given by $\boldsymbol{j}_{\mathrm{drift}} = \sigma \mathbf{E}$. The sum of the drift and diffusion current density $\boldsymbol{j} = \boldsymbol{j}_{\mathrm{drift}} + \boldsymbol{j}_{\mathrm{diffusion}}$ is

$$\boldsymbol{j} = \sigma \mathbf{E} + eD\nabla n, \tag{2.24}$$

where $\boldsymbol{j}_{\mathrm{diffusion}} = eD\nabla n$ is the diffusion current density obtained from Eq. (2.23). σ is the electrical conductivity and $e = 1.602 \times 10^{-19}$ C is the elementary charge. Since $N(E_F)\nabla \mu^c = \nabla n$, the gradient in the electrochemical potential $\mu = \mu^c - e\phi$ is

$$\nabla \mu = e\boldsymbol{E} + \frac{\nabla n}{N(E_F)}. \tag{2.25}$$

Thus, for $\nabla \mu = 0$, the total current density

$$\boldsymbol{j} = \left(\sigma - e^2 N(E_F)D\right)\boldsymbol{E}, \tag{2.26}$$

must be zero and thus one obtains the Einstein relation:

$$\sigma = e^2 N(E_F)D. \tag{2.27}$$

Because of Eqs. (2.25) and (2.27), we can write

$$\boldsymbol{j} = \frac{\sigma}{e}\nabla \mu. \tag{2.28}$$

This relation expresses the fact that the driving force for a current in this system is the gradient of the electrochemical potential $\nabla \mu$.

Next, we consider the spin degree of freedom. The driving force for a diffusion or drift spin current is the gradient of the difference in the spin-dependent electrochemical potential μ_σ for spin up ($\sigma =\uparrow$) and spin down ($\sigma =\downarrow$). The current density $\boldsymbol{j}_\sigma$ for spin channel σ ($\sigma =\uparrow, \downarrow$) is expressed as

$$\boldsymbol{j}_\sigma = \frac{\sigma_\sigma}{e}\nabla \mu_\sigma, \tag{2.29}$$

where $\mu_\sigma = \mu_\sigma^c - e\phi$ is the spin-dependent electrochemical potential. Here, we introduce a charge current $\boldsymbol{j}_c = \boldsymbol{j}_\uparrow + \boldsymbol{j}_\downarrow$ and a spin current $\boldsymbol{j}_s = \boldsymbol{j}_\uparrow - \boldsymbol{j}_\downarrow$, which are rewritten as

$$\boldsymbol{j}_c = \frac{1}{e}\nabla(\sigma_\uparrow \mu_\uparrow + \sigma_\downarrow \mu_\downarrow), \tag{2.30}$$

$$\boldsymbol{j}_s = \frac{1}{e}\nabla(\sigma_\uparrow \mu_\uparrow - \sigma_\downarrow \mu_\downarrow). \tag{2.31}$$

In nonmagnetic metals or semiconductors, the electrical conductivity is spin-independent, $\sigma_\uparrow = \sigma_\downarrow = (1/2)\sigma_N$, and thus $\boldsymbol{j}_s = (\sigma_N/2e)\nabla(\mu_\uparrow - \mu_\downarrow)$.

The continuity equation for charge is

$$\frac{d}{dt}\rho = -\text{div}\boldsymbol{j}_c. \tag{2.32}$$

The continuity equation for spins can be written as

$$\frac{d}{dt}M_z = -\text{div}\boldsymbol{j}_s + T_z, \tag{2.33}$$

where M_z is the z-component of magnetization. z is defined as the quantization axis. T_z represents spin relaxation, which can be written as $T_z = e(n_\uparrow - \bar{n}_\uparrow)/\tau_{\uparrow\downarrow} - e(n_\downarrow - \bar{n}_\downarrow)/\tau_{\downarrow\uparrow}$ using the single-pole relaxation approximation.

$\bar{n}_\sigma$ is the equilibrium carrier density with spin σ, and $\tau_{\sigma\sigma'}$ is the scattering time of an electron from spin state σ to σ'. Note that the detailed balance principle imposes that $N_\uparrow/\tau_{\uparrow\downarrow} = N_\downarrow/\tau_{\downarrow\uparrow}$, so that in equilibrium no net spin scattering takes place, where N_σ denotes the spin-dependent density of states at the Fermi energy. This implies that, in general, in a ferromagnet, $\tau_{\uparrow\downarrow}$ and $\tau_{\downarrow\uparrow}$ are not the same. In the equilibrium condition, $d\rho/dt = dM_z/dt = 0$, substituting Eqs. (2.30), (2.31), and $N_\uparrow/\tau_{\uparrow\downarrow} = N_\downarrow/\tau_{\downarrow\uparrow}$ into Eqs. (2.32) and (2.33), we have

$$\nabla^2(\sigma_\uparrow\mu_\uparrow + \sigma_\downarrow\mu_\downarrow) = 0, \tag{2.34}$$

$$\nabla^2(\mu_\uparrow - \mu_\downarrow) = \frac{1}{\lambda^2}(\mu_\uparrow - \mu_\downarrow). \tag{2.35}$$

Equation (2.35) is known as the spin diffusion equation. $\lambda = \sqrt{D\tau_{sf}}$ is the spin diffusion length. Here, $D = D_\uparrow D_\downarrow(N_\uparrow + N_\downarrow)/(N_\uparrow D_\uparrow + N_\downarrow D_\downarrow)$ is the spin-averaged diffusion constant, where D_σ is the spin-dependent diffusion constant. The spin relaxation time τ_{sf} is given by $1/\tau_{sf} = 1/\tau_{\uparrow\downarrow} + 1/\tau_{\downarrow\uparrow}$.

Now, we consider a simple example of a spin current in a ferromagnetic/nonmagnetic (F/N) junction with a charge current passing through the interface as shown in Fig. 2.3(a). The general solution of Eqs. (2.34) and (2.35) is

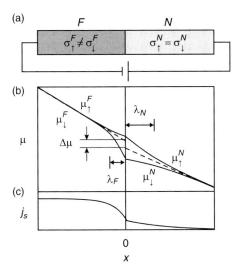

FIG. 2.3. (a) A ferromagnetic/nonmagnetic (F/N) junction with a current passing through the interface. (b) The spatial variation of the electrochemical potential $\mu_\sigma^{F(N)}$ for spin-up and spin-down electrons. (c) The spatial variation of a spin current j_s.

$$\mu_\uparrow^F = A_F + B_F x + \frac{C_F}{\sigma_\uparrow^F} \exp\left(\frac{x}{\lambda_F}\right) + \frac{D_F}{\sigma_\uparrow^F} \exp\left(-\frac{x}{\lambda_F}\right), \tag{2.36}$$

$$\mu_\downarrow^F = A_F + B_F x - \frac{C_F}{\sigma_\downarrow^F} \exp\left(\frac{x}{\lambda_F}\right) - \frac{D_F}{\sigma_\downarrow^F} \exp\left(-\frac{x}{\lambda_F}\right), \tag{2.37}$$

$$\mu_\uparrow^N = A_N + B_N x + \frac{C_N}{\sigma_\uparrow^N} \exp\left(\frac{x}{\lambda_N}\right) + \frac{D_N}{\sigma_\uparrow^N} \exp\left(-\frac{x}{\lambda_N}\right), \tag{2.38}$$

$$\mu_\downarrow^N = A_N + B_N x - \frac{C_N}{\sigma_\downarrow^N} \exp\left(\frac{x}{\lambda_N}\right) - \frac{D_N}{\sigma_\downarrow^N} \exp\left(-\frac{x}{\lambda_N}\right), \tag{2.39}$$

where $\mu_\sigma^{F(N)}$, $\lambda_{F(N)}$, and $\sigma_\sigma^{F(N)}$ are the electrochemical potential, the spin diffusion length, and the electrical conductivity for the $F(N)$ layer, respectively. From Eq. (2.29), the current density $j_\sigma^{F(N)}$ is

$$j_\sigma^{F(N)} = \frac{\sigma_\sigma^{F(N)}}{e} \frac{\partial}{\partial x} \mu_\sigma^{F(N)}. \tag{2.40}$$

In the F layer, the electrical conductivity is spin dependent and thus $\sigma_\uparrow^F + \sigma_\downarrow^F = \sigma_F$. In contrast, in the N layer, the electrical conductivity is spin independent: $\sigma_\uparrow^N = \sigma_\downarrow^N = \sigma_N/2$. The coefficients $A_{F(N)}$, $B_{F(N)}$, $C_{F(N)}$, and $D_{F(N)}$ are determined by boundary conditions. Without loss of generality, we can define the first boundary conditions as

$$\mu_\uparrow^F(x = -\infty) = \mu_\downarrow^F(x = -\infty), \tag{2.41}$$

$$\mu_\uparrow^N(x = \infty) = \mu_\downarrow^N(x = \infty). \tag{2.42}$$

These conditions yield $D_F = 0$ and $C_N = 0$. An applied charge current density j_c is

$$j_\uparrow^F + j_\downarrow^F = j_\uparrow^N + j_\downarrow^N = j_c, \tag{2.43}$$

which gives $B_F = ej_c/(\sigma_\uparrow^F + \sigma_\downarrow^F) = ej_c/\sigma_F$ and $B_N = ej_c/(\sigma_\uparrow^N + \sigma_\downarrow^N) = ej_c/\sigma_N$. At the F/N interface, the boundary conditions representing the continuity of $\mu_\sigma^{F(N)}$ and the conservation of $j_\sigma^{F(N)}$ are

$$\mu_\sigma^F(x = 0) = \mu_\sigma^N(x = 0), \tag{2.44}$$

$$j_\sigma^F(x = 0) = j_\sigma^N(x = 0). \tag{2.45}$$

Setting $A_F = 0$ and using these boundary conditions, one finds

$$\mu_\uparrow^F = \frac{ej_c}{\sigma_F} x - \frac{ej_c P \lambda_N (1 - P^2) \sigma_F}{2\sigma_\uparrow^F \sigma_N \left(1 + (1 - P^2)\dfrac{\sigma_F \lambda_N}{\sigma_N \lambda_F}\right)} \exp\left(\frac{x}{\lambda_F}\right), \tag{2.46}$$

$$\mu_\downarrow^F = \frac{ejc}{\sigma_F}x + \frac{ejc P \lambda_N (1-P^2)\sigma_F}{2\sigma_\downarrow^F \sigma_N \left(1 + (1-P^2)\dfrac{\sigma_F \lambda_N}{\sigma_N \lambda_F}\right)} \exp\left(\frac{x}{\lambda_F}\right), \qquad (2.47)$$

$$\mu_\uparrow^N = \frac{ejc P^2 \lambda_N}{\sigma_N \left(1 + (1-P^2)\dfrac{\sigma_F \lambda_N}{\sigma_N \lambda_F}\right)} + \frac{ejc}{\sigma_N}x$$

$$- \frac{ejc P \lambda_N}{\sigma_N \left(1 + (1-P^2)\dfrac{\sigma_F \lambda_N}{\sigma_N \lambda_F}\right)} \exp\left(-\frac{x}{\lambda_N}\right), \qquad (2.48)$$

$$\mu_\downarrow^N = \frac{ejc P^2 \lambda_N}{\sigma_N \left(1 + (1-P^2)\dfrac{\sigma_F \lambda_N}{\sigma_N \lambda_F}\right)} + \frac{ejc}{\sigma_N}x$$

$$+ \frac{ejc P \lambda_N}{\sigma_N \left(1 + (1-P^2)\dfrac{\sigma_F \lambda_N}{\sigma_N \lambda_F}\right)} \exp\left(-\frac{x}{\lambda_N}\right), \qquad (2.49)$$

where $P = (\sigma_\uparrow^F - \sigma_\downarrow^F)/(\sigma_\uparrow^F + \sigma_\downarrow^F)$ is the spin polarization of the F layer. Figure 2.3(b) shows the spatial variation of the electrochemical potential for spin-up and spin-down electrons with a current through a F/N interface. In the N layer, a spin current j_s driven by $\nabla(\mu_\uparrow^N - \mu_\downarrow^N)$ flows from the interface toward the inside of the N layer as shown in Fig. 2.3(c). The decay length of j_s is characterized by the spin diffusion length λ_N. In the F layer, a spin-polarized current is suppressed near the interface ($\sim \lambda_F$) due to the back flow of spin-polarized electrons induced by the spin accumulation at the interface.

The spin polarization of the current at the interface $\alpha = (j_\uparrow^N - j_\downarrow^N)/(j_\uparrow^N + j_\downarrow^N) = (j_\uparrow^F - j_\downarrow^F)/(j_\uparrow^F + j_\downarrow^F)$ is obtained as

$$\alpha = P \frac{1}{1 + (1-P^2)\dfrac{\sigma_F \lambda_N}{\sigma_N \lambda_F}}. \qquad (2.50)$$

Note that the spin polarization α of a current injected into the N layer is different from the bulk polarization P of the F layer.

Although $\mu_\uparrow$ and $\mu_\downarrow$ are continuous at the interface, the slope of the electrochemical potentials can be discontinuous at the interface (see the dotted lines in Fig. 2.3b). This voltage drop at the interface $\Delta\mu$ gives the spin-coupled interface resistance $R_s = \Delta\mu/(ejc)$:

$$R_s = P^2 \frac{\lambda_N \sigma_N^{-1}}{1 + (1-P^2)\dfrac{\sigma_F \lambda_N}{\sigma_N \lambda_F}}. \qquad (2.51)$$

The above equations show that the magnitude of the spin polarization and the spin-coupled resistance contain the same factor $(\sigma_F \lambda_N)/(\sigma_N \lambda_F)$. In many cases, the spin diffusion length of F is much shorter than that of N, $\lambda_F \ll \lambda_N$, and, in this case, λ_F is a limiting factor to obtain a large spin polarization. This problem becomes serious when a ferromagnetic metal is used to inject spin-polarized currents into semiconductors. In this case, the electrical conductivity, $\sigma_N \ll \sigma_F$, drastically limits the polarization. This problem is known as the conductivity mismatch problem. A way to overcome the conductivity mismatch problem of spin injection into a semiconductor is to use a ferromagnetic semiconductor as a spin source. Another way is to insert a spin-dependent interface resistance at a metal–semiconductor interface.

There are some methods for experimentally detecting pure spin currents, spin currents without accompanying charge currents. One direct method is the utilization of the inverse spin Hall effect, a method which was demonstrated first by spin pumping [5] and nonlocal techniques [5, 7]. The details are discussed in Chapter 14 and 15. In semiconductors, an optical method was also demonstrated [8]. As an alternative way, one can infer spin-current generation indirectly by measuring spin accumulation.

References

[1] Ashcroft, N. W. and Mermin, N. D. (1976). *Solid State Physics*. Brooks Cole, Pacific Grove.

[2] Kittel, C. (2004). *Introduction to Solid State Physics*; 8th edition. John Wiley, New York.

[3] Ibach, H. and Lüth, H. (2009). *Solid-State Physics: An Introduction to Principles of Materials*; 4th edition. Springer, New York.

[4] Takahashi, S. and Maekawa, S. (2008). *J. Phys. Soc. Jpn.*, **77**, 031009.

[5] Saitoh, E., Ueda, M., Miyajima, H., and Tatara, G. (2006). *Appl. Phys. Lett.*, **88**, 182509.

[6] Kimura, T., Otani, Y., Sato, T., Takahashi, S., and Maekawa, S. (2007). *Phys. Rev. Lett.*, **98**, 156601.

[7] Valenzuela, S. O. and Tinkham, M. (2006). *Nature*, **442**, 176.

[8] Werake, L. K. and Zhao, H. (2010). *Nature Phys.*, **6**, 875.

3 Exchange spin current

E. Saitoh and K. Ando

A spin current is carried also by a spin wave, a collective excitation of magnetization in magnets. In this section, we first rewrite the exchange interaction in magnets by introducing the concept of exchange spin current and then formulate a spin-wave spin current.

3.1 Magnetic order and exchange interaction

State of matters can be classified into several types in terms of magnetic properties. In paramagnetic and diamagnetic states, matter has no magnetic order and exhibits zero magnetization in the absence of external magnetic fields. By applying a magnetic field, matter in a paramagnetic state exhibits magnetization parallel to the external field while that in a diamagnetic state exhibits magnetization antiparallel to the field. The other types of material states contain magnetic order. In ferromagnetic states, the permanent magnetic moments of atoms or ions align parallel to a certain direction and the matter exhibits finite magnetization even in the absence of external magnetic fields. Antiferromagnetic states refer to states in which the permanent magnetic moments align antiparallel and cancel each other out and the net magnetization is zero in the absence of magnetic fields. In ferrimagnets, the moments align antiparallel but the cancellation is not perfect and net magnetization appears.

The interaction that aligns spins is called the exchange interaction. One typical model for the exchange interaction is Heisenberg's Hamiltonian: $H = -J \sum s_i \cdot s_j$, s_i represents the spin operator of an atom or an ion at the position labeled by i. J is the interaction coefficient; for $J > 0$, parallel alignment of s_i and s_j reduces the energy. When this energy reduction is greater than the thermal fluctuation energy, a ferromagnetic state can appear. The summation runs over all the combination of nearest neighboring i and j.

3.2 Exchange spin current

3.2.1 Landau–Lifshitz–Gilbert equation

In this section, we derive an equation which describes spin or magnetization dynamics. Consider a system described by the Hamiltonian

$$\mathcal{H} = -\boldsymbol{M} \cdot \boldsymbol{H}_{\text{eff}}, \tag{3.1}$$

which describes the fact that a spin s tends to align parallel to the external magnetic field H due to Zeeman's interaction. The magnetization M satisfies the following commutation relation of angular momentum

$$[M_i, M_j] = i\gamma\hbar\epsilon_{ijk}M_k. \tag{3.2}$$

The dynamics of M is described by a Heisenberg equation of motion [1]

$$\frac{d\boldsymbol{M}}{dt} = -\frac{i}{\hbar}[\boldsymbol{M}, \mathcal{H}]. \tag{3.3}$$

Substituting Eq. (3.2) into this equation, the following result is obtained.

$$\frac{d\boldsymbol{M}}{dt} = -\gamma\boldsymbol{M} \times \boldsymbol{H}. \tag{3.4}$$

This equation describes the dynamics of an isolated spin magnetic moment; a spin keeps undergoing a precession motion around the magnetic field H, as shown in Fig. 3.1. However, as shown in Eq. (3.1), the energy is minimized when the magnetic moment is aligned parallel to the external magnetic field. Therefore the precession motion should be relaxed to this energy minimized state before too long. This relaxation is due to the interaction of the spin with the environmental degrees of freedom, such as conduction electrons and/or lattice vibrations in the matter. This relaxation is often taken into consideration by adding a term called the Gilbert term,

$$\boldsymbol{D} = -\frac{\alpha}{M}\boldsymbol{M} \times \frac{d\boldsymbol{M}}{dt} \tag{3.5}$$

to the equation of motion. Note that the Gilbert term is always directed toward the magnetic field direction, or the precession axis. The equation of motion then becomes

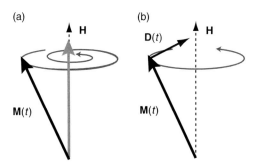

FIG. 3.1. (a) Concept of spin precession and decay. (b) Directions of Gilbert term $\boldsymbol{D}(t)$.

$$\frac{d\boldsymbol{M}}{dt} = -\gamma \boldsymbol{M} \times \boldsymbol{H} + \frac{\alpha}{M} \boldsymbol{M} \times \frac{d\boldsymbol{M}}{dt}, \qquad (3.6)$$

which is called the Landau–Lifshitz–Gilbert (LLG) equation.

Next, we consider general interactions acting on a single spin magnetic moment. For small-angle spin dynamics, the interaction can be introduced into the Landau–Lifshitz–Gilbert equation simply by replacing $\boldsymbol{H}$ with an effective magnetic field $\boldsymbol{H}_{\text{eff}}$ as follows

$$\boldsymbol{H}_{\text{eff}} = -\frac{\delta E_i(\boldsymbol{S})}{\delta \boldsymbol{s}_i}. \qquad (3.7)$$

$E_i(\boldsymbol{S})$ is the energy on the i site electron as a function of the spin direction of the system. $\boldsymbol{H}_{\text{eff}}$ describes the total interactions acting on the spin including external magnetic fields, magnetic anisotropy, and the exchange interaction.

3.2.2 Rewriting the Landau–Lifshitz–Gilbert equation

We are now in a position to consider the ferromagnetic interaction described by Heisenberg's Hamiltonian for a ferromagnet

$$E_i = -2J \sum_j \boldsymbol{s}_i \cdot \boldsymbol{s}_j \quad (J > 0). \qquad (3.8)$$

Let us apply the continuum approximation to $\boldsymbol{s}_i$ to rewrite $\boldsymbol{s}$ as a field value $\boldsymbol{s}(\boldsymbol{r})$ where $\boldsymbol{r}$ represents the position vector. When $\boldsymbol{s}_i = \boldsymbol{s}(\boldsymbol{r})$, a neighboring $\boldsymbol{s}_j$ is written as $\boldsymbol{s}(\boldsymbol{r} + \boldsymbol{a})$ where $\boldsymbol{a}$ is the displacement vector of the j site measured from the i site. $\boldsymbol{s}(\boldsymbol{r} + \boldsymbol{a})$ is expanded as

$$\boldsymbol{s}(\boldsymbol{r} + \boldsymbol{a}) = \boldsymbol{s}(\boldsymbol{r}) + \frac{\partial \boldsymbol{s}(\boldsymbol{r})}{\partial \boldsymbol{r}} \cdot \boldsymbol{a} + \frac{1}{2} \frac{\partial^2 \boldsymbol{s}(\boldsymbol{r})}{\partial \boldsymbol{r}^2} \boldsymbol{a}^2 + \cdots. \qquad (3.9)$$

The second term of the expansion vanishe in Eq. (3.8) due to the summation of i and j since there are the same atoms or ions at $r = a$ and $r = -a$ and, via the summation, the second terms of the expansion for these atoms are canceled out. Therefore, the third term is the dominant term and we will neglect the higher-order terms. We then obtain

$$\boldsymbol{H}_{\text{eff}} = -2Ja^2 \frac{\partial^2 \boldsymbol{s}(\boldsymbol{r})}{\partial \boldsymbol{r}^2} \equiv A\nabla^2 \boldsymbol{M}(\boldsymbol{r}). \qquad (3.10)$$

A is called the spin stiffness constant.

Then the Landau–Lifshitz–Gilbert equation in which exchange interaction is taken into consideration becomes

$$\frac{\partial}{\partial t} \boldsymbol{M}(\boldsymbol{r}) = -A\gamma \boldsymbol{M}(\boldsymbol{r}) \times \nabla^2 \boldsymbol{M}(\boldsymbol{r}) + \frac{\alpha}{M} \boldsymbol{M}(\boldsymbol{r}) \times \frac{\partial}{\partial t} \boldsymbol{M}(\boldsymbol{r}). \qquad (3.11)$$

Next, let us rewrite this equation in the form of a continuity equation. By using a mathematical formula of vector analysis, $\boldsymbol{A} \times \nabla^2 \boldsymbol{A} = \mathrm{div}(\boldsymbol{A} \times \nabla \boldsymbol{A})$, the Landau–Lifshitz–Gilbert equation for exchange-interacting spins is rewritten as

$$\frac{\partial}{\partial t} \boldsymbol{M}(\boldsymbol{r}) = -\mathrm{div}\left[A\gamma \boldsymbol{M}(\boldsymbol{r}) \times \nabla \boldsymbol{M}(\boldsymbol{r})\right] + \frac{\alpha}{M} \boldsymbol{M}(\boldsymbol{r}) \times \frac{\partial}{\partial t} \boldsymbol{M}(\boldsymbol{r}). \tag{3.12}$$

For now, we neglect the Gilbert relaxation term for simplicity, say,

$$\frac{\partial}{\partial t} \boldsymbol{M}(\boldsymbol{r}) = -\mathrm{div}\left[A\gamma \boldsymbol{M}(\boldsymbol{r}) \times \nabla \boldsymbol{M}(\boldsymbol{r})\right]. \tag{3.13}$$

This equation has the same form as a continuity equation. In fact, by defining the current $\boldsymbol{j}_s$ as $\boldsymbol{j}_s = A\gamma \boldsymbol{M} \times \nabla \boldsymbol{M}$, the equation becomes

$$\frac{\partial \boldsymbol{M}}{\partial t} = -\mathrm{div}\boldsymbol{j}_s, \tag{3.14}$$

which represents the conservation rule of spin angular momentum. $\boldsymbol{j}_s$ is interpreted as a flow of spin, say, a spin current, called an exchange spin current or a magnetization current. Since the above is a simple rewriting of the Landau–Lifshitz–Gilbert equation and the exchange interaction, it means that the exchange interaction between spins is equivalent to a flow of an exchange spin currents $\boldsymbol{j}_s$. Then, we restore the Gilbert damping: $\dot{\boldsymbol{M}} = -\mathrm{div}\boldsymbol{j}_s + \frac{\alpha}{M} \boldsymbol{M} \times \frac{\partial}{\partial t} \boldsymbol{M}$

The steady state described by this equation is obtained by $\partial \boldsymbol{M}/\partial t = 0$ as

$$\mathrm{div}\boldsymbol{j}_s = \mathrm{div}(A\gamma \boldsymbol{M} \times \nabla \boldsymbol{M}) = 0, \tag{3.15}$$

which implies a uniform alignment of magnetization.

In this way, exchange interaction can be rewritten in terms of the exchange spin current, which derives a ferromagnetic order in cooperation with the damping. In a uniform magnetization state, there are no exchange spin currents. In a steady nontrivial magnetic structure, for instance in a magnetic domain wall, the torque due to the exchange spin current is balanced by the magnetic anisotropy torque.

3.3 Spin-wave spin current

The exchange spin current can be driven in an nonequilibrium manner by exciting spin waves, or magnons. A spin wave is an elementary excitation from magnetically ordered states, which can be generated by, for instance, applying a microwave. At finite temperature, a spin current is generated also as a thermal fluctuation.

3.3.1 *Spin-wave formulation*

Now we consider low-energy excitations from a ferromagnetic ground state. Let us assume that spins are coupled with nearest neighbor spins via the exchange interaction. If one of spins is tilted against the ground state direction, the

(a)

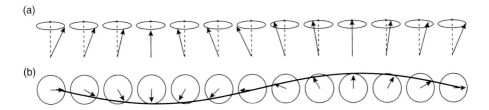

(b)

FIG. 3.2. Spin wave of a one-atomic chain in (a) side view and (b) top view.

neighboring spins tend to follow this tilt and the whole system will start to perform a collective motion just as a linear chain of masses connected by springs as shown in Fig. 3.2; the masses become the magnetic moments of the spins and the role of the springs is taken by the exchange interaction. These collective excitations of spins are the spin waves.

In a simple situation where only nearest neighbor interactions are important and all nearest exchange interactions are equal, the Hamiltonian is

$$\mathcal{H} = -2J \sum_{\langle i,j \rangle} \boldsymbol{s}_i \cdot \boldsymbol{s}_j - g\mu_{\mathrm{B}} H \sum_i s_{iz}. \tag{3.16}$$

The second term represents the Zeeman energy. We will further assume $s = 1/2$ and $J > 0$. Equation (3.16) can be written in a convenient form using the spin raising and lowering operators for the ith spin site:

$$s_i^+ = s_{ix} + i s_{iy}, \tag{3.17}$$

$$s_i^- = s_{ix} - i s_{iy}. \tag{3.18}$$

Now consider a state $|s, M\rangle$, which is an eigenstate of the spin operators $\boldsymbol{s}_i^2$ and s_{iz}: $\boldsymbol{s}_i^2 |s, M\rangle = s(s+1) |s, M\rangle$ and $s_{iz} |s, M\rangle = M |s, M\rangle$. This yields

$$s_i^+ |s, M\rangle = [s(s+1) - M(M+1)]^{1/2} |s, M+1\rangle, \tag{3.19}$$

$$s_i^- |s, M\rangle = [s(s+1) - M(M-1)]^{1/2} |s, M-1\rangle. \tag{3.20}$$

In terms of these operators, the Hamiltonian becomes

$$\mathcal{H} = -2J \sum_{\langle i,j \rangle} \left[\frac{1}{2} \left(s_i^+ s_j^- + s_i^- s_j^+ \right) + s_{iz} s_{jz} \right] - g\mu_B H \sum_i s_{iz}. \tag{3.21}$$

The dynamics of $\boldsymbol{s}_j$ is obtained using the Heisenberg equation of motion

$$\frac{d\boldsymbol{s}_j}{dt} = \frac{i}{\hbar} [\mathcal{H}, \boldsymbol{s}_j] = -\frac{1}{\hbar} \left(\boldsymbol{\epsilon}_j \times \boldsymbol{s}_j \right), \tag{3.22}$$

where

$$\epsilon_j = 2J \sum_{i \neq j} s_i + g\mu_{\mathrm{B}} H \qquad (3.23)$$

is the effective magnetic field acting on s_j. The second relation of Eq. (3.22) can be obtained from the commutation relations for spin operators.

When the external magnetic field is $H = (0, 0, H)$, each spin is aligned along the z-axis. In the case of low-energy excited states, the deviation of the spin from the z-direction is small and the change of the z-component of s_i is a small quantity of second order. Then one can approximate $s_{jz} \simeq s$. Substituting Eq. (3.23) with $H = (0, 0, H)$ into Eq. (3.22), we obtain

$$\hbar \frac{ds_{jx}}{dt} = -2Js \sum_i (s_{iy} - s_{jy}) + g\mu_{\mathrm{B}} H s_{jy}, \qquad (3.24)$$

$$\hbar \frac{ds_{jy}}{dt} = -2Js \sum_i (s_{jx} - s_{ix}) - g\mu_{\mathrm{B}} H s_{jx}. \qquad (3.25)$$

Using $s_j^{\pm} = s_{jx} \pm i s_{jy}$, we have

$$\hbar \frac{ds_j^{\pm}}{dt} = \pm i \left[-2Js \sum_i (s_j^{\pm} - s_i^{\pm}) - g\mu_{\mathrm{B}} H s_j^{\pm} \right]. \qquad (3.26)$$

The local change of spins propagates through the whole spin system through the first term on the right-hand side. The coupled motion of neighboring spins can be decoupled by exploiting the periodicity with the Bloch representation:

$$s_{\boldsymbol{k}}^{\pm} = \frac{1}{\sqrt{N}} \sum_j e^{-i\boldsymbol{k} \cdot \boldsymbol{R}_j} s_j^{\pm}. \qquad (3.27)$$

With these normal coordinates, we arrive at

$$\frac{\hbar}{i} \frac{ds_{\boldsymbol{k}}^{\pm}}{dt} = \left[2Js \sum_i \left(1 - e^{-i\boldsymbol{k} \cdot (\boldsymbol{R}_i - \boldsymbol{R}_j)} \right) + g\mu_{\mathrm{B}} H \right] s_{\boldsymbol{k}}^{\pm}. \qquad (3.28)$$

Assuming $s_{\boldsymbol{k}}^{-} \propto \delta s_{\boldsymbol{k}} e^{i\omega_{\boldsymbol{k}} t + i\alpha}$, we find the eigenfrequency of Eq. (3.27) as

$$\hbar \omega_{\boldsymbol{k}} = 2JsZ (1 - \gamma_{\boldsymbol{k}}) + g\mu_{\mathrm{B}} H, \qquad (3.29)$$

where Z is the number of nearest neighbors and $\gamma_{\boldsymbol{k}} = (1/Z) \sum_{\boldsymbol{R}_i - \boldsymbol{R}_j} e^{i\boldsymbol{k} \cdot (\boldsymbol{R}_i - \boldsymbol{R}_j)}$. The above discussion shows that the whole spin configuration of a crystal behaves as an oscillatory motion with frequency $\omega_{\boldsymbol{k}}$ and wavevector $\boldsymbol{k}$. This collective mode is the spin wave that is mediated by the exchange interaction. This collective mode corresponds to a coherent precession of the individual spins around the direction of the ferromagnetic orientation. It is completely analogous to the lattice modes in a solid.

For a cubic lattice where the nearest neighbors are along the $\pm x$, $\pm y$, and $\pm z$ axes at a distance a, $\gamma_{\boldsymbol{k}} = (1/3)\left(\cos k_x a + \cos k_y a + \cos k_z a\right)$. Therefore, for small k, we have a quadratic dispersion relation for spin waves:

$$\hbar\omega_{\boldsymbol{k}} = g\mu_B H + 2Jsa^2 k^2. \tag{3.30}$$

This shows a quadratic dependence on the wavevector around the minimum at $\boldsymbol{k} = 0$.

3.3.2 *Spin current carried by a spin wave*

Here we show that spin-wave propagation carries spin angular momentum. In the following, the Gilbert damping term is neglected for simplicity. We can rewrite the LLG equation as

$$\frac{\partial}{\partial t}\boldsymbol{M}(\boldsymbol{r}, t) = \gamma\boldsymbol{H}_{\text{eff}} \times \boldsymbol{M}(\boldsymbol{r}, t) - \mathrm{div}\boldsymbol{j}^{M_\alpha}(\boldsymbol{r}, t). \tag{3.31}$$

Here, $\boldsymbol{j}^{M_\alpha}$ is the exchange spin current, defined above, whose components are

$$j_\beta^{M_\alpha} = \frac{D}{M_s}[\boldsymbol{M} \times \nabla_\beta \boldsymbol{M}]_\alpha. \tag{3.32}$$

The z-component of the LLG equation (3.31) gives a continuity equation for exchange spin currents: $\partial M_z/\partial t + \mathrm{div}\boldsymbol{j}^{M_z} = 0$, which represents spin angular momentum conservation, when $\alpha = 0$.

Here we consider an exchange spin current carried when a spin wave is excited. We introduce a spin-wave wavefunction $\psi(\boldsymbol{r}, t) = M_+(\boldsymbol{r}, t) = M_x(\boldsymbol{r}, t) + iM_y(\boldsymbol{r}, t)$ and its complex conjugate $\psi^*(\boldsymbol{r}, t)$. The z-component of the exchange spin current is written as

$$j_\beta^{M_z} = \frac{1}{2i}\frac{D}{M_s}\left[\psi^*(\boldsymbol{r}, t)\nabla_\beta\psi(\boldsymbol{r}, t) - \psi(\boldsymbol{r}, t)\nabla_\beta\psi^*(\boldsymbol{r}, t)\right]. \tag{3.33}$$

By introducing creation and annihilation operators $(b_q^\dagger, b_q)$ of spin-wave excitations (magnons) with frequency ω_q and wave number q by $\psi = M_+ = \sqrt{2/M_s}\sum_q b_q e^{i\boldsymbol{q}\cdot\boldsymbol{r}}$ and $\psi^* = M_- = \sqrt{2/M_s}\sum_q b_q^\dagger e^{-i\boldsymbol{q}\cdot\boldsymbol{r}}$, the exchange spin current is expressed as

$$j_x^{M_z} = \sum_{p,q} \nu_q n_q, \tag{3.34}$$

where $\nu_q = \partial\omega_q/\partial q = 2Dq$ is the spin-wave group velocity and $n_q = \langle b_q^\dagger b_q\rangle$ is the number of spin waves. Equation (3.34) means that, when the numbers of excited spin waves are different between q and $-q$ in k-space, a nonzero net exchange spin current is carried by the spin waves: a spin-wave spin current. Such a spin-wave property was observed in Ref. [2].

References

[1] Chikazumi, S. (2009). *Physics of Ferromagnetism (International Series of Monographs on Physics)*; 2nd edition. Oxford University Press, New York.

[2] Kajiwara, Y., Harii, K., Takahashi, S., Ohe, J., Uchida, K., Mizuguchi, M., Umezawa, H., Kawai, H., Ando, K., Takanashi, K., Maekawa, S., and Saitoh, E. (2010). *Nature (London)*, **464**, 262–266.

4 Topological spin current

E. Saitoh

In the previous chapter, we showed that the exchange interaction among spins can be rewritten by introducing an equilibrium exchange spin current. In this section, another type of equilibrium spin current is discussed, which is a topological spin current. Topological spin currents are driven by topological band structure and classified into bulk and surface topological spin currents.

4.1 Bulk topological spin current

Electrons in crystals are confined onto electron-band manifolds and their motions are sometimes affected by this confinement. This contribution can be argued in terms of Berry's phase. Here we go over a standard method to treat this problem [1], a method combining the equations of motion and the Boltzmann equation for semi-classical electrons in a band. This method considers a wavepacket of electrons and tracks its motion, assuming that its position and momentum are defined with moderate accuracy without violating the uncertainty principle. We obtain the following equations assuming that the band index n does not change because interband transitions of electrons do not happen under a weak external perturbation

$$\frac{d\boldsymbol{r}}{dt} = \frac{\partial \epsilon_n(\boldsymbol{k})}{\partial \boldsymbol{k}}, \tag{4.1}$$

$$\frac{d\boldsymbol{p}}{dt} = -e\left(\boldsymbol{E} + \hbar\frac{d\boldsymbol{r}}{dt} \times \boldsymbol{B}\right). \tag{4.2}$$

Simple semiclassical motion of an electron is determined without knowledge of the wavefunctions. However, exact motions in solids must be modified reflecting the electron-band curvature, which is represented by the connection of a wavefunction in k-space. This is due to the band structure which causes the Hilbert space to be projected into electron-band manifolds. The operator for the position of the electrons $\boldsymbol{r} = (x_j) = (x, y, z)$ is canonically conjugate with the wavenumber vector $\boldsymbol{k} = (k_j)$ and satisfies the commutation relation

$$[x_i, k_j] = i\delta_{i,j}, \tag{4.3}$$

thus

$$\boldsymbol{r} = i\nabla_{\boldsymbol{k}} \tag{4.4}$$

holds. Due to the curvature of the electron-band manifold, this position operator should be generalized into the gauge covariant derivative using the "vector potential" in $\boldsymbol{k}$-space expressing the connection due to the curvature

$$a_{nj}(\boldsymbol{k}) = -i\langle n\boldsymbol{k}|\nabla_k|n\boldsymbol{k}\rangle \tag{4.5}$$

and becomes

$$\boldsymbol{r} = i\nabla_{\boldsymbol{k}} - \boldsymbol{a}_n(\boldsymbol{k}), \tag{4.6}$$

and a nontrivial noncommutation relationship holds:

$$[x_i, x_j] = -i\epsilon_{ij}\frac{\partial}{\partial k_i}a_{nj}(\boldsymbol{k}). \tag{4.7}$$

This noncommutation relationship modifies the equation of motion as

$$\dot{x}_\mu = -\frac{i}{\hbar}[x_\mu, \mathcal{H}] = -\frac{i}{\hbar}[x_\mu, k_\nu]\frac{\partial}{\partial k_\mu}\mathcal{H} - \frac{i}{\hbar}[x_\mu, x_\nu]\frac{\partial}{\partial x_\nu}\mathcal{H}$$

$$= \frac{\partial}{\hbar\partial k_\mu}\epsilon_n(\boldsymbol{k}) - \frac{\partial}{\partial k_\mu}a_{n\nu}(\boldsymbol{k})\frac{\partial}{\partial x_\nu}\frac{V(\boldsymbol{r})}{\hbar}. \tag{4.8}$$

The second term on the right-hand side is a new term called the anomalous velocity. This term generates an electric current (topological current), even under thermal equilibrium. The anomalous velocity is one mechanism for the spin Hall effect and the anomalous Hall effect.

Taking this effect into consideration, the semiclassical equation of motion of electrons in solids becomes [1]

$$\frac{d\boldsymbol{r}}{dt} = \frac{\partial\epsilon_n(\boldsymbol{k})}{\hbar\partial\boldsymbol{k}} + \frac{d\boldsymbol{k}}{dt}\times\boldsymbol{b}_n(\boldsymbol{k}), \tag{4.9}$$

$$\frac{d\boldsymbol{k}}{dt} = -e\left(-\frac{\partial\phi(\boldsymbol{r})}{\partial\boldsymbol{r}} + \frac{d\boldsymbol{r}}{dt}\times\boldsymbol{B}(\boldsymbol{r})\right). \tag{4.10}$$

In these equations, the duality of r and k is obvious; using the field $\boldsymbol{b}_n(\boldsymbol{k})$ instead of the magnetic field $\boldsymbol{B}(\boldsymbol{r})$ and using $\epsilon_n(\boldsymbol{k})$ instead of the electrostatic potential $\phi(\boldsymbol{r})$ results in a dual relationship between the equations of motion for $\boldsymbol{r}$ and $\boldsymbol{k}$. The second term on the right-hand side is the anomalous velocity, and the Hall current can be obtained by adding the states occupied by electrons

$$j_x = -e\sum_{\boldsymbol{k},n} f\big(\epsilon_n(\boldsymbol{k})\big)b_{nz}(\boldsymbol{k})\dot{k}_y. \tag{4.11}$$

If an electric field exists in the y-direction only, $\dot{k}_y = -eE_y$, thus

$$\sigma_{xy} = e^2\sum_{\boldsymbol{k},n} f\big(\epsilon_n(\boldsymbol{k})\big)b_{nz}(\boldsymbol{k}). \tag{4.12}$$

In paramagnetic or diamagnetic metals, when the gauge field $\boldsymbol{b}_n(\boldsymbol{k})$ is spin dependent, electrons with different spins have different velocities; an electric current is converted into a spin current of conduction electrons. Such a spin current is a bulk topological spin current.

4.2   Surface topological spin current

The spin currents discussed above are basically those that flow in bulk. Finally, the other type of spin current is introduced very quickly: a surface (edge) spin current, which is limited near surfaces (edges) of a three (two)-dimensional system and flows along the surfaces (edges).

This surface spin current is known to appear in topological insulators. In topological insulators, the bulk is insulating but the surface or edge is electrically conducting due to the surface or edge state: an electronic state localized at the surface/edge. In such a system, the spin degeneracy of the surface (edge) state is lifted except for the $k = 0$ point and the surface (edge) states of wavevector k and $-k$ have opposite spin. The situation means the state accompanies a spin current in an equilibrium state even without external perturbation.

There are two-dimensional and three-dimensional topological insulators. In two-dimensional topological insulators, spin-flip scattering in the edge states is predicted to be significantly suppressed due to the absence of spin degeneracy. The details will be discussed in Chapter 17.

References

[1] Xiao, D., Chang, M. C., and Niu, Q. (2010). *Rev. Mod. Phys.*, **82**, 1950.
[2] Morrish, A. H. (1980). *The Physical Principles of Magnetism*. Robert E. Krieger, New York.

5 Spin polarization in magnets

K. Takanashi and Y. Sakuraba

5.1 Spin polarization in ferromagnets

The exchange splitting between up- and down-spin bands in ferromagnets unexceptionally generates spin-polarized electronic states at the Fermi energy (E_F). The quantity of spin polarization P in ferromagnets is one of the important parameters for application in spintronics since a ferromagnet having a higher P is able to generate larger various spin-dependent effects such as the magnetoresistance effect, spin transfer torque, spin accumulation, and so on.

Usually, P is defined as

$$P = \frac{D_\uparrow(E_F) - D_\downarrow(E_F)}{D_\uparrow(E_F) + D_\downarrow(E_F)} \tag{5.1}$$

where $D_{\uparrow(\downarrow)}(E_F)$ is the density of states (DOS) for the up- (down-) spin channel at the Fermi level. A typical transition ferromagnet has two components of electric structure near E_F: narrow d-bands that are highly spin polarized due to the exchange energy and broad s-bands with low degree of spin polarization due to hybridization with d-bands. Thus, if the orbital character at the Fermi surface of a ferromagnet is d-like, naturally P expressed by Eq. (5.1) will be high. However, when it comes to spin-dependent metallic transport or tunneling, the expression for P written in Eq. (5.1) is not sufficient to describe the actual spin polarization of electric carriers, because carrier mobility, effective mass, and tunneling probability, which differs absolutely in orbital character, must be taken into consideration. Thus, the spin polarization of a conduction electron P_c, considering a single transport mode at E_F, is often written as

$$P_C = \frac{D_\uparrow(E_F)\nu_{F\uparrow} - D_\downarrow(E_F)\nu_{F\downarrow}}{D_\uparrow(E_F)\nu_{F\uparrow} + D_\downarrow(E_F)\nu_{F\downarrow}} \tag{5.2}$$

where $\nu_{F\uparrow(\downarrow)}$ denotes the Fermi velocity of the up- (down-) spin band. In other words, Eq. (5.2) represents the spin asymmetry of the conductivity $(\sigma_\uparrow \sigma_\downarrow)/(\sigma_\uparrow + \sigma_\downarrow)$, which is identical to the value of β defined in Valet and Fert's model for giant-magnetoresistive devices [1]. In the case of tunneling, the tunneling spin polarization P_T, considering a single transport mode at E_F, is expressed as follows,

$$P_T = \frac{D_\uparrow(E_F)|T_\uparrow|^2 - D_\downarrow(E_F)|T_\downarrow|^2}{D_\uparrow(E_F)|T_\uparrow|^2 + D_\downarrow(E_F)|T_\downarrow|^2} \qquad (5.3)$$

where $T_{\uparrow(\downarrow)}$ are the spin-dependent tunneling matrix elements for the up- (down-) spin band. The tunneling effect is sensitive to the barrier/electrode interface, so the meaning of $D_{\uparrow(\downarrow)}(E_F)$ in Eq. (5.3) is the density of states at the barrier interface. It is easily seen from Eqs. (5.1)–(5.3), that the quantities and even the signs of P_C and P_T are often different from those of P. It is important to notice that, obtaining high P_C or P_T, and not P, is required to enhance spin-dependent transport properties. A material with a large P originated by narrow d-states at E_F in one spin channel often shows poorer spin-dependent transport properties than we expected because of the large effective mass and small tunneling probability of d-electrons.

5.2 Half-metallic ferromagnets

As mentioned in Section 5.1, the spin polarization of a ferromagnet is one of the important parameters for various spintronic phenomena. However, the spin polarizations of general $3d$ transition metals or alloys are generally below 0.6, which limits the size of spin-dependent effects. Thus, "half-metals" are attracting much interest as an ideal source of spin(-polarized) current and spin-dependent scattering because they possess perfectly spin-polarized conduction electrons, i.e. $P = 1$, due to the energy band gap in either the up- or down-spin channel at the Fermi level. For example, according to the conventional model for tunneling magnetoresistance (TMR) proposed by Julliere [2], the TMR ratio is expressed as $2P_L P_R/(1-P_L P_R)$, where $P_{L(R)}$ is the spin polarization of the left (right) ferromagnetic electrode. Thus, the TMR ratio becomes infinite in the ideal case of using half-metals for both electrodes. Schmidt *et al.* predicted that the conductance mismatching problem, which is a main obstacle to injecting spin-polarized current into a semiconducting material from a ferromagnetic metal, can be solved by using a half-metal as a spin-injector to the semiconductor [3]. Other various kinds of spin-dependent phenomena like spin accumulation, the spin Hall effect, and spin-torque induced phenomena can also be largely enhanced using half-metals. As candidates for half-metals, some Heusler compounds [4–6], zinc-blende structure materials [7], and magnetic oxides (CrO_2 [8] and Fe_3O_4 [9]) etc. were predicted to have a half-metallic electric structure by first-principles calculations. However, although in several candidates of half-metals the high spin polarization reflecting half-metallicity was clearly observed at low temperature (LT), the evidence of half-metallicity has never been confirmed at room temperature (RT) so far.

Some of the Heusler compounds having $L2_1$-or $C1_b$-structures are promising candidates showing half-metallicity at RT because of their high Curie temperatures and chemical stability. The $L2_1$-structure consists of four fcc sublattices (general chemical formula is X_2YZ), and the $C1_b$-structure has one unoccupied sublattice for X atoms (i.e. XYZ), which are often called full- and half-(semi-) Heusler compounds, respectively. In 1983 de Groot *et al.* showed by first-principles calculation that one of the half-Heusler compounds, NiMnSb has a semiconducting gap at the Fermi level in only the down-spin channel leading to 100% spin polarization [4]. The other half-Heusler compounds with XMnSb or XCrSb composition such as PdMnSb and NiCrSb, etc. were also predicted to have a half-metallic electronic structure. Some of the full-Heusler compounds are other choices of half-metals; Co_2MnX (X = Al, Si, Ge, etc.) is one of the popular compounds because of their high Curie temperatures of 600–1000 K [5,6]. Other quaternary compounds such as $Co_2(Cr,Fe)Al$ [10, 11], $Co_2Fe(Al,Si)$ [12–14], and $Co_2(Mn,Fe)Si$ [15] are also promising candidates. Apart from the Co-based full-Heusler compounds, Fe_2CrAl [16] and Mn_2VAl [17], etc. were predicted to have half-metallicity with small magnetization.

In half- and full-Heuser-based half-metals, the number of occupied valence spin-down states given by the number of spin-down bands (N_{down}) is 9 and 12, respectively. Thus, it is well known that the total magnetic moments (M_t) of the half-metallic half- and full-Heusler compounds must be an integer number following the simple Slater–Pauling rule: $M_t = Z_t - 2N_{down}$, where Z_t is the total number of valence electrons. For example, Z_t in Co_2MnSi is 29, thus $M_t = 29 - 24 = 5$ μ_B.

Figure 5.1 shows the spin- and element-resolved density of states in Co_2MnSi with perfect $L2_1$-structure. It is noteworthy that the half-metallic energy gap in the Co-based full-Heusler compounds is derived from the d-states of Co. More concretely, the conduction and valence edges of the gap are formed by the anti-bonding (t_{1u} and e_u, respectively) states generated from the hybridization of the nearest Co-Co as shown in Fig. 5.1 [7]. Thus, it is easily predicted that the half-metallicity is very fragile against chemical disordering and surface/interface termination involving Co atoms. Picozzi *et al.* reported that the creation of a Co anti-site destroys the half-metallicity in Co_2MnSi and Co_2MnGe due to the formation of in-gap states [18]. The disappearance of half-metallicity was also predicted at the Co-terminated [001] surface/interface in contrast to the preservation of half-metallicity at the Mn-Si termination [19,20]. Therefore the control of the chemical ordering and the terminated surface/interface is a critical issue in obtaining the half-metallicity in half-metallic Heusler compounds.

In experiments, previous studies did not indicate the half-metallicity in half-Heusler compounds so far; a maximum value of 58% for the spin polarization was obtained by a point-contact Andreev reflection method [20]. The superconducting tunnel junction and the magnetic tunnel junction with NiMnSb electrodes did not show a high spin polarization, suggesting half-metallicity even at LT [21,22]. Although nearly 100% polarization was reported on NiMnSb(100) sputtered thin films using spin-polarized inverse photoemission, these measurement were k resolved and therefore do not directly demonstrate the half-metallic character across the entire Brillouin zone [23,24]. On the other hand, recent extensive stud-

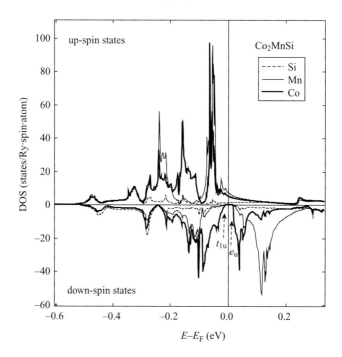

FIG. 5.1. Calculated element and spin-resolved DOS in Co$_2$MnSi on the basis of first principles.

ies for the full-Heusler compounds clearly confirmed their half-metallic nature. The MTJ with a Co$_2$MnSi/Al-O/Co$_2$MnSi structure showed a giant TMR ratio of 560% at 2 K, indicating a high spin polarization of 86% for Co$_2$MnSi electrodes [25]. In current-perpendicular-to-plane giant magnetoresistive (CPP-GMR) devices with Co$_2$MnSi, Co$_2$FeAl$_{0.5}$Si$_{0.5}$, and Co$_2$FeGa$_{0.5}$Ge$_{0.5}$, large MR ratios over 30% were observed at RT, which is one order of magnitude larger than those reported in devices with normal transition metals [26–29]. The details will be mentioned in Section 5.4

5.3 Experimental techniques for spin-polarization measurement

5.3.1 *Point-contact Andreev reflection (PCAR)*

Point-contact Andreev reflection (PCAR) is the simplest approach to measure P_C for a metal by just making a metallic point contact between the sample and a superconductor (SC) [20]. This method is convenient since it requires no magnetic field and no special constraints on the sample; thin films, single crystals, and foils of several metals have been successfully measured [20]. The principle of the method is depicted in Fig. 5.2. A metallic contact allows coherent two-particle transfer at the interface between the metal and the SC. The conversion between

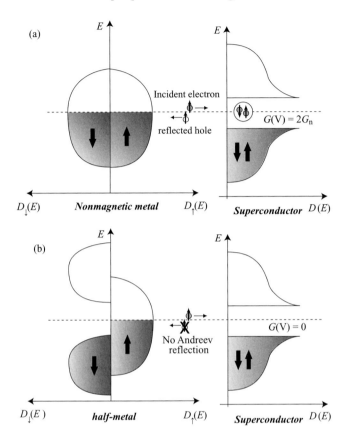

FIG. 5.2. Schematic illustrations of supercurrent conversion at the supercon-
ductor/metal interface due to Andreev reflection for a nonmagnetic metal
($P_C = 0$, (a)) and a half-metal ($P_C = 1$, (b)).

Cooper pairs and the single-particle charge carriers of the metal at the interface
(called Andreev reflection) gives information about P_C. The electron entering the
SC from the metal must condense and proceed as part of the supercurrent, thus
it becomes a member of a pair. Because a superconducting pair is composed
of a spin-up and spin-down electron, the other electron having an opposite
spin direction is obtained from the metal for the formation of the pair. Thus
a hole is left behind and propagates away from the interface. Fig. 5.2(a) shows
the case of a superconducting contact with a nonmagnetic metal, i.e. $P_C = 0$.
The Andreev reflected holes double the normal-state conductance G_n of the
applied voltages eV< Δ, where Δ is the superconducting gap at the interface.
In contrast, the conductance G become zero within Δ in the case of contact
with a half-metal since there is no supercurrent conversion at the interface due

to the lack of electrons with one spin direction. For the quantitative analysis of P_C, the Blonder–Tinkham–Klapwijk (BTK) theory modified by considering spin polarization is applied to fit an observed G-V curve, where the interfacial scattering at the point contact is taken into account as a scattering parameter Z [30, 31]. A ballistic contact with no scattering has $Z = 0$, whereas a tunnel junction corresponds to the limit $Z \to \infty$. The intrinsic P_C is obtained at the limit of ballistic contact, thus, P_C is usually estimated by extrapolating Z to 0 in the experimentally observed Z dependence of P_C. Although the P_C for CrO_2 and LSMO films was found to be high enough over 90%, suggesting half-metallicity [20,32,33], large spin-polarization has never been observed by PCAR in the half-metallic candidates of Heusler compounds, showing typically 50–70% [34–37]. Since there are still ambiguities in the analysis of G-V curve especially in the understanding of the scattering parameter Z [38], the model for the analysis is being gradually improved for more reliable determination of P_C [39].

5.3.2 *Superconducting tunneling spectroscopy (STS)*

Spin-polarized tunneling spectroscopy in a superconductor (SC)/tunnel barrier (I)/ ferromagnetic metal (FM) junction, called superconducting tunneling spectroscopy (STS), is a traditional and powerful technique to measure the tunneling spin polarization P_T using a quasi-particle tunneling from SC, which was first developed by Meservey and Tedrow [40]. As shown in Fig. 5.3(b), in an applied magnetic field $H = 0$, the peaks of the tunnel conductance $(\mathrm{d}I/\mathrm{d}V)$ indicate that the superconducting gap edges appear at the same voltage (ΔV) for the spin-up and spin-down electron channels, thus only two peaks are observed. When H is applied in parallel to the film plane, the quasi-particle DOS is energetically separated by $2\mu_B H$ according to their spin orientation, which is a phenomenon known as Zeeman splitting (Fig. 5.3a). An electron passes from/to these Zeeman split quasi-particle states, which gives rise to two split tunnel conductance $(\mathrm{d}I/\mathrm{d}V)$ peaks for each spin-up and spin-down electron channel as shown in Fig. 5.3(c). From the shape of the $\mathrm{d}I/\mathrm{d}V$–V curve the tunneling spin polarization P_T near E_F can be quantitatively analyzed. The details of the method are well described in the review paper by Meservey and Tedrow [41].

Popular SC materials for this measurement are Al, or Al with slight Cu or Si impurities [42] since the spin–orbit interaction in Al is small enough to observe a splitting of the spin-up and down channels in the $\mathrm{d}I/\mathrm{d}V$–V curve. However, a big disadvantage of Al is low superconducting transition temperature (T_c) below ~ 2.5 K, thus measurements must be made at even lower temperatures, typically below ~ 0.4 K. Yang *et al.* used NbN as an alternative superconductor and successfully observed a splitting of the peak even at 1.2 K due to the high T_c of NbN ~ 16 K [43].

It is well known that tunneling spin polarization P_T cannot be explained by the total DOS at E_F in a ferromagnet, since it is necessary to consider the mobility of electrons (sp-like or d-like) at E_F [44], interfacial bonding, and

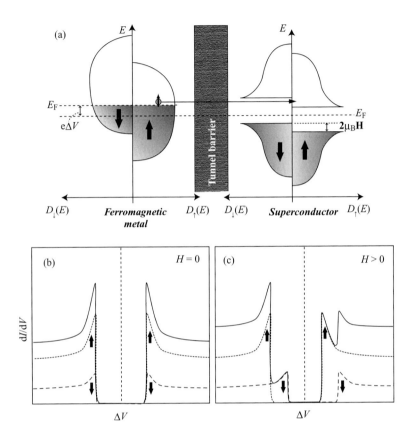

FIG. 5.3. (a) Electron DOS of the ferromagnet and quasi-particle DOS of the
SC electrode when the magnetic field H and bias voltage $\Delta V (= V_{\mathrm{FM}} - V_{\mathrm{SC}})$
are applied. A Zeeman splitting of $2\mu_{\mathrm{B}}H$ is induced by the external magnetic
field. (b) $dI/dV - \Delta V$ curve at $H = 0$ and (c) $H > 0$.

wavefunction symmetry, including the spin-detector side, the material and the
thickness of the tunneling barrier [45,46], and various other extrinsic factors
that affect spin transport such as magnetic impurities at the interface/inside the
barrier. For example, there is a large DOS for the minority d band at E_{F} with
near -100% for P, but the observed P_{T} in Ni was always positive, 23% [41] and
46% [42]. In the case of a $\mathrm{La_{0.7}Sr_{0.3}MnO_3/SrTiO_3/Co}$ junction [47], the observed
inverse TMR ratio was interpreted by a negative P_{T} of Co, whereas a positive
P_{T} was always measured in $\mathrm{Co/SrTiO_3/Al}$ and $\mathrm{Co/Al\text{-}O/Al}$ junctions, which
indicated the importance of the spin-detector and tunneling barrier materials

[42,48]. The details of tunneling spin polarization are well described in the recent review paper by Miao *et al.* [49].

5.3.3 *Spin-resolved photoemission spectroscopy (SP-PES)*

Photoemission spectroscopy (PES) is a general technique to investigate the electric structure of valence states or inner shells by detecting photoelectrons excited from those occupied states by incident photon with the energy in the region from ultraviolet light to X-rays. The information about band dispersion is also obtainable by angle-resolved detection of emitted photoelectrons. In order to obtain spin resolution, a Mott detector [50,51] or spin low-energy diffraction (SPLEED) detector [52], etc. are used to count photoelectrons with spin information. Previously, the low efficiency of spin detection was a critical problem to investigate a detailed spin-resolved electric structure with high energy resolution, in contrast to the excellent energy resolution of spin-unresolved photoemission spectroscopy $\sim$ a few meV, but recent improvements of the instruments is gradually making it possible to obtain higher energy resolution of 8–30 meV in spin-resolved photoemission spectroscopy (SP-PES) [53,54]. The strong surface sensitivity of photoemission spectroscopy is also very attractive to study the spin splitting due to large spin–orbit interaction by no space reversal asymmetry at the surface or interface.

The spin polarization (P) obtained from SP-PES is basically the spin polarization of the total DOS, which is different from both P_C and P_T observed by PCAR and STS. P at the surface of various ferromagnets has been studied by spin-resolved PES for three decades [50,51]. An almost fully spin-polarized state at E_F was claimed in several half-metallic materials: a P of 100% and 95% was observed in LSMO [55] and CrO_2 [56], respectively, by SP-PES spin-resolved inverse PES measurements also found a half-metallic P of 100% for LSMO [57] at 100 K and NiMnSb at 300 K [58]. However, there is skepticism of the claims of half-metallicity from these SR-PES measurements [24], because finite-temperature effects leading to the population of spin minority states near E_F would be most significant at wavevectors away from $k_{11} = 0$, thus they may not be observed by SR-PES at normal emission if the sample is single-crystalline or is polycrystalline with texture growth [23,55–58].

5.4 Magnetoresistive devices with half-metals

5.4.1 *Magnetic tunnel junctions with half-metals*

One of the most attractive applications of a half-metal is its use as FM electrodes for magnetic tunnel junctions (MTJs) because the TMR ratio is drastically enhanced when P_T becomes near unity as is easily expected from Julliere's formula [2]. A pioneering result in MTJ with half-metallic electrodes was reported for the $La_{0.7}Sr_{0.3}MnO_3$(LSMO)/$SrTiO_3$/LSMO structure, where a giant TMR ratio of 1800% ($P_T = 95\%$ for LSMO) was observed at 4.2 K [58]. However, the TMR ratio perfectly disappeared at RT because of low Curie temperature of

LSMO ($T_C \sim$ 350–370 K). In MTJs with other half-metallic oxides, CrO_2 and Fe_3O_4, large TMR ratios, suggesting a half-metallicity, have never been observed [59,60], whereas, large TMR ratios reflecting half-metallicity were observed in MTJs with various half-metallic candidates in Heusler compounds; a giant TMR ratio of 570% was observed in the $Co_2MnSi/Al-O/Co_2MnSi$ MTJ at 2 K, indicating a P_T of 86% for Co_2MnSi electrodes [25]. A recent trend is to combine a half-metallic Heusler electrode and a MgO crystalline barrier, where the multiplied enhancement effects of the TMR ratio, i.e. half-metallicity and spin-filter effect of the MgO barrier, are theoretically expected [61]. Tezuka *et al.* reported a large MR ratio of 832% at 9 K in the $Co_2FeAl_{0.5}Si_{0.5}/MgO/Co_2FeAl_{0.5}Si_{0.5}$-MTJ [62], and Taira *et al.* realized a giant TMR ratio of 1804% at 2 K in the Co-Mn-Si/MgO/Co-Mn-Si MTJ [63]. A critical issue in half-metallic Heusler-based MTJs is the rapid reduction of the TMR ratio with temperature as shown in Fig. 5.4, which is not simply understood because the Curie temperatures of those materials are much higher than RT (e.g. $T_C \sim$ 985 K for Co_2MnSi). There is still a controversy about the origin of this large temperature dependence;

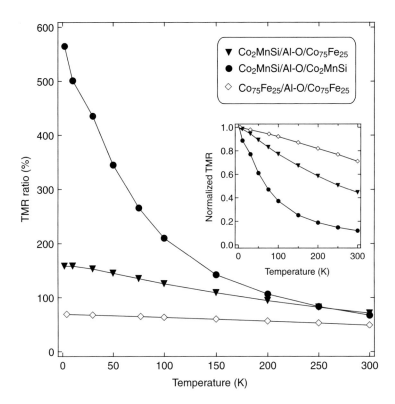

FIG. 5.4. Temperature dependence of the TMR ratio in the Co_2MnSi-based MTJs and CoFe-based MTJ. The inset shows the data normalized at 2 K.

Mavropoulos *et al.* suggested that interface states appearing in the minority-spin gap can contribute to the tunneling conductance in the antiparallel state through spin-mixing processes such as magnon excitations and inelastic scattering at RT [64]. Sakuma *et al.* theoretically predicted the reduction of the exchange energy of the Co layer terminated at the MgO barrier in the (001)-oriented CMS/MgO structure, suggesting large magnon excitation at the CMS/MgO interface [19]. It is suggested as a solution of the temperature dependence to insert another thin ferromagnetic layer at the barrier interface to suppress the creation of interface states [65] and improve the exchange energy at the interface. Although Tsunegi *et al.* reported a slight improvement of the temperature dependence of the TMR ratio by inserting a thin CoFeB layer at the CMS/MgO interface [66], a striking breakthrough to solve the problem at the interface seems to be required to realize high P_T reflecting the half-metallicity at RT.

5.4.2 *Current-perpendicular-to-plane magnetoresistive device with half-metals*

When an electric current flows in the direction perpendicular to the film with a stacking structure of FM layers separated by nonmagnetic metal (NM) layers, spin-dependent electron scattering at the FM/NM interfaces and inside the FM layer generates an MR effect, which is called the current-perpendicular-to-plane giant magnetoresistive (CPP-GMR) effect. According to the prediction by Mavropoulos [64], the half-metallicity of the electrodes can be fully exploited in the CPP-GMR structure in contrast to MTJ, because spin-dependent scattering not only at the interface but also inside the FM layer contributes to the MR effect, that is, the interface states in the half-metallic gap play almost no significant role in the CPP-GMR structure. In the (001)-Co_2MnSi/Ag/ Co_2MnSi epitaxial device, a large MR ratio of 36% was reported at RT, which was one order of magnitude higher than that observed in normal $3d$-transition FM based CPP-GMR devices [26,27]. Spin asymmetries of resistance at the interface (γ) and in the FM layer (β) have been analyzed on the basis of the Valet–Ferts model [1]. γ and β are expressed as $(R_\downarrow - R_\uparrow/R_\downarrow + R_\uparrow)$ and $(\rho_\downarrow - \rho_\uparrow/\rho_\downarrow + \rho_\uparrow)$, respectively, where $R_{\uparrow(\downarrow)}$ and $\rho_{\uparrow(\downarrow)}$ indicate the interface resistance and the resistivity for up (down) spin electrons. As a result of the analysis, a large γ over 0.8 at the Co_2MnSi/Ag interface was found, and was explained by the good Fermi surface matching at the (001)-Co_2MnSi/Ag interface according to this first-principles calculation [27]. Nakatani *et al.* also observed a large MR ratio of 34% in the $Co_2FeAl_{0.5}Si_{0.5}$/Ag/ $Co_2FeAl_{0.5}Si_{0.5}$ epitaxial device and a large β of 0.71–0.78, indicating a high spin polarization of conduction electrons [28]. β can also be expressed as $(\sigma_\uparrow - \sigma_\downarrow)/(\sigma_\downarrow - \sigma_\uparrow)$, which corresponds to the definition of P_C. Thus, in principle, the P_C observed by PCAR should be equal to the evaluated β in CPP-GMR devices. In other words, an FM material showing high P_C by PCAR is promising as electrodes for a CPP-GMR device to enhance the MR effect. Takahashi *et al.* investigated P_C for $Co_2Fe(Ga_xGe_{1-x})$ with a different x by PCAR and found the highest P_C of 0.69 at $x = 0.5$. They also prepared the CPP-GMR device using $Co_2Fe(Ga_{0.5}Ge_{0.5})$ and observed a large MR ratio of 41.7% at RT [29]. Although the estimated β of 0.77 for $Co_2Fe(Ga_{0.5}Ge_{0.5})$ deviated from the P_C observed by PCAR because of the ambiguity in the estimation of P_C in PCAR (mentioned in Section 5.3.1), this result seems to support the identity between P_C and β.

References

[1] T. Valet and A. Fert, *Phys. Rev. B*, 48 (1993) 7099.

[2] M. Julliere, *Phys. Lett. A* 54 (1975) 225.

[3] G. Schmidt, D. Ferrand, L. W. Molenkamp, A. T. Filip, and B. J. van Wees, *Phys. Rev. B*, 62 (2000) R4790.

[4] R. A. de Groot, F. M. Mueller, P. G. van Engen, and K. H. J. Buschow, *Phys. Rev. Lett.* 50 (1983) 2024.

[5] I. Galanakis, P. H. Dederichs, and N. Papanikolaou, *Phys. Rev. B* 66 (2002) 134428.

[6] S. Ishida, S. Fujii, S. Kashiwagi, and S. Asano, *J. Phys. Soc. Jpn.* 64 (1995) 2152.

[7] I. Gakanakis, *Phys. Rev. B* 66 (2002) 012406.

[8] K. Schwart, *J. Phys. F: Met. Phys.* 16 (1986) L211.

[9] Z. Zhang and S. Satpathy, *Phys. Rev. B*, 19 (1991) 44319.

[10] Y. Miura, K. Nagao, and M. Shirai, *Phys. Rev. B* 69 (2004) 144413.

[11] V. N. Antonov, H. A. Durr, Yu. Kucherenko, L. V. Bekenov, and A. N. Yaresko, *Phys. Rev. B* 72 (2005) 054441.

[12] N. Tezuka, N. Ikeda, S. Sugimoto, and K. Inomata, *Jpn. J. Appl. Phys., Part 2*, 46 (2007) L454.

[13] N. Tezuka, N. Ikeda, F. Mitsuhashi, and S. Sugimoto, *Appl. Phys. Lett.* 94 (2009) 162504.

[14] R. Shan, H. Sukegawa, W. H. Wang, M. Kodzuka, T. Furubayashi, T. Ohkubo, S. Mitani, K. Inomata, and K. Hono, *Phys. Rev. Lett.* 102 (2009) 246601.

[15] T. Kubota, S. Tsunegi, M. Oogane, S. Mizukami, T. Miyazaki, H. Naganuma, Y. Ando, *Appl. Phys. Lett.* 94 (2009) 122504.

[16] I. Galanakis, P. H. Dederichs, and N. Papanikolaou, *Phys. Rev. B* 66 (2002) 174429.

[17] R. Weht and W. E. Pickett, *Phys. Rev. B* 60 (1999) 13006.

[18] S. Picozzi, A. Continenza, and A. J. Freeman, *Phys. Rev. B* 69 (2004) 094423.

[19] A. Sakuma, Y. Toga, and H. Tsuchiura, *J. Appl. Phys.* 105 (2009) 07C910.

[20] R. J. Soulen Jr., J. M. Byers, M. S. Osofsky, B. Nadgorny, T. Ambrose, S. F. Cheng, P. R. Broussard, C. T. Tanaka, J. Nowak, J. S. Moodera, A. Barry, and J. M. D. Coey, *Science* 282 (1998) 85.

[21] C. T. Tanaka, J. Nowak, and J. S. Moodera, *J. Appl. Phys.* 86 (1999) 6239.

[22] C. T. Tanaka, J. Nowak, and J. S. Moodera, *J. Appl. Phys.* 81 (1997) 5515.

[23] D. Ristoiu, J. P. Nozières, C. N. Borca, T. Komesu, H.-K. Jeong, and P. A. Dowben, *Europhys. Lett.* 49 (2000) 624.

[24] P. A. Dowben and R. Skomski, *J. Appl. Phys.* 95 (2004) 7453.

[25] Y. Sakuraba, M. Hattori, M. Oogane, Y. Ando, H. Kato, A. Sakuma, and T. Miyazaki, *Appl. Phys. Lett.* 88 (2006) 192508.

[26] T. Iwase, Y. Sakuraba, S. Bosu, K. Saito, S. Mitani, and K. Takanashi, *Appl. Phys. Express* 2 (2009) 063003.

[27] Y. Sakuraba, K. Izumi, Y. Miura, K. Futasukawa, T. Iwase, S. Bosu, K. Saito, K. Abe, M. Shirai, and K. Takanashi, *Phys. Rev. B* 82 (2010) 094444.

[28] T. M. Nakatani, T. Furubayashi, S. Kasai, H. Sukegawa, Y. K. Takahashi, S. Mitani, and K. Hono, *Appl. Phys. Lett.* 96 (2010) 212501.

[29] Y. K. Takahashi, A. Srinivasan, B. Varaprasad, A. Rajanikanth, N. Hase, T. M. Nakatani, S. Kasai, T. Furubayashi, and K. Hono, *Appl. Phys. Lett.* 98 (2011) 152501.

[30] G. E. Blonder, M. Tinkham, and T. M. Klapwijk, *Phys. Rev. B* 25 (1982) 4515.

[31] G. J. Strijkers, Y. Ji, F. Y. Yang, C. L. Chien, and J. M. Byers, *Phys. Rev. B* 63 (2001) 104510.

[32] Y. Ji, G. J. Strijkers, F. Y. Yang, C. L. Chien, J. M. Byers, A. Anguelouch, G. Xiao, and A. Gupta, *Phys. Rev. Lett.* 86 (2001) 5585.

[33] B. Nadgorny, I. I. Mazin, M. Osofsky, R. J. Soulen, Jr., P. Broussard, R. M. Stroud, D. J. Singh, V. G. Harris, A. Arsenov, and Ya. Mukovskii, *Phys. Rev. B* 63 (2001) 184433.

[34] T. M. Nakatani, A. Rajanikanth, Z. Gercsi, Y. K. Takahashi, K. Inomata, and K. Hono, *J. Appl. Phys.* 102 (2007) 033916.

[35] A. Rajanikanth, Y. K. Takahashi, and K. Hono, *J. Appl. Phys.* 101 (2007) 023901.

[36] S. V. Karthik, A. Rajanikanth, Y. K. Takahashi, T. Okhkubo, and K. Hono, *Appl. Phys. Lett.* 89 (2006) 052505.

[37] B. S. D. Ch. S. Varaprasad, A. Rajanikanth, Y. K. Takahashi, and K. Hono, *Act. Mater.* 57 (2009) 2702.

[38] G. T. Woods, R. J. Soulen Jr., I. Mazin, B. Nadgorny, M. S. Osofsky, J. Sanders, H. Srikanth, W. F. Egelhoff, and R. Datle, *Phys. Rev. B* 70 (2004) 054416.

[39] T. Löfwander, R. Grein, and M. Eschrig, *Phys. Rev. Lett.* 105 (2010) 207001.

[40] P. M. Tedrow and R. Meservey, *Phys. Rev. Lett.* 26 (1971) 192.

[41] R. Meservey and P. M. Tedrow, *Phys. Rep.* 238 (1994) 173.

[42] D. J. Monsma and S. S. P. Parkin, *Appl. Phys. Lett.* 77 (2000) 720.

[43] H. Yang, S.-H. Yang, C. Kaiser, and S. Parkin, *Appl. Phys. Lett.* 88 (2006) 182501.

[44] J. W. Gadzuk, *Phys. Rev.* 182 (1969) 416.

[45] M. Mnzenberg and J. S. Moodera, *Phys. Rev. B* 70 (2004) 060402(R).

[46] W. H. Butler, X.-G. Zhang, and T. C. Schulthess, *Phys. Rev. B* 63 (2002) 054416.

[47] J. M. de Teresa, A. Barthélémy, A. Fert, J. P. Contour, R. Lyonnet, F. Montaigne, P. Seneor, and A. Vaures, *Phys. Rev. Lett.* 82 (1999) 4288.

[48] A. Thomas, J. S. Moodera, and B. Satpati, *J. Appl. Phys.* 97 (2005) 10C908.

[49] G-X Miao, M. Munzenberg, and J. S. Moodera, *Rep. Prog. Phys.* 74 (2011) 036501.

[50] R. Raue , H. Hopster and E. Kisker, *Rev. Sci. Instrum.* 55 (1984) 383.

[51] E. Kisker, R. Clauberg, and W. Gudat, *Rev. Sci. Instrum.* 53 (1982) 1137.

[52] G.-C. Wang, *Phys. Rev. B* 23 (1981) 1761.

[53] T. Okuda, Y. Takeichi, Y. Maeda, A. Harasawa, I. Matsuda, T. Kinoshita, and A. Kakizaki, *Rev. Sci. Instrum.* 79 (2008) 123117.

[54] S. Souma, A. Takayama, K. Sugawara, T. Sato, and T. Takahashi, *Rev. Sci. Instrum.* 81 (2010) 095101.

[55] J.-H. Park, E. Vescovo, H.-J. Kim, C. Kwon, R. Ramesh, and T. Venkatesan, *Nature (London)* 392 (1998) 794.

[56] K. P. Kämper, W. Schmitt, G. Güntherodt, R. J. Gambino, and R. Ruf, *Phys. Rev. Lett.* 59 (1987) 2788.

[57] R. Bertacco, M. Portalupi, M. Marcon, L. Duo, F. Ciccacci, M. Bowen, J.-P. Contour, and A. Barthelemy, *J. Magn. Magn. Mater.* 242–245 (2002) 710.

[58] M. Bowen, M. Bibes, A. Barthelemy, J.-P. Contour, A. Anane, Y. Lemaitre, and A. Fert, *Appl. Phys. Lett.* 82 (2003) 233.

[59] A. Gupta, X. W. Li, and G. Xiao, *Appl. Phys. Lett.* 78 (2001) 1894.

[60] F. Greullet, E. Snoeck, C. Tiusan, M. Hehn, D. Lacour, O. Lenoble, C. Magen, and L. Calmels, *Appl. Phys. Lett.* 92 (2010) 053508.

[61] Y. Miura, H. Uchida, Y. Oba, K. Nagao, and M. Shirai, *J. Phys.: Cond. Matt.* 19 (2007) 365228.

[62] N. Tezuka, N. Ikeda, F. Mitsuhashi, and S. Sugimoto: *Appl. Phys. Lett.* 94 (2009) 162504.

[63] T. Taira, H.-X. Liu, S. Hirata, K.-i. Matsuda, T. Uemura, and M. Yamamoto, 55th Annual Conference on Magnetism & Magnetic Materials, Abstracts (CD-ROM), pp. 118–119, BH-10, Atlanta, Georgia, USA, November 14–18, 2010.

[64] P. Mavropoulos, M. Ležai, and S. Blügel, *Phys. Rev. B* 72 (2005) 174428.

[65] Y. Miura, H. Uchida, Y. Oba, K. Abe, and M. Shirai, *Phys. Rev. B* 78 (2008) 064416.

[66] S. Tsunegi, Y. Sakuraba, M. Oogane, H. Naganuma, K. Takanashi, and Y. Ando: *Appl. Phys. Lett.* 94 (2009) 252503.

6 Optically induced and detected spin current

A. Hirohata

6.1 Introduction

6.1.1 *Optical generation of spins*

An alternative method of injecting spin-polarized electrons into a nonmagnetic semiconductor is photoexcitation. This method uses circularly polarized light, whose energy needs to be the same as, or slightly larger than, the semiconductor band-gap, to excite spin-polarized electrons. This process will introduce a spin-polarized electron–hole pair, which can be detected as electrical signals. Such an optically induced spin-polarized current can only be generated in a direct band-gap semiconductor due to the selection rule described below. This introduction of circularly polarized light can also be used for spin-polarized scanning tunneling microscopy (spin STM) as shown in Fig. 6.1.

6.1.2 *Spin polarization in GaAs*

In a direct band-gap semiconductor, such as GaAs, which is widely used in spintronics, the valence band maximum and the conduction band minimum are aligned at the Γ-point as shown in Fig. 6.2(a). This point is the center of the Brillouin zone ($\mathbf{k} = 0$), indicating that the only transition induced by a photon of energy $h\nu$ occurs at Γ [1,2]. For GaAs, the energy gap is measured to be $E_g = 1.43\,\mathrm{eV}$ at room temperature (RT). The valence band (p-symmetry) splits into a four-fold degenerate $P_{3/2}$ state at Γ_8 and a two-fold degenerate $P_{1/2}$ state at Γ_7, which lies at an energy $\Delta = 0.34\,\mathrm{eV}$ below $P_{3/2}$. The $P_{3/2}$ band consists of two-fold degenerate bands; heavy hole and light hole sub-bands. On the other hand, the conduction band (s-symmetry) is a two-fold degenerate $S_{1/2}$ state at Γ_6.

When $h\nu = E_g$, circularly polarized light excites electrons from $P_{3/2}$ to $S_{1/2}$. According to the selection rule ($\Delta m_j = \pm 1$), two transitions for each photon helicity (right, σ^+, and left, σ^-, circular) are possible. However, the relative transition probabilities for the light and heavy holes need to be taken into account in order to estimate the net spin polarization (see Fig. 6.2b). For example, if electrons are excited only from the valence band maximum (Γ_8) by circularly polarized light, three times more spins are excited from $m_j = \pm 3/2$ than from $m_j = \pm 1/2$ states. Although the maximum spin polarization is expected to be 50% in theory, the maximum observed experimentally is $\sim 40\%$ at the threshold

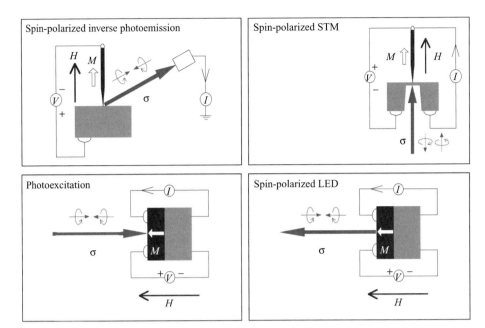

FIG. 6.1. Major experimental techniques with circularly polarized photons;
spin-polarized inverse photoemission, spin-polarized scanning tunneling
microscopy (spin STM), photoexcitation and spin-polarized light-emitting
diode (spin LED).

as shown in Fig. 6.3. This can be explained due to experimental limitations, such
as spin depolarization in the GaAs layer and the interfaces [1,3].

For $E_g + \Delta < h\nu$, the polarization decreases with increasing $h\nu$ due to the
mixture of light and heavy hole states with the split-off valence band states,
which have the opposite spin orientation. Such interband absorption occurs only
through the spin–orbit interaction, since the electric field of the exciting light
only influences the electron orbital motion. For $E_g + \Delta \ll h\nu$, the spin–orbit
interaction becomes negligible and spin depolarization during the cascade process
can dominate the process. Therefore photoexcited spin-polarized electrons are
absent.

6.1.3 Photoexcitation model

The helicity-dependent photocurrent I is measured by modulating the photon
helicity from right (σ^+) to left (σ^-). The two helicity values correspond to opposite spin angular momentum values of the incident photon, and the helicity gives
rise to opposite spin polarizations of electrons photoexcited in the GaAs [1–3].
The magnetization $(\mathbf{M})$ in the ferromagnet is aligned perpendicular $(H = 1.8\,\text{T})$
or in the plane $(H = 0)$ by using an external field. For σ parallel to $\mathbf{M}$

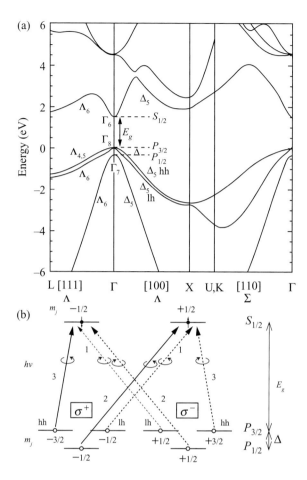

Fig. 6.2. (a) Schematic band structure of GaAs in the vicinity of the Γ-point (center of the Brillouin zone) in **k**-space. The energy gap E_g between the conduction band and the valence band for both heavy and light holes is shown. The spin–orbit splitting Δ also exists. (b) Schematic diagram of the allowed transitions for right (σ^+, solid lines) and left (σ^-, dashed lines) circularly polarized light in GaAs. The selection rule is $\Delta m_j = +1$ for σ^+ and $\Delta m_j = -1$ for σ^-. The numbers near the arrows represent the relative transition probabilities. The magnetic quantum numbers are also indicated at the corresponding energy levels. The heavy and light holes are abbreviated to hh and lh, respectively.

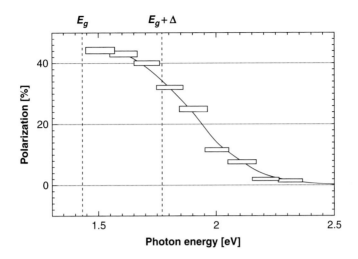

F$_{\text{IG}}$. 6.3. Photoemission spectrum of spin polarization from GaAs+CsOCs at $T \leq 10\,\text{K}$. The experimental data are shown as boxes including experimental errors [1].

(or antiparallel), the electrons in the ferromagnet and the semiconductor share the same spin quantization axis, while for σ perpendicular to $\mathbf{M}$, the two possible spin states created by the circularly polarized light are equivalent when projected along the magnetization direction in the ferromagnet (see Fig. 6.4). Consequently, in the remanent state ($\sigma \perp \mathbf{M}$), when $\mathbf{M}$ is orthogonal to the photoexcited spin polarization, both up and down spin-polarized electrons in the semiconductor can flow into the ferromagnet. At perpendicular saturation ($\sigma//\mathbf{M}$), on the other hand, the up-spin electron current from the semiconductor is filtered due to the spin-split density of states (DOS) at the Fermi level E_F of the ferromagnet [4–6], i.e. only minority spin electrons contribute to the transmitted current from the semiconductor to the ferromagnet. Spin filtering is therefore turned on ($\sigma//\mathbf{M}$) and off ($\sigma \perp \mathbf{M}$) by controlling the relative axes of σ and $\mathbf{M}$, and is detected as the helicity-dependent photocurrent I. Hence, the helicity-dependent photocurrents I^0 and I^n correspond to the magnetization configurations $\sigma \perp \mathbf{M}$ (see Fig. 6.4a) and $\sigma//\mathbf{M}$ respectively (see Fig. 6.4b).

Different transport mechanisms, such as hole diffusion into the ferromagnet, thermionic emission of electrons over the AlGaAs barrier, electron tunneling across the AlGaAs barrier, will contribute to the unpolarized photocurrent, depending upon the applied bias voltage. Significant spin filtering effects are expected to occur at reverse bias for the case of spin-dependent hole transport and at forward bias for the case of spin-dependent electron transport. The difference in the helicity-dependent photocurrent ΔI is a superposition of magneto-optical (ΔI_{MCD}) and spin filtering (ΔI_{SF}) effects:

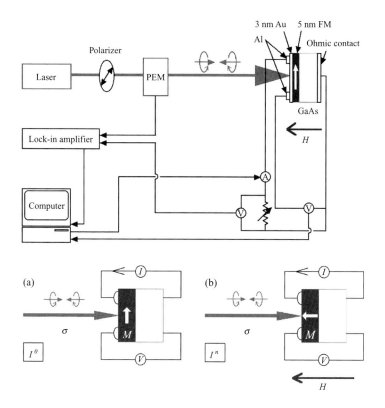

FIG. 6.4. Schematic configuration of the polar photoexcitation setup. The laser light is linearly polarized at 45° to the modulator axis pointing along the normal to the sample plane. Right/left circular light is produced using a photoelastic modulator (PEM). The photocurrent is measured via $I–V$ methods combined with lock-in techniques. The magnetization **M** in the FM and the photon helicity σ are shown with the field H applied normal to the sample. Two configurations, (a) without (I^0) and (b) with (I^n) a magnetic field are shown in the inset.

$$\Delta I = \Delta I_{\mathrm{SF}} + \Delta I_{\mathrm{MCD}}, \qquad (6.1)$$

with ΔI_{MCD} being proportional to the unpolarized photocurrent ($\Delta I_{\mathrm{MCD}} = \alpha I_{\mathrm{ph}}$).

The well-defined structure allows a clear separation of all these contributions [7]. As seen in Fig. 6.5, a significant difference between the bias dependence of the unpolarized photocurrent and the helicity-dependent photocurrent is only observed at forward bias (0.4–0.8 V), where electron tunneling occurs. The bias dependence of both currents match each other closely at reverse bias. This latter finding shows that spin-dependent hole transport does not play an important

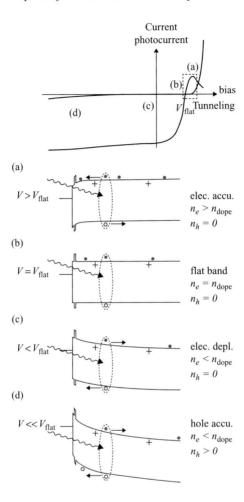

FIG. 6.5. Schematic diagrams of band bending and the corresponding current–voltage $(I–V)$ and photocurrent–voltage characteristics in the case of (a) forward bias electron accumulation, (b) forward bias flat band configuration, (c) zero bias electron depletion, and (d) reverse bias hole accumulation [7].

role and that the helicity-dependent photocurrent at reverse bias arises mainly from magnetic circular dichroism (MCD). The spin filtering efficiency can then be quantified in terms of an effective polarization P_{eff}:

$$P_{\text{eff}} = (\Delta I - \alpha I_{\text{ph}})/2I_{\text{ph}}. \qquad (6.2)$$

6.2   Optical spin injection

6.2.1   *Photoexcitation*

The possibility of detecting a spin-polarized current through thin film tunnel junctions of both $Co/Al_2O_3/GaAs$ and Co/τ-$MnAl/AlAs/GaAs$ induced by photoexcitation was first discussed by Prins *et al.* [8]. For the former structure, a spin-dependent tunneling current was observed while only MCD signals were seen in the latter structure. In their experiment, a sample with a 2 nm Al_2O_3 tunneling barrier showed the largest helical asymmetry of the photoexcited current of approximately 1.2% at 1.5 eV (near the GaAs band-gap). Accordingly, many studies of spin-dependent tunneling through metal/oxide/semiconductor (MOS) junctions have been carried out, e.g. Co (or $Ni)/Al_2O_3/p$-GaAs [11], with the intention of realizing optically pumped spin-polarized scanning tunneling microscopy (spin STM) as described in Section 6.2.3. These results are summarized in Table 6.1.

6.2.2   *Schottky diodes*

By depositing a ferromagnetic metal layer directly onto a semiconductor substrate a Schottky barrier is formed intrinsically at the interface. Since the barrier is formed at the surface region of the semiconductor the electron flow can be prevented depending on the bending shape of the barrier resulting in current rectification. The Schottky barrier acts as an intrinsic tunneling barrier for electrons traveling across the interface. This offers a way to overcome the conductance mismatch, which fundamentally reduces the spin polarization of the current across a ferromagnet/semiconductor interface [15,16].

RT spin filtering of spin-polarized electrons has been systematically investigated at the ferromagnet/semiconductor interface in forward bias [9,17]. The bias and GaAs doping density dependence of spin-filterer signals suggest that

Table 6.1 List of recent optical spin injection studies.

Structures	Spin polarization	Refs.
Ferromagnet/semiconductor hybrid structures:		
$\quad$ $Co/Al_2O_3/p$-GaAs	$\sim 1.2\%$ (RT)	[8]
$\quad$ (NiFe, Co and Fe)/n-GaAs	$\sim 2 \pm 1\%$ (RT)	[9]
$\quad$ (FeCo and Fe)/GaAs QW	$\sim 0.5\% \sim$ MCD (10 K)	[10]
MOS junctions:		
$\quad$ (Ni and Co)/Al_2O_3/p-GaAs	~ 2.5 and 1.0% (RT)	[11]
$\quad$ Epitaxial Fe/GaAs	$-4 \sim +4\%$ (RT)	[21]
Spin STM:		
$\quad$ Ni STM tip/GaAs	$< 10\%$ (RT)	[12]
$\quad$ p-GaAs STM tip/Co/mica	$\sim 10\%$ (RT)	[13]
$\quad$ p-GaAs STM tip/NiFe/Si	$\sim 7\%$ (RT)	[14]

electron tunneling is the spin-dependent transport mechanism. Further proof of this picture has been achieved by temperature-dependent measurements of band gap engineered NiFe/AlGaAs/GaAs structures [18]. Spin-dependent effects were only observed in the bias and temperature range where electron tunneling occurs. In addition, strong optical magnetocurrent effects at RT have been observed in spin-valve/semiconductor structures. The difference in the optical magnetocurrent obtained between parallel and antiparallel spin-valve configurations was extremely large (up to 2400%) [19]. This indicated that spin-dependent electron transport across spin-valve structures was determined by the relative spin alignment of the ferromagnetic layers and the initial spin polarization of the photoexcited electrons. The spin filtering effect can be used for future spintronic devices such as optically assisted magnetic sensors [20].

The photon energy dependence of the optical magnetocurrent also proved that the photoexcited electrons tunnel into the ferromagnet ballistically as shown in Fig. 6.6 [21]. This result shows the possibility of tuning the spin polarization at the epitaxial Fe/GaAs interface using the interfacial resonant states. Dery and Sham proposed a spin switch by using spin filtering through localized electron states in a heavily doped semiconductor [22]. In their model, s conduction electrons tunnel through a Schottky barrier and carry positive spin polarization into a ferromagnet. Localized d electrons transfer negative spin polarization as the spin DOS for the down spins at E_F is larger than that for the up spins. Chantis et $al.$ also predicted negative spin polarization in a range of bias voltages across a Fe/GaAs (8 ML)/bcc Cu structure due to the formation

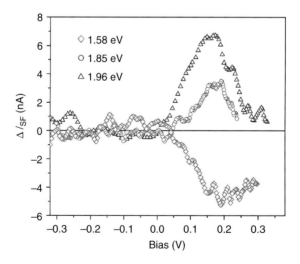

FIG. 6.6. Bias dependence of a photoexcited current for an epitaxial Fe/GaAs(001) interface for different photon energies [21].

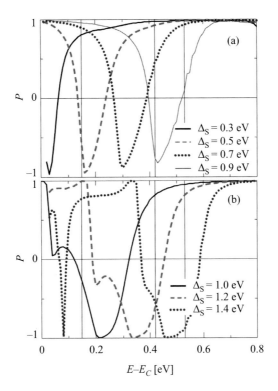

FIG. 6.7. Calculated results of the spin polarization of the tunnel conductance as
a function of energy for (a) Fe-As and (b) Fe-Ga contacts with the depletion
layer thickness of 200 ML and various values of the Schottky barrier height
Δ_S. The vertical lines correspond to the energy $E - E_C = E_{\mathrm{ph}} - E_{\mathrm{gap}}$ for
three photon energies E_{ph} used in experiments [24].

of interface resonant states for down spin electrons formed at the interface to
the Fe layer [23]. Since the spin polarization is found to be dependent upon
the photoexcitation energy, a new model has been proposed by Honda *et al.*
as shown in Fig. 6.7 [24]. It is shown that band matching of resonant interface
states within the Schottky barrier defines the sign of the spin polarization of the
electrons transported through the barrier. The predictions agree very well with
experimental results including those for the tunneling of photoexcited electrons
[21] and suggest that spin polarization (from -100% to 100%) is dependent
on the Schottky barrier height. They also suggest that the sign of the spin
polarization can be controlled with a bias voltage.

6.2.3 *Spin-polarized scanning tunneling microscopy (spin STM)*

Spin STM was proposed in 1993 by Molotkov [25] and Laiho and Reittu [26].
This technique used a direct band-gap semiconductor tip. This is expected

to be able to image surface magnetic configurations with almost atomic resolution.

Spin-polarized electron tunneling from a Ni STM tip into a GaAs substrate was first demonstrated by Alvarado and Renaud [27]. The Ni tip was magnetized by an electromagnet and was used as a spin injector. The tip scans over the sample surface in its measurement state. Spin-polarized electron tunneling through the vacuum was detected as circularly polarized electroluminescence (EL) signals, which change by $\sim 30\%$ at RT. This value correspondeds to the minority electron spin polarization of Ni(001) at the Fermi level. This suggests that the minority spin electrons provide the dominant contribution to the tunneling current.

After the first photoexcitation measurement by Prins *et al.* [8], modulated circularly polarized light was used to excite spin-polarized electrons in a semiconductor (e.g. GaAs). Although optically excited electrons are scattered mainly at the semiconductor surface with back illumination [28], Sueoka *et al.* demonstrated the possibility of detecting spin-polarized signals by scanning a Ni STM tip over a GaAs film with circularly polarized light shone through an AlGaAs membrane [12]. Suzuki *et al.* also performed a similar observation by scanning a *p*-GaAs STM tip over a Co film with back illumination through a mica/Au/ Co film and obtained magnetic domain images [13]. The GaAs tip was fabricated using photolithography and anisotropic etching to prevent limitation due to facets {105}. A three-monolayer (ML) Co film exhibited perpendicular magnetization and showed less than the MCD effect of 0.14%, which was much smaller than the observed polarization response of about 10%. Polarization modulation response images of the spin STM showed very good agreement with magnetic force microscopy (MFM) images. In order to avoid the MCD effect and possible light scattering through the sample structures, Kodama *et al.* introduced photon helicity into a GaAs tip in the vicinity of the sample which is the equivalent of front illumination [14]. They detected a change of approximately 7% in $I\text{--}V$ curves between right and left circularly polarized light irradiated onto NiFe films.

6.3 Optical spin detection

6.3.1 *Spin-polarized light-emitting diodes (spin LED)*

By inverting the photoexcitation process one can detect the spin polarization of an electrically injected current using an optical method. For this case, a highly spin-polarized ferromagnet is necessary to inject a large degree of spin polarization as an electrical current. Since a dilute magnetic semiconductor (DMS) shows a large Zeeman splitting and ferromagnetism [29], a DMS can be used as a spin aligner to inject spin-polarized carriers, i.e. spin-polarized electrons or holes into a semiconductor. This is an alternative method to avoid the interfacial conductance mismatch [15]. Spin polarization of the injected carriers is detected optically through circularly polarized EL from the semiconductor. Such structures are called spin-polarized light-emitting diodes (spin LEDs). With ferromagnetic *p*-GaMnAs as a spin aligner, spin-polarized hole injection was reported at low temperature [30]. At forward bias, spin-polarized holes from the p-GaMnAs layer as well as unpolarized electrons from the n-GaAs layer were injected into the InGaAs quantum well (QW), so that the recombination of the spin-polarized holes created circularly polarized EL emission from the QW. However, as the spin relaxation time for the holes is much shorter than that for the electrons [31], the spin-polarization signal through the recombination process in the GaAs was very small (about $\pm 1\%$) [30]. Using paramagnetic n-BeMnZnSe as a spin aligner, highly efficient electron spin injection has been achieved with an applied field of ~ 3 T (spin polarization in EL $\sim 90\%$) [32]. This is because the spin diffusion length of the electrons has been reported to be above 100 μm in GaAs [33]. Similar results were obtained using CdMnTe [34], ZnMnSe [35,36], ZnSe [37], and MnGe [38] but only at low temperatures (typically $T < 80$ K). Since RT ferromagnetism has been predicted in several DMS compounds [39] but not yet observed, spin injection at RT with a DMS may be achievable in the near future. These results are summarized in Table 6.2.

Table 6.2 List of recent optical spin detection studies.

Structures	Spin polarization	Refs.
Spin LED (spin-polarized *electron* injection):		
BeMgZnSe+BeMnZnSe/n-AlGaAs/i-GaAs QW/.../p-GaAs	$\sim 42\%$ (<5 K)	[32]
CdMnTe/CdTe	$\sim 30\%$ (5 K)	[34]
n-ZnMnSe/AlGaAs/GaAs QW/AlGaAs	$\sim 83\%$ (4.5 K)	[35] [36]
Fe/GaAs/InAs QW/GaAs	$\sim 2\%$ (25 K)	[40]
Fe/AlGaAs/GaAs QW/GaAs	$\sim 13\%$ (4.5 K) $\sim 8\%$ (240 K)	[41]
NiFe+CoFe/AlO$_x$/AlGaAs/GaAs QW/GaAs	$\sim 9.2\%$ (80 K)	[42]
FeCo/AlO$_x$/AlGaAs/GaAs QW/.../p-GaAs	$\sim 21\%$ (80 K) $\sim 16\%$ (300 K)	[43]
CoFe/MgO/AlGaAs/GaAs QW/.../p-GaAs	$\sim 57\%$ (100 K) $\sim 47\%$ (290 K)	[44]
Co$_{2.4}$Mn$_{1.6}$Ga/n-AlGaAs/.../InAs QW/GaAs	$\sim 13\%$ (5 K)	[45]
Fe/n-GaAs	$\sim 30\%$ (4 K)	[46]
Fe/Al$_2$O$_3$/n-Si/.../GaAs QW	$\sim 5.5\%$ (20 K) $\sim 3\%$ (125 K)	[49]
Spin LED (spin-polarized *hole* injection): p-GaMnAs/GaAs/InAs QW	$\sim 1\%$ (<31 K)	[30]
Spin STM: Ni STM tip/GaAs	$\sim 30\%$ (RT)	[27]

6.3.2 *Schottky diodes*

A ferromagnet/semiconductor Schottky diode consisting of an Fe (20 nm)/
GaAs/InGaAs QW LED structure was also used to measure circularly polarized EL by Zhu *et al.* [40]. Spin injection from the Fe to the GaAs was
achieved with an efficiency of about 2% at 25 K which was found to be
independent of temperature. However, the right and left circularly polarized
EL intensity did not show a clear difference. Therefore, by examining the
tails of the Gaussian-like EL intensity distributions, a heavy-hole excitation
contribution was estimated. Hanbicki *et al.* [41] performed a similar experiment
with an Fe (12.5 nm)/AlGaAs/GaAs QW LED and observed a spin injection
efficiency of 30%. They clearly observed a significant difference between the
right and left circular EL intensity. The spin polarization was estimated to
be 13% at 4.5 K (8% at 240 K). Taking the spin relaxation time in the

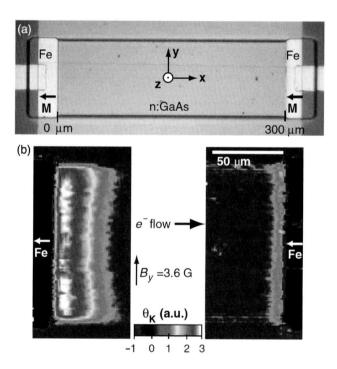

FIG. 6.8. (A) Photomicrograph of the lateral ferromagnet/semiconductor device
used for electron spin injection, transport, accumulation, and detection.
(B) Images of Kerr rotation angle θ_K ($\propto S_z$) near the source and drain
contacts. $V_b = +0.4$ V. The region of spin accumulation near the drain contact
also exhibits positive θ_K indicating that both the injected and accumulated
spin polarizations are antiparallel to $\boldsymbol{M}$ [46].

QW into account, they reported a small temperature dependence in the spin injection efficiency which was consistent with spin-polarized electron tunneling theory.

Crooker *et al.* measured a spin polarization of 32% at an Fe/GaAs Schottky junction as shown in Fig. 6.8 [46]. In Fe/GaAs/Fe junctions more up-spin electrons were injected on one side and more down-spins are ejected from the other end by flowing current across the junction. This indicates that positive spin polarization was achieved in reversed bias and negative polarization in forward bias. This is similar to the observation that both positive and negative spin polarization have been measured in an Fe/GaAs Schottky junction by introducing spin-polarized electrons by circularly polarized photoexcitation [21].

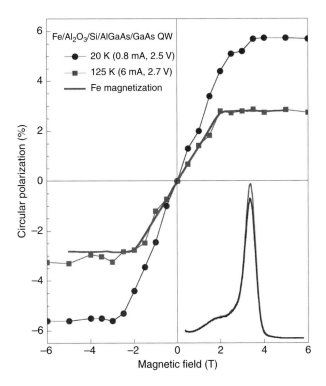

FIG. 6.9. Magnetic-field dependence of P_{circ} for the GaAs quantum-well free exciton in the electroluminescence spectra from the Fe/Al$_2$O$_3$/Si/ AlGaAs/GaAs quantum well/AlGaAs n–i–p structure. P_{circ} tracks the magnetization and majority electron spin orientation of the Fe film shown as the solid line. The electroluminescence spectra at 3 T and 20 K are shown in the inset, analyzed for $\sigma+$ (higher) and $\sigma-$ (lower) polarization. The main peak is the free exciton at 1.54 eV from the 10 nm GaAs quantum well [49].

6.3.3 *Spin injection into Si*

Since Si has an indirect band-gap poor spin injection is expected. It has long been believed that intrinsic spin polarization in Si is typically a few percent at RT. However, present nano-electronic devices predominantly depend on Si-based technology indicating the importance of spin injection into Si with high efficiency. A junction of $Co/Al_2O_3/Si$ has been used to demonstrate that the product (RA) can be tuned over eight orders of magnitude by inserting an ultrathin Gd layer which has a lower work function against Si [47]. Such tunability in the product RA is very useful to realize a spin MOS field effect transistor (FET), which requires a narrow RA window against the Si doping density. Recently, spin injection into Si has been successfully demonstrated by Jonker *et al.* in $Fe/Al_2O_3/n$-Si with an LED structure underneath [48,49]. As seen in Fig. 6.9, circular light polarization of 5.6% at 20 K (2.8% at 125 K) was measured indicating an injected spin polarization of approximately 30% in Si. This experiment has opened the door to Si spintronics, which possesses a significant advantage for implementation of spintronics into current Si-based nano-electronics.

Acknowledgments

The author would like to thank Prof. Kevin O'Grady for proof-reading the manuscript. This work was partially supported by EPSRC and JST PRESTO programs.

References

[1] D. T. Pierce and F. Meier, *Phys. Rev. B* **13**, 5484 (1976).

[2] S. Adachi, *GaAs and Related Materials* (World Scientific, Singapore, 1994).

[3] F. Meier and B. P. Zakharchenya, *Optical Orientation* (North-Holland Physics Publishing, Amsterdam, 1984).

[4] I. I. Mazin, *Phys. Rev. Lett.* **83**, 1427 (1999).

[5] B. Nadgorny, R. J. Soulen, Jr., M. S. Osofsky, I. I. Mazin, G. Laprade, R. J. M. van de Veerdonk, A. A. Smits, S. F. Cheng, E. F. Skelton, and S. B. Qadri, *Phys. Rev. B* **61**, R3788 (2000).

[6] C. Li, A. J. Freeman, and C. L. Fu, *J. Magn. Magn. Mater.* **75**, 53 (1988).

[7] S. E. Andersen, S. J. Steinmuller, A. Ionescu, G. Wastlbauer, C. M. Guertler, and J. A. C. Bland, *Phys. Rev. B* **68**, 073303 (2003).

[8] M. W. J. Prins, H. van Kempen, H. van Leuken, R. A. de Groot, W. van Roy, and J. de Boeck, *J. Phys.: Condens. Matter* **7**, 9447 (1995).

[9] A. Hirohata, S. J. Steinmueller, W. S. Cho, Y. B. Xu, C. M. Guertler, G. Wastlbauer, J. A. C. Bland, and S. N. Holmes, *Phys. Rev. B* **66**, 035330 (2002).

[10] A. F. Isakovic, D. M. Carr, J. Strand, B. D. Schultz, C. J. Palmstrøm, and P. A. Crowell, *Phys. Rev. B* **64**, R161304 (2001).

[11] K. Nakajima, S. N. Okuno, and K. Inomata, *Jpn. J. Appl. Phys.* **37**, L919 (1998).

[12] K. Sueoka, K. Mukasa, and K. Hayakawa, *Jpn. J. Appl. Phys.* **32**, 2989 (1993).

[13] Y. Suzuki, W. Nabhan, R. Shinohara, K. Yamaguchi, and K. Mukasa, *J. Magn. Magn. Mater.* **198–199**, 540 (1999).

[14] H. Kodama, T. Uzumaki, M. Oshiki, K. Sueoka, and K. Mukasa, *J. Appl. Phys.* **83**, 6831 (1999).

[15] G. Schmidt, D. Ferrand, L. W. Molenkamp, A. T. Filip, and B. J. van Wees, *Phys. Rev. B* **62**, R4790 (2000).

[16] E. I. Rashba, *Phys. Rev. B* **62**, 16267 (2000).

[17] A. Hirohata, Y. B. Xu, C. M. Guertler, and J. A. C. Bland, *J. Appl. Phys.* **87**, 4670 (2000).

[18] J. A. C. Bland, A. Hirohata, Y. B. Xu, C. M. Guertler, and S. N. Holmes, *IEEE Trans. Magn.* **36**, 2827 (2000).

[19] S. J. Steinmuller, T. Trypiniotis, W. S. Cho, A. Hirohata, W. S. Lew, C. A. F. Vaz, and J. A. C. Bland, *Phys. Rev. B* **69**, 153309 (2004).

[20] J. A. C. Bland and A. Hirohata, "An optically addressed spin-polarized diode sensor," *Basic British Patent (AU8608701) and PCT/GB01/04088 (WO 02/23638 A2)*.

[21] H. Kurebayashi, S. J. Steinmuller, J. B. Laloë, T. Trypiniotis, S. Easton, A. Ionescu, J. R. Yates, and J. A. C. Bland, *Appl. Phys. Lett.* **91**, 102114 (2007).

[22] H. Dery and L. J. Sham, *Phys. Rev. Lett.* **98**, 46602 (2007).

[23] A. N. Chantis, K. D. Belashchenko, D. L. Smith, E. Y. Tsymbal, M. van Schilfgaarde, and L. C. Albers, *Phys. Rev. Lett.* **99**, 196603 (2007).

[24] S. Honda, H. Itoh, J. Inoue, H. Kurebayashi, T. Trypiniotis, C. H. W. Barnes, A. Hirohata, and J. A. C. Bland, *Phys. Rev. B* **78**, 245316 (2008).

[25] S. N. Molotkov, *Surf. Sci.* **287/288**, 1098 (1993).

[26] R. Laiho and H. J. Reittu, *Surf. Sci.* **289**, 363 (1993).

[27] S. F. Alvarado and P. Renaud, *Phys. Rev. Lett.* **68**, 1387 (1992).

[28] S. M. Sze, *Physics of Semiconductor Devices, 2nd ed.* (John Wiley, New York, 1981).

[29] H. Ohno, *Science* **281**, 951 (1998).

[30] Y. Ohno, D. K. Young, B. Beschoten, F. Matsukura, H. Ohno, and D. D. Awschalom, *Nature* **402**, 790 (1999).

[31] M. E. Flatté and J. M. Byers, *Phys. Rev. Lett.* **84**, 4220 (2000).

[32] R. Fiederling, M. Keim, G. Reuscher, W. Ossau, G. Schmidt, A. Waag, and L. W. Molenkamp, *Nature* **402**, 787 (1999).

[33] J. M. Kikkawa and D. D. Awschalom. *Nature*, **397**, 139 (1999).

[34] M. Oestreich, J. Hübner, D. Hägele, P. J. Klar, W. Heimbrodt, W. W. Rühle, D. E. Ashenford, and B. Lunn, *Appl. Phys. Lett.* **74**, 1251 (1999).

[35] B. T. Jonker, Y. D. Park, B. R. Bennett, H. D. Cheong, G. Kioseoglou, and A. Petrou, *Phys. Rev. B* **62**, 8180 (2000).

[36] B. T. Jonker, A. T. Hanbicki, Y. D. Park, G. Itskos, M. Furis, G. Kioseoglou, A. Petrou, and X. Wei, *Appl. Phys. Lett.* **79**, 3098 (2001).

[37] I. Malajovich, J. M. Kikkawa, D. D. Awschalom, J. J. Berry, and N. Samarth, *Phys. Rev. Lett.* **84**, 1015 (2000).

[38] Y. D. Park, A. T. Hanbicki, S. C. Erwin, C. S. Hellberg, J. M. Sullivan, J. E. Mattson, T. F. Ambrose, A. Wilson, G. Spanos, and B. T. Jonker, *Science* **295**, 651 (2002).

[39] T. Dietl, H. Ohno, F. Matsukura, J. Cibert, and D. Ferrand, *Science* **287**, 1019 (2000).

[40] H. J. Zhu, M. Ramsteiner, H. Kostial, M. Wassermeier, H.-P. Schönher, and K. H. Ploog, *Phys. Rev. Lett.* **87**, 016601 (2001).

[41] A. T. Hanbicki, B. T. Jonker, G. Itskos, G. Kioseoglou, and A. Petrou, *Appl. Phys. Lett.* **80**, 1240 (2002).

[42] V. F. Motsnyi, J. de Boeck, J. Das, W. van Roy, G. Borghs, E. Goovaerts, and V. I. Safarov, *Appl. Phys. Lett.* **81**, 265 (2002).

[43] V. F. Motsnyi, P. van Dorpe, W. van Roy, E. Goovaerts, V. I. Safarov, G. Borghs, and J. de Boeck, *Phys. Rev. B* **68**, 245319 (2003).

[44] X. Jing, R. Wang, R. M. Shelby, R. M. Macfarlane, S. R. Bank, J. S. Harris, and S. S. P. Parkin, *Phys. Rev. Lett.* **94**, 056601 (2005).

[45] M. C. Hickey, C. D. Damsgaard, I. Farrer, S. N. Holmes, A. Husmann, J. B. Hansen, C. S. Jacobsen, D. A. Ritchie, R. F. Lee, and G. A. C. Jones, *Appl. Phys. Lett.* **86**, 252106 (2005).

[46] S. A. Crooker, M. Furis, X. Lou, C. Adelmann, D. L. Smith, C. J. Palmström, and P. A. Crowell, *Science* **309**, 2191 (2005).

[47] B.-C. Min, K. Motohashi, J. C. Lodder, and R. Jansen, *Nature Mater.* **5**, 817 (2006).

[48] O. H. J. van't Erve, G. Kioseoglou, A. T. Hanbicki, C. H. Li, B. T. Jonker, R. Mallory, M. Yasar, and A. Petrou, *Appl. Phys. Lett.* **84**, 4334 (2004).

[49] B. T. Jonker, G. Kioseoglou, A. T. Hanbicki, C. H. Li, and P. E. Thompson, *Nature Phys.* **3**, 542 (2007).

7 Theory of spinmotive forces in ferromagnetic structures

S. E. Barnes

7.1 Introduction

Electromotive forces $\mathcal{E}$, as predicted by Faraday's 1831 law of electromagnetic induction [1], power everything from locomotives to light bulbs. The force exists by virtue of the charge e of an electron. This chapter is dedicated to the introduction of *spinmotive forces* which reflect the magnetic moment of an electron, that is, its spin charge $\pm\hbar/2$. This motive force reflects the energy conservation requirements of the spin-torque transfer (STT) process which is at the heart of *spintronics*. The very existence of spin was established in the Stern–Gerlach experiment using spin-dependent forces. However, the forces relevant to that experiment are proportional to the gradient of the Zeeman potential energy $\Phi_0 = -\vec{\mu} \cdot \vec{B}$, where $\vec{\mu}$ is the magnetic moment, and as such are conservative. Such forces, as do the nonconservative forces of interest here, change sign with the spin quantum number $\sigma_z = \pm 1$. They would exist even if an electron was not charged and *do* exist for uncharged excitations such as magnons or phonons. Such forces are of particular importance in ferromagnetic materials where the spinmotive force usually drives an electronic charge current due to the higher mobility of the majority electrons. This new motive force describes the conversion of magnetic to electrical energy which exists within a *spin battery* [2].

7.2 Spin Faraday's law

The usual Faraday's law can be written as [1]:

$$\mathcal{E} = -\frac{d\Phi}{dt} = -\frac{\hbar}{e}\frac{d\gamma}{dt} \tag{7.1}$$

and relates the emf $\mathcal{E} \equiv \oint_C \vec{E} \cdot d\ell$, the line integral of the electric field $\vec{E}$ on some contour C, to the rate of change of the magnetic flux $\Phi = \int_S \vec{B} \cdot d\vec{S}$ due to a magnetic field $\vec{B}$, where the surface S closes on C. The second version on the right-hand side is equivalent if γ is the Aharonov–Bohm phase [3] γ_e and which is the Berry phase [4] associated with the electronic charge. The generalization of Faraday's law to include nonconservative spin forces, as proposed by Barnes and Maekawa [5], corresponds to this same expression but with $\gamma = \gamma_e + \gamma_s$ which is the sum of the Aharonov–Bohm and spin Berry γ_s phases. The latter phase is independent of the charge but changes sign with σ_z and thereby leads to a spinmotive force. This expression is identical to one proposed by A. Stern [6]. Another lineage [7, 8] stemming from Volovik [9] leads to the same result. A result due to Berger [10] is cited [7] for a domain wall as being equivalent to this latter approach. These all differ, in the definition of γ_s, from what is presented here. This will be illustrated by considering an experiment proposed by Stern [6] and in connection with the anomalous Hall effect [11, 12, 13]. Both the Aharonov–Bohm [14] and spin [15] Berry phase are measurable in the solid state context. There are several experimental observations which confirm the existence of an smf [2, 7, 16, 17]. Other theoretical approaches exist [18, 19, 20, 21].

7.2.1 *Magnetic momentum, spin electric, and magnetic fields*

At a more formal level, following Feynman [22], Faraday's law is a consequence of the definition of the momentum

$$\vec{p} = m\vec{v} + e\vec{A} \tag{7.2}$$

conjugate to the position $\vec{r}$ of an electron charge e (< 0). This momentum is equal to the mechanical momentum, $m\vec{v}$, to which is added the momentum $e\vec{A}$ which reflects a momentum associated with all the fields or excitations to which the particle is coupled. The usual $e\vec{A}$ corresponds to an electromagnetic field. Here is added $e\vec{A}_s$ to reflect the *linear* momentum of the magnetic order parameter $\vec{M}$. The mathematical properties reflect those of these fields, so that, while the usual $e\vec{A}$ is just a complex number corresponding to a U(1) gauge theory, $e\vec{A}_s$, derives from a SU(2) theory and is a linear combination of Pauli matrices. The potential, which is the time-like component of the four-vector potential, $\Phi_0 = eA_0 = eA_{0e} + eA_{0s}$ has similar algebraic properties. The *physically meaningful* electric and magnetic fields are defined by the Lorentz force $\vec{F} = e(\vec{E} + \vec{v} \times \vec{B})$ and this is determined by calculating $m\frac{d\vec{v}}{dt}$, either classically or here quantum mechanically via Ehrenfest's theorem [23]. The results, in compact form, are:

$$B_i = -\frac{im^2}{e\hbar}\epsilon_{ijk}[v_j, v_k]; \quad i = \{x, y, z\}; \quad \text{and} \quad E_i = -\frac{im^2}{e\hbar}[v_0, v_i] \tag{7.3}$$

where $i\hbar\partial_t = mv_0 + eA_0$, and where $\partial_t \equiv \frac{\partial}{\partial t}$. Again for the usual U(1) electromagnetic theory, these reduce to the familiar $\vec{E} = -\vec{\nabla}\Phi_0 - \partial\vec{A}/\partial t$ and $\vec{B} = \nabla \times \vec{A}$ and Faraday's law in differential form, the equivalent of Eq. (7.1), is recovered since $\vec{\nabla} \times \vec{E} = -(\partial\vec{A}/\partial t)\vec{\nabla} \times \vec{A} = -\partial\vec{B}/\partial t$. While the result is still true when the SU(2) fields are included, this requires the specific expressions for $e\vec{A}_s$ and eA_{0s}. That these involve Pauli matrices is of great importance to the evaluation of the commutators in Eq. (7.3).

7.2.2 *Calculation of magnetic momentum for adiabatic electrons*

Here it is assumed that the magnetization $\vec{M}$ (direction $\hat{m}$) changes sufficiently slowly that an electron's spin adiabatically follows the local direction of $\hat{m}$.

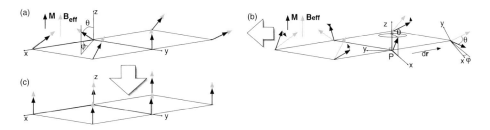

FIG. 7.1. (a) Stationary spin structure with definitions of θ and ϕ. Such a structure does not produce a spin magnetic field $\vec{B}_s$ and there is no spin flux. A field $\vec{B}_s$ only occurs when there is magnetic energy which might be transferred to the electrons and this implies the dynamic situation shown in (b). The pure gauge transformation $U_{\theta,\phi}$ which takes (a) to the uniform system (c) does not lead to a finite $\vec{B}_s$. The similar transformation $U_{\hat{n}}$ between (b) and (a) *does* generate the field $\vec{B}_s$.

Even then, the definition of $e\vec{A}_s$ is not unique; the physical spin fields given by Eqs. (7.3) are unchanged[1] by a *pure gauge transformation*. Imagine the common situation in which electrons move adiabatically through a *stationary* spin texture, as illustrated in Fig. 7.1(a). It is usually imagined, in connection with the anomalous Hall effect [11, 12, 13], that this produces a static $\vec{B}_s$. However this problem can be diagonalized by a pure gauge transformation, which therefore results in $\vec{B}_s = 0$. The relevant rotation $U_{\theta,\phi} = e^{-i\phi s_z/\hbar} e^{-i\theta s_y/\hbar} e^{i\phi s_z/\hbar}$ aligns $\vec{m}$ along a reference $\hat{z}$ direction, i.e. takes Fig. 7.1(a) to Fig. 7.1(c). For the majority electrons, define the wavefunction as $|\psi\rangle = U_{\theta,\phi}|\uparrow\rangle$, and substitute this into the full time-dependent Schrodinger equation

$$i\hbar\partial_t|\psi\rangle = H(\vec{p}, \vec{r})|\psi\rangle \tag{7.4}$$

to obtain $(i\hbar\partial_t - e\vec{A}^0)|\uparrow\rangle = H(\vec{p} - e\vec{A}_s, \vec{r})|\uparrow\rangle$, where the vector potential

$$e\vec{A}_s^0 = (i\hbar U_{\theta,\phi}^\dagger \vec{\nabla} U_{\theta,\phi}) = s_z(\cos\theta - 1)\vec{\nabla}\phi + (s_y\vec{\nabla}\theta - s_x\sin\theta\vec{\nabla}\phi), \tag{7.5}$$

for which the evaluation of $B_i = -\frac{im^2}{e\hbar}\epsilon_{ijk}[v_j, v_k]$ explicitly gives *zero*.[2] The result of this pure gauge transformation is a uniform magnet, as illustrated.

---

[1] The spin magnetic field $\vec{B}_s$ is technically known as the "spin Berry curvature" since it causes particles with a spin charge, i.e. a magnetic moment, to have curved paths. It is common to insist that this is defined by $\vec{B}_s = \vec{\nabla} \times \vec{A}_s$, which as a mathematical exercise is perfectly admissible. The resulting force $\frac{1}{2}(\vec{v} \times \vec{B}_s + \vec{B}_s \times \vec{v})$ is however not that acting on a real particle such as an electron. The curvature given by the full expression Eqs. (7.3) is unique in that it gives the measurable physical forces.

[2] It is usual to drop the off-diagonal parts $(s_y\vec{\nabla}\theta + s_x\sin\theta\vec{\nabla}\phi)$ since they couple states which differ in energy by the large exchange splitting $\sim J_0$. This "U(1) projection" [9] results in a

Magnetic textures which produce finite fields $\vec{E}_s$ and $\vec{B}_s$ are *time dependent*. The first step, which defines the meaningful $e\vec{A}_s$, is to eliminate this time dependence making the transformation Fig. 7.1(b) to Fig. 7.1(a), followed by a second, pure gauge transformation $U_{\theta,\phi}$, which turns Fig. 7.1(a) into Fig. 7.1(c) and renders the system uniform.

The first step, Fig. 7.1(b) to Fig. 7.1(a), aligns the local direction of the $\vec{m}$, not along a fixed axis, but rather along the direction of $\vec{B}_{\text{eff}}$ which would, and will, appear *in the Landau–Lifshitz equations*. Interest is in determining $\vec{A}_s$ and A_{0s} and which involve derivatives. This considerably simplifies the task since, at the point P where the fields are to be evaluated, at $t = 0$, the reference system of axes can be chosen such that $\vec{B}_{\text{eff}} = B_{\text{eff}}\hat{\mathbf{z}}$ with $\vec{m}$ in the z–x plane. At P, and $t = 0$, a y-rotation θ corresponding to U_θ brings $\vec{m}$ along the original direction of $\vec{B}_{\text{eff}}$. Following this, stationary and uniform, rotation the exchange interaction $-J\vec{S} \cdot \vec{s}$, where $\vec{S}$ is the order parameter and $\vec{s}$ the spin operator of some specific electron, is diagonal, i.e. $-J\vec{S} \cdot \vec{s}$ reduces to $-JSs_z$. At a neighbouring point $-J\vec{S} \cdot \vec{s} \approx -JSs_z - JS[s_x d\theta + s_y \sin\theta d\phi]$.[3] This defines $U_{\hat{\mathbf{n}}}$ by $U_{\hat{\mathbf{n}}} s_z U_{\hat{\mathbf{n}}}^\dagger = s_z + s_y d\theta - s_x \sin\theta d\phi$, where $U_{\hat{\mathbf{n}}} = 1 + \vec{t}_s \cdot d\vec{r}$ which then defines $\vec{t}_s = \frac{1}{i\hbar}(-s_y \vec{\nabla}\theta + s_x \sin\theta\vec{\nabla}\phi)$ as the generator of displacements in space. The wave-function in the vicinity of this point is now, e.g. for majority electrons $|\psi\rangle = U_\theta U_{\hat{\mathbf{n}}}|\uparrow\rangle$. The results for $e\vec{A}_s = i\hbar\vec{t}_s$ and the equivalent, $e\vec{A}_{0s}$, for displacements in time, are:

$$e\vec{A}_s = \left[-\sin\theta\vec{\nabla}\phi\, s_x + \vec{\nabla}\theta\, s_y\right]; \qquad e\vec{A}_{0s} = -\left[-\sin\theta\partial_t\phi\, s_x + \partial_t\theta\, s_y\right], \quad (7.6)$$

As it is for the U(1) theory, $U_{\hat{\mathbf{n}}} = \exp\left\{i\frac{e}{\hbar}\int^{\vec{r}} d\vec{r}' \cdot \vec{A}_s(\vec{r}')\right\}$, while unitary, involves a *path-dependent* integral $\int^{\vec{r}} d\vec{r}' \cdot \vec{A}_s(\vec{r}')$, i.e. $U_{\hat{\mathbf{n}}}$, unlike $U_{\theta,\phi}$, is not analytic.

7.2.3 The spin fields $\vec{E}_s$ and $\vec{B}_s$ and Faraday's law

Substituting into Eqs. (7.2), the spin derived fields are,

$$\vec{B}_s = \sigma_z \frac{\hbar}{2e}\sin\theta\, \vec{\nabla}\theta \times \vec{\nabla}\phi, \quad (7.7)$$

and

$$\vec{E}_s = \sigma_z\left[-\frac{\hbar}{2e}\sin\theta(\vec{\nabla}\theta\partial_t\phi - \vec{\nabla}\phi\partial_t\theta) - \frac{g\mu_B}{2e}\vec{\nabla}(\hat{\mathbf{m}} \cdot \vec{B}_{\text{eff}})\right]. \quad (7.8)$$

Note that here θ and ϕ are defined by Fig. 7.1(b) whereas usually they are defined by Fig. 7.1(a). In the continuum limit, these fields explicitly obey Faraday's law Eq. (7.1).

---

finite diagonal $\vec{B}_s = \vec{\nabla} \times \vec{A}_s = s_z \sin\theta\vec{\nabla}\theta \times \vec{\nabla}\phi$. Necessarily, this and the associated $\vec{E}_s$, have the opposite signs to the results obtained below.

[3]The term involving s_z is second order in infinitesimals.

7.2.4 *Landau–Lifshitz equations*

The Landau–Lifshitz (or Bloch) equations are an emergent property of the adiabatic approximation. They correspond to the, obvious, requirement that there be no *transverse* effective magnetic field acting upon the conduction electrons. The effective Hamiltonian following the $U_{\hat{n}}$ rotation is

$$H' = \frac{1}{2m}\left(\vec{p} - e\vec{A}_s\right)^2 + e(\Phi'_0 + A_{0s}) \tag{7.9}$$

where, since the *total* S_z is conserved, the spin dependent Zeeman potential $\Phi'_0 = g\mu_B B \sin\theta s_x$ is off-diagonal. This H' contains spin off-diagonal terms, $-e\vec{v}\cdot\vec{A}_s - g\mu_B \sin\theta B_{\text{eff}} s_x + eA_0$. That, on average, the electrons see no effective *transverse* magnetic field is the *requirement* that the slow magnetic dynamics have been well separated from the fast transport dynamics of the charge degrees of freedom. Explicitly this average, over unpaired electrons, is

$$\vec{v}_s \cdot \left[\sin\theta\vec{\nabla}\phi\, s_x - \vec{\nabla}\theta\, s_y\right] - Pg\mu_B \sin\theta B_{\text{eff}} s_x + P\sin\theta\left[\partial_t\phi\, s_x - \partial_t\theta\, s_y\right] = 0 \tag{7.10}$$

where $P\vec{v}_s = \vec{j}_\uparrow - \vec{j}_\downarrow$ is the spin current and $P = \langle\sigma\rangle = n_\uparrow - n_\downarrow$ the average polarization, and, requiring the coefficients of *both* s_x and s_y be zero, results in the expected Landau–Lifshitz equations, but without relaxation:

$$\frac{d\phi}{dt} + \vec{v}_s\cdot\vec{\nabla}\phi = \frac{\gamma B_{\text{eff}}}{\hbar}; \quad \text{and} \quad \frac{d\theta}{dt} + \vec{v}_s\cdot\vec{\nabla}\theta = 0. \tag{7.11}$$

The terms proportional to $\vec{v}_s$ correspond to the STT effect. As will be seen by example below, in fact, these terms *do* include relaxation when dissipative effects are reflected in $\vec{v}_s$.

7.2.5 *Spin-torque-transfer (STT), spin valves, and magnetic tunnel junctions (MTJs)*

The electron motion is now no longer adiabatic. The standard theory [24, 25, 26], is illustrated in Fig. 7.2(a). The ↑-electron component $\cos\frac{\theta}{2}$ of an incident electron is transmitted while most of the ↓-amplitude $\sin\frac{\theta}{2}$ is reflected. However, corresponding to the thin arrow of Fig. 7.2(a), there is a, usually neglected, spin-flip process which is, in fact, solely responsible for the STT effect. The spin wavefunction at the surface of the magnet is $\cos\frac{\theta}{2}|\uparrow\rangle|S, S_z = S\rangle + \sin\frac{\theta}{2}|\downarrow\rangle|S, S_z = S\rangle$. The total transverse and longitudinal angular momentum components are $\frac{\hbar}{2}\sin\theta$ and $\hbar(S + \frac{1}{2}\cos\theta)$. The electron leaving this free layer is parallel to the magnetization and the *changes* in the angular momenta are $\frac{\hbar}{2}\sin\theta$ and $-\frac{\hbar}{2}(1 - \cos\theta)$, i.e. the magnetization not only rotates by becomes *shorter*. The state with a *definite* $S_z = \hbar\cos\theta$ is proportional to $\int d\phi e^{-iS_z\phi/\hbar}|\theta, \phi\rangle$ where $|\theta, \phi\rangle = U_{\theta,\phi}|S, S_z = S\rangle$ is the maximal spin state rotated by θ and ϕ. It can

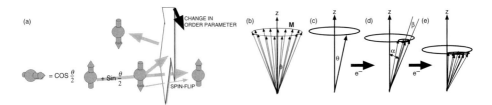

FIG. 7.2. (a) Cartoon version of the STT effect. Added here is the spin-flip channel. (b) State with S_z fixed but ϕ uncertain. (c) Initial state $|\theta\phi\rangle$. (b) The exaggerated result after one electron, (c) after two. There are changes in θ, i.e. the angle to the z-axis and in the length.

be pictured, Fig. 7.2(b), as a state[4] with a definite θ but indefinite ϕ. Given a beginning maximum spin state $|\theta, \phi\rangle$, Fig. 7.2(c), the passage of a first and second electron results in the states depicted in Fig. 7.2(d) and Fig. 7.2(e). In Fig. 7.2(d) there is a rotation by $\delta\theta = \frac{1}{2S}\sin\theta$ *plus* an opening of an angle $\delta\gamma = \frac{1}{2S}(1 - \cos\theta)$. These are double for Fig. 7.2(e). The resulting rate of rotation is

$$\frac{\partial\theta}{\partial t} = \frac{I}{2eS}\sin\theta, \tag{7.12}$$

where I is the current *in this majority-majority channel*. This is different from $\partial_t\theta = \frac{I}{eS}\tan\frac{\theta}{2}$, the famous result of Slonczewski [25] for the half-metal case. It is posited, at least for slow dynamics of $\vec{m}$, that relaxation of the angle γ is *much* faster than that of θ, due to Gilbert damping.

For MTJs, the tunneling barrier dominates the resistance and it is expected that,[5] $I = \cos^2\frac{\theta}{2}V/R$, where R is the resistance in the parallel configuration, thus the Slonczewski [25] result $\partial_t\theta = \frac{I}{eS}\tan\frac{\theta}{2}$ gives $\partial_t\theta = \frac{V}{2ReS}\sin\theta$. It is $\partial_t\theta/V$, rather than $\partial_t\theta/I$, which has a $\sin\theta$ dependence. This leads to the definition of *torkance*. This difference between angular dependence as a function of I or V is evident in first-principles calculations for the spin-torque transfer in real materials [27, 28, 29] but which are based on the Slonczewski approach and which do not account for changes in the order parameter *length*. Although the $\sin\theta$ dependence is emphasized, experiment [30] shows little difference between $(\partial\theta/\partial t)/V$ and $(\partial\theta/\partial t)/I$, both of which have an approximately $\sin\theta$ dependence. This reflects a magnetoresistance with a θ dependence much weaker than would be expected.

---

[4]More generally a state can be defined by $\int d\phi\rho(\theta,\phi)|\theta,\phi\rangle$ where $\rho(\theta,\phi)$ describes an "ensemble", even though this is a "pure" state.

[5]There is no fixed layer so in [25] only half of the angular momentum is given to each layer and $\partial_t\theta = \frac{I}{2eS}\tan\frac{\theta}{2}$. This approach implies a conductivity $\sigma(\theta) = \sigma_0\cos^2\frac{\theta}{2}$, whereas here $\sigma(\theta)$ is constant. The zero voltage magneto-resistance reflects the smf, see below.

7.2.6 Spinmotive force (smf), spin valves and MTJs

An smf arises when the field $\vec{B}_{\mathrm{eff}}$ is added. Assume for simplicity that this is aligned along the z-axis. This is a common experimental situation. Relative to this z-axis the amplitudes are, $\sin\frac{\theta}{2}$ for the spin-flip, and $\cos\frac{\theta}{2}$ for the regular channel. It follows $\partial_t m = \frac{2I}{eS}\sin^2\frac{\theta}{2}$. Reflecting the conservation of energy, as is illustrated in Fig. 7.3(a), there is an smf

$$\mathcal{E}_s = -\frac{\hbar}{e}\frac{\partial\phi}{\partial t} = \frac{g\mu_B B}{e} \tag{7.13}$$

but *only* in the spin-flip channel. Corresponding to the current model, Fig.7.3(a), the resistances in the spin-flip and regular channels are inversely proportional to the conductivities $\sigma_{\mathrm{sf}} = \sigma\sin^2\frac{\theta}{2}$ and $\sigma_r = \sigma\cos^2\frac{\theta}{2}$, where σ is the conductivity for the parallel (P) orientation when $\theta = 0$. Of the power $\mathcal{E}_s I \sin^2\frac{\theta}{2}$ dissipated in the spin-flip channel only the part $\frac{1}{4}\mathcal{E}_s I \sin^2\theta = \mathcal{E}_s I \cos^2\frac{\theta}{2}\sin^2\frac{\theta}{2}$ arises from the Zeeman energy $-\frac{1}{2}g\mu_B BS\cos\theta$ and rotations described by Eq. (7.12). The difference $\frac{1}{4}\mathcal{E}_s I(1-\cos\theta)^2$, for temperature $T = 0$, is dissipated *directly* as heat as indicated in Fig. 7.3(a), i.e. the spin-flip channel splits into two with the part $\frac{1}{4}I\sin^2\theta$ causing the rotations implied by Eq. (7.12) and with $\frac{1}{4}I(1-\cos\theta)^2$ corresponding to length changes and producing heat which adds to that produced by the regular channel.

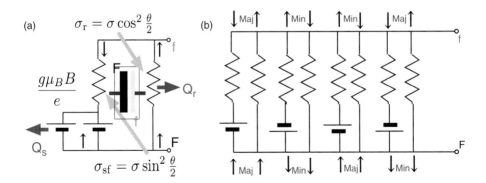

FIG. 7.3. (a) Shown are the *two*, i.e. spin-flip and regular conduction channels for half-metals. A smf of $g\mu_B B/e$ occurs in only the spin-flip channel. The two batteries in this channel reflect separation between the work done to change the Zeeman energy through rotations and that which is done to change the length of M and which is lost as heat as indicated by Q_s. The capital F (lower case f) corresponds to the "Fixed" ("free") layer. (b) For other than half-metals there is a total of eight channels. The sense of the battery and the conductance are deduced by comparison with the half-metal case. For simplicity the splitting of the batteries is not shown.

Even for half-metals, the presence of the smf changes the behavior of the device. In order for electrons to pass down the spin-flip channel, which is uniquely responsible of the spin-torque transfer, it is necessary that there is a voltage drop of at least $\mathcal{E}_s = \frac{\gamma B}{e}$ across the *regular* channel, implying a critical voltage for the onset of the effect. In addition there must be sufficient current in this channel to overcome Gilbert damping. Perhaps certain MTJs are the closest to this case. In Fig. 7.4(b) are shown the P and AP, IV branches of such a junction [31]. While the zero-voltage magnetoresistance, i.e. the difference in slope at the origin, is large, ideally $\sigma_{V=0} = \sigma_r = \sigma \cos^2 \frac{\theta}{2}$, this is not the case for larger voltages for which the slopes become almost equal. There is rather an almost parallel displacement by a voltage $\Delta V \approx 0.25\,\mathrm{V}$, as indicated. This is not directly $\mathcal{E}_s$ since a very large voltage drop across the oxide is needed to produce the appropriate $\vec{E}$ in the metal-free layer. The actual $\mathcal{E}_s$ is perhaps 10^{-4} times smaller. There is also not a step along the voltage axis, as the equivalent circuit would suggest, since the width of resonance of the uniform mode, which determines $\partial_t \phi$ through Eq. (7.11), with relaxation included, implies the channel opens over a range of $\vec{E}$ reflecting this width which is apparently quite large for this junction [31].

In a realistic model, there are four types of channel: majority/majority, majority/minority, minority/majority, and minority/minority, each with two sub-channels, to make a total of eight channels of which four contain an smf $\mathcal{E}_s = \pm \frac{\gamma B}{e}$. Not shown in the figure, for simplicity, is that each spin-flip channel also splits into two corresponding to the rotation and length change parts, i.e. the grand total is 16 channels. For other channels, which lead to relaxation

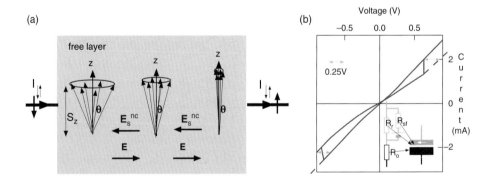

FIG. 7.4. (a) A current $I_{\downarrow\uparrow}$ flowing in the spin-flip channel produces a texture and a spin electric field $\vec{E}_s$ which is opposed by a regular $\vec{E}$ electric field resulting in a voltage across the free layer. (b) The IV curve for a MTJ with the P and AP branches shown [31]. An arrow shows the displacement of the curves by approximately 0.25 V at larger voltages and the insert shows the effective circuit with a large series resistance due to the tunnel barrier.

and which *enhance* Gilbert damping, there is a driving smf even without an external voltage, that is, currents will flow in an external circuit as the magnet relaxes from a high to a low level state. This possibility is that of a spin battery [2].

7.2.7 *The spin electric field $\vec{E}_s$*

In the single-electron picture, the rotation of the spin current inside the free layer corresponds to virtual excitations of electrons. In a realistic theory it is rather magnons which are created at the surface and this rotation reflects rather virtual magnons. These virtual excitations, in turn, give the free layer a texture which is ultimately responsible for the coupling between the electron charge current and the order parameter. It is this magnon texture which implies a spin electric field $\vec{E}_s$. It is necessary to adapt the earlier theory to this new situation. Not only does the order parameter not have a constant length, now this length also has a texture, i.e. the magnons are partially incoherent and there is an average over a ϕ ensemble. There is no gradient in ϕ. The nonconservative $\vec{E}_s^{\rm nc} = \sigma \frac{\hbar}{2e}(\vec{\nabla}\cos\theta)\partial_t\phi$, Eq. (7.8), does not depend upon the absolute value of ϕ. Accounting for the ϕ average: $\vec{E}_s^{\rm nc} = \sigma \frac{\hbar}{2e}[\vec{\nabla}m + \vec{\nabla}\cos\theta]\partial_t\phi$ where the first term reflects changes in the *length* and the second, with a redefined θ, the remaining texture due *rotations* of the order parameter. Here interest is in uniform rotations for which θ is a constant, i.e. $\vec{E}_s^{\rm nc} = \sigma \frac{\hbar}{2e}(\vec{\nabla}m)\partial_t\phi$. While a constant ∇m costs energy, it is not reflected in spin-wave energies and in particular the uniform mode.[6] Consider, a current I in the x-direction, a half-magnet without relaxation, near the AP configuration. This configuration isolates the spin-flip channel. Continuity requires $\hbar S\partial_t m = \frac{\hbar I}{e} = -T\frac{\hbar I}{e}\partial_x m$ where T is the thickness. The result $\partial_x m = -\frac{1}{T}$ is independent of the current and, as illustrated in Fig. 7.4(a), gives a *constant*:

$$\vec{E}_s^{\rm nc} = -\frac{\hbar}{T}\frac{\hbar}{2e}\partial_t\phi\,\hat{x} = -\frac{\mathcal{E}_s}{T}\hat{x} \qquad (7.14)$$

in the *spin-flip channel*. There is evidently no such spin electric field in the regular channel. As is also illustrated in Fig. 7.4(a), in order to have a current in the spin-flip channel, and hence to have a STT effect, it is necessary for the charge electric field E to exceed the spin E_s. Since the regular and spin-flip channels are

---

[6]This result has to do with the existence of evanescent magnons which decay in a linear manner. When the length mS of the magnetization is no longer a constant, the magnon energy is of the form: $\frac{1}{2}AS^2[(\nabla m)^2 + m^2((\nabla\theta)^2 + \sin^2\theta(\nabla\phi)^2)]$, where A is the intersite exchange. With the STT effect, for a constant current in the z-direction, the relevant *Bloch* equations of motion, without relaxation, are $\partial_t\phi + v\partial_z\phi = \partial_z^2 m + \partial_z^2\theta + \cos\theta(\nabla\phi)^2$, $\partial_t\theta + v\partial_z\theta = 0$, and for the length $\partial_t m + \partial_z vm = 0$. The latter is just the equation of continuity for the z-component of the magnetization. For a uniform mode, that is $\partial_z\theta = 0$, the energy, determined by $\partial_t\phi$, is unchanged provided $\partial_z m$ is a *constant*.

in parallel, this implies there is enough current in the regular channel in order to have a voltage $V > g\mu_B B_{\text{eff}}/e$ across the free layer.

It might seem strange that there is a gradient in the magnetization for one channel and not for another. The present approach is a semiclassical picture which represents evanescent magnons.[6] An electron in the spin-flip channel is accompanied by a cloud of virtual magnons which carry angular momentum and energy into the body of the free layer and the gradient in m reflects this. There is evidently no such cloud in the regular channel.

7.3   Examples of spinmotive forces

7.3.1   *A plain Néel domain wall*

This is the simplest example. The spin-motive force is easily deduced by the requirements of the conservation of energy [32]. It is a perfect example which illustrates Eq. (7.8) for $\vec{E}_s$ and which here simplifies to $\vec{E}_s = \vec{E}_s^{\text{nc}} + \vec{E}_s^{\text{c}}$ where

$$\vec{E}_s^{\text{nc}} = -\sigma_z \frac{\hbar}{2e} \sin\theta (\vec{\nabla}\theta)(\partial_t\phi) \quad \text{and} \quad \vec{E}_s^{c} = -\frac{g\mu_B}{2e} B(\vec{\nabla}m), \tag{7.15}$$

since $\vec{\nabla}\phi = 0$ and $\vec{B}_{\text{eff}} = B\hat{x}$. The nonconservative, i.e. motive force term: $\vec{E}_s^{\text{nc}} = \sigma_z \frac{\hbar}{2e}(\partial_x \cos\theta)(\partial_t\phi)\hat{x}$ is the exact opposite of $\vec{E}_s^{c} = -\sigma\frac{\hbar}{2e}\frac{g\mu_B}{2e}B(\partial_x \cos\theta)\hat{x}$, due to the gradient of the conservative Zeeman energy. This follows since $m = \cos\theta$ and $\hbar\partial_t\phi = g\mu_B B$. What is physically measured, by a voltmeter, is the *potential difference* $\Phi(x = +\infty) - \Phi(x = -\infty) = \sigma_z g\mu_B B/e$, since the effective potential seen by the conduction electrons is $e\Phi = (g\mu_B B/2)\cos\theta$ and $\theta = 0$ at $x = +\infty$ and $\theta = \pi$ at $z = -\infty$. The resulting spinmotive force is $\mathcal{E}_s = g\mu_B B/e \approx 100\,\mu\text{V/T}$. This prediction [5] is confirmed experimentally [7] but analyzed differently.

While it is the potential difference which is measurable, this is *maintained* by the nonconservative force without which it would collapse. If, for example, the wall was locally pinned, the effective field seen by the wall is $g\mu_B B_{\text{eff}} = \hbar\frac{\partial\phi}{\partial t} = 0$, and $\vec{E}_s^{nc} = 0$. Now the finite $\vec{E}_s^{c}$ would cause electrons to flow, and charge would accumulate, until the chemical potentials are constants. At this point, the regular U(1) electric field is $\vec{E} + \vec{E}_s^{c} = 0$, and there is no measurable voltage across the sample.

In other analyses [7], the situation is inverted, and a U(1) field $\vec{E}$ arises in the dynamic situation. This $\vec{E}$ is determined by solving the Laplace equation with $\vec{\nabla} \cdot \vec{E}_s^{c}$ as a source term. The difference is easily tested, first because the sign of the predicted voltage is opposite, and also by using ferromagnetic contacts to measure the potential difference. If the contacts are half-metals then the contact chemical potentials will align with the majority/minority bands of the sample with the same spin direction. If the contacts have antiparallel magnetizations, and the magnetization at both ends is in the same sense as the sample, see Fig. 7.5, then they align themselves with the majority chemical potential. When

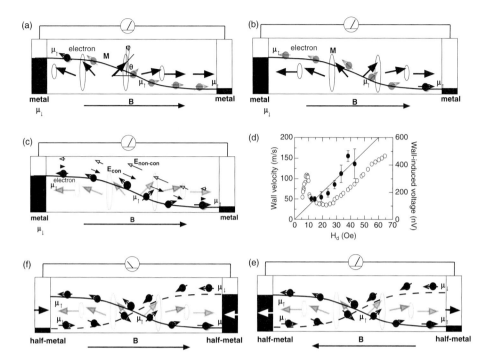

FIG. 7.5. In (a) and (b) is illustrated the displacement of the wall with passage of an electron. As the electron adiabatically rotates, its Zeeman, potential energy decreases but that of the magnet increases due to its displacement in the sense opposite to the direction of $\vec{B}$. The spin electric field $\vec{E}_s^{nc}$ reflects the transfer of this latter energy to the conduction electron to constitute a smf while the Zeeman energy corresponds to a potential. (c) Illustrates that locally $\vec{E}_s^{nc}$ cancels $\vec{E}_s^c$. That there is a net force which overcomes the resistance implies $E_s^c < E_s^{nc}$. (d) Experimental results of Beach in [7]. With half-metal *contacts*, (e) and (f), the voltage changes sign with the reversal of their magnetization, as shown; other theories do not predict such a change.

the contact magnetizations are both reversed, the alignment is with the minority bands and the measured voltage changes sign. If instead voltages reflect the U(1) field $\vec{E}$, there is no such sign reversal.

7.3.2 *Reciprocal relations*

This problem is also an excellent illustration of reciprocal relationships. The relationships which exist between the STT effect and the smf arise from current–current couplings and follow from Newton's third law. Onsager relationships are between transport coeffecients and do not apply here. This contrasts with what is to be found in the next chapter, and elsewhere [33], which explains the essential nature of such transport relationships to structure of "magneto-electronic circuit theory". For a domain wall located at x, the Zeeman potential energy is $e\Phi = -g\mu_B BnA\, x$, where n is the number density and A the cross-sectional area and x the position of the wall. With a charge q on its positive plate, a battery corresponds to a potential energy $q\mathcal{E}$. The Lagrangian is then

$$L = -(enA\dot{x} - \dot{q})\frac{\hbar}{e}\phi - g\mu_B BnAx - q\mathcal{E} = (enA\dot{x} - \dot{q})\frac{\hbar}{e}\phi - H \qquad (7.16)$$

where H is the Hamiltonian. Corresponding to $enA\dot{x}\phi$, x and the angle ϕ, are the conjugate co-ordinate and momentum, ultimately corresponding to the spin commutation rules. Lastly $\dot{q}\frac{\hbar}{e}\phi$ is the average over the wall of $-\vec{v}\cdot\vec{A}_s$, where $\vec{A}_s$ is given by Eq. (7.5), i.e. the current $i = \dot{q}$ is proportional to the negative of the conduction electron momentum and $\frac{\hbar}{e}\phi$ that of the magnetic system. This current–current term *alone* generates both the smf and the STT. Dissipation is introduced by a Rayleigh dissipation function $\mathcal{R} = \frac{1}{2}R\dot{q}^2$. The three Euler–Lagrange equations of motion are

$$\hbar\dot{\phi} = g\mu_B B; \qquad \dot{x} = \frac{1}{enA}\dot{q}; \qquad \text{and} \qquad \frac{\hbar}{e}\dot{\phi} + \mathcal{E} = iR. \qquad (7.17)$$

The third-law reciprocal relationship is between the STT term $\frac{\hbar}{enA}\dot{q}$ in the second equation and the smf term $\mathcal{E}_s = -\frac{\hbar}{e}\dot{\phi}$ in the third. The only transport coeffecient R cannot have an Onsager relationship since it appears in the single circuit equation. Eliminating the current, the equivalents of the Landau–Lifshitz equations are $enAv = \frac{\mathcal{E}}{R} + \frac{1}{R}\frac{\hbar}{e}\dot{\phi}$ and $\hbar\dot{\phi} = g\mu_B B$, and contain a relaxation term corresponding to $\alpha = \beta$, despite the absence of nonadiabatic processes deemed [34] responsible for a finite β.

7.3.3 The spinmotive force and magnons

Consider magnons of wavevector $\vec{k}$, generated at one side of a dynamic domain wall, and detected at the other side. In the absence of forces, i.e. since the net $\vec{E}_s = 0$, $\vec{p} = \hbar\vec{k}$ is unchanged by the passage through the wall, but due to the change in sign in the energy $\hbar\omega_{\vec{k}}^{\pm} = g\mu_B(B_a \pm B) + (\hbar k)^2/2m_s$ there *is* a frequency shift such that $\hbar\Delta\omega_{\vec{k}} = 2g\mu_B B$. (For a pinned wall $\vec{E}_s^{nc} = 0$ and there *is* an acceleration and no Doppler shift.) This makes a clear prediction relative to approaches in which, *for the electrons,* $\vec{E}_s \neq 0$, but which, in equilibrium, is cancelled by the U(1), i.e. charge field $\vec{E}$.

Accompanying a charge current, for a conducting magnet, there is an electron transport spin current flowing in or out of a domain wall, depending upon its sense "head to head" or "tail to tail." The equivalent Doppler shift of magnon energies, due to *charge* spin currents, has already been detected [35].

7.3.4 *Spin forces and Doppler shifts for phonons*

The generation and detection of surface acoustic waves (SAW) is a well-proven technique [36] which is somewhat easier than the detection of $\vec{q} \neq 0$ magnons. Phonons are also chargeless excitations subject to the spin electric field $\vec{E}_s$ with some small effective charge e_p. This charge, e_p, reflects what tradition would identify with electron–phonon or magnon–phonon drag effects, but which are more correctly identified as spinmotive forces, since first they can cause currents to flow in external circuits and secondly because there is no electric charge to be identified with the forces. It is also of interest that, in contrast with magnons, phonons are not subject to a "chemical potential" shift due to an applied magnetic field $\vec{B}$, and hence do not feel either the charge field $\vec{E}$, *or* the spin $\vec{E}_s^c$. They however *are* accelerated by $\vec{E}_s^{nc}$ and hence have a Doppler shift in frequency, but now due to the change in $\vec{k}$. Together, such experiments offer the possibility of determining all of the fields $\vec{E}_e$, $\vec{E}_s^c$, and $\vec{E}_s^{nc}$.

7.3.5 *Voltage steps and magnetoresistance*

In the experiments of Sun *et al.* [37], on a number of spin valves, with perpendicular fields large compared with the demagnetization field $4\pi M_s$, there are voltage steps ΔV at currents I_c for which $\Delta V = \delta R I_c$ and where δR is approximately the low-current magnetoresistance. The steps $\Delta V \approx 4.5\mu_B(B + B_a)$ occur when the system switches from P to AP.

These results can be understood in terms of the multichannel model, Fig. 7.3(c). When the low-field, low-current magnetoresistance is measured, in the P configuration the resistance is $R_P < R_{AP}$ and these are in series with a resistance, R, which is perhaps 10 times larger that R_{AP}. It follows that $\delta R \approx R_{AP}$. For large currents and fields, in order to maintain the system in the AP configuration, it must be that the voltage across R_{AP} is greater that $\hbar\omega_s/e$ where $\hbar\omega_s \approx 2\mu_B(B + B_a)$ and where the demagnetization field now adds to B, i.e. there must be a voltage $\delta R I_c \approx 2\mu_B(B + B_a)/e$ across the free layer in order to open the spin-flip channel and have a STT to maintain the negative-temperature AP state. However since it is in series with a resistance R the voltage across the device is larger by a factor of $R/\delta R$.

This picture differs from that based on Slonczewski's theory [25, 37] where the Gilbert damping constant α is of prime importance. It is suggested [37] that the observations are due to the domination of the extra α' spin-pumping term to this constant. The present model contains this effect. At finite small angles there is a current through the regular R_P channel and which constitutes a rather large contribution to α'. That this relaxation is not important is surprising but is consistent on FMR experiments [38, 39, 40, 41] of bi- and tri-layer system very similar to those used for spin valves.

7.3.6 *Static magnetic vortices*

Such a vortex is illustrated in Fig. 7.6(a). They form for disks of suitable aspect ratio. Here the concept of spin magnetic field $\vec{B}_s$ is very relevant, and the correct definition of $e\vec{A}_s$ is central. If the U(1) projection is assumed [9, 7, 8], $\vec{B}_s = -s_z \sin\theta \vec{\nabla}\theta \times \vec{\nabla}\phi$, and passing though such a static vortex, is a unit of magnetic flux. If this was the case it might be used to produce circulating currents reflecting the smf equation $\mathcal{E}_s = -d\gamma_s/dt$, in what is an adaptation of the original smf experiment suggested by Stern [6]. The spin orientation at a given distance from the vortex center is shown in Fig. 7.6(b) and corresponds to that suggested for this experiment [6]. If a vertical rf field is imposed, the angle θ these spins make to the vertical will be modulated and it is imagined this will modulate the spin Berry phase γ_s and produce a circulating current. For spin one-half, in the static situation, the spin Berry phase is given by $\gamma_s = \Omega/2$ where Ω is the angle subtended by the spin trajectory on a unit sphere. Thus, in Fig. 7.6(b), as an electron passes around the loop, in spin space it has the trajectory shown in Fig. 7.6(d) and subtends a solid angle $\Omega(t) = 2\pi(1 - \cos\theta(t))$. When the angle θ is modulated, so is the static flux and predicated is a $\mathcal{E}_s = \pi(\hbar/2e)\sin\theta(t)d\theta/dt$ and a circulating ac spin and charge current, which might be rather easily detected. Similar circulating currents, in the core region, would be expected for the vortex as a whole. That $\vec{B}_s = 0$ for the static system implies no circulating currents. A finite flux also results in a measurable anomalous Hall effect [11, 12, 13] or measurable force $\vec{v} \times \vec{B}_s$, in Lorentz microscopy.

7.3.7 *Field-driven magnetic vortices*

A gyrating vortex, if its radius of gyration R exceeds the core radius r_c, *does* appear to carry a unit of magnetic flux $(2\pi\hbar/e)$. Consider two points separated by roughly twice the core radius. For a velocity v the unit flux will pass by the points in a time $2r_c/v$ so the rate of change is $|\mathcal{E}_s| = |d\Phi_s/dt| \sim (2\pi\hbar/e)v/2r_c$ which is the result obtained more directly below. The circular motion, Fig. 7.6(c), of such a vortex reflects a radial restoring force $\vec{f}$ directed towards the center, which, in turn, corresponds to a tangential $\vec{B}_{\text{eff}}$. Consider a line through the centre of the vortex and parallel to $\vec{B}_{\text{eff}}$. Along this line the orientation of $\hat{m}$ is the same as in a Bloch wall. As the vortex moves by a distance $v\Delta t = 2r_c$, twice the core radius, the local $\hat{m}$, initially just in front of the core, reverses its direction as the vortex passes by, thereby rotating by $\Delta\phi = \pi$, that is, $\frac{d\phi}{dt} = \frac{\pi}{2r_c}v = \gamma B_{\text{eff}}/\hbar$ determining B_{eff}. However along this direction $\theta = \pi/2$, a constant, and there is no tangential $\vec{E}_s$. In contrast, along a radial direction through the core, $\hat{m}$ nearest the center is tangential in one sense $(\theta = 0)$ while the more distant $\hat{m}$ has the opposite sense $(\theta = \pi)$. The magnetization direction $\hat{m}$ close to and inside the core precesses about $\hat{n}$ resulting in a finite radial $\vec{E}_s = \hat{r}\frac{\hbar}{2e}\frac{d\cos\theta}{dy}\frac{d\phi}{dt}$ such that $e\mathcal{E}_s = \gamma B_{\text{eff}} = \frac{\pi\hbar}{2r_c}v$ is proportional to the velocity v and reflects the passage of a unit flux. For frequency f, $v = 2\pi fR$ and finally $\mathcal{E}_s \sim \left(\frac{\hbar\omega}{e}\right)\left(\frac{R}{2r_c}\right)$. Compared

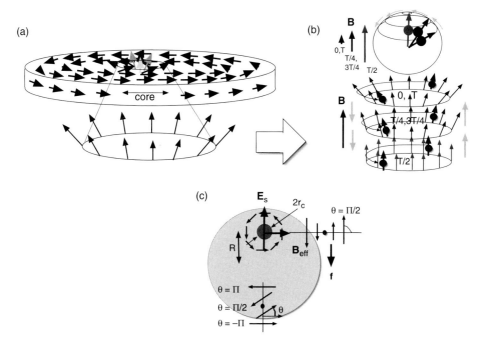

FIG. 7.6. (a) Static magnetic vortex structure. The circulating magnetization has two senses, i.e. chiralities. There is no surface magnetic charge minimizing the dipole energy at the expense of the exchange energy. Near the center, the magnetization must turn out of the plane in order to reduce the exchange energy. This core has two senses, either pointing up, as in the figure, or down. Not too far from the core the orientation of the magnetization is as shown in the projection. (b) This texture is that envisaged by Stern [6]. It is imagined that a periodic vertical field $\vec{B}$ will twist the magnetization to and from the vertical as shown. Here T is the period. For a circulating electron with a spin which follows the magnetization adiabatically its spin direction maps out in spin space a solid angle Ω which is a function of time. This in turn implies a time-varying spin Berry phase $\gamma_s = \Omega/2$ and a smf which causes a circulating current. However this is predicated on the insistence that a static spin texture has a spin Berry phase. The current SU(2) theory *does not* predict a circulating current. (c) While a *static* magnetic vortex does *not* carry a unit of magnetic flux, one which rotates due to the effects of a driving rf field which causes the radius of motion R to exceed that of the core radius r_c, *does* appear to carry such a flux. In this situation there is a considerable "magnification" of the smf compared to the value $\hbar\omega/e$, as explained in the text.

to a photoelectric effect or a simple domain wall for which the voltage is $\frac{\hbar\omega}{e}$ ($\approx 0.3\mu$V, for 80 MHz), an enhancement by $\left(\frac{R}{2r_c}\right)$, the vortex orbital radius in units of the core diameter, is expected, and observed [17].

7.4 Ferromagnetic resonance (FMR) and spinmotive forces

Ferromagnetic resonance occurs when the sample is subject to a magnetic field with a radio frequency (rf) ω_{rf} which is close to that $\omega_{q=0}$ of the uniform (or magnetostatic) modes. Driving such a resonance, in structures of interest, where a magnet is in contact with a metal, produces an smf which leads to relaxation or drives a current in an external circuit and constitutes an externally driven "spin battery" [42].

7.4.1 Field-driven FMR

Assume for simplicity that $\omega_{\mathrm{rf}} = \omega_{q=0}$ whence, as shown in Fig. 7.7(a), the rf field provides a torque which opposes relaxation. The steady state $\sin\theta = b_{\mathrm{rf}}/\alpha B_{\mathrm{eff}}$ corresponds to a coherent state $|n_{\vec{q}=0}\rangle = \prod_n e^{(n_{\vec{q}=0}/S)^{1/2}e^{i\phi}b_n^\dagger}|\rangle$ with a macroscopic occupation $n_{\vec{q}=0} \sim S\sin^2\frac{\theta}{2}$ of the uniform mode. Thermal fluctuations, Fig. 7.7(a), imply a θ_T such that $g\mu_B B_{\mathrm{eff}}(1 - \cos\theta_T) \sim k_B T$ but for a coherent state $|\theta_T\rangle$ for which ϕ is random. The FMR state is then $|n_{\vec{q}=0}\rangle = \prod_n e^{(n_{\vec{q}=0}/S)^{1/2}e^{i\phi}b_n^\dagger}|\theta_T\rangle$.

7.4.2 STT-FMR and a-FMR resonance

With STT-FMR a dc current, in a suitable sense, provides a torque, Fig. 7.7(a), essentially equivalent to that of a rf field. For small currents this leads to small-angle FMR, but above a critical current can jump to a large angle. When $\theta > \pi/2$ the anisotropy field changes sign and with this the sense of precession. Now the STT is in the same sense as the relaxation-fluctuations.

Because the individual spins have a finite magnitude s, at finite temperature T, there can be macroscopically occupied coherent states of the form $|n_{\vec{q}}^a\rangle = \prod_n e^{(n_{\vec{q}}^a/S)^{1/2}e^{i\phi}b_n}|\theta_T\rangle$ which now involves the magnon *destruction* operator. This "anti"-ferromagnetic resonance (a-FMR), often described as "thermally excited FMR," occurs for the opposite sign of current and small angles, and is actually well characterized [43] in MTJs. Driving the a-FMR necessarily produces (or is in response to) a heat current. This implies spin Seebeck and Peltier effects.

7.4.3 The spin Seebeck effect and the smf

Consider a single ferromagnetic thin film subject to a heat gradient, as shown in Fig. 7.8(a). A suitable spin current drives the FMR or a-FMR and this produces an $\vec{E}_s$ and a heat current. In the reciprocal situation, a heat current drives an FMR and an a-FMR and produces a $\vec{E}_s$ and an smf. It is the a-FMR which absorbs heat and is driven at the hot end and the FMR which ejects heat

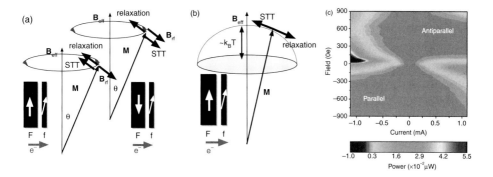

FIG. 7.7. (a) The two cases in which, for zero temperature, the STT and $\vec{B}_{\text{eff}}$ effects either add, leading to a reduction in the effective relaxation and a reduction in the effective rate of relaxation, or the inverse when they act in the opposite directions. For half-metals the relative orientation of the Fixed (F) and free (f) layers along with the direction of electron flow are shown. (b) The situation at finite temperature. Thermal fluctuations would have the magnetization somewhere close to the edge of the shaded circle. This reflects the equipartition theorem. When the resonance is driven by a suitable sense of the electron current, as indicated, this corresponds to an a-FMR. As usual the STT opposes relaxation but since the precession angle is smaller than that which corresponds to thermal fluctuations, relaxation-fluctuation actually causes there to be an *increase* in energy (but not free energy), implying the system actually absorbs heat energy. (c) The experimental data of Deac *et al.* [43]. The contour plot shows the rf power in the field–current plane. The positive current power corresponds to an a-FMR or "thermally excited FMR."

which operates at the cold end. Such a state has a constant $\vec{\nabla}m$ (and hence constant $\vec{E}_s$). If then the dipole fields are considered as perturbations, the result of a temperature gradient is a $n = 1$ magnetostatic mode, also illustrated in Fig. 7.8(a). In this picture the Pt strips used to detect the spin Seeback effect [44] reflect spin currents which flow into the a-FMR and out of the FMR.

7.4.4 *FMR with thermal gradients*

Given an FMR driven by an external field, the constant $\vec{\nabla}m$, constant $\vec{E}_s$, state ejects heat and hence has an increased amplitude at the cold end and a reduced one at the hot end; put differently it is a linear combination of the uniform $n = 0$ and $n = 1$ magnetostatic modes, Fig. 7.8(b). There is some tentative evidence [45] for this effect, Fig. 7.8(c).

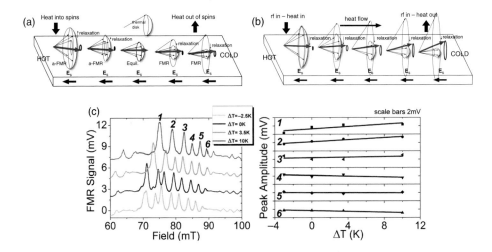

FIG. 7.8. (a) The spin electric field $\vec{E}_s$ associated with an odd "thermally excited" magnetostatic mode driven by the temperature gradient. There are several arrows which correspond to the thermally excited magnetization with a random phase and which lies near the "thermal cone." The coherent amplitude shown by the large gray arrow arises because of magnetization which is excited to greater angles indicated by the single arrow on the larger circle. Relaxation is towards the thermal cone, i.e. towards smaller angles, which ejects heat. Only thermal magnetization arises in the center and the grey coherent part has zero amplitude. The a-FMR is involved, at the high temperatures for which the grey coherent magnetization is out of phase with the cold part. This magnetization is now created by moving magnetization to lower angles, indicated by the smaller inner circle, and which relaxes to larger angle and bringing heat energy *into* the magnetic system. (b) The coherent magnetization increases steadily for FMR with a smallish temperature gradient and this is generated by magnetization which is shifted to greater angles. (c) Relative to the case without a heat current, the detected FMR increases at the cold end, but decreases at the hot end, i.e. there is a change in sign of the signal with the sense of the heat current, as shown experimentally for the uniform mode "1."

7.5 Spinmotive forces, magnons, and phonons

7.5.1 *The $s = 1$ nature of magnons and phonons*

Both magnons and phonons have no charge. They see only the spin electric $\vec{E}_s$ and magnetic $\vec{B}_s$ fields. Conventional wisdom would have the $s = 1$ nature of phonons reflecting the polarization. This is not so at least in the present context of spin-derived forces and has interesting consequences. For the

Holstein–Primakoff magnon boson b the Lagrangian density is $\mathcal{L} = i\hbar b^\dagger \frac{\partial}{\partial t} b - \mathcal{H}$ where $\mathcal{H}$ is the appropriate Hamiltonian density and *defines* Bose statistics. *However*, the $s = 1$ nature of the magnons corresponds to $s_z = s - b^\dagger b$, and has nothing to do with polarization. The "spin" Berry phase corresponds to the boson *number* and can exist, e.g. for longitudinal phonons. Relevant rather is the coupling between the *momenta* $e\vec{A}$ of the electrons, magnons, and phonons, e.g. electron–phonon *drag*.

7.5.2 *Realization of smf effects with magnons and phonons*

Imagine a dc spin-motive force produced by a magnon spin current. As illustrated in Fig. 7.9(a), consider magnons which propagate in the x-direction in an insulating magnet such as YIG. The addition of a metal capping layer will cause an additional damping in much the same manner as it does for the FMR. As a consequence, as illustrated, the amplitude θ will decrease as the magnons pass under the metal capping. Thus there exists a nonconservative $E_s^{\text{nc}} = \frac{\hbar}{e}(\partial_z \cos\theta)\partial_t \phi$ where $\partial_t \phi = \omega_{\vec{k}}$, the angular frequency of the magnon. It is predicted that a longitudinal voltage appears across the metal cap, as shown.

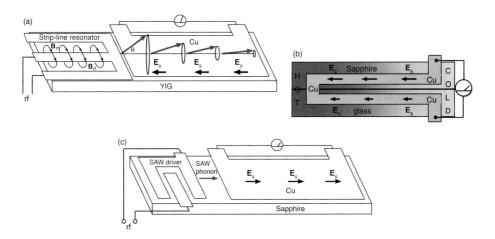

FIG. 7.9. (a) Magnons are produced, in say insulating YIG, by the strip-line resonator and attenuated by the metal, say Cu, since they share momentum. This leads to an $\vec{E}_s$ and a voltage. (b) In a phonon thermocouple the co-momentum of the metal (Cu) with now the phonons leads to a phonon Seebeck effect since that of the Cu cancels. (c) Phonons produced by surface acoustic modes produced by electrostatic coupling will also result in an $\vec{E}_s$ and a voltage.

Possible is a phonon thermocouple. As shown in Fig. 7.9(b) imagine a piece of a good thermal conductor such as sapphire glued to a poor conductor, say, glass and lay a continuous thin film of some good metal, say Cu or Ag, to form a "U", as shown. Even though one end is hotter than the other, the net Seebeck effect cancels in the metal layer. However there will be a larger heat current in the sapphire than the glass and hence a larger smf induced in one metal leg than the other. A net voltage will appear across the terminals and which might be made to drive an electrical current, and hence constitutes a nonconservative motive force.

Sapphire is a material with an exceptionally long phonon mean-free-path. In the dispersion relationship [46], there is a local minimum with a small "mass" at about 5 THz corresponding to a temperature ~ 150 K. A large temperature gradient and the momentum–momentum coupling can lead to spontaneous $\vec{E}_s$ associated with this 5 THz phonon which would correspond to the production of such phonons by thermal effects.

As with magnons, a more basic experiment is to induce a dc motive force in a metal capping layer due to phonons propagating in an insulator, as illustrated in Fig. 7.9(c). This experiment has been performed [47] in a different geometry some time ago and interpreted in terms of electron–phonon momentum coupling but using a different language.

References

[1] J. D. Jackson, *Classical Electrodynamics* (John Wiley, New Jersey, 1998).

[2] P. N. Hai, S. Ohya, M. Tanaka, S. E. Barnes, and S. Maekawa, Nature **458**, 489 (2009).

[3] Y. Aharonov and D. Bohm, Phys. Rev. **115**, 485 (1959).

[4] M. V. Berry, Proc R. Soc. Lond. A **392**, 45 (1984).

[5] S. E. Barnes and S. Maekawa, Phys. Rev. Lett. **98**, 246601 (2007).

[6] A. Stern, Phys. Rev. Lett. **68**, 1022 (1992).

[7] S. A. Yang, G. S. D. Beach, C. Knutson, D. Xiao, Q. Niu, M. Tsoi, and J. L. Erskine, Phys. Rev. Lett. **102**, 067201 (2009); S. A. Yang, G. S. D. Beach, C. Knutson, D. Xiao, Z. Zhang, M. Tsoi, Q. Niu, A. H. MacDonald, and J. L. Erskine, Phys. Rev. B **82**, 067201 (2009).

[8] Y. Tserkovnyak and M. Mecklenburg, Phys. Rev. B **77**, 134407 (2008).

[9] G. E. Volovik, J. Phys. C **20**, L83 (1987).

[10] L. Berger, Phys. Rev. B **33**, 1572 (1986).

[11] W. Koshibae, M. Yamanaka, M. Oshikawa, and S. Maekawa, Phys. Rev. Lett. **82**, 2119 (1999).

[12] Y. Taguchi, Y. Oohara, H. Yoshizawa, N. Nagaosa, and Y. Tokura, Science **291**, 2573 (2001).

[13] N. Nagaosa, J. Phys. Soc. Jpn. **75**, 042001 (2006).

[14] R. A Webb, S. Washburn, C. P. Umbach, and R. B. Laibowitz, Phys. Rev. Lett. **54**, 26962699 (1985).

[15] M. J. Yang, C. H. Yang, and Y. B. Lyanda-Geller, Europhys. Lett. **66**, 826 (2004).

[16] Y. Yamane, K. Sasage, T. An, K. Harii, J.-I. Ohe, J. Ieda, S. E. Barnes, E. Saitoh, and S. Maekawa, Phys. Rev. Lett. **107**, 236602 (2011).

[17] K. Tanabe, D. Chiba, J. Ohe, S. Kasai, H. Kohno, S. E. Barnes, S. Maekawa, K. Kobayashi, and T. Ono, Nat. Commun. **3**, 845 (2011).

[18] W. M. Saslow, Phys. Rev. B **76**, 184434 (2007).

[19] R. A. Duine, Phys. Rev. B **77**, 014409 (2008).

[20] M. Stamenova, T. N. Todorov, and S. Sanvito, Phys. Rev. B **77**, 054439 (2008).

[21] S. S.-L. Zhang and S. Zhang, Phys. Rev. B **82**, 184423 (2010).

[22] see: F. J. Dyson, Am. J. Phys. **58** 209 (1990).

[23] see, e.g., J. J. Sakurai, *Modern Quantum Mechanics* (Benjamin/Cummings, 1985).

[24] see, e.g., D. C. Ralph and M. A. Stiles, J. Magn. Magn. Mater. **320**, 1190 (2008).

[25] J. C. Slonczewski, J. Magn. Magn. Mater. **159**, L1 (1996).

[26] L. Berger, Phys. Rev. B **54**, 9353 (1996).

[27] C. Heiliger and M. D. Stiles, Phys. Rev. Lett. **100** 186805 (2008).

[28] I. Theodonis, N. Kioussis, A. Kalitsov, M. Chshiev, and W. H. Butler, Phys. Rev. Lett. **97**, 237205 (2006).

[29] P. M. Haney, D. Waldron, R. A. Duine, A. S. Nunez, H. Guo, and A. H. Mac-Donald, Phys. Rev. B **76**, 024404 (2007).

[30] J. C. Sankey, T-T. Cui, J. Sun, J. Slonczewski, R. A. Buhrman, and D. C Ralph, Nat. Phys. **4**, 67 (2008); see, in addition the supplemental information.

[31] T. Devolder, J. Hayakawa, K. Ito, H. Takahashi, S. Ikeda, P. Crozat, N. Zerounian, Joo-Von Kim, C. Chappert, and H. Ohno, Phys. Rev. Lett. **100**, 057206 (2008).

[32] S. E. Barnes, J. Ieda, and S. Maekawa, Appl. Phys. Lett. **89**, 122507 (2006).

[33] K. M. D. Hals, A. Brataas, and Y. Tserkovnyak, Euro. Phys. Lett. **90**, 47002 (2010).

[34] S. Zhang and Z. Li, Phys. Rev. Lett. **93**, 127204 (2004).

[35] V. Vlaminck and M. Bailleul, Science **322**, 410 (2008).

[36] C. K. Campbell, Proc. IEEE **77**, 1453 (1989).

[37] J. Z. Sun, B. Ozyilmaz, W. Chen, M. Tsoi, and A. D. Kent, J. Appl. Phys. **97**, 10714 (2005).

[38] R. Urban, G. Woltersdorf, and B. Heinrich, Phys. Rev. Lett. **87**, 217204 (2001).

[39] S. Mizukami, Y. Ando, and T. Miyazaki, J. Magn. Magn. Mater. **226**, 1640 (2001); ibid. **239**, 42 (2002).

[40] B. Heinrich, Y. Tserkovnyak, G. Woltersdorf, A. Brataas, R. Urban, and G. E. W. Bauer, Phys. Rev. Lett. **90**, 187601 (2003).

[41] H. Hurdequint, V. Castel., *et al.* Private communication.

[42] A. Brataas, Y. Tserkovnyak, G. E. W. Bauer, and B. I. Halperin, Phys. Rev. B **66**, 060404(R) (2002).

[43] A. M. Deac, A. Fukushima, H. Kubota, H. Maehara, Y. Suzuki, S. Yuasa, Y. Nagamine, K. Tsunekawa, D. D. Djayaprawira, and N. Watanabe, Nature Phys. **4**, 803, (2008).

[44] K. Uchida, S. Takahashi, K. Harii, J. Ieda, W. Koshibae, K. Ando, S. Maekawa, and E. Saitoh, Nature **455**, 778 (2008).

[45] E. Papa, M. Abi, G. Boero, S. E. Barnes, and J.-Ph. Ansermet, unpublished.

[46] H. Bialas and H. J. Stolz, Z. Phy. B **21**, 319 (1975).

[47] H. Karl, W. Dietsche, A. Fischer, and K. Ploog, Phys. Rev. Lett. **61**, 2360 (1988).

8 Spin pumping and spin transfer

A. Brataas, Y. Tserkovnyak, G. E. W. Bauer, and P. J. Kelly

8.1 Introduction

8.1.1 *Technology pull and physics push*

The interaction between electric currents and the magnetic order parameter in conducting magnetic micro- and nanostructures has developed into a major sub-field in magnetism [1]. The main reason is the technological potential of magnetic devices based on transition metals and their alloys that operate at ambient temperatures. Examples are current-induced tunable microwave generators (spin-torque oscillators) [2, 3], and nonvolatile magnetic electronic architectures that can be randomly read, written, or programmed by current pulses in a scalable manner [4]. The interaction between currents and magnetization can also cause undesirable effects such as enhanced magnetic noise in read heads made from magnetic multilayers [5]. While most research has been carried out on metallic structures, current-induced magnetization dynamics in semiconductors [6] or even insulators [7] has been pursued as well.

Physicists have been attracted in large numbers to these issues because on top of the practical aspects the underlying phenomena are so fascinating. Berger [8] and Slonczewski [9] are in general acknowledged to have started the whole field by introducing the concept of current-induced magnetization dynamics by the transfer of spin. The importance of their work was fully appreciated only after experimental confirmation of the predictions in multilayered structures [10, 11]. The reciprocal effect, i.e. the generation of currents by magnetization dynamics now called *spin pumping*, was expected long ago [12, 13], but it took some time before Tserkovnyak *et al.* [14, 15] developed a rigorous theory of spin pumping for magnetic multilayers, including the associated increased magnetization damping [16].

8.1.2 *Discrete versus homogeneous*

Spin-transfer torque and spin pumping in magnetic metallic multilayers are by now relatively well understood and the topic has been covered by a number of review articles [15, 19, 20]. It can be understood very well in terms of a time-dependent extension of magneto-electronic circuit theory [19, 21], which corresponds to the assumption of spin diffusion in the bulk and quantum mechanical boundary conditions at interfaces. Random matrix theory [22] can be shown to be equivalent to circuit theory [19, 23, 24]. The technologically important current-induced switching in magnetic tunnel junctions has recently been the focus of attention [25]. Tunnel junctions limit the transport such that circuit issues are less important, whereas the quantum-mechanical nature of the tunneling process becomes essential. We will not review this issue in more detail here.

The interaction of currents and magnetization in continuous magnetization textures has also attracted much interest, partly due to possible applications such as nonvolatile shift registers [26]. From a formal point of view the physics of current–magnetization interaction in a continuum poses new challenges as compared to heterostructures with atomically sharp interfaces. In magnetic textures such as magnetic domain walls, currents interact over length-scales corresponding to the wall widths that are usually much longer than even the transport mean-free path. Issues of the in-plane *vs.* magnetic-field-like torque [27] and the spin-motive force in moving magnetization textures [28] took some time to get sorted out, but the understanding of the complications associated with continuous textures has matured by now. There is now general consensus about the physics of current-induced magnetization excitations and magnetization-dynamics induced currents [29, 30]. Nevertheless, the similarities and differences of spin torque and spin pumping in discrete and continuous magnetic systems has to our knowledge never been discussed in a coherent fashion. It has also only recently been realized that both phenomena are directly related, since they reflect identical microscopic correlations according to the Onsager reciprocity relations [31–33].

8.1.3   *This chapter*

In this chapter, we (i) review the basic understandings of spin-transfer torque *vs.* spin pumping and (ii) knit together our understanding of both concepts for heterogeneous and homogeneous systems. We discuss the general phenomenology guided by Onsager's reciprocity in the linear response regime [34]. We will compare the in- and out-of-plane spin-transfer torques at interfaces as governed by the real and imaginary parts of the so-called spin-mixing conductances with that in textures, which are usually associated with the adiabatic torque and its dissipative correction [27], usually described by a dimensionless factor β in order to stress the relation with the Gilbert damping constant α. We argue that the spin pumping phenomenon at interfaces between magnets and conductors is identical to the spin-motive force due to magnetization texture dynamics such as moving domain walls [28]. We emphasize that spin pumping is on a microscopic level identical to the spin-transfer torque, thus arriving at a significantly simplified conceptual picture of the coupling between currents and magnetization. We also point out that we are not limited to a phenomenological description relying on fitting parameters by demonstrating that the material dependence of crucial parameters such as α and β can be computed from first principles.

8.2 Phenomenology

In this section we explain the basics physics of spin pumping and spin-transfer torques, introduce the dependence on material and externally applied parameters, and prove their equivalence in terms of Onsager's reciprocity theorem.

8.2.1 *Mechanics*

On a microscopic level electrons behave as wave-like fermions with quantized intrinsic angular momentum. However, in order to understand the electron wavepackets at the Fermi energy in high-density metals and the collective motion of a large number of spins at not too low temperatures classical analogues can be useful.

Spin-transfer torque and spin pumping are on a fundamental level mechanical phenomena that can be compared with the game of billiards, which is all about the transfer of linear and angular momenta between the balls and cushions. A skilled player can use the cue to transfer velocity and spin to the billiard ball in a controlled way. The path of the spinning ball is governed by the interaction with the reservoirs of linear and angular momentum (the cushions and the felt/baize) and with other balls during collisions. A ball that for instance hits the cushion at normal angle with top or bottom spin will reverse its rotation and translation, thereby transferring twice its linear and angular moment to the frame of the billiard table.

Since the work by Barnett [35] and Einstein and de Haas [36] almost a century ago, we know that magnetism is caused by the magnetic moment of the electron, which is intimately related with its mechanical angular momentum. How angular momentum transfer occurs between electrons in magnetic structures can be imagined mechanically: just replace the billiard balls by spin-polarized electrons and the cushion by a ferromagnet. Good metallic interfaces correspond to a cushion with high friction. The billiard ball reverses angular and linear momentum, whereas the electron is reflected with a spin flip. While the cushion and the billiard table absorb the angular momentum, the magnetization absorbs the spin angular momentum. The absorbed spins correspond to a torque that, if it exceeds a critical value, will set the magnetization into motion. Analogously, a time-dependent magnetization injects net angular momentum into a normal metal contact. This "spin pumping" effect, i.e. the main topic of this chapter, can also be visualized mechanically: a billiard ball without spin will pick up angular momentum under reflection if the cushion is rotating along its axis.

8.2.2 *Spin-transfer torque and spin pumping*

Ferromagnets do not easily change the modulus of the magnetization vector due to large exchange energy costs. The low-energy excitations, so-called spin waves or magnons, only modulate the magnetization direction with respect to the equilibrium magnetization configuration. In this regime the magnetization dynamics of ferromagnets can be described by the Landau–Lifshitz–Gilbert (LLG) equation,

$$\dot{\mathbf{m}} = -\gamma \mathbf{m} \times \mathbf{H}_{\text{eff}} + \tilde{\alpha}\mathbf{m} \times \dot{\mathbf{m}}, \tag{8.1}$$

where $\mathbf{m}(\mathbf{r}, t)$ is a unit vector along the magnetization direction, $\dot{\mathbf{m}} = \partial\mathbf{m}/\partial t$, $\gamma = g^*\mu_B/\hbar > 0$ is (minus) the gyromagnetic ratio in terms of the effective g-factor and the Bohr magneton μ_B, and $\tilde{\alpha}$ is the Gilbert damping tensor that determines the magnetization dissipation rate. Under isothermal conditions the effective magnetic field $\mathbf{H}_{\text{eff}} = -\delta F[\mathbf{m}]/\delta(M_s\mathbf{m})$ is governed by the magnetic free energy F and M_s is the saturation magnetization. We will consider both spatially homogeneous and inhomogeneous situations. In the former case, the magnetization is constant in space (macrospin), while the torques are applied at the interfaces. In the latter case, the effective magnetic field $\mathbf{H}_{\text{eff}}$ also includes a second-order spatial gradient arising from the (exchange) rigidity of the magnetization and torques as well as motive forces that are distributed in the ferromagnet.

Equation (8.1) can be rewritten in the form of the Landau–Lifshitz (LL) equation:

$$\left(1 + \tilde{\alpha}^2\right)\dot{\mathbf{m}} = -\gamma\mathbf{m} \times \mathbf{H}_{\text{eff}} - \gamma\tilde{\alpha}\mathbf{m} \times (\mathbf{m} \times \mathbf{H}_{\text{eff}}). \tag{8.2}$$

Additional torques due to the coupling between currents and magnetization dynamics should be added to the right-hand side of the LLG or LL equation, but some care should be exercised in order to keep track of dissipation in a consistent manner. In our approach the spin pumping and spin-transfer torque contributions are most naturally added to the LLG equation (8.1), but we will also make contact with the LL equation (8.2) while exploring the Onsager reciprocity relations.

In the remaining part of this section we describe the extensions of the LLG equation due to spin-transfer and spin-pumping torques for discrete and bulk systems in Sections 8.2.2.1 and 8.2.2.2, respectively. In the next section we demonstrate in more detail how spin-pumping and spin-transfer torque are related by Onsager reciprocity relations for both discrete and continuous systems.

8.2.2.1 *Discrete systems* Berger and Slonczewski predicted that in spin-valve structures with current perpendicular to the interface planes (CPP) a dc current can excite and even reverse the relative magnetization of magnetic layers separated by a normal metal spacer [8, 9]. The existence of this phenomenon has been amply confirmed by experiments [10, 11, 20, 37–41]. We can understand current-induced magnetization dynamics from first principles in terms of the coupling of spin-dependent transport with the magnetization. In a ferromagnetic metal majority and minority electron spins often have very different electronic structures. Spins that are polarized noncollinear with respect to the magnetization direction are not eigenstates of the ferromagnet, but can be described as a coherent linear combination of majority and minority electron spins at the given energy shell. If injected at an interface, these states precess on time- and length-scales that depend on the orbital part of the wavefunction. In high electron-density transition metal ferromagnets like Co, Ni, and Fe a large number of wavevectors are available at the Fermi energy. A transverse spin current injected from a diffuse reservoir generates a large number of wavefunctions oscillating with different wavelength that lead to efficient destructive interference or decoherence of the spin momentum. Beyond a transverse magnetic coherence length, which in these materials is around 1 nm, a transversely polarized spin current cannot persist. [21] This destruction of transverse angular momentum is per definition equal to a torque. Slonczewski's spin-transfer torque is therefore equivalent to the absorption of a spin current at the interface between a normal metal and a ferromagnet whose magnetization is transverse to the spin current polarization. Each electron carries an electric charge $-e$ and an angular momen- tum of $\pm\hbar/2$. The loss of transverse spin angular momentum at the normal metal–ferromagnet interface is therefore $\hbar\left[\mathbf{I}_s - (\mathbf{I}_s \cdot \mathbf{m})\,\mathbf{m}\right]/(2e)$, where the spin current $\mathbf{I}_s$ is measured in units of an electrical current, e.g. in amperes. In the macrospin approximation the torque has to be shared with all magnetic moments or $M_s\mathcal{V}$ of the ferromagnetic particle or film with volume $\mathcal{V}$. The torque on magnetization equals the rate of change of the total magnetic moment of the magnet $\partial\left(\mathbf{m}M_s\mathcal{V}\right)_{\mathrm{stt}}/\partial t$, which equals the spin current absorption [9]. The rate of change of the magnetization direction therefore reads:

$$\boldsymbol{\tau}_{\mathrm{stt}} = \left(\frac{\partial\mathbf{m}}{\partial t}\right)_{\mathrm{stt}} = -\frac{\gamma\hbar}{2eM_s\mathcal{V}}\mathbf{m}\times(\mathbf{m}\times\mathbf{I}_s).\qquad(8.3)$$

We still need to evaluate the spin current that can be generated, e.g. by the inverse spin Hall effect in the normal metal or optical methods. Here we concentrate on the layered normal metal–ferromagnet systems in which the current generated by an applied bias is polarized by a second highly coercive magnetic layer as in the schematic of Fig. 8.1. Magneto-electronic circuit theory is especially suited to handle such a problem [21]. For simplicity we disregard here extrinsic dissipation of spin angular momentum due to spin–orbit coupling and disorder, which can be taken into account when the need arises [43, 44]. We allow for a nonequilibrium magnetization or spin accumulation $\mathbf{V}_N^{(s)}$ in the normal metal layer. $\mathbf{V}_N^{(s)}$ is a vector pointing in the direction of the local net magnetization, whose modulus $V_N^{(s)}$ is the difference between the differences in electric potentials (or electrochemical potentials divided by $2e$) of both spin species. Including the charge accumulation $V_N^{(c)}$ (local voltage), the potential experienced by a spin-up (spin-down) electron along the direction of the spin accumulation in the normal metal is $V_N^\uparrow = V_N^{(c)} + V_N^{(s)}$ $\left(V_N^\downarrow = V_N^{(c)} - V_N^{(s)}\right)$. Inside a ferromagnet, the spin accumulation must be aligned to the magnetization direction $\mathbf{V}_F^{(s)} = \mathbf{m}V_F^{(s)}$. Since $V_F^{(s)}$ does not directly affect the spin-transfer

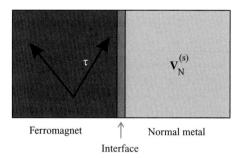

FIG. 8.1. Illustration of the spin-transfer torque in a layered normal metal–ferromagnet system. A spin accumulation $\mathbf{V}_N^{(s)}$ in the normal metal induces a spin-transfer torque $\boldsymbol{\tau}_{\mathrm{stt}}$ on the ferromagnet.

torque at the interface we disregard it for convenience here (see Ref. [19] for a complete treatment), but retain the charge accumulation $V_F^{(c)}$. We can now compute the torque at the interface between a normal metal and a ferromagnet arising from a given spin accumulation $\mathbf{V}_N^{(s)}$. Ohm's law for the spin-current projections aligned ($I_\uparrow$) and anti-aligned ($I_\downarrow$) to the magnetization direction then read [21, 42] (positive currents correspond to charge flowing from the normal metal towards the ferromagnet)

$$I_\uparrow = G_\uparrow \left[\left(V_N^{(c)} - V_F^{(c)} \right) + \mathbf{m} \cdot \left(\mathbf{V}_N^{(s)} - \mathbf{m} V_F^{(c)} \right) \right], \qquad (8.4)$$

$$I_\downarrow = G_\downarrow \left[\left(V_N^{(c)} - V_F^{(c)} \right) - \mathbf{m} \cdot \left(\mathbf{V}_N^{(s)} - \mathbf{m} V_F^{(c)} \right) \right]. \qquad (8.5)$$

where $G_\uparrow$ and $G_\downarrow$ are the spin-dependent interface conductances. The total charge current $I^{(c)} = I_\uparrow + I_\downarrow$, is continuous across the interface, $I_N^{(c)} = I_F^{(c)} = I^{(c)}$. The (longitudinal) spin current defined by Eqs. (8.4) and (8.5) $(I_\uparrow - I_\downarrow)\,\mathbf{m}$ is polarized along the magnetization direction. The transverse part of the spin current can be written as the sum of two vector components in the space spanned by the $\mathbf{m}, \mathbf{V}_N^{(s)}$ plane as well as its normal. The total spin current on the normal metal side close to the interface reads [19, 21]:

$$\mathbf{I}_N^{(s,\mathrm{bias})} = (I_\uparrow - I_\downarrow)\,\mathbf{m} - 2G_\perp^{(R)} \mathbf{m} \times \left(\mathbf{m} \times \mathbf{V}_N^{(s)} \right) - 2G_\perp^{(I)} \left(\mathbf{m} \times \mathbf{V}_N^{(s)} \right), \quad (8.6)$$

where $G_\perp^{(R)}$ and $G_\perp^{(I)}$ are two independent transverse interface conductances. $\mathbf{I}_N^{(s,\mathrm{bias})}$ is driven by the external bias $\mathbf{V}_N^{(s)}$ and should be distinguished from the pumped spin current addressed below. (R) and (I) refer to the real and imaginary parts of the microscopic expression for these "spin mixing" interface conductances $G_{\uparrow\downarrow} = G_\perp^{(R)} + iG_\perp^{(I)}$.

The transverse components are absorbed in the ferromagnet within a very thin layer. Detailed calculations show that transverse spin-current absorption in the ferromagnet happens within a nanometer of the interface, where disorder suppresses any residual oscillations that survived the above-mentioned destructive interference in ballistic structures [45]. Spin transfer in transition metal based multilayers is therefore an interface effect, except in ultrathin ferromagnetic films [46]. As discussed above, the divergence of the transverse spin current at the interface gives rise to the torque

$$\boldsymbol{\tau}_{\mathrm{stt}}^{\mathrm{(bias)}} = -\frac{\gamma\hbar}{eM_s\mathcal{V}} \left[G_\perp^{(R)} \mathbf{m} \times \left(\mathbf{m} \times \mathbf{V}_N^{(s)} \right) + G_\perp^{(I)} \left(\mathbf{m} \times \mathbf{V}_N^{(s)} \right) \right]. \tag{8.7}$$

Adding this torque to the Landau–Lifshitz–Gilbert equation leads to the Landau–Lifshitz–Gilbert–Slonczewski (LLGS) equation

$$\dot{\mathbf{m}} = -\gamma\mathbf{m} \times \mathbf{H}_{\mathrm{eff}} + \boldsymbol{\tau}_{\mathrm{stt}}^{\mathrm{(bias)}} + \alpha\mathbf{m} \times \dot{\mathbf{m}}. \tag{8.8}$$

The first term in Eq. (8.7) is the (Slonczewski) torque in the $\left(\mathbf{m}, \mathbf{V}_N^{(s)} \right)$ plane, which resembles the Landau–Lifshitz damping in Eq. (8.2). When the spin accumulation $\mathbf{V}_N^{(s)}$ is aligned with the effective magnetic field $\mathbf{H}_{\mathrm{eff}}$, the Slonczewski torque effectively enhances the damping of the ferromagnet and stabilizes the magnetization motion towards the equilibrium direction. On the other hand, when $\mathbf{V}_N^{(s)}$ is antiparallel to $\mathbf{H}_{\mathrm{eff}}$, this torque opposes the damping. When exceeding a critical value it leads to precession or reversal of the magnetization. The second term in Eq. (8.7) proportional to $G_\perp^{(I)}$ modifies the magnetic field torque and precession frequency. While the in-plane torque leads to dissipation of the spin accumulation, the out-of-plane torque induces a precession of the spin accumulation in the ferromagnetic exchange field along $\mathbf{m}$. It is possible to implement the spin-transfer torque into the Landau–Lifshitz equation, but the conductance parameters differ from those in Eq. (8.7).

Since spin currents can move magnetizations, it is natural to consider the reciprocal effect, *viz.* the generation of spin currents by magnetization motion. It was recognized in the 1970s that spin dynamics is associated with spin currents in normal metals. Barnes [47] studied the dynamics of localized magnetic moments embedded in a conducting medium. He showed that the dynamic susceptibility in diffuse media is limited by the spin-diffusion length. Janossy and Monod [12] and Silsbee *et al.* [13] postulated a coupling between dynamic ferromagnetic magnetization and spin accumulation in adjacent normal metals in order to explain that microwave transmission through normal metal foils is enhanced by a coating with a ferromagnetic layer. The scattering theory for spin currents induced by magnetization dynamics was developed by Tserkovnyak *et al.* [14] on the basis of the theory of adiabatic quantum pumping [48], hence the name "spin pumping." Theoretical results were confirmed by the agreement of the spin-pumping induced increase of the Gilbert damping with experiments by Mizukami

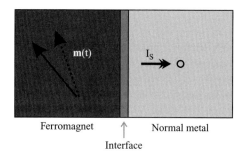

FIG. 8.2. Spin pumping in normal metal–ferromagnet systems. A dynamical magnetization "pumps" a spin current $\mathbf{I}^{(s)}$ into an adjacent normal metal.

et al. [16]. A schematic picture of spin pumping in normal|ferromagnetic systems is shown in Fig. 8.2. At not too high excitations and temperatures, the ferromagnetic dynamics conserves the modulus of the magnetization $M_s\mathbf{m}$. Conservation of angular momentum then implies that the spin current $\mathbf{I}_N^{(s,\text{pump})}$ pumped out of the ferromagnet has to be polarized perpendicularly to $\mathbf{m}$, *viz.* $\mathbf{m} \cdot \mathbf{I}_N^{(s,\text{pump})} = 0$. Furthermore, the adiabatically pumped spin current is proportional to $|\dot{\mathbf{m}}|$. Under these conditions, therefore, [14, 15]

$$\frac{e}{\hbar}\mathbf{I}_N^{(s,\text{pump})} = G_{\perp}^{\prime(R)}\left(\mathbf{m} \times \dot{\mathbf{m}}\right) + G_{\perp}^{\prime(I)}\dot{\mathbf{m}}, \qquad (8.9)$$

where $G_{\perp}^{\prime R}$ and $G_{\perp}^{\prime I}$ are two transverse conductances that depend on the materials. Here the sign is defined to be negative when $\mathbf{I}_N^{(s,\text{pump})}$ implies loss of angular momentum for the ferromagnet. For $|\dot{\mathbf{m}}| \neq 0$, the right-hand side of the LLGS equation (8.8) must be augmented by Eq. (8.9). The leakage of angular momentum leads e.g. to an enhanced Gilbert damping [16].

Onsager's reciprocity relations dictate that conductance parameters in thermodynamically reciprocal processes must be identical when properly normalized. We prove below that spin-transfer torque (8.7) and spin pumping (8.9) indeed belong to this category and must be identical, *viz.* $G_{\perp}^{(R)} = G_{\perp}^{\prime(R)}$ and $G_{\perp}^{(I)} = G_{\perp}^{\prime(I)}$. Spin-transfer torque and spin pumping are therefore opposite sides of the same coin, at least in the linear response regime. Since spin-mixing conductance parameters governing both processes are identical, an accurate measurement of one phenomenon is sufficient to quantify the reciprocal process. Magnetization dynamics induced by the spin-transfer torque are not limited to macrospin excitations and experiments are carried out at high current levels that imply heating and other complications. On the other hand, spin pumping can be directly detected by the line width broadening of FMR spectra of thin multilayers. In the absence of two-magnon scattering phenomena and a sufficiently strong static magnetic field, FMR excites only the homogeneous macrospin mode, allowing the measurement of the transverse conductances $G_{\perp}'^{(R)}$ and, in principle, $G_{\perp}'^{(I)} \cdot G_{\perp}'^{(I)}$. Experimental results and first-principles calculations [14, 15] agree quantitatively well. Rather than attempting to measure these parameters by current-induced excitation measurements, the values $G_{\perp}'^{(R)}$ and $G_{\perp}'^{(I)}$ should be inserted, concentrating on other parameters when analyzing these more complex magnetization phenomena. Finally we note that spin-mixing conductance parameters can be derived as well from static magnetoresistance measurements in spin valves [46] or by detecting the spin current directly by the inverse spin Hall effect [78, 79].

8.2.2.2 *Continuous systems* The coupling effects between (spin-polarized) electrical currents and magnetization dynamics also exist in magnetization textures of bulk metallic ferromagnets. Consider a magnetization that adiabatically varies its direction in space. The dominant contribution to the spin-transfer torque can be identified as a consequence of violation of angular momentum conservation: in a metallic ferromagnet, a charge current is spin polarized along the magnetization direction to leading order in the texture gradients. In the bulk, i.e. separated from contacts by more than the spin-diffusion length, the current polarization is $P = (\sigma_\uparrow - \sigma_\downarrow)/(\sigma_\uparrow + \sigma_\downarrow)$, in terms of the ratio of the conductivities for majority and minority electrons, where we continue to measure spin currents in units of electric currents. We first disregard spin-flip processes that dissipate spin currents to the lattice. To zeroth order in the gradients, the spin current $\mathbf{j}^{(s)}$ flowing in a specified (say x-) direction at position $\mathbf{r}$ is polarized along the local magnetization, $\mathbf{j}^{(s)}(\mathbf{r}) = \mathbf{m}(\mathbf{r})j^{(s)}(\mathbf{r})$. The gradual change of the magnetization direction corresponds to a divergence of the angular momentum of the itinerant electron subsystem, $\partial_x \mathbf{j}^{(s)} = j^{(s)}\partial_x \mathbf{m} + \mathbf{m}\partial_x j^{(s)}$, where the latter term is aligned with the magnetization direction and does not contribute to the magnetization torque. The former change of spin current does not leave the electron system but flows into the magnetic order, thus inducing a torque on the magnetization. This process does not cause any dissipation and the torque is reactive, as can be seen as well from its time reversal symmetry. To first order in the texture gradient, or adiabatic limit, and for arbitrary current directions [49, 50]

$$\tau_{\text{stt}}^{(\text{bias})}(\mathbf{r}) = \frac{g^* \mu_B P}{2e M_s} (\mathbf{j} \cdot \nabla)\,\mathbf{m}\,, \tag{8.10}$$

where $\mathbf{j}$ is the charge current density vector and the superscript "bias" indicates that the torque is induced by a voltage bias or electric field. From symmetry arguments another torque should exist that is normal to Eq. (8.10), but still perpendicular to the magnetization and proportional to the lowest order in its gradient. Such a torque is dissipative, since it changes sign under time reversal. For isotropic systems, we can parameterize the out-of-plane torque by a dimensionless parameter β such that the total torque reads [27, 51],

$$\tau_{\text{stt}}^{(\text{bias})}(\mathbf{r}) = \frac{g^* \mu_B}{2e M_s} \sigma P \left[(\mathbf{E} \cdot \nabla)\,\mathbf{m} + \beta \mathbf{m} \times (\mathbf{E} \cdot \nabla)\,\mathbf{m} \right], \tag{8.11}$$

where we have used Ohm's law, $\mathbf{j} = \sigma \mathbf{E}$. In the adiabatic limit, i.e. to first order in the gradient of the magnetization $\partial_i m_j$, the spin-transfer torque Eq. (8.11) describes how the magnetization dynamics is affected by currents in isotropic ferromagnets.

Analogous to discrete systems, we may expect a process reciprocal to (8.11) in ferromagnetic textures similar to spin pumping at interfaces. Since we are now operating in a ferromagnet, a pumped spin current is transformed into a charge current. To leading order a time-dependent texture is expected to pump a current proportional to the rate of change of the magnetization direction and the gradient of the magnetization texture. For isotropic systems, we can express the expected charge current as

$$j_i^{(\text{pump})} = \frac{\hbar}{2e} \sigma P' \left[\mathbf{m} \times \partial_i \mathbf{m} + \beta' \partial_i \mathbf{m} \right] \cdot \dot{\mathbf{m}}, \tag{8.12}$$

where P' is a polarization factor and β' an out-of-plane contribution. Note that we have here been assuming a strong spin-flip rate so that the spin-diffusion length is much smaller than the typical length of the magnetization texture. Volovik considered the opposite limit of weak spin dissipation and kept track of currents in two independent spin bands [49]. In that regime he derived the first term in (8.12), proportional to P' and proved that $P = P'$. This result was re-derived by Barnes and Maekawa [28]. The last term, proportional to the β-factor, was first discussed by Duine [52] for a mean-field model, demonstrating that $\beta = \beta'$. More general textures and spin relaxation regimes were treated by Tserkovnyak and Mecklenburg [31]. In the following we demonstrate by the Onsager reciprocity relations that the coefficients appearing in the spin-transfer torques (8.11) are identical to those in the pumped current (8.12), i.e. $P = P'$ and $\beta = \beta'$.

The proposed relations for the spin-transfer torques and pumped current in continuous systems form a local relationship between torques, current, and electric and magnetic fields. For ballistic systems, this is not satisfied since the current at one spatial point depends on the electric field in the whole sample or global voltage bias and not just on the local electric field. The local assumption also breaks down in other circumstances. The long-range magnetic dipole interaction typically breaks a ferromagnet into uniform domains. The magnetization gradually changes in the region between the domains, the domain wall. When the domain wall width is smaller than the phase coherence length or the mean free path, one should replace the local approach by a global strategy for magnetization textures in which the dynamics is characterized by one or more dynamic (soft) collective coordinates $\{\xi_a(\tau)\}$ that are allowed to vary (slowly) in time

$$\mathbf{m}(\mathbf{r}\tau) = \mathbf{m}_{\text{st}}(\mathbf{r}; \{\xi_a(\tau)\}), \tag{8.13}$$

where $\mathbf{m}_{\text{st}}$ is a static description of the texture. In order to keep the discussion simple and transparent we disregard thermoelectric effects, which can be important in principle [53]. The thermodynamic forces are $-\partial F/\partial \xi_a$, where F is the free energy as well as the bias voltage across the sample V. In linear response the rate of change of the dynamic collective coordinates and the charge current in the system are related to the thermodynamic forces $-\partial F/\partial \xi$ and V by a response matrix

$$\begin{pmatrix} \dot{\xi} \\ I \end{pmatrix} = \begin{pmatrix} \tilde{L}_{\xi\xi} & \tilde{L}_{\xi I} \\ \tilde{L}_{I\xi} & \tilde{L}_{II} \end{pmatrix} \begin{pmatrix} -\partial F/\partial \xi \\ V \end{pmatrix}, \tag{8.14}$$

where $\tilde{L}_{\xi V}$ describes the bias voltage-induced torque and $\tilde{L}_{I\xi}$ the current pumped by the moving magnetization texture. These expressions are general and include, e.g. effects of spin–orbit interaction. Onsager's reciprocity relations imply $\tilde{L}_{I\xi_i}\{\mathbf{m}, \mathbf{H}\} = \tilde{L}_{\xi_i I}\{-\mathbf{m}, -\mathbf{H}\}$ or $\tilde{L}_{I\xi_i}\{\mathbf{m}, \mathbf{H}\} = -\tilde{L}_{\xi_i I}\{-\mathbf{m}, -\mathbf{H}\}$ depending on how the collective coordinates transform under time reversal. The coefficient $\tilde{L}_{I\xi}$ can be easily expressed in terms of the scattering theory of adiabatic pumping as discussed below. This strategy was employed to demonstrate for (Ga,Mn) As that the spin–orbit interaction can enable a torque arising from a pure charge current bias in Ref. [43] and to compute β in Ref. [32].

8.2.2.3 *Self-consistency: Spin battery and enhanced Gilbert damping* We have discussed two reciprocal effects: torque induced by charge currents (voltage or electric field) on the magnetization and the current induced by a time-dependent magnetization. These two effects are not independent. For instance, in layered systems, when the magnetization precesses, it can pump spins into adjacent normal metal. The spin pumping affects magnetization dynamics depending on whether the spins return into the ferromagnet or not. When the adjacent normal metal is a good spin sink, this loss of angular momentum affects the magnetization dynamics by an enhanced Gilbert damping. In the opposite limit of little or no spin relaxation in an adjacent conductor of finite size, the pumped steady-state spin current is canceled by a diffusion spin current arising from the build-up of spin accumulation potential in the adjacent conductor. The build-up of the spin accumulation can be interpreted as a spin battery [54]. Similarly, in magnetization textures, the dynamic magnetization pumps currents that in turn exert a torque on the ferromagnet.

In the spin battery the total spin current in the normal metal consists of the diffusion-driven Eq. (8.6) and the pumped Eq. (8.9) spin currents [54]. When there are no other intrinsic time-scales in the transport problem (e.g. instantaneous diffusion) and in the steady state, conservation of angular momentum dictates that the total spin current in the normal metal must vanish,

$$\mathbf{I}_N^{(s,\text{bias})} + \mathbf{I}_N^{(s,\text{pump})} = 0,$$

which from Eqs. (8.6) and (8.9) results in a spin accumulation, which can be called a spin-battery bias or spin-motive force:

$$eV_N^{(s)} = \hbar \mathbf{m} \times \dot{\mathbf{m}}. \qquad (8.15)$$

This is a manifestation of Larmor's theorem [15]. In diffusive systems, the diffusion of the pumped spins into the normal metal takes a finite amount of time. When the typical diffusion time is longer than the typical precession time, the ac component averages out to zero [54]. In this regime, the spin-battery bias is constant and determined by

$$\left[eV_N^{(s)}\right]^{(\mathrm{DC})} = \int_{\tau_\mathrm{p}} \frac{dt}{\tau_\mathrm{p}} \mathbf{m} \times \hbar\dot{\mathbf{m}}, \qquad (8.16)$$

where τ_p is the precession period. Without spin-flip processes, the magnitude of the steady-state spin bias is governed by the FMR frequency of the magnetization precession $eV_N^{(s)} = \hbar\omega_\mathrm{FMR}$ and is independent of the interface properties. Spin-flip scattering in the normal metal reduces the spin bias $eV_N^{(s)} < \hbar\omega_\mathrm{FMR}$ in a nonuniversal way [15, 54]. The loss of spin angular momentum implies a damping torque on the ferromagnet. Asymmetric spin-flip scattering rates in adjacent left and right normal metals can also induce a charge potential difference resulting from the spin battery, which has been measured. [55, 56] The spin-battery effect has also been measured via the spin Hall effect in Ref. [57].

In the opposite regime, when spins relax much faster than their typical injection rate into the adjacent normal metal, (8.3), the net spin current is well described by the spin-pumping mechanism. According to Eq. (8.9), in which primes may be removed because of the Onsager reciprocity,

$$\tau_\mathrm{stt}^{(\mathrm{pump})} = \frac{\gamma\hbar^2}{2e^2 M_s \mathcal{V}} \left[G_\perp^{(R)}\mathbf{m} \times \dot{\mathbf{m}} + G_\perp^{(I)}\dot{\mathbf{m}}\right]. \qquad (8.17)$$

We use the superscript "pump" to clarify that this torque arises from the emission of spins from the ferromagnet. The first term in Eq. (8.17) is equal to the Gilbert damping term in the LLG equation (8.1). This implies that the spin pumping into an adjacent conductor maximally enhances the Gilbert damping by

$$\alpha_\mathrm{stt}^{(\mathrm{pump})} = \frac{\gamma\hbar^2}{2e^2 M_s \mathcal{V}} G_\perp^{(R)}. \qquad (8.18)$$

This damping is proportional to the interface conductance $G_\perp^{(R)}$ and thus the normal metal–ferromagnet surface area as well as inversely proportional to the volume of the ferromagnet and therefore scales as $1/d_F$, where d_F is the thickness of the ferromagnetic layer. The transverse conductance per unit area agrees well with theory [15]. The microscopic expression for $G_\perp^{(R)} > 0$ and therefore $\alpha_\mathrm{stt}^{(\mathrm{pump})} > 0$. The second term on the right-hand side of Eq. (8.17), modifies the gyromagnetic ratio and ω_FMR. For conventional ferromagnets like Fe, Ni, and Co, $G_\perp^{(I)} \ll G_\perp^{(R)}$ by near cancellation of positive and negative contributions in momentum space. In these systems $G_\perp^{(I)}$ is much smaller than $G_\perp^{(R)}$ and the effects of $G_\perp^{(I)}$ might therefore be difficult to observe.

A similar argument leads us to expect an enhancement of the Gilbert damping in magnetic textures. By inserting the pumped current Eq. (8.12) into the torque Eq. (8.11) in place of $\sigma \mathbf{E}$, we find a contribution caused by the magnetization dynamics [58–60]

$$\tau_{\text{stt}}^{(\text{drift})}(\mathbf{r}) = \frac{\gamma \hbar^2}{4e^2 M_s} P^2 \sigma \left[([\mathbf{m} \times \partial_i \mathbf{m} + \beta \partial_i \mathbf{m}] \cdot \dot{\mathbf{m}}) \right.$$

$$\left. + \beta \mathbf{m} \times ([\mathbf{m} \times \partial_i \mathbf{m} + \beta \partial_i \mathbf{m}] \cdot \dot{\mathbf{m}}_i) \right] \partial_i \mathbf{m}, \qquad (8.19)$$

which gives rise to additional dissipation of the order $\gamma \hbar^2 P^2 \sigma / 4e^2 M_s \lambda_w^2$, where λ_w is the typical length-scale for the variation of the magnetization texture such as the domain wall width or the radius of a vortex. Equation (8.19) inserted into the LLG equation also renormalizes the gyromagnetic ratio by an additional factor β. The additional dissipation becomes important for large gradients as in narrow domain walls and close to magnetic vortex centers [58, 60].

Finally, we point out that the fluctuation–dissipation theorem dictates that equilibrium spin-current fluctuations associated with spin pumping by thermal fluctuations must lead to magnetization dissipation. This connection was worked out in Ref. [61].

8.2.3   Onsager reciprocity relations

The Onsager reciprocity relations express fundamental symmetries in the linear response matrix relating thermodynamic forces and currents. In normal metal–ferromagnetic heterostructures, a spin accumulation in the normal metal in contact with a ferromagnet can exert a torque on the ferromagnet, see Eq. (8.7). The reciprocal process is spin pumping: a precessing ferromagnet induces a spin current in the adjacent normal metal as described by Eq. (8.9). Both these effects are nonlocal since the spin-transfer torque on the ferromagnet arises from the spin accumulation potential in the normal metal and the pumped spin current in the normal metal is a result of the collective magnetization dynamics. In bulk ferromagnets, a current (or electric field) induces a spin-transfer torque on a magnetization texture. The reciprocal pumping effect is now an electric current (or emf) generated by the texture dynamics. In the next two subsections we provide technical details of the derivation of the Onsager reciprocity relations under these circumstance

8.2.3.1   *Discrete systems*   As an example of a discrete system, we consider a normal metal–ferromagnet bilayer without any spin–orbit interaction (see Ref. [43] for a more general treatment that takes spin-flip processes into account) and under isothermal conditions (the effects of temperature gradients are discussed in Refs. [33, 62, 63]). The spin-transfer physics is induced by a pure spin accumulation in the normal metal, whose creation does not concern us here. The central ingredients for Onsager's reciprocity relations are the thermodynamic variables with associated forces and currents that are related by a linear response matrix [34]. In order to uniquely define the linear response, currents J and forces X have to be normalized such that $\dot{F} = \sum XJ$. This is conventionally done by the rate of change of the free energy in the nonequilibrium situation in terms of currents and forces [34].

Let us consider first the electronic degrees of freedom. In the normal metal reservoir of a constant spin accumulation $\mathbf{V}_N^{(s)}$ the rate of change of the free energy F_N in terms of the total spin $\mathbf{s}_N$ (in units of electric charge e) reads

$$\dot{F}_N = -\dot{\mathbf{s}}_N \cdot \mathbf{V}_N^{(s)}. \tag{8.20}$$

This identifies $\mathbf{V}_N^{(s)}$ as a thermodynamic force that induces spin currents $\mathbf{I}_s = \dot{\mathbf{s}}_N$, which is defined to be positive when leaving the normal metal. In the ferromagnet, all spins are aligned along the magnetization direction $\mathbf{m}$. The associated spin accumulation potential $V_F^{(s)}$ can only induce a contribution to the longitudinal part of the spin current, e.g. a contribution to the spin current along the magnetization direction $\mathbf{m}$. In our discussion of the Onsager reciprocity relations, we will set this potential to zero for simplicity and disregard associated change in the free energy, but it is straightforward to include the effects of a finite $V_F^{(s)}$[19].

Next, we address the rate of change of the free energy related to the magnetic degrees of freedom in the ferromagnet,

$$\dot{F}(\mathbf{m}) = -M_s \mathcal{V}\mathbf{H}_{\text{eff}} \cdot \dot{\mathbf{m}},$$

where $F(\mathbf{m})$ is the magnetic free energy. The total magnetic moment $M_s \mathcal{V}\mathbf{m}$ is a thermodynamic quantity and the effective magnetic field $\mathbf{H}_{\text{eff}} = -\partial F/\partial(M_s \mathcal{V}\mathbf{m})$ is the thermodynamic force that drives the magnetization dynamics $\dot{\mathbf{m}}$.

In linear response, the spin current $\mathbf{I}_s = \dot{\mathbf{s}}$ and magnetization dynamics $M_s \mathcal{V}\dot{\mathbf{m}}$ are related to the thermodynamic forces as

$$\begin{pmatrix} M_s \mathcal{V}\dot{\mathbf{m}} \\ \mathbf{I}_N^{(s)} \end{pmatrix} = \begin{pmatrix} \tilde{L}^{(mm)} & \tilde{L}^{(ms)} \\ \tilde{L}^{(sm)} & \tilde{L}^{(ss)} \end{pmatrix} \begin{pmatrix} \mathbf{H}_{\text{eff}} \\ \mathbf{V}_N^{(s)} \end{pmatrix}, \tag{8.21}$$

where $\tilde{L}^{(mm)}$, $\tilde{L}^{(ms)}$, $\tilde{L}^{(sm)}$, and $\tilde{L}^{(ss)}$ are 3×3 tensors in, e.g. a Cartesian basis for the spin and magnetic moment vectors. Onsager discovered that microscopic time-reversal (anti-)symmetry leads to relations between the off-diagonal components of these linear-response matrices. Both the magnetization in the ferromagnet and the spin-accumulation in the normal metal are antisymmetric under time reversal leading to the reciprocity relations

$$L_{ij}^{(sm)}(\mathbf{m}) = L_{ji}^{(ms)}(-\mathbf{m}). \tag{8.22}$$

Some care should be taken when identifying the Onsager symmetries in spin accumulation-induced magnetization dynamics. Specifically, the LLGS equation (8.8) cannot simply be combined with the linear response relation (8.21) and Eq. (8.22). Only the Landau–Lifshitz–Slonczewski (LL) Eq. (8.2) directly relates $\dot{\mathbf{m}}$ to $\mathbf{H}_{\text{eff}}$ as required by Eq. (8.21). In terms of the 3×3 matrix $\tilde{O}$ e.g.

$$\tilde{O}_{ij}(\mathbf{m}) = \sum_k \epsilon_{ikj} m_k, \qquad (8.23)$$

where $\epsilon_{ijk} = \frac{1}{2}(j-i)(k-i)(k-j)$ is the Levi-Civita tensor, $\mathbf{m} \times \mathbf{H}_{\text{eff}} = \tilde{O}\mathbf{H}_{\text{eff}}$, and the LLGS equation (8.8) can be written as

$$\left(1 - \alpha \tilde{O}\right) \dot{\mathbf{m}} = \tilde{O}\left(-\gamma \mathbf{H}_{\text{eff}}\right) + \boldsymbol{\tau}_{\text{stt}}. \qquad (8.24)$$

By Eq. (8.21), the pumped current in the absence of spin accumulation ($\mathbf{V}_N^{(s)} = 0$) is $\mathbf{I}_N^{(s)} = \tilde{L}^{(sm)}\mathbf{H}_{\text{eff}}$. Then, by Eq. (8.9), $\mathbf{I}_N^{(s)} = \tilde{X}^{(sm)}\dot{\mathbf{m}}$, where the 3×3 tensor $\tilde{X}^{(sm)}$ has components

$$\tilde{X}_{ij}^{(sm)}(\mathbf{m}) = -\frac{\hbar}{e}\left[G_{\perp}'^{(R)}\sum_n \epsilon_{inj} m_n + G_{\perp}'^{(I)}\sum_{nkl}\epsilon_{ink} m_n \epsilon_{klj} m_k\right]. \qquad (8.25)$$

From the LLG equation (8.24) for a vanishing spin accumulation ($\mathbf{V}_N^{(s)} = 0$) and thus no bias-induced spin-transfer torque ($\boldsymbol{\tau}_{\text{stt}}^{(\text{bias})} = 0$), the pumped spin current can be expressed as $\mathbf{I}_N^{(s)} = \tilde{X}^{(sm)}\tilde{O}\left[1 - \alpha\tilde{O}\right]^{-1}(-\gamma\mathbf{H}_{\text{eff}})$, which identifies the linear response coefficient $\tilde{L}^{(sm)}$ in terms of $\tilde{X}^{(sm)}$ as

$$\tilde{L}^{(sm)} = -\gamma \tilde{X}^{(sm)}\tilde{O}\left[1 - \alpha\tilde{O}\right]^{-1}. \qquad (8.26)$$

Using the Onsager relation (8.22) and noticing that $\tilde{O}_{ij}(\mathbf{m}) = \tilde{O}_{ji}(-\mathbf{m})$ and $\tilde{X}_{ij}^{(sm)}(\mathbf{m}) = \tilde{X}_{ji}^{(sm)}(-\mathbf{m})$

$$\tilde{L}^{(ms)} = -\gamma \left[1 - \alpha\tilde{O}\right]^{-1}\tilde{O}\tilde{X}^{(sm)}. \qquad (8.27)$$

The rate of change of the magnetization by the spin accumulation therefore becomes

$$\dot{\mathbf{m}}_{\text{stt}} = \frac{1}{M_s \mathcal{V}}\tilde{L}^{(ms)}\mathbf{V}_N^{(s)}$$

$$= -\frac{\gamma}{M_s \mathcal{V}}\left[1 - \alpha\tilde{O}\right]^{-1}\tilde{O}X^{(sm)}\mathbf{V}_N^{(s)}. \qquad (8.28)$$

Furthermore, the LLGS equation (8.24) in the absence of an external magnetic field reads $\left[1 - \alpha\tilde{O}\right]\dot{\mathbf{m}}_{\text{stt}} = \tau_{\text{stt}}^{(\text{drift})}$. Inserting the phenomenological expression for the spin-transfer torque (8.7), we identify the linear response coefficient $\tilde{L}^{(ms)}$:

$$
\begin{aligned}
\tau_{\text{stt}}^{(\text{drift})} &= -\frac{\gamma}{M_s \mathcal{V}}\tilde{O}X^{(sm)}\mathbf{V}_N^{(s)} \\
&= \frac{\gamma}{M_s \mathcal{V}e}\left[G_{\perp}^{\prime(R)}\mathbf{m} \times \left(\mathbf{m} \times \mathbf{V}_N^{(s)}\right) + G_{\perp}^{\prime(I)}\left(\mathbf{m} \times \mathbf{V}_N^{(s)}\right)\right].
\end{aligned}
\tag{8.29}
$$

This agrees with the phenomenological expression (8.7) when

$$
G_{\perp}^{\prime(R)} = G_{\perp}^{(R)}; \quad G_{\perp}^{\prime(I)} = G_{\perp}^{(I)}.
\tag{8.30}
$$

Spin pumping as expressed by Eq. (8.9) is thus reciprocal to the spin-transfer torque as described by Eq. (8.7). In Section 8.3.1.1 these relations are derived by first principles from quantum-mechanical scattering theory, resulting in. $G_{\perp}^{\prime(R)} = G_{\uparrow\downarrow} = (e^2/h)\sum_{nm}\left[\delta_{nm} - r_{nm}^{\uparrow}\left(r_{nm}^{\uparrow}\right)^*\right]$ for a narrow constriction, where $r_{nm}^{\uparrow}$ ($r_{nm}^{\downarrow}$) is the reflection coefficient for spin-up (spin-down) electrons from waveguide mode m to waveguide mode n. For layered systems with a constant cross section the microscopic expressions of the transverse (mixing) conductances should be renormalized by taking into account the contributions from the Sharvin resistances [23, 81], which increases the conductance by roughly a factor of two and is important for a quantitatively comparison between theory and experiments [15, 19].

8.2.3.2 *Continuous systems* The Onsager reciprocity relations also relate the magnetization torques and currents in the magnetization texture of bulk magnets. Following Refs. [31, 32], the rate of change of the free energy related to the electronic degrees of freedom in the ferromagnet is $\dot{F}_F = -\int d\mathbf{r}\dot{q}V$, where q is the charge density and $eV = \mu$ is the chemical potential. Inserting charge conservation, $\dot{q} + \nabla \cdot \mathbf{j} = 0$ and by partial integration,

$$
\dot{F}_F = -\int d\mathbf{r}\mathbf{j} \cdot \mathbf{E}
\tag{8.31}
$$

which identifies charge as a thermodynamic variable, while the electric field $\mathbf{E} = \nabla V$ is a thermodynamic force which drives the current density $\mathbf{j}$. For the magnetic degrees of freedom, the rate of change of the free energy (or entropy) is

$$
\dot{F}_m = -M_s\int d\mathbf{r}\dot{\mathbf{m}}(\mathbf{r}) \cdot \mathbf{H}_{\text{eff}}(\mathbf{r}).
\tag{8.32}
$$

Just like for discrete systems, $\mathbf{H}_{\text{eff}}(\mathbf{r})$, is the thermodynamic force and $M_s\mathbf{m}$ is the thermodynamic variable to which it couples. In a local approximation the (linear) response depends only on the force at the same location:

$$\begin{pmatrix} M_s \dot{\mathbf{m}} \\ \mathbf{j} \end{pmatrix} = \begin{pmatrix} \tilde{L}^{(mm)} & \tilde{L}^{(mE)} \\ \tilde{L}^{(Em)} & \tilde{L}^{(EE)} \end{pmatrix} \begin{pmatrix} M_s \mathbf{H}_{\text{eff}} \\ \mathbf{E} \end{pmatrix}, \tag{8.33}$$

where $\tilde{L}^{(mm)}$, $\tilde{L}^{(mj)}$, $\tilde{L}^{(jm)}$, and $\tilde{L}^{(jj)}$ are the local response functions. Onsager's reciprocity relations dictate again that

$$\tilde{L}_{ji}^{(jm)}(\mathbf{m}) = \tilde{L}_{ij}^{(mj)}(-\mathbf{m}). \tag{8.34}$$

Starting from the expression for current pumping (8.12), we can determine the linear response coefficient $\tilde{L}^{(Em)}$ from

$$\left[\tilde{L}^{(Em)} \left[1 - \alpha \tilde{O} \right] \tilde{O}^{-1} \right]_{ij} = -\gamma \frac{\hbar}{2e} \sigma P' \left[\epsilon_{jkl} m_k \partial_i m_l + \beta' \partial_i m_j \right], \tag{8.35}$$

where the operator $\tilde{O}$ is introduced in the same way as for discrete systems (8.23) to transform the LLG equation into the LL form (8.24). According to Eq. (8.34)

$$\left[\tilde{O}^{-1} \left[1 - \alpha \tilde{O} \right] \tilde{L}^{(mj)} \right]_{ij} = -\gamma \frac{\hbar}{2e} \sigma P' \left[\epsilon_{ikl} m_k \partial_j m_l - \beta' \partial_j m_i \right]. \tag{8.36}$$

The change in the magnetization induced by an electric field is then $M_s \dot{\mathbf{m}}_{\text{stt}}^{(\text{bias})} = \tilde{L}^{(mj)} \mathbf{E}$ so that the spin-transfer torque due to a drift current $\boldsymbol{\tau}_{\text{stt}}^{(\text{bias})} = \left[1 - \alpha \tilde{O} \right] \dot{\mathbf{m}}_{\text{stt}}^{(\text{bias})}$ can be written as

$$\boldsymbol{\tau}_{\text{stt}}^{(\text{bias})} = -\frac{\gamma \hbar}{2e M_s} \sigma P' \epsilon_{imn} m_m \left[\epsilon_{nkl} m_k E_j \partial_j m_l - \beta' E_j \partial_j m_n \right], \tag{8.37}$$

$$\boldsymbol{\tau}_{\text{stt}}^{(\text{bias})} = \gamma \frac{g^* \mu_B}{2e M_s} \sigma P' \left[(\mathbf{E} \cdot \nabla) \mathbf{m} + \beta' \mathbf{m} \times \mathbf{E} \cdot \nabla \mathbf{m} \right]. \tag{8.38}$$

This result agrees with the phenomenological expression for the pumped current (8.12) when $P = P'$ and $\beta = \beta'$. Therefore, the pumped current and the spin-transfer torque in continuous systems are reciprocal processes. The pumped current can be formulated as the response to a spin-motive force [28].

In small systems and thin wires, the current-voltage relation is not well represented by a local approximation. A global approach based on collective coordinates as outlined around Eq. (8.13) is then a good choice to keep the computational effort in check. Of course, the Onsager reciprocity relations between the pumped current and the effective current-induced torques on the magnetization then hold as well [32].

8.3 Microscopic derivations

8.3.1 *Spin-transfer torque*

8.3.1.1 *Discrete systems—Magneto-electronic circuit theory* Physical properties across a scattering region can be expressed in terms of the region's scattering matrix, which requires a separation of the system into reservoirs, leads, and a

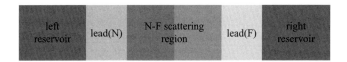

FIG. 8.3. Schematic of how transport between a normal metal and a ferromagnet
is computed by scattering theory. The scattering region, which may contain
the normal metal–ferromagnet interface and diffusive parts of the normal
metal as well as ferromagnet, is attached to real or fictious leads that are in
contact with a left and right reservoir. In the reservoirs, the distributions of
charges and spins are assumed to be equilibrated and known via the charge
potential and spin accumulation bias.

scattering region, see Fig. 8.3. In the lead with index α, the field operator for spin
s-electrons at longitudinal and transverse coordinates $(x,\boldsymbol{\rho})$ and time t is [62]

$$\hat{\Psi}_\alpha^{(s)} = \int \frac{d\epsilon}{\sqrt{2\pi}} \sum_n \left[v_\alpha^{(ns)} \right]^{-1/2} \varphi_\alpha^{(ns)}(\boldsymbol{\varrho}) e^{-i\epsilon_\alpha^{(nks)}t/\hbar} \left[e^{ikx} \hat{a}_\alpha^{(ns)}(\epsilon) + e^{-ikx} \hat{b}_\alpha^{(ns)}(\epsilon) \right]$$

(8.39)

in terms of the annihilation operators $\hat{a}_\alpha^{(ns)}$ $(\hat{b}_\alpha^{(ns)})$ for particles incident on
(outgoing from) the scattering region in transverse waveguide modes with orbital
quantum number n and spin quantum number s ($s = \uparrow$ or $s = \downarrow$). Furthermore, the
transverse wavefunction is $\varphi_\alpha^{(ns)}(\boldsymbol{\varrho})$, the transverse coordinate $\boldsymbol{\varrho}$, the longitudinal
coordinate along the waveguide is x, and $v_\alpha^{(ns)}$ is the longitudinal velocity for
waveguide mode ns. The positive definite momentum k is related to the energy ϵ
by $\hbar k = (2m\epsilon)^{1/2}$. The annihilation operators for incident and outgoing electrons
are related by the scattering matrix

$$\hat{b}_\alpha^{(ns)}(\epsilon) = \sum_{\beta m s'} S_{\alpha\beta}^{(nsms')}(\epsilon) \hat{a}_\beta^{(ms')}(\epsilon).$$ (8.40)

In the basis of the leads ($\alpha = N$ (normal metal) or $\alpha = F$ (ferromagnet)), the
scattering matrix is

$$S = \begin{pmatrix} r & t \\ t' & r' \end{pmatrix},$$

where r (t) is a matrix of the reflection (transmission) coefficients between the
waveguide modes for an electron incident from the left. Similarly, r' and t'
characterize processes where the electron is incident from the right.

In terms of the field operators defined by Eq. (8.39) and the scattering matrix
Eq. (8.40), at low frequencies, the spin current that flows in the normal metal
$\alpha = N$ in the direction towards the scattering region is

$$\mathbf{I}_\alpha^{(s)}(t) = \frac{e}{h} \int_{-\infty}^{\infty} d\epsilon_1 \int_{-\infty}^{\infty} d\epsilon_2 \sum_{\beta\gamma} \sum_{nml} \sum_{\sigma\sigma'} \exp(i\,(\epsilon_1 - \epsilon_2)\,t/\hbar)$$

$$\mathbf{A}_{\alpha\beta,\alpha\gamma}^{(nm,nl),(\sigma,\sigma')}(\epsilon_1, \epsilon_2) \hat{a}_\beta^{(m\sigma)\dagger}(\epsilon_1) \hat{a}_\gamma^{(l\sigma')}(\epsilon_2), \tag{8.41}$$

where

$$\mathbf{A}_{\alpha\beta,\alpha\gamma}^{(nm,nl)(\sigma,\sigma')}(\epsilon_1, \epsilon_2) = \sum_{ss'} \Big[\delta_{\alpha\beta}\delta^{(nm)}\delta^{(s\sigma)}\delta_{\alpha\gamma}\delta^{(nl)}\delta^{(s'\sigma')}$$

$$- S_{\alpha\beta}^{(ns,m\sigma)*}(\epsilon_1) S_{\alpha\gamma}^{(ns',l\sigma')}(\epsilon_2) \Big] \boldsymbol{\sigma}^{(ss')}$$

and $\boldsymbol{\sigma}^{(ss')}$ is a vector of the 2×2 Pauli matrices that depends on the spin indices s and s' of the waveguide mode. The charge current can be found in a similar way. We are interested in the expectation value of the spin current (8.41) when the system is driven out of equilibrium. In *equilibrium*, the expectation values are

$$\left\langle \hat{a}_\alpha^{(ns)\dagger}(\epsilon) \hat{a}_\beta^{(ms')}(\epsilon') \right\rangle_{\mathrm{eq}} = \delta(\epsilon - \epsilon')\delta_{\alpha\beta}\delta^{(ss')}\delta^{(nm)} f_{\mathrm{FD}}(\epsilon), \tag{8.42}$$

where $f_{\mathrm{FD}}(\epsilon)$ is the Fermi–Dirac distribution of electrons with energy ϵ. A non-equilibrium spin accumulation in the normal metal reservoir is not captured by the local equilibrium ansatz in Eq. (8.42), however. A spin accumulation in the normal metal reservoir can still be postulated when spin-flip dissipation is slow compared to all other relevant time-scales. We assume the normal metal and ferromagnet have an isotropic distribution of spins in orbital space, and for clarity consider no charge bias. The expectation for the number of charges and spins in the waveguide describing normal metal leads attached to the normal reservoirs are

$$\left\langle \hat{a}_N^{(ns)\dagger}(\epsilon) \hat{a}_N^{(ms')}(\epsilon') \right\rangle = \delta(\epsilon - \epsilon') \Big[\delta^{(mn)}\delta^{(ss')} f_{\mathrm{FD}}(\epsilon) + \delta^{(mn)} f_N^{(s's)}(\epsilon) \Big]. \tag{8.43}$$

The spin accumulation $\mathbf{V}_N^{(s)}$ is related to the 2×2 out-of-equilibrium distribution matrix $f_N^{(s's)}(\epsilon)$ by

$$\boldsymbol{\sigma}^{(ss')} \cdot \mathbf{V}_N^{(s)} = \int_{-\infty}^{\infty} d\epsilon\, f_N^{(ss')}(\epsilon)/e. \tag{8.44}$$

For the spin-transfer physics, a bias voltage in the ferromagnet does not contribute since it only gives rise to a charge current and a longitudinal spin current. As in the previous section, we therefore set this voltage to zero for simplicity, so that in the ferromagnetic lead attached to the ferromagnetic reservoir

$$\left\langle \hat{a}_F^{(ns)\dagger}(\epsilon) \hat{a}_F^{(ms')}(\epsilon') \right\rangle = \delta(\epsilon - \epsilon')\delta^{(ms)}\delta^{(s's)} f_{\mathrm{FD}}(\epsilon). \tag{8.45}$$

Furthermore, the expectation values of the cross-correlations remain zero also out-of-equilibrium, $\left\langle \hat{a}_N^{(ns)\dagger}(\epsilon)\hat{a}_F^{(ms')}(\epsilon') \right\rangle = 0$ because we assume that phase coherence is broken in the leads. The spin current in lead α is then

$$\mathbf{I}_\alpha^{(s)}(t) = \frac{e}{h}\int_{-\infty}^{\infty} d\epsilon \sum_{nmlss'\sigma\sigma'} \left[\delta^{(nm)}\delta^{(s\sigma)}\delta^{(nl)}\delta^{(s'\sigma')} - r_{NN}^{(ns,m\sigma)*}r_{NN}^{(ns',l\sigma')}\right]\boldsymbol{\sigma}^{(\sigma'\sigma)}f^{(\sigma'\sigma)}.$$

(8.46)

Without spin-flip scattering, the reflection coefficient can be expressed as

$$r_{NN}^{nsm\sigma} = \left(r_{NN}^{nm,\uparrow} + r_{NN}^{nm,\downarrow}\right)\delta^{(s\sigma)}/2 + \mathbf{m}\cdot\boldsymbol{\sigma}_{s\sigma}\left(r_{NN}^{nm,\uparrow} - r_{NN}^{nm,\downarrow}\right)/2 \qquad (8.47)$$

which can be represented in spin space as

$$r_{NN}^{nsm\sigma} = r_{NN}^{nm,(c)}1 + r_{NN}^{nm,(s)}\mathbf{m}\cdot\boldsymbol{\sigma} \qquad (8.48)$$

since the scattering matrix can be decomposed into components aligned and anti-aligned with the magnetization direction. These matrices only depend on the orbital quantum numbers (n and m). Using the representation of the out-of-equilibrium spin density in terms of the spin accumulation (8.11) [21],

$$\mathbf{I}_N^{(s)} = (G_\uparrow + G_\downarrow)\,\mathbf{m}\left(\mathbf{m}\cdot\mathbf{V}_N^{(s)}\right) - 2G_\perp^{(R)}\mathbf{m}\times\left(\mathbf{m}\times\mathbf{V}_N^{(s)}\right) - 2G_\perp^{(I)}\left(\mathbf{m}\times\mathbf{V}_N^{(s)}\right)$$

(8.49)

in agreement with (8.6) when there is no bias voltage in the ferromagnet ($V_F = 0$) which we have assumed for clarity here. We identify the microscopic expressions for the conductances [21] associated with spins aligned and anti-aligned with the magnetization direction

$$G_\uparrow = \frac{e^2}{h}\sum_{nm}\left[\delta_{nm} - \left|r_{NN}^{nm,\uparrow}\right|^2\right], \qquad (8.50)$$

$$G_\downarrow = \frac{e^2}{h}\sum_{nm}\left[\delta_{nm} - \left|r_{NN}^{nm,\downarrow}\right|^2\right], \qquad (8.51)$$

and the transverse (complex-valued) spin-mixing conductance

$$G_\perp = \frac{e^2}{h}\sum_{nm}\left[\delta_{nm} - r_{NN}^{nm,\uparrow}r_{NN}^{nm,\downarrow*}\right]. \qquad (8.52)$$

These results are valid when the transmission coefficients are small such that currents do not affect the reservoirs. Otherwise, the transverse conductance parameters should be renormalized by taking into account the Sharvin resistances, as described above [23, 81]. In the limit we consider here, the expression for the spin current depends only on the reflection coefficients for transport from the normal metal towards the ferromagnet and not on the transmission coefficients for propagation from the normal metal into the ferromagnet. This follows from our assumption that the ferromagnet is thicker than the transverse coherence length as well as our disregard of the spin accumulation in the ferromagnet. Both assumptions can be easily relaxed if necessary [15, 19].

8.3.1.2 *Continuous systems* Spin torques in continuous spin textures can be studied by either quantum kinetic theory, [65] imaginary-time [66] and functional Keldysh [67] diagrammatic approaches, or the scattering-matrix formalism [32]. The latter is particularly powerful when dealing with nontrivial band structures with strong spin–orbit interactions, while the others give complementary insight, but are mostly limited to simple model systems. When the magnetic texture is sufficiently smooth on the relevant length-scales (the transverse spin coherence length and, in special cases, the spin–orbit precession length) the spin torque can be expanded in terms of the local magnetization and current density as well as their spatial-temporal derivatives. An example is the phenomenological Eq. (8.11) for the electric-field-driven magnetization dynamics of an isotropic ferromagnet. While the physical meaning of the coefficients is clear, the microscopic origin and magnitude of the dimensionless parameter β has still to be clarified.

The solution of the LLG equation (8.1) appended by these spin torques depends sensitively on the relationship between the dimensionless Gilbert damping constant α and the dissipative spin-torque parameter β: the special case $\beta/\alpha = 1$ effectively manifests Galilean invariance [68] while the limits $\beta/\alpha \gg 1$ and $\beta/\alpha \ll 1$ are regimes of qualitatively distinct macroscopic behavior. The ratio β/α determines the onset of the ferromagnetic current-driven instability [65] as well as the Walker threshold [69] for the current-driven domain-wall motion [51], and both diverge as $\beta/\alpha \to 1$. The subthreshold current-driven domain-wall velocity is proportional to β/α [27], while $\beta/\alpha = 1$ at a special point, at which the effect of a uniform current density **j** on the magnetization dynamics is eliminated in the frame of reference that moves with velocity $\mathbf{v} \propto \mathbf{j}$, which is of the order of the electron drift velocity [70]. Although the exact ratio β/α is a system-dependent quantity, some qualitative aspects not too sensitive to the microscopic origin of these parameters have been discussed in relation to metallic systems[65, 66, 68, 71]. However, these approaches fail for strongly spin–orbit coupled systems such as dilute magnetic semiconductors [32].

Let us outline the microscopic origin of β for a simple toy model for a ferromagnet. In Ref. [65], we developed a self-consistent mean-field approach, in which itinerant electrons are described by a single-particle Hamiltonian

$$\hat{\mathcal{H}} = [\mathcal{H}_0 + U(\mathbf{r}, t)]\,\hat{1} + \frac{\gamma\hbar}{2}\hat{\boldsymbol{\sigma}} \cdot (\mathbf{H} + \mathbf{H}_{\text{xc}})\,(\mathbf{r}, t) + \hat{\mathcal{H}}_\sigma\,, \qquad (8.53)$$

where the unit matrix $\hat{1}$ and a vector of Pauli matrices $\hat{\boldsymbol{\sigma}} = (\hat{\sigma}_x, \hat{\sigma}_y, \hat{\sigma}_z)$ form a basis for the Hamiltonian in spin space. $\mathcal{H}_0$ is the crystal Hamiltonian including kinetic and potential energy. U is the scalar potential consisting of disorder and applied electric-field contributions. The total magnetic field consists of the applied, **H**, and exchange, $\mathbf{H}_{\text{xc}}$, fields that, like U, are parametrically time dependent. Finally, the last term in the Hamiltonian, $\hat{\mathcal{H}}_\sigma$, accounts for spin-dephasing processes, e.g. due to quenched magnetic disorder or spin–orbit scattering associated with impurity potentials. This last term is responsible for low-frequency dissipative processes affecting dimensionless parameters α and β in the collective equation of motion.

In the time-dependent spin-density-functional theory [72–74] of itinerant ferromagnetism, the exchange field $\mathbf{H}_{xc}$ is a functional of the time-dependent spin-density matrix

$$\rho_{\alpha\beta}(\mathbf{r}, t) = \langle \hat{\Psi}_\beta^\dagger(\mathbf{r}) \hat{\Psi}_\alpha(\mathbf{r}) \rangle_t, \tag{8.54}$$

where $\hat{\Psi}$'s are electronic field operators, which should be computed self-consistently as solutions of the Schrödinger equation for $\hat{\mathcal{H}}$. The spin density of conducting electrons is given by

$$\mathbf{s}(\mathbf{r}) = \frac{\hbar}{2} \mathrm{Tr} \left[\hat{\boldsymbol{\sigma}} \hat{\rho}(\mathbf{r}) \right]. \tag{8.55}$$

We focus on low-energy magnetic fluctuations that are long ranged and transverse and restrict our attention to a single parabolic band. Consideration of more realistic band structures is also in principle possible from this starting point [75]. We adopt the adiabatic local-density approximation (ALDA, essentially the Stoner model) for the exchange field:

$$\gamma\hbar\mathbf{H}_{xc}[\hat{\rho}](\mathbf{r}, t) \approx \Delta_{xc}\mathbf{m}(\mathbf{r}, t), \tag{8.56}$$

with direction $\mathbf{m} = -\mathbf{s}/s$ locked to the time-dependent spin density (8.55).

In another simple model of ferromagnetism, the so-called s-d model, conducting s electrons interact with the exchange field of the d electrons that are assumed to be localized to the crystal lattice sites. The d-orbital electron spins account for most of the magnetic moment. Because d-electron shells have large net spins and strong ferromagnetic correlations, they are usually treated classically. In a mean-field s-d description, therefore, conducting s orbitals are described by the same Hamiltonian (8.53) with an exchange field (8.56). The differences between the Stoner and s-d models for the magnetization dynamics are subtle and rather minor. In the ALDA/Stoner model, the exchange potential is (on the scale of the magnetization dynamics) instantaneously aligned with the total magnetization. In contrast, the direction of the unit vector $\mathbf{m}$ in the s-d model corresponds to the d magnetization, which is allowed to be slightly misaligned with the s magnetization, transferring angular momentum between the s and d magnetic moments. Since most of the magnetization is carried by the latter, the external field $\mathbf{H}$ couples mainly to the d spins, while the s spins respond to and follow the time-dependent exchange field (8.56). As Δ_{xc} is usually much larger than the external (including demagnetization and anisotropy) fields that drive collective magnetization dynamics, the total magnetic moment will always be very close to $\mathbf{m}$. A more important difference of the philosophy behind the two models is the presumed shielding of the d orbitals from external disorder. The reduced coupling with dissipative degrees of freedom would imply that their dynamics are more coherent. Consequently, the magnetization damping has to originate from the disorder experienced by the itinerant s electrons. As in the case of the itinerant ferromagnets, the susceptibility has to be calculated self-consistently with the magnetization dynamics parametrized by $\mathbf{m}$. For more details on this model, we refer to Refs. [76] and [65]. With the above differences in mind, the following discussion is applicable to both models. The Stoner model is more appropriate for transition-metal ferromagnets because of the strong hybridization between d and s, p electrons. For dilute magnetic semiconductors with deep magnetic impurity states the s-d model appears to be a better choice.

The single-particle itinerant electron response to electric and magnetic fields in Hamiltonian (8.53) is all that is needed to compute the magnetization dynamics microscopically. Stoner and s-d models have to be distinguished only at the final stages of the calculation, when we self-consistently relate $\mathbf{m}(\mathbf{r}, t)$ to the electron spin response. The final result for the simplest parabolic-band Stoner model with isotropic spin-flip disorder comes down to the torque (8.11) with $\alpha \approx \beta$. The latter is proportional to the spin-dephasing rate τ_σ^{-1} of the itinerant electrons:

$$\beta \approx \frac{\hbar}{\tau_\sigma \Delta_{\mathrm{xc}}}. \tag{8.57}$$

The derivation assumes $\omega, \tau_\sigma^{-1} \ll \Delta_{\mathrm{xc}}/\hbar$, which is typically the case in real materials sufficiently below the Curie temperature. The s-d model yields the same result for β, Eq. (8.57), but the Gilbert damping constant

$$\alpha \approx \eta\beta \tag{8.58}$$

is reduced by the ratio η of the itinerant to the total angular momentum when the d-electron spin dynamics is not damped. (Note that Eq. (8.58) is also valid for the Stoner model since then $\eta = 1$.)

These simple model considerations shed light on the microscopic origins of dissipation in metallic ferromagnets as reflected in the α and β parameters. In Section 8.4 we present a more systematic, first-principles approach based on the scattering-matrix approach, which accesses the material dependence of both α and β with realistic electronic band structures.

8.3.2 Spin pumping

8.3.2.1 *Discrete systems* When the scattering matrix is time dependent, the energy of outgoing and incoming states does not have to be conserved and the scattering relation (8.40) needs to be appropriately generalized [77]. We will demonstrate here how this is done in the limit of slow magnetization dynamics, i.e. adiabatic pumping. When the time dependence of the scattering matrix $\hat{S}_{\alpha\beta}^{(nm)}[X_i(t)]$ is parameterized by a set of real-valued parameters $X_i(t)$, the pumped spin current in excess of its static bias-driven value (8.49) is given by [14]

$$\mathbf{I}_\alpha^s(t) = e \sum_i \frac{\partial \mathbf{n}_\alpha}{\partial X_i} \frac{dX_i(t)}{dt}, \tag{8.59}$$

where the "spin emissivity" vector by the scatterer into lead α is [80]

$$\frac{\partial \mathbf{n}_\alpha}{\partial X_i} = \frac{1}{2\pi} \mathrm{Im} \sum_\beta \sum_{mn} \sum_{ss'\sigma} \frac{\partial S_{\alpha\beta}^{(ms,n\sigma)*}}{\partial X_i} \hat{\boldsymbol{\sigma}}^{(ss')} S_{\alpha\beta}^{(ms',n\sigma)}. \tag{8.60}$$

Here, $\hat{\boldsymbol{\sigma}}^{(ss')}$ is again the vector of Pauli matrices. In the case of a magnetic monodomain insertion and in the absence of spin–orbit interactions, the spin-dependent scattering matrix between the normal-metal leads can be written in terms of the respective spin-up and spin-down scattering matrices:[21]

$$S_{\alpha\beta}^{(ms,ns')}[\mathbf{m}] = \frac{1}{2} S_{\alpha\beta}^{(mn)\uparrow} \left(\delta^{(ss')} + \mathbf{m} \cdot \hat{\boldsymbol{\sigma}}^{(ss')} \right) + \frac{1}{2} S_{\alpha\beta}^{(mn)\downarrow} \left(\delta^{(ss')} - \mathbf{m} \cdot \hat{\boldsymbol{\sigma}}^{(ss')} \right). \tag{8.61}$$

Here, $\mathbf{m}(t)$ is the unit vector along the magnetization direction and $\uparrow$ ($\downarrow$) are spin orientations defined along (opposite) to $\mathbf{m}$.

Spin pumping due to magnetization dynamics $\mathbf{m}(t)$ is then found by substituting Eq. (8.61) into Eqs. (8.60) and (8.59). After straightforward algebra:[14]

$$\mathbf{I}_\alpha^s(t) = \left(\frac{\hbar}{e} \right) \left(G_\perp^{(R)} \mathbf{m} \times \frac{d\mathbf{m}}{dt} + G_\perp^{(I)} \frac{d\mathbf{m}}{dt} \right). \tag{8.62}$$

As before, we assume here a sufficiently thick ferromagnet, on the scale of the transverse spin-coherence length. Note that the spin pumping is expressed in terms of the same complex-valued mixing conductance $G_\perp = G_\perp^{(R)} + iG_\perp^{(I)}$ as the dc current (8.49), in agreement with the Onsager reciprocity principle as found on phenomenological grounds in Section 8.2.3.

Charge pumping is governed by expressions similar to Eqs. (8.59) and (8.60), subject to the following substitution: $\hat{\boldsymbol{\sigma}} \to \delta$ (Kronecker delta). Finite charge pumping by monodomain magnetization dynamics into normal-metal leads, however, requires a ferromagnetic analyzer or finite spin–orbit interactions and appropriately reduced symmetries, as discussed in Refs. [43, 82–84].

An immediate consequence of the pumped spin current (8.62) is an enhanced Gilbert damping of the magnetization dynamics [14]. Indeed, when the reservoirs are good spin sinks and spin backflow can be disregarded, the spin torque associated with the spin current (8.62) into the αth lead, as dictated by the conservation of the spin angular momentum, Eq. (8.3), contributes (cf. Eq. (8.18)):

$$\alpha' = g^* \frac{\hbar \mu_B}{2e^2} \frac{G_\perp^{(R)}}{M_s \mathcal{V}} \tag{8.63}$$

to the Gilbert damping of the ferromagnet in Eq. (8.1). Here, $g^* \sim 2$ is the g factor of the ferromagnet, $M_s \mathcal{V}$ its total magnetic moment, and μ_B is the Bohr magneton. For simplicity, we neglected $G_\perp^{(I)}$, which is usually not important for intermetallic interfaces. If we disregard energy relaxation processes inside the ferromagnet, which would drain the associated energy dissipation out of the electronic system, the enhanced energy dissipation associated with the Gilbert damping is associated with heat flows into the reservoirs. Phenomenologically, the dissipation power follows from the magnetic free energy F and the LLG Eq. (8.1) as

$$P \equiv -\partial_{\mathbf{m}} F_m \cdot \dot{\mathbf{m}} = M_s \mathcal{V} \mathbf{H}_{\text{eff}} \cdot \dot{\mathbf{m}} = \frac{\alpha M_s \mathcal{V}}{\gamma} \dot{\mathbf{m}}^2 \tag{8.64}$$

or, more generally, for anisotropic damping (with, for simplicity, an isotropic gyromagnetic ratio), by

$$P = \frac{M_s \mathcal{V}}{\gamma} \dot{\mathbf{m}} \cdot \tilde{\alpha} \cdot \dot{\mathbf{m}}. \tag{8.65}$$

Heat flows can be also calculated microscopically by the scattering-matrix transport formalism. At low temperatures, the heat-pumping rate into the αth lead is given by [85–87]

$$I_\alpha^E = \frac{\hbar}{4\pi} \sum_\beta \sum_{mn} \sum_{ss'} \left| \dot{S}_{\alpha\beta}^{(ms,ns')} \right|^2 = \frac{\hbar}{4\pi} \sum_\beta \text{Tr} \left(\hat{\dot{S}}_{\alpha\beta}^\dagger \hat{\dot{S}}_{\alpha\beta} \right), \tag{8.66}$$

where the carets denote scattering matrices with suppressed transverse-channel indices. When the time dependence is entirely due to the magnetization dynamics, $\dot{S}_{\alpha\beta}^{(ms,ns')} = \partial_{\mathbf{m}} S_{\alpha\beta}^{(ms,ns')} \cdot \dot{\mathbf{m}}$. Utilizing again Eq. (8.61), we find for the heat current into the αth lead:[88]

$$I_\alpha^E = \dot{\mathbf{m}} \cdot \tilde{G}_\alpha \cdot \dot{\mathbf{m}}, \tag{8.67}$$

in terms of the dissipation tensor [88]

$$\tilde{G}_\alpha^{ij} = \frac{\gamma^2 \hbar}{4\pi} \text{Re} \sum_\beta \text{Tr} \left(\frac{\partial \hat{S}_{\alpha\beta}^\dagger}{\partial m_i} \frac{\partial \hat{S}_{\alpha\beta}}{\partial m_j} \right). \tag{8.68}$$

In the limit of vanishing spin-flip in the ferromagnet, meaning that all dissipation takes place in the reservoirs, we find

$$\tilde{G}_\alpha^{ij} = \frac{\gamma^2 \hbar}{4\pi} \text{Re} \sum_\beta \text{Tr} \left(\frac{\partial \hat{S}_{\alpha\beta}^\dagger}{\partial m_i} \frac{\partial \hat{S}_{\alpha\beta}}{\partial m_j} \right) = \gamma^2 \frac{1}{2} \left(\frac{\hbar}{e} \right)^2 G_\perp^{(R)} \delta_{ij}. \tag{8.69}$$

Equating this I_α^E with P above, we obtain a microscopic expression for the Gilbert damping tensor $\tilde{\alpha}$:

$$\tilde{\alpha} = g^* \frac{\hbar \mu_B}{2e^2} \frac{G_\perp^{(R)}}{M_s \mathcal{V}} \overset{\leftrightarrow}{1}, \tag{8.70}$$

which agrees with Eq. (8.63). Indeed, in the absence of spin–orbit coupling the damping is necessarily isotropic. While Eq. (8.63) reproduces the additional Gilbert damping due to the interfacial spin pumping, Eq. (8.69) is more general, and can be used to compute bulk magnetization damping, as long as it is of a purely electronic origin [88, 89].

8.3.2.2 *Continuous systems* As has already been noted, spin pumping in continuous systems is the Onsager counterpart of the spin-transfer torque discussed in Section 8.3.1.2 [31]. While a direct diagrammatic calculation for this pumping is possible [52], with results equivalent to those of the quantum-kinetic description of the spin-transfer torque outlined above, we believe that the scattering-matrix formalism is the most powerful microscopic approach [32]. The latter is particularly suitable for implementing parameter-free computational schemes that allow a realistic description of material-dependent properties.

An important example is pumping by a moving domain wall in a quasi-one-dimensional ferromagnetic wire. When the domain wall is driven by a weak magnetic field, its shape remains to a good approximation unaffected, and only its position $r_w(t)$ along the wire is needed to parameterize its slow dynamics. The electric current pumped by the sliding domain wall into the αth lead can then be viewed as pumping by the r_w parameter, which leads to [80]

$$I_\alpha^c = \frac{e \dot{r}_w}{2\pi} \operatorname{Im} \sum_\beta \operatorname{Tr}\left(\frac{\partial \hat{S}_{\alpha\beta}}{\partial r_w} \hat{S}_{\alpha\beta}^\dagger \right). \tag{8.71}$$

The total heat flow into both leads induced by this dynamics is according to Eq. (8.66)

$$I^E = \frac{\hbar \dot{r}_w^2}{4\pi} \sum_{\alpha\beta} \operatorname{Tr}\left(\frac{\partial \hat{S}_{\alpha\beta}^\dagger}{\partial r_w} \frac{\partial \hat{S}_{\alpha\beta}}{\partial r_w} \right). \tag{8.72}$$

Evaluating the scattering-matrix expressions on the right-hand side of the above equations leads to microscopic magnetotransport response coefficients that describe the interaction of the domain wall with electric currents, including spin transfer and pumping effects.

These results lead to microscopic expressions for the phenomenological response [32] of the domain-wall velocity $\dot{r}_w$ and charge current I^c to a voltage V and magnetic field applied along the wire H:

$$\begin{pmatrix} \dot{r}_w \\ I^c \end{pmatrix} = \begin{pmatrix} L_{ww} & L_{wc} \\ L_{cw} & L_{cc} \end{pmatrix} \begin{pmatrix} 2AM_sH \\ V \end{pmatrix}, \tag{8.73}$$

subject to appropriate conventions for the signs of voltage and magnetic field and assuming a head-to-head or tail-to-tail wall such that the magnetization outside of the wall region is collinear with the wire axis. $2AM_sH$ is the thermodynamic force normalized to the entropy production by the magnetic system, where A is the cross-sectional area of the wire. We may therefore expect the Onsager's symmetry relation $L_{cw} = L_{wc}$. When a magnetic field moves the domain wall in the absence of a voltage $I^c = (L_{cw}/L_{ww})\dot{r}_w$, which, according to Eq. (8.71), leads to the ratio L_{cw}/L_{ww} in terms of the scattering matrices. The total energy dissipation for the same process is $I^E = \dot{r}_w^2/L_{ww}$, which, according to Eq. (8.72), establishes a scattering-matrix expression for L_{ww} alone. By supplementing these equations with the standard Landauer-Büttiker formula for the conductance

$$G = \frac{e^2}{h}\text{Tr}\left(\hat{S}_{12}^\dagger \hat{S}_{12}\right), \tag{8.74}$$

valid in the absence of domain-wall dynamics, we find L_{cc} in the same spirit since $G = L_{cc} - L_{wc}^2/L_{ww}$. Summarizing, the phenomenological response coefficients in Eq. (8.73) read [32]:

$$L_{ww}^{-1} = \frac{\hbar}{4\pi}\sum_{\alpha\beta}\text{Tr}\left(\frac{\partial \hat{S}_{\alpha\beta}^\dagger}{\partial r_w}\frac{\partial \hat{S}_{\alpha\beta}}{\partial r_w}\right), \tag{8.75}$$

$$L_{cw} = L_{wc} = L_{ww}\frac{e}{2\pi}\text{Im}\sum_{\beta}\text{Tr}\left(\frac{\partial \hat{S}_{\alpha\beta}}{\partial r_w}\hat{S}_{\alpha\beta}^\dagger\right), \tag{8.76}$$

$$L_{cc} = \frac{e^2}{h}\text{Tr}\left(\hat{S}_{12}\hat{S}_{12}^\dagger\right) + \frac{L_{wc}^2}{L_{ww}}. \tag{8.77}$$

When the wall is sufficiently smooth, we can model spin torques and pumping by the continuum theory based on the gradient expansion in the magnetic texture, Eqs. (8.11) and (8.12). Solving for the magnetic-field and current-driven dynamics of such domain walls is then possible using the Walker ansatz [69, 90]. Introducing the domain-wall width λ_w:

$$\alpha = \frac{\gamma\lambda_w}{2AM_sL_{ww}} \quad \text{and} \quad \beta = -\frac{e\lambda_w}{\hbar PG}\frac{L_{wc}}{L_{ww}}. \tag{8.78}$$

When the wall is sharp the adiabatic approximation underlying the leading-order gradient expansion breaks down. These relations can still be used as definitions of the effective domain-wall α and β. As such, these could be distinct from the bulk values that are associated with smooth textures. This is relevant for dilute magnetic semiconductors, for which the adiabatic approximation easily breaks down [32]. In transition-metal ferromagnets, on the other hand, the adiabatic approximation is generally perceived to be a good starting point, and we may expect the dissipative parameters in Eq. (8.78) to be comparable to their bulk values discussed in Section 8.3.1.2.

8.4 First-principles calculations

We have shown that the essence of spin pumping and spin transfer can be captured by a small number of phenomenological parameters. In this section we address the material dependence of these phenomena in terms of the (reflection) mixing conductance $G_\perp$, the dimensionless Gilbert damping parameter α, and the out-of-plane torque parameter β.

For discrete systems the (reflection) mixing conductance $G_\perp$ was studied theoretically by Xia *et al.* [91], Zwierzycki *et al.* [45] and Carva *et al.* [92]. $G_\perp$ describes the spin current flowing in response to an externally applied spin accumulation $e\mathbf{V}_s$ that is a vector with length equal to half of the spin-splitting of the chemical potentials $e|\mathbf{V}_s| = e(V_\uparrow - V_\downarrow)/2$. It also describes the spin torque exerted on the moment of the magnetic layer [9, 21, 45, 91–94]. Consider a spin accumulation in a normal metal N, which is in contact with a ferromagnet on the right magnetized along the z-axis. The spin current incident on the interface is proportional to the number of incident channels in the left lead, $\mathbf{I}_{\text{in}}^N = 2G_N^{\text{Sh}}\mathbf{V}_s$, while the reflected spin current is given by

$$\mathbf{I}_{\text{out}}^N = 2 \begin{pmatrix} G_N^{\text{Sh}} - G_\perp^{(R)} & -G_\perp^{(I)} & 0 \\ G_\perp^{(I)} & G_N^{\text{Sh}} - G_\perp^{(R)} & 0 \\ 0 & 0 & G_N^{\text{Sh}} - \frac{G_\uparrow + G_\downarrow}{2} \end{pmatrix} \mathbf{V}_s, \qquad (8.79)$$

where G_σ are the conventional Landauer–Büttiker conductances. The real and imaginary parts of $G_N^{\text{Sh}} - G_\perp = (e^2/h)\sum_{mn} r_{mn}^\uparrow r_{mn}^{\downarrow\star}$ are related to the components of the reflected transverse spin current and can be calculated by considering a single N–F interface [91]. When the ferromagnet is a layer with finite thickness d sandwiched between normal metals, the reflection mixing conductance depends on d and it is necessary to consider also the transmission mixing conductance $(e^2/h)\sum_{mn} t'^\uparrow_{mn} t'^{\downarrow\star}_{mn}$. In Ref. [45], both reflection and transmission mixing conductances were calculated for Cu–Co–Cu and Au–Fe–Au sandwiches as a function of magnetic layer thickness d. The real and imaginary parts of the transmission mixing conductance and the imaginary part of the reflection mixing conductance were shown to decay rapidly with increasing d implying that the absorption of the transverse component of the spin current occurs within a few monolayers of the N–F interface for ideal lattice matched interfaces. When a minimal amount of interface disorder was introduced the absorption increased. The limit $G_\perp \to G_N^{\text{Sh}}$ corresponds to the situation where all of the incoming transverse polarized spin current is absorbed in the magnetic layer. The torque is then proportional to the Sharvin conductance of the normal metal. This turns out to be the situation for all but the thinnest (few monolayers) and cleanest

Co and Fe magnetic layers considered by Zwierzycki *et al.* [45] However, when there is nesting between Fermi surface sheets for majority and minority spins so that both spins have the same velocities over a large region of reciprocal space, then the transverse component of the spin current does not damp so rapidly and $G_\perp$ can continue to oscillate for large values of d. This has been found to occur for ferromagnetic Ni in the (001) direction [92].

Equation (8.17) implies that the spin pumping renormalizes both the Gilbert damping parameter α and the gyromagnetic ratio γ of a ferromagnetic film embedded in a conducting nonmagnetic medium. However, in view of the results discussed in the previous paragraph, we conclude that the main effect of the spin pumping is to enhance the Gilbert damping. The correction is directly proportional to the real part of the reflection mixing conductance and is essentially an interface property. Oscillatory effects are averaged out for realistic band structures, especially in the presence of disorder. $G_\perp^{(R)}$ determines the damping enhancement of a single ferromagnetic film embedded in a perfect spin-sink medium and is usually very close to G_N^{Sh} for intermetallic interfaces [91, 93].

8.4.1 *Alpha*

We begin with a discussion of the small-angle damping measured as a function of temperature using ferromagnetic resonance (FMR). There is general agreement that spin–orbit coupling and disorder are essential ingredients in any description of how spin excitations relax to the ground state. In the absence of intrinsic disorder, one might expect the damping to increase monotonically with temperature in clean magnetic materials and indeed, this is what is observed for Fe. Heinrich *et al.* [95] developed an explicit model for this high-temperature behavior in which itinerant s electrons scatter from localized d moments and transfer spin angular momentum to the lattice via spin–orbit interaction. This $s - d$ model results in a damping that is inversely proportional to the electronic relaxation time, $\alpha \sim 1/\tau$, i.e. is *resistivity*-like. However, at low temperatures, both Co and Ni exhibit a sharp rise in damping as the temperature decreases. The so-called breathing Fermi surface model was proposed [96–98] to describe this low-temperature *conductivity*-like damping, $\alpha \sim \tau$. In this model the electronic population lags behind the instantaneous equilibrium distribution due to the precessing magnetization and requires dissipation of energy and angular momentum to bring the system back to equilibrium.

Of the numerous microscopic models that have been proposed [99] to explain the damping behaviour of metals, only the so-called "torque correlation model" (TCM) [100] is qualitatively successful in explaining the nonmonotonic damping observed for hcp Co that results from conductivity-like and resistivity-like behaviors at low and high temperatures, respectively. The central result of the TCM is the expression

$$\tilde{G} = \frac{g^2 \mu_B^2}{\hbar} \sum_{n,m} \int \frac{d\mathbf{k}}{(2\pi)^3} \left| \langle n, \mathbf{k} | [\sigma_-, \hat{\mathcal{H}}_{so}] | m, \mathbf{k} \rangle \right|^2 W_{n,m}(\mathbf{k}) \qquad (8.80)$$

for the damping. The commutator $[\sigma_-, \hat{\mathcal{H}}_{so}]$ describes a torque between the spin and orbital moments that arises as the spins precess. The corresponding matrix elements in (8.80) describe transitions between states in bands n and m induced by this torque whereby the crystal momentum $\mathbf{k}$ is conserved. Disorder enters in the form of a phenomenological relaxation time τ via the spectral overlap

$$W_{n,m}(\mathbf{k}) = -\frac{1}{\pi} \int A_n(\varepsilon, \mathbf{k}) A_m(\varepsilon, \mathbf{k}) \frac{df}{d\varepsilon} d\varepsilon \qquad (8.81)$$

where the electron spectral function $A_n(\varepsilon, \mathbf{k})$ is a Lorentzian centered on the band n, whose width is determined by the scattering rate. For intraband transitions with $m = n$, integration over energy yields a spectral overlap which is proportional to the relaxation time, like the conductivity. For interband transitions with $m \neq n$, the energy integration leads to a spectral overlap that is roughly inversely proportional to the relaxation time, like the resistivity.

To interpret results obtained with the TCM, Gilmore *et al.* [100–104] used an effective field approach expressing the effective field about which the magnetization precesses in terms of the total energy

$$\mu_0 \mathbf{H}^{\mathrm{eff}} = -\frac{\partial E}{\partial \mathbf{M}} \qquad (8.82)$$

and then approximated the total energy by a sum of single-particle eigenvalues $E \sim \sum_{n,\mathbf{k}} \varepsilon_{n\mathbf{k}} f_{n\mathbf{k}}$, so that the effective field naturally splits into two parts

$$\mathbf{H}^{\mathrm{eff}} = \frac{1}{\mu_0 M} \sum_{n,\mathbf{k}} \left[\frac{\partial \varepsilon_{n\mathbf{k}}}{\partial \mathbf{m}} f_{n\mathbf{k}} + \varepsilon_{n\mathbf{k}} \frac{\partial f_{n\mathbf{k}}}{\partial \mathbf{m}} \right] \qquad (8.83)$$

the first of which corresponds to the breathing Fermi surface model, intraband transitions and conductivity-like behavior while the second term could be related to interband transitions and resistivity-like behaviour. Evaluation of this model for Fe, Co, and Ni using first-principles calculations to determine $\varepsilon_{n\mathbf{k}}$ including spin–orbit coupling yields results for the damping α in good qualitative and reasonable quantitative agreement with the experimental observations [101].

In spite of this real progress, the TCM has disadvantages. As currently formulated, the model can only be applied to periodic lattices. Extending it to handle inhomogeneous systems such as ferromagnetic substitutional alloys like permalloy ($Ni_{80}Fe_{20}$), magnetic multilayers or heterojunctions, disordered materials or materials with surfaces, is far from trivial. The TCM incorporates disorder in terms of a relaxation time parameter τ and so suffers from the same disadvantages as all transport theories similarly formulated, namely, that it is difficult to relate microscopically measured disorder unambiguously to a given value of τ. Indeed, since τ in general depends on the incoming and scattered band index n, the wavevector $\mathbf{k}$, as well as the spin index, assuming a single value for it is a gross simplification. An improved theoretical framework would allow us to study not only crystalline materials such as the ferromagnetic metals Fe, Co, and Ni and substitutional disordered alloys such as permalloy (Py), but also amorphous materials and configurations such as magnetic heterojunctions, multilayers, thin films, etc. which become more important and are more commonly encountered as devices are made smaller.

The scattering–theoretical framework discussed in Section 8.3.2 satisfies these requirements and has recently been implemented by extending a first-principles scattering formalism [105, 106] based upon the local spin density approximation (LSDA) of density functional theory (DFT) to include noncollinearity, spin–orbit coupling (SOC), and chemical or thermal disorder on equal footings [89]. Relativistic effects are included by using the Pauli Hamiltonian. To calculate the scattering matrix, a "wavefunction matching" (WFM) scheme [105–107] has been implemented with a minimal basis of tight-binding linearized muffin-tin orbitals (TB-LMTOs) [108, 109]. Atomic-sphere-approximation (ASA) potentials [108, 109] are calculated self-consistently using a surface Green's function (SGF) method also implemented [110] with TB-LMTOs.

8.4.1.1 *NiFe alloys*

The flexibility of the scattering–theoretical formulation of transport can be demonstrated with an application to NiFe binary alloys [89]. Charge and spin densities for binary alloy A and B sites are calculated using the coherent potential approximation (CPA) [111] generalized to layer structures [110]. For the transmission matrix calculation, the resulting spherical potentials are distributed at random in large lateral supercells (SC) subject to maintenance of the appropriate concentration of the alloy [105, 106]. Solving the transport problem using lateral supercells makes it possible to go beyond effective medium approximations such as the CPA. As long as one is only interested in the properties of bulk alloys, the leads can be chosen for convenience and Cu leads with a single scattering state for each value of crystal momentum, $\mathbf{k}_\parallel$, are very convenient. The alloy lattice constants are determined using Vegard's law and the lattice constants of the leads are made to match. Though NiFe is fcc only for the concentration range $0 \le x \le 0.6$, the fcc structure is used for all values of x.

To illustrate the methodology, we begin by calculating the electrical resistivity of $Ni_{80}Fe_{20}$. In the Landauer–Büttiker formalism, the conductance can be expressed in terms of the transmission matrix t as $G = (e^2/h)Tr\left\{tt^\dagger\right\}$ [112, 113]. The resistance of the complete system consisting of ideal leads sandwiching a layer of ferromagnetic alloy of thickness L is $R(L) = 1/G(L) = 1/G_{\mathrm{Sh}} + 2R_{\mathrm{if}} + R_{\mathrm{b}}(L)$ where $G_{\mathrm{Sh}} = \left(2e^2/h\right)N$ is the Sharvin conductance of each lead with N conductance channels per spin, R_{if} is the interface resistance of a single N–F interface, and $R_{\mathrm{b}}(L)$ is the bulk resistance of a ferromagnetic layer of thickness L [81, 106]. When the ferromagnetic slab is sufficiently thick, Ohmic behavior is recovered whereby $R_{\mathrm{b}}(L) \approx \rho L$ as shown in the inset to Fig. 8.4 and the bulk resistivity ρ can be extracted from the slope of $R(L)$. For currents parallel and perpendicular to the magnetization direction, the resistivities are different and have to be calculated separately. The average resistivity is given

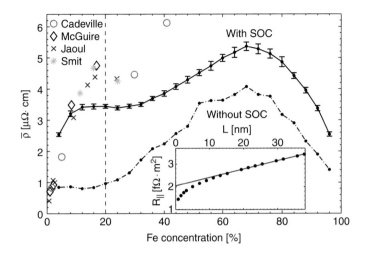

FIG. 8.4. Calculated resistivity as a function of the concentration x for fcc $Ni_{1-x}Fe_x$ binary alloys with (solid line) and without (dashed-dotted line) SOC. Low-temperature experimental results are shown as symbols [114–117]. The composition $Ni_{80}Fe_{20}$ is indicated by a vertical dashed line. Inset: resistance of Cu–$Ni_{80}Fe_{20}$–Cu as a function of the thickness of the alloy layer. Dots indicate the calculated values averaged over five configurations while the solid line is a linear fit.

by $\bar{\rho} = (\rho_\parallel + 2\rho_\perp)/3$, and the anisotropic magnetoresistance ratio (AMR) by $(\rho_\parallel - \rho_\perp)/\bar{\rho}$.

For $Ni_{80}Fe_{20}$ we find values of $\bar{\rho} = 3.5 \pm 0.15$ μOhm cm and AMR $= 19 \pm 1\%$, compared to experimental low-temperature values in the range 4.2–4.8 μOhm-cm for $\bar{\rho}$ and 18% for AMR [114]. The resistivity calculated as a function of x is compared to low-temperature literature values [114–117] in Fig. 8.4. The overall agreement with previous calculations is good [118, 119]. In spite of the smallness of the SOC, the resistivity of Py is underestimated by more than a factor of 4 when it is omitted, underlining its importance for understanding transport properties.

Assuming that the Gilbert damping is isotropic for cubic substitutional alloys and allowing for the enhancement of the damping due to the F–N interfaces [14, 45, 120, 121], the total damping in the system with a ferromagnetic slab of thickness L can be written $\tilde{G}(L) = \tilde{G}_{if} + \tilde{G}_b(L)$ where we express the bulk damping in terms of the dimensionless Gilbert damping parameter α $\tilde{G}_b(L) = \alpha \gamma M_s(L) = \alpha \gamma \mu_s A L$, where μ_s is the magnetization density and A is the cross-section. The results of calculations for $Ni_{80}Fe_{20}$ are shown in the inset to Fig. 8.5. The intercept at $L = 0$, $\tilde{G}_{if}$, allows us to extract the damping enhancement [45] but here we focus on the bulk properties and leave consideration of the material dependence of the interface enhancement for later study. The value of

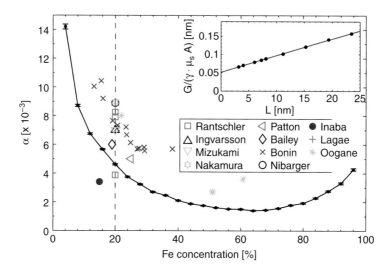

Fig. 8.5. Calculated zero-temperature (solid line) and experimental room temperature (symbols) values of the Gilbert damping parameter as a function of the concentration x for fcc $Ni_{1-x}Fe_x$ binary alloys [120–131]. Inset: total damping of Cu–$Ni_{80}Fe_{20}$–Cu as a function of the thickness of the alloy layer. Dots indicate the calculated values averaged over five configurations while the solid line is a linear fit.

α determined from the slope of $\tilde{G}(L)/(\gamma\mu_s A)$ is 0.0046 ± 0.0001 that is at the lower end of the range of values 0.004–0.013 measured at room temperature for Py [120–131].

Fig. 8.5 shows the Gilbert damping parameter as a function of x for $Ni_{1-x}Fe_x$ binary alloys in the fcc structure. From a large value for clean Ni, it decreases rapidly to a minimum at $x \sim 0.65$ and then grows again as the limit of clean *fcc* Fe is approached. Part of the decrease in α with increasing x can be explained by the increase in the magnetic moment per atom as we progress from Ni to Fe. The large values of α calculated in the dilute alloy limits can be understood in terms of conductivity-like enhancement at low temperatures [132, 133] that has been explained in terms of intraband scattering [100–102, 104]. The trend exhibited by the theoretical $\alpha(x)$ is seen to be reflected by experimental results obtained at room temperature. In spite of a large spread in measured values, these seem to be systematically larger than the calculated values. Part of this discrepancy can be attributed to an increase in α with temperature [122, 134].

Calculating α for the end members, Ni and Fe, of the substitutional alloy $Ni_{1-x}Fe_x$ presents a practical problem. In these limits there is no scattering whereas in experiment there will always be some residual disorder at low temperatures, and at finite temperatures, electrons will scatter from the thermally displaced ions. We introduce a simple "frozen thermal disorder" scheme to study

Ni and Fe and simulate the effect of temperature via electron–phonon coupling by using a random Gaussian distribution of ionic displacements $\mathbf{u}_i$, corresponding to a harmonic approximation. This is characterized by the root-mean-square (RMS) displacement $\Delta = \sqrt{\langle |\mathbf{u}_i|^2 \rangle}$ where the index i runs over all atoms. Typical values will be of the order of a few hundredths of an angstrom. We will not attempt to relate Δ to a real lattice temperature here.

We calculate the total resistance $R(L)$ and Gilbert damping $\tilde{G}(L)$ for thermally disordered scattering regions of variable length L and extract the resistivity ρ and damping α from the slopes as before. The results for Ni are shown as a function of the RMS displacement in Fig. 8.6. The resistivity is seen to increase monotonically with Δ underlining the correlation between Δ and a real temperature. For large values of Δ, α saturates for Ni in agreement with experiment [132] and calculations based on the torque-correlation model [101, 103, 104] where no concrete scattering mechanism is attached to the relaxation time τ. The absolute value of the saturated α is about 70% of the observed value. For small values of Δ, the Gilbert damping increases rapidly as Δ decreases. This sharp rise corresponds to the experimentally observed conductivity-like behavior at low temperatures and confirms that the scattering formalism can reproduce this feature.

8.4.2 Beta

To evaluate expressions (8.78) for the out-of-plane spin-torque parameter β given in Section 8.3.2 requires modeling domain walls (DW) in the scattering region sandwiched between ideal Cu leads. A head-to-head Néel DW is introduced inside the permalloy region by rotating the local magnetization to follow

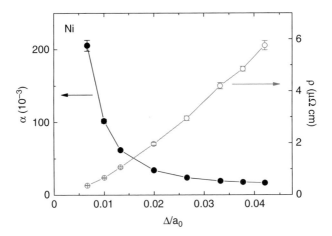

FIG. 8.6. Calculated Gilbert damping and resistivity for fcc Ni as a function of the relative RMS displacement with respect to the corresponding lattice constant, $a_0 = 3.524$ Å.

the Walker profile, $\mathbf{m}(z) = [f(z), 0, g(z)]$ with $f(z) = \cosh^{-1}[(z - r_w)/\lambda_w]$ and $g(z) = -\tanh[(z - r_w)/\lambda_w]$ as shown schematically in Fig. 8.7(a). r_w is the DW center and λ_w is a parameter characterizing its width. In addition to the Néel wall, we also study a rotated Néel wall with magnetization profile $\mathbf{m}(z) = [g(z), 0, f(z)]$ sketched in Fig. 8.7(b) and a Bloch wall with $\mathbf{m}(z) = [g(z), f(z), 0]$ sketched in Fig. 8.7(c).

The effective Gilbert damping constant α of permalloy in the presence of all three DWs calculated using (8.78) is shown in Fig. 8.8. For different types of DWs, α is identical within the numerical accuracy indicating that the Gilbert damping is isotropic due to the strong impurity scattering [103]. In the adiabatic limit, α saturates to the same value (the dashed line in Fig. 8.8) calculated for bulk permalloy using (8.68). It implies that the DWs in permalloy have little effect on the magnetization relaxation and the strong impurity scattering is the dominant mechanism to release energy and magnetization. This is in contrast to DWs in (Ga,Mn) As where Gilbert damping is mostly contributed by the reflection of the carriers from the DW [30]. At $\lambda_w < 5$ nm, the nonadiabatic reflection of conduction electrons due to the rapidly varying magnetization direction becomes significant and results in a sharp rise in α for narrow DWs.

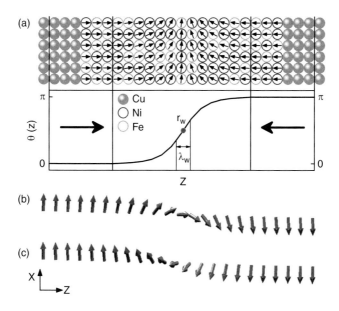

FIG. 8.7. (a) Sketch of the configuration of a Néel DW in Py sandwiched by two Cu leads. The arrows denote local magnetization directions. The curve shows the mutual angle between the local magnetization and the transport direction (z-axis). (b) Magnetization profile of a rotated Néel wall. (c) Magnetization profile of a Bloch wall.

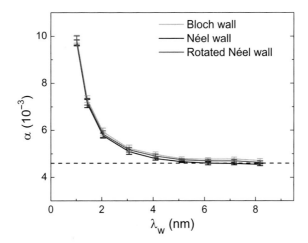

FIG. 8.8. Calculated effective Gilbert damping constant α for Py DWs as a function of λ_w. The dashed line shows the calculated α for bulk Py with the magnetization parallel to the transport direction [89].

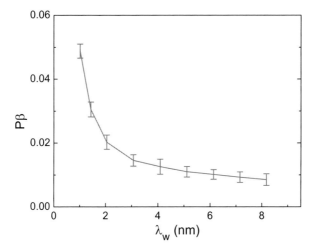

FIG. 8.9. Calculated out-of-plane spin torque parameter $P\beta$ for a permalloy DW as a function of λ_w.

The out-of-plane torque is formulated as $\beta(\hbar\gamma P/2eM_s)\mathbf{m} \times (\mathbf{j} \cdot \nabla)\mathbf{m}$ in the Landau–Lifshitz–Gilbert (LLG) equation under a finite current density $\mathbf{j}$. In principle, the current polarization P is required to determine β. Since the spin-dependent conductivities of permalloy depend on the angle between the current and the magnetization, P is not well defined for magnetic textures. Instead, we calculate the quantity $P\beta$, as shown in Fig. 8.9 for a Bloch DW. For $\lambda_w < 5$ nm, $P\beta$ decreases quite strongly with increasing λ_w corresponding to an expected nonadiabatic contribution to the out-of-plane torque. This arises from the spin-flip scattering induced by the rapidly varying magnetization in narrow DWs [135] and does not depend on the specific type of DW. For $\lambda_w > 5$ nm, which one expects to be in the adiabatic limit, $P\beta$ decreases slowly to a constant value [27, 32, 51, 66, 75, 135–143]. It is unclear what length-scale is varying so slowly. Unfortunately, the spread of values for different configurations is quite large for the last data point and our best estimate of $P\beta$ for a Bloch DW in permalloy is ~ 0.008. Taking the theoretical value of $P \sim 0.7$ for permalloy [89], our best estimate of β is a value of ~ 0.01.

8.5   Theory versus experiments

Spin-torque induced magnetization dynamics in multilayers and its reciprocal effect, spin pumping, are experimentally well established and quantitatively understood within the framework described in this chapter, and need not be discussed further here [15, 20]. Recent FMR experiments also confirm the spin-pumping contribution to the enhanced magnetization dissipation [144]. Spin pumping occurs in magnetic insulators as well [7, 145].

The parameters that control the current-induced dynamics of continuous textures are much less well known. Most experiments are carried out on permalloy (Py). It is a magnetically very soft material with large domain wall widths of the order of 100 nm. Although the adiabatic approximations appears to be a safe assumption in Py, many systems involve vortex domain walls with large gradients in the wall center, and, therefore, possibly sizable nonadiabatic corrections. An effective description for such vortex dynamics has been constructed in Ref. [60], where it was shown, in particular, that self-consistent quadratic corrections to damping (which stem from self-pumped currents inducing back-action on the magnetic order) is generally nonnegligible in transition-metal ferromagnets.

Early experimental studies [146, 147] for the torque-supplemented [Eq. (8.11)] LLG equation describing current-driven domain-wall motion in magnetic wires reported values of the β/α ratios in Py close to unity, in agreement with simple Stoner-model calculations. However, much larger values $\beta/\alpha \sim 8$ were extracted from the current-induced oscillatory motion of domain walls [148]. The inequality $\beta \neq \alpha$ was also inferred from a characteristic transverse-to-vortex wall structure transformation, although no exact value of the ratio was established [149]. In Ref. [150], vanadium doping of Py was shown to enhance β up to nearly 10α, with little effect on α itself. Even larger ratios, $\beta/\alpha \sim 20$, were found for magnetic vortex motion by an analysis of their displacement as a function of an applied dc current in disk structures [151, 152].

Eltschka *et al.* [153] reported on a measurement of the dissipative spin-torque parameter β entering Eq. (8.11), as manifested by thermally activated motion of transverse and vortex domain walls in Py. They found the ratio $\beta_v/\beta_t \sim 7$ for the vortex *vs* transverse wall, attributing the larger β to high magnetization gradients in the vortex wall core. Their ratio $\beta_t/\alpha \sim 1.3$ turns out to be close to unity, where α is the bulk Gilbert damping. The importance of large spin-texture gradients on the domain-wall and vortex dynamics was theoretically discussed in Refs. [58, 60].

The material dependence of the current-induced torques is not yet well investigated. A recent study on CoNi and FePt wires with perpendicular magnetization found $\beta \approx \alpha$, in spite of the relatively narrow domain walls in these materials [154]. Current-induced domain-wall dynamics in dilute magnetic semiconductors [155] generally exhibit similar phenomenology, but a detailed discussion, especially of the domain wall creep regime that can be accessed in these systems, is beyond the scope of this review.

Finally, the first term in the spin-pumping expression (8.12) has been measured by Yang *et al.* [156] for a domain wall moved by an applied magnetic field above the Walker breakdown field. These experiments confirmed the existence of pumping effects in magnetic textures, which are Onsager reciprocals of spin torques and thus expected on general grounds. Similar experiments carried out below the Walker breakdown would also give direct access to the β parameter.

8.6 Conclusions

A spin-polarized current can excite magnetization dynamics in ferromagnets via spin-transfer torques. The reciprocal phenomenon is spin pumping where a dynamic magnetization pumps spins into adjacent conductors. We have discussed how spin-transfer torques and spin pumping are directly related by Onsager reciprocity relations.

In layered normal metal–ferromagnet systems, spin-transfer torques can be expressed in terms of two conductance parameters governing the flow of spins transverse to the magnetization direction and the spin accumulation in the normal metal. In metallic systems, the field-like torque is typically much smaller than the effective energy gain/damping torque, but in tunnel systems they might become comparable. Spin pumping is controlled by the same transverse conductance parameters as spin-transfer torques, the magnetization direction, and its rate of change. It can lead to an enhanced magnetization dissipation in ultrathin ferromagnets or a build-up of spins, a spin battery, in normal metals where the spin-flip relaxation rate is low.

Spin-transfer torque and spin-pumping phenomena in magnetization textures are similar to their counterparts in layered normal metal–ferromagnet systems. A current becomes spin polarized in a ferromagnet and this spin-polarized current in a magnetization texture gives rise to a reactive torque and a dissipative torque in the lowest gradient expansion. The reciprocal pumping phenomena can be viewed as an electromotive force; the dynamic magnetization texture pumps a spin current that in turn is converted to a charge current or voltage by the giant magnetoresistance effect. Naturally, the parameters governing the spin-transfer torques and the pumping phenomena are also the same in continuously textured ferromagnets.

When the spin–orbit interaction becomes sufficiently strong, additional effects arise in the coupling between the magnetization and itinerant electrical currents. A charge potential can then by itself induce a torque on the ferromagnet and the reciprocal phenomenon is that a precessing ferromagnet can induce a charge current in the adjacent media. The latter can be an alternative way to carry out FMR measurements on small ferromagnets by measuring the induced voltage across a normal metal–ferromagnet–normal metal device.

These phenomena are well known and we have reviewed them in a unified physical picture and discussed the connection between these and some experimental results.

Acknowledgments

We are grateful to Jørn Foros, Bertrand I. Halperin, Kjetil M. D. Hals, Alexey Kovalev, Yi Liu, Hans Joakim Skadsem, Anton Starikov, Zhe Yuan, and Maciej Zwierzycki for discussions and collaboration.

This work was supported in part by EU-ICT-7 contract no. 257159 MACALO – Magneto Caloritronics, DARPA, NSF under Grant No. DMR-080965, the FOM foundation, and DFG Priority Program SpinCaT.

References

[1] S.D. Bader and S.S.P. Parkin, *Spintronics*, Ann. Rev. Conden. Matter Phys. **1**, 71 (2010).

[2] T.J. Silva and W.H. Rippard, *Developments in nano-oscillators based upon spin-transfer point-contact devices*, J. Magn. Magn. Mater. **320**, 1260 (2008).

[3] P.M. Braganca, B.A. Gurney, B.A. Wilson, J.A. Katine, S. Maat, and J.R. Childress, *Nanoscale magnetic field detection using a spin torque oscillator*, Nanotechnology **21**, 235202 (2010).

[4] S. Matsunaga, K. Hiyama, A. Matsumoto, S. Ikeda, H. Hasegawa, K. Miura, J. Hayakawa, T. Endoh, H. Ohno, and T. Hanyu, *Standby-power-free compact ternary content-addressable memory cell chip using magnetic tunnel junction devices*, Appl. Phys. Expr. **2**, 023004 (2009).

[5] K. Nagasaka, *CPP-GMR technology for magnetic read heads of future high-density recording systems*, J. Magn. Magn. Mater. **321**, 508 (2009).

[6] D.D. Awschalom and M. Flatté. *Challenges for semiconductor spintronics,* Nature Physics **3**, 153 (2007).

 [7] Y. Kajiwara, K. Harii, S. Takahashi, J. Ohe, K. Uchida, M. Mizuguchi, H. Umezawa, H. Kawai, K. Ando, K. Takanashi, S. Maekawa, and E. Saitoh, *Transmission of electrical signals by spin-wave interconversion in a magnetic insulator,* Nature **464**, 262 (2010).

 [8] L. Berger, *Emission of spin waves by a magnetic multilayer traversed by a current,* Phys. Rev. B **54**, 9353 (1996).

 [9] J.C. Slonczewski, *Current-driven excitation of magnetic multilayers,* J. Magn. Magn. Mater. **159**, L1 (1996).

[10] M. Tsoi, A.G.M. Jansen, J. Bass, W.C. Chiang, M. Seck, V. Tsoi, and P. Wyder, *Excitation of a magnetic multilayer by an electric current,* Phys. Rev. Lett. **81**, 493 (1998).

[11] E.B. Myers, D.C. Ralph, J.A. Katine, R.N. Louie, and R.A. Buhrman, *Current-induced switching in magnetic multilayer devices,* Science **285**, 867 (1999).

[12] A. Janossy and P. Monod, *Spin waves for single electrons in paramagnetic metals,* Phys. Rev. Lett. **37**, 612 (1976).

[13] R.H. Silsbee, A. Janossy, and P. Monod, *Coupling between ferromagnetic and conduction-spin-resonance modes at a aferromagnet-normal-metal interface,* Phys. Rev. B **19**, 4382 (1979).

[14] Y. Tserkovnyak, A. Brataas, and G.E.W. Bauer, *Enhanced Gilbert damping in thin ferromagnetic films,* Phys. Rev. Lett. **88**, 117601 (2002).

[15] Y. Tserkovnyak, A. Brataas, G.E.W. Bauer, and B.I. Halperin, *Nonlocal magnetization dynamics in ferromagnetic heterostructures,* Rev. Mod. Phys. 77, 1375 (2005).

[16] S. Mizukami, Y. Ando, and T. Miyazaki, *The Study on Ferromagnetic Resonance Linewidth for NM/80NiFe/NM (NM=Cu, Ta, Pd and Pt) Films,* Jpn. J. Appl. Phys. **40**, 580 (2001); S. Mizukami, Y. Ando, and T. Miyazaki, *Effect of spin diffusion on Gilbert damping for a very thin permalloy layer in Cu/permalloy/Cu/Pt film,* Phys. Rev. B **66**, 104413 (2002).

[17] R. Urban, G. Woltersdorf, and B. Heinrich, *Gilbert damping in Single and Multilayer Ultrathin Films: Role of Interfaces in Nonlocal Spin Dynamics,* Phys. Rev. Lett. **87**, 217204 (2001).

[18] B. Heinrich, Y. Tserkovnyak, G. Woltersdorf, A. Brataas, R. Urban, and G. E. W. Bauer, *Dynamic Exchange Coupling in Magnetic Bilayers,* Phys. Rev. Lett. **90**, 187601 (2003).

[19] A. Brataas, G.E.W. Bauer, and P.J. Kelly, *Non-collinear magnetoelectronics,* Phys. Rep. **427**, 157 (2006).

[20] D.C. Ralph and M.D. Stiles, *Spin transfer torques,* J. Magn. Magn. Materials **320**, 1190 (2008).

[21] A. Brataas, Yu.V. Nazarov, and G.E.W. Bauer, *Finite-element theory of transport in ferromagnet-normal metal systems,* Phys. Rev. Lett. **84**, 2481 (2000); *Spin-transport in multi-terminal normal metal-ferromagnet systems with non-collinear magnetizations,* Eur. Phys. J. B **22**, 99 (2001).

[22] X. Waintal, E.B. Myers, P.W. Brouwer, and D.C. Ralph, *Role of spin-dependent interface scattering in generating current-induced torques in magnetic multilayers,* Phys. Rev. B **62**, 12317 (2000).

[23] G.E.W. Bauer, Y. Tserkovnyak, D. Huertas-Hernando, and A. Brataas, *Universal angular magnetoresistance and spin torque in ferromagnetic/normal metal hybrids,* Phys. Rev. B **67**, 094421 (2003).

[24] V.S. Rychkov, S. Borlenghi, H. Jaffres, A. Fert, and X. Waintal, *Spin torque and waviness in magnetic multilayers: A bridge between Valet–Fert theory and quantum approaches,* Phys. Rev. Lett. **103**, 066602 (2009).

[25] J.Z. Sun and D.C. Ralph, *Magnetoresistance and spin-transfer torque in magnetic tunnel junctions,* J. Magn. Magn. Mater. **320**, 1227 (2008).

[26] S.S.P. Parkin, M. Hayashi, and L. Thomas, *Magnetic domain-wall racetrack memory,* Science **320**, 190 (2008).

[27] S. Zhang and Z. Li, *Roles of nonequilibrium conduction electrons on the magnetization dynamics of ferromagnets,* Phys. Rev. Lett. **93**, 127204 (2004).

[28] S.E. Barnes and S. Maekawa, *Generalization of Faraday's law to include nonconservative spin forces,* Phys. Rev. Lett. **98**, 246601 (2007).

[29] G. Tatara, H. Kohno, and J. Shibata, *Microscopic approach to current-driven domain wall dynamics,* Phys. Rep. **468**, 213 (2008).

[30] G.S.D. Beach, M. Tsoi, and J.L. Erskine, *Current-induced domain wall motion,* J. Magn. Magn. Mater. **320**, 1272 (2008).

[31] Y. Tserkovnyak and M. Mecklenburg, *Electron transport driven by nonequilibrium magnetic textures,* Phys. Rev. B **77**, 134407 (2008).

[32] K.M.D. Hals, A.K. Nguyen, and A. Brataas, *Intrinsic coupling between current and domain wall motion in (Ga,Mn)As,* Phys. Rev. Lett. **102**, 256601 (2009).

[33] G.E.W. Bauer, S. Bretzel, A. Brataas, and Y. Tserkovnyak, *Nanoscale magnetic heat pumps and engines,* Phys. Rev. B **81**, 024427 (2010).

[34] S.R. de Groot, *Thermodynamics of Irreversible Processes,* Interscience, New York, 1952.

[35] S.J. Barnett, *Magnetization by rotation,* Phys. Rev. **6**, 239 (1915); S.J. Barnett, *Gyromagnetic and electron-inertia effects,* Rev. Mod. Phys. **7**, 129 (1935).

[36] A. Einstein and W.J. de Haas, *Experimenteller Nachweis der Ampereschen Molekularströme,* Deutsche Physikalische Gesellschaft, Verhandlungen **17**, 152 (1915).

[37] J. Grollier, V. Cros, A. Hamzic, J.M. George, H. Jaffres, A. Fert, G. Faini, J. Ben Youssef, and H. Legall, *Spin-polarized current induced swithing in Co/Cu pillars,* Appl. Pys. Lett. **78**, 3663 (2001).

[38] S.I. Kiselev, J.C. Sankey, I.N. Krivorotov, N.C. Emley, R.J. Schoelkopf, R.A. Buhrman, and D.C. Ralph, *Microwave oscillations of a nanomagnetic driven by a spin-polarized current,* Nature **425**, 380 (2003).

[39] B. Ozyilmaz, A.D. Kent, D. Monsma, J.Z. Sun, M.J. Rooks, and R.H. Koch, *Current-induced magnetization reversal in high magnetic field in Co/Cu/Co nanopillars*, Phys. Rev. Lett. **91**, 067203 (2003).

[40] I.N. Krivorotov, N.C. Emley, J.C. Sankey, S.I. Kiseev, D.C. Ralphs, and R.A. Buhrman, *Time-domain measurements of nanomagnet dynamics driven by spin-transfer torques*, Science 307, **228** (2005).

[41] Y.T. Cui, G. Finocchio, C. Wang, J.A. Katine, R.A. Buhrman, and D.C. Ralph, *Single-shot time-domain studies of spin-torque-driven switching in magnetic tunnel junctions*, Phys. Rev. Lett. **104**, 097201 (2010).

[42] A. Brataas, Y. Tserkovnyak, and G.E.W. Bauer, *Current-induced macrospin versus spin wave excitations in spin valves*, Phys. Rev. B **73**, 014408 (2006).

[43] K.M.D. Hals, A. Brataas, and Y. Tserkovnyak, *Scattering theory of charge-current-induced magnetization dynamics*, EPL **90**, 4702 (2010).

[44] A.A. Kovalev, A. Brataas, and G.E.W. Bauer, *Spin-tranfer in diffusive ferromagnet-normal metal systems with spin-flip scattering*, Phys. Rev. B **66**, 224424 (2002).

[45] M. Zwierzycki, Y. Tserkovnyak, P.J. Kelly, A. Brataas, and G.E.W. Bauer, *First-principles study of magnetization relaxation enhancement and spin transfer in thin magnetic films*, Phys. Rev. B **71**, 064420 (2005).

[46] A.A. Kovalev, G.E.W. Bauer, and A. Brataas, *Perpendicular spin valves with ultrathin ferromagnetic layers: Magnetoelectronic circuit investigation of finite-size effects*, Phys. Rev. B **73**, 054407 (2006).

[47] S.E. Barnes, *The effect that finite lattic spacing has upon the ESR Bloch equations*, J. Phys. F: Met. Phys. **4**, 1535 (1974).

[48] M. Büttiker, H. Thomas, and A. Prétre, *Current partition in multiprobe conductors in the presence of slowly oscillating external potentials*, Z. Phys. B **94**, 133 (1994).

[49] G.E. Volovik, *Linear momentum in ferromagnets*, J. Phys. C, **L83** (1987).

[50] G. Tatara and H. Kohno, *Theory of current-driven domain wall motion: spin transfer versus momentum transfer,* Phys. Rev. Lett. **92**, 086601 (2004).

[51] A. Thiaville, *Micromagnetic understanding of current-driven domain wall motion in patterned nanowires, EPL **69**, 990 (2005).*

[52] R.A. Duine, *Spin pumping by a field-driven domain wall*, Phys. Rev. B **77**, 014409 (2008).

[53] G.E.W. Bauer, A.H. MacDonald, and S. Maekawa, *Spin caloritronics*, Solid State Comm. **150**, 459 (2010).

[54] A. Brataas, Y. Tserkovnyak, G.E.W. Bauer, and B.I. Halperin, *Spin battery operated by ferromagnetic resonance,* Phys. Rev. B **66**, 060404 (2002).

[55] M.V. Costache, M. Sladkov, S.M. Watts, C.H. van der Wal, and B.J. van Wees, *Electrical detection of spin pumping due to the precessing magnetization of a single ferromagnet,* Phys. Rev. Lett. **97**, 216603 (2006); M.V. Costache, S.M. Watts, C.H. van der Wal, and B.J. van Wees, *Electrical*

detection of spin pumping: dc voltage generated by ferromagnetic resonacne at ferroamgnet/nonmagnet contact, Phys. Rev. B **78**, 064423 (2008).

[56] X. Wang, G.E.W. Bauer, B.J. van Wees, A. Brataas, and Y. Tserkovnyak, *Voltage generation by ferromagnetic resonancen at a nonmagnetic to ferromagnet contact,* Phys. Rev. Lett. **97**, 216602 (2006).

[57] K. Ando, S. Takahashi, J. Ieda, H. Kurebayashi, T. Trypiniotis, C.H.W. Barnes, S. Maekawa, and E. Saitoh, *Electrically tunable spin injector free from the impedance mismatch problem,* Nature Materials, Advance Online Publicatino, 26. June 2011.

[58] J. Foros, A. Brataas, Y. Tserkovnyak, and G.E.W. Bauer, *Current-induced noise and damping in nonuniform ferromagnets,* Phys. Rev. B **78**, 140402 (2008).

[59] S. Zhang and S.S.-L. Zhang, *Generalization of the Landau–Lifshitz–Gilbert equation for conducting ferromagnets,* Phys. Rev. Lett. **102**, 086601 (2009).

[60] C.H. Wong and Y. Tserkovnyak, *Dissipative dynamics of magnetic solitons in metals,* Phys. Rev. B **81**, 060404 (2010).

[61] J. Foros, A. Brataas, Y. Tserkovnyak, and G.E.W. Bauer, *Magnetization noise in magnetoelectronic nanostructures,* Phys. Rev. Lett. **95**, 016601 (2005).

[62] M. Hatami, G.E.W. Bauer, Q. Zhang, and J. Kelly, *Thermal spin-transfer torque in magnetoelectronic devices,* Phys. Rev. Lett. **99**, 066603 (2007).

[63] A.A. Kovalev and Y. Tserkovnyak, *Thermoelectric spin transfer in textured magnets,* Phys. Rev. B 80, 100408 (2009).

[64] M. Büttiker, *Scattering theory of current and intensity noise correlations in conductors and waveguides,* Phys. Rev. B **46**, 12485 (1992).

[65] Y. Tserkovnyak, H.J. Skadsem, A. Brataas, and G.E.W. Bauer, *Current-induced magnetization dynamics in disordered itinerant ferromagnets,* Phys. Rev. B **74**, 144405 (2006).

[66] H. Kohno and G. Tatara, *Microscopic calculation of spin torques in disordered ferromagnets,* J. Phys. Soc. Jpn, **75**, 113706 (2006).

[67] R.A. Duine, A.S. Núñez, J. Sinova, and A.H. MacDonald, *Functional Keldysh theory of spin torques,* Phys. Rev. B **75**, 214420 (2007).

[68] S.E. Barnes and S. Maekawa, *Current-spin coupling for ferromagnetic domain walls in fine wires,* Phys. Rev. Lett. **95**, 107204 (2005).

[69] N.L. Schryer and L.R. Walker, *The motion of 180 degree domain walls in uniform dc magnetic fields,* J. Appl. Phys. **45**, 5406 (1974).

[70] Y. Tserkovnyak, A. Brataas, and G.E. Bauer, *Theory of current-driven magnetization dynamics in inhomogeneous ferromagnets,* J. Magn. Magn. Mater. **320**, 1282 (2008).

[71] H.J. Skadsem, Y. Tserkovnyak, A. Brataas, and G.E.W. Bauer, *Magnetization damping in a local-density approximation,* Phys. Rev. B **75**, 094416 (2007).

[72] E. Runge and E.K.U. Gross, *Density-functional theory for time-dependent systems*, Phys. Rev. Lett. **52**, 997 (1984).

[73] K. Capelle, G. Vignale, and B.L. Györffy, *Spin currents and spin dynamics in time-dependent density-functional theory,* Phys. Rev. Lett. **87**, 206403 (2001).

[74] Z. Qian and G. Vignale, *Spin dynamics from time-dependent spin-density-functional theory,* Phys. Rev. Lett. **88**, 056404 (2002).

[75] I. Garate, K. Gilmore, M.D. Stiles, and A.H. MacDonald, *Nonadiabatic spin-transfer torque in real materials*, Phys. Rev. B, **79**, 104416 (2009).

[76] Y. Tserkovnyak, G.A. Fiete, and B.I. Halperin, *Mean-field magnetization relaxation in conducting ferromagnets*, Appl. Phys. Lett. **84**, 5234 (2004).

[77] M. Büttiker, H. Thomas, and A. Prêtre, *Current partition in multiprobe conductors in the presence of slowly oscillating external potentials*, Z. Phys. B **94**, 133 (1994).

[78] E. Saitoh, M. Ueda, H. Miyajima, and G. Tatara, *Conversion of spin current into charge current at room temperature: inverse spin-Hall effect*, Appl. Phys. Lett. **88**, 182509 (2006).

[79] F.D. Czeschka, L. Dreher, M.S. Brandt, M. Weiler, M. Althammer, I.-M. Imort, G. Reiss, A. Thomas, W. Schoch, W. Limmer, H. Huebl, R. Gross, and S.T.B. Goennenwein, *Scaling behavior of the spin pumping effect in ferromagnet/platinum bilayers,* Phys. Rev. Lett. **107**, 046601 (2011).

[80] P.W. Brouwer, *Scattering approach to parametric pumping*, Phys. Rev. B **58**, R10135 (1998).

[81] K.M. Schep, J.B.A.N. van Hoof, P.J. Kelly, G.E.W. Bauer, and J.E. Inglesfield, *Interface resistances of magnetic multilayers*, Phys. Rev. B **56**, 10805 (1997).

[82] A. Chernyshov, M. Overby, X. Liu, J.K. Furdyna, Y. Lyanda-Geller, and L.P. Rokhinson, *Evidence for reversible control of magnetization in a ferromagnetic material by means of spin–orbit magnetic field*, Nature Phys. **5**, 656 (2009).

[83] A. Manchon and S. Zhang, *Theory of nonequilibrium intrinsic spin torque in a single nanomagnet*, Phys. Rev. B **78**, 212405 (2008).

[84] I. Garate and A.H. MacDonald, *Influence of a transport current on magnetic anisotropy in gyrotropic ferromagnets*, Phys. Rev. B **80**, 134403 (2009).

[85] J.E. Avron, A. Elgart, G.M. Graf, and L. Sadun. *Optimal quantum pumps*, Phys. Rev. Lett. **87**, 236601 (2001).

[86] M. Moskalets and M. Büttiker, *Dissipation and noise in adiabatic quantum pumps*, Phys. Rev. B **66**, 035306 (2002).

[87] M. Moskalets and M. Büttiker, *Floquet scattering theory of quantum pumps,* Phys. Rev. B **66**, 205320 (2002).

[88] A. Brataas, Y. Tserkovnyak, and G.E.W. Bauer, *Scattering theory of Gilbert damping*, Phys. Rev. Lett. **101**, 037207 (2008).

[89] A.A. Starikov, P.J. Kelly, A. Brataas, Y. Tserkovnyak, and G.E.W. Bauer, *A unified first-principles study of gilbert damping, spin-flip diffusion and resistivity in transition metal alloys*, Phys. Rev. Lett. **105**, 236602 (2010).

[90] Z. Li and S. Zhang, *Domain-wall dynamics driven by adiabatic spin-transfer torques*, Phys. Rev. B **70**, 024417 (2004).

[91] K. Xia, P.J. Kelly, G.E.W. Bauer, A. Brataas, and I. Turek, *Spin torques in ferromagnetic/normal-metal structures*, Phys. Rev. B **65**, 220401 (2002).

[92] K. Carva and I. Turek, *Spin-mixing conductances of thin magnetic films from first principles*, Phys. Rev. B **76**, 104409 (2007).

[93] M.D. Stiles and A. Zangwill, *Anatomy of spin-transfer torque*, Phys. Rev. B **66**, 014407 (2002).

[94] A. Brataas, G. Zaránd, Y. Tserkovnyak, and G.E.W. Bauer, *Magnetoelectronic spin echo*, Phys. Rev. Lett. **91**, 166601 (2003).

[95] B. Heinrich, D. Fraitová, and V. Kamberský, *The influence of s-d exchange on relaxation of magnons in metals*, Phys. Stat. Sol. B **23**, 501–507 (1967).

[96] V. Kamberský, *Ferromagnetic resonance in iron whiskers*, Can. J. Phys. **48**, 1103 (1970).

[97] V. Korenman and R.E. Prange, *Anomalous damping of spin waves in magnetic metals*, Phys. Rev. B **6**, 2769 (1972).

[98] J. Kuneš and V. Kamberský, *First-principles investigation of the damping of fast magnetization precession in ferromagnetic 3d metals*, Phys. Rev. B **65**, 212411 (2002).

[99] B. Heinrich, *Spin relaxation in magnetic metallic layers and multilayers*, in *Ultrathin Magnetic Structures III* (J.A.C. Bland and B. Heinrich, eds.), Springer, New York, 2005, pp. 143–210.

[100] V. Kamberský, *On ferromagnetic resonance damping in metals*, Czech. J. Phys. **26**, 1366–1383 (1976).

[101] K. Gilmore, Y.U. Idzerda, and M.D. Stiles, *Identification of the dominant precession-damping mechanism in Fe, Co, and Ni by first-principles calculations*, Phys. Rev. Lett. **99**, 027204 (2007).

[102] K. Gilmore, Y.U. Idzerda, and M.D. Stiles, *Spin–orbit precession damping in transition metal ferromagnets*, J. Appl. Phys. **103**, 07D303 (2008).

[103] K. Gilmore, M.D. Stiles, J. Seib, D. Steiauf, and M. Fähnle, *Anisotropic damping of the magnetization dynamics in Ni, Co, and Fe*, Phys. Rev. B **81**, 174414 (2010).

[104] V. Kamberský, *Spin–orbital Gilbert damping in common magnetic metals*, Phys. Rev. B **76**, 134416 (2007).

[105] K. Xia, P.J. Kelly, G.E.W. Bauer, I. Turek, J. Kudrnovský, and V. Drchal, *Interface resistance of disordered magnetic multilayers*, Phys. Rev. B **63**, 064407 (2001).

[106] K. Xia, M. Zwierzycki, M. Talanana, P.J. Kelly, and G. E. W. Bauer, *First-principles scattering matrices for spin-transport*, Phys. Rev. B **73**, 064420 (2006).

[107] T. Ando, *Quantum point contacts in magnetic fields*, Phys. Rev. B **44**, 8017 (1991).

[108] O.K. Andersen, *Linear methods in band theory*, Phys. Rev. B **12**, 3060 (1975).

[109] O.K. Andersen, Z. Pawlowska, and O. Jepsen, *Illustration of the linear-muffin-tin-orbital tight-binding representation: Compact orbitals and charge density in Si*, Phys. Rev. B **34**, 5253 (1986).

[110] I. Turek, V. Drchal, J. Kudrnovský, M. Šob, and P. Weinberger, *Electronic structure of disordered alloys, surfaces and interfaces*, Kluwer, Boston, 1997.

[111] P. Soven, *Coherent-potential model of substitutional disordered alloys*, Phys. Rev. **156**, 809 (1967).

[112] M. Büttiker, Y. Imry, R. Landauer, and S. Pinhas, *Generalized many-channel conductance formula with application to small rings*, Phys. Rev. B **31**, 6207 (1985).

[113] S. Datta, *Electronic Transport in Mesoscopic Systems*, Cambridge University Press, Cambridge, 1995.

[114] J. Smit, *Magnetoresistance of ferromagnetic metals and alloys at low temperatures*, Physica **17**, 612–627 (1951).

[115] T.R. McGuire and R.I. Potter, *Anisotropic magnetoresistance in ferromagnetic 3d alloys*, IEEE Trans. Mag. **11**, 1018 (1975).

[116] O. Jaoul, I. Campbell, and A. Fert, *Spontaneous resistivity anisotropy in Ni alloys*, J. Magn. and Magn. Mater. **5**, 23 (1977).

[117] M.C. Cadeville and B. Loegel, *On the transport properties in concentrated Ni-Fe alloys at low temperatures*, J. Phys. F: Met. Phys. **3**, L115 (1973).

[118] J. Banhart, H. Ebert, and A. Vernes, *Applicability of the two-current model for systems with strongly spin-dependent disorder*, Phys. Rev. B **56**, 10165 (1997).

[119] J. Banhart and H. Ebert, *First-principles theory of spontaneous-resistance anisotropy and spontaneous hall effect in disordered ferromagnetic alloys*, Europhys. Lett. **32**, 517 (1995).

[120] S. Mizukami, Y. Ando, and T. Miyazaki, *Ferromagnetic resonance linewidth for NM/80NiFe/NM films (NM=Cu, Ta, Pd and Pt)*, J. Magn. and Magn. Mater. **226–230**, 1640 (2001).

[121] S. Mizukami, Y. Ando, and T. Miyazaki, *The study on ferromagnetic resonance linewidth for NM/80NiFe/NM (NM = Cu, Ta, Pd and Pt) films*, Jpn. J. Appl. Phys. **40**, 580 (2001).

[122] W. Bailey, P. Kabos, F. Mancoff, and S. Russek, *Control of magnetization dynamics in $Ni_{81}Fe_{19}$ thin films through the use of rare-earth dopants*, IEEE Trans. Mag. **37**, 1749 (2001).

[123] C.E. Patton, Z. Frait, and C.H. Wilts, *Frequency dependence of the parallel and perpendicular ferromagnetic resonance linewidth in Permalloy films, 2-36 GHz*, J. Appl. Phys. **46**, 5002 (1975).

[124] S. Ingvarsson, G. Xiao, S.S.P. Parkin, and R.H. Koch, *Tunable magneti-zation damping in transition metal ternary alloys*, Appl. Phys. Lett. **85**, 4995 (2004).

[125] H. Nakamura, Y. Ando, S. Mizukami, and H. Kubota, *Measurement of magnetization precession for $NM/Ni_{80}Fe_{20}/NM$ (NM= Cu and Pt) using time-resolved Kerr effect*, Jpn. J. Appl. Phys. **43**, L787 (2004).

[126] J.O. Rantschler, B.B. Maranville, J.J. Mallett, P. Chen, R.D. McMichael, and W.F. Egelhoff, *Damping at normal metal/Permalloy interfaces*, IEEE Trans. Mag. **41**, 3523 (2005).

[127] R. Bonin, M.L. Schneider, T.J. Silva, and J.P. Nibarger, *Dependence of magnetization dynamics on magnetostriction in NiFe alloys*, J. Appl. Phys. **98**, 123904 (2005).

[128] L. Lagae, R. Wirix-Speetjens, W. Eyckmans, S. Borghs, and J. de Boeck, *Increased Gilbert damping in spin valves and magnetic tunnel junctions*, J. Magn. and Magn. Mater. **286**, 291 (2005).

[129] J.P. Nibarger, R. Lopusnik, Z. Celinski, and T.J. Silva, *Variation of magnetization and the Landé g factor with thickness in Ni-Fe films*, Appl. Phys. Lett. **83**, 93 (2003).

[130] N. Inaba, H. Asanuma, S. Igarashi, S. Mori, F. Kirino, K. Koike, and H. Morita, *Damping constants of Ni-Fe and Ni-Co alloy thin films*, IEEE Trans. Mag. **42**, 2372 (2006).

[131] M. Oogane, T. Wakitani, S. Yakata, R. Yilgin, Y. Ando, A. Sakuma, and T. Miyazaki, *Magnetic damping in ferromagnetic thin films*, Jpn. J. Appl. Phys. **45**, 3889 (2006).

[132] S.M. Bhagat and P. Lubitz, *Temperature variation of ferromagnetic relax-ation in the 3d transition metals*, Phys. Rev. B **10**, 179 (1974).

[133] B. Heinrich, D.J. Meredith, and J.F. Cochran, *Wave number and temperature-dependent Landau–Lifshitz damping in nickel*, J. Appl. Phys. **50**, 7726 (1979).

[134] D. Bastian and E. Biller, *Damping of ferromagnetic resonance in Ni-Fe alloys*, Phys. Stat. Sol. A **35**, 113 (1976).

[135] J. Xiao, A. Zangwill, and M.D. Stiles, *Spin-transfer torque for continuously variable magnetization*, Phys. Rev. B **73**, 054428 (2006).

[136] J.-P. Adam, N. Vernier, J. Ferré, A. Thiaville, V. Jeudy, A. Lemaître, L. Thevenard, and G. Faini, *Nonadiabatic spin-transfer torque in (Ga,Mn)As with perpendicular anisotropy*, Phys. Rev. B **80**, 193204 (2009).

[137] C. Burrowes, A.P. Mihai, D. Ravelosona, J.-V. Kim, C. Chappert, L. Vila, A. Marty, Y. Samson, F. Garcia-Sanchez, L.D. Buda-Prejbeanu, I. Tudosa, E.E. Fullerton, and J.-P. Attané, *Non-adiabatic spin-torques in narrow magnetic domain walls*, Nature Physics **6**, 17 (2010).

[138] M. Eltschka, M. Wötzel, J. Rhensius, S. Krzyk, U. Nowak, M. Kläui, T. Kasama, R.E. Dunin-Borkowski, L.J. Heyderman, H.J. van Driel, and R.A. Duine, *Nonadiabatic spin torque investigated using thermally*

activated magnetic domain wall dynamics, Phys. Rev. Lett. **105**, 056601 (2010).

[139] M. Hayashi, L. Thomas, C. Rettner, R. Moriya, and S.S.P. Parkin, *Dynamics of domain wall depinning driven by a combination of direct and pulsed currents*, Appl. Phys. Lett. **92**, 162503 (2008).

[140] S. Lepadatu, J.S. Claydon, C.J. Kinane, T.R. Charlton, S. Langridge, A. Potenza, S.S. Dhesi, P.S. Keatley, R.J. Hicken, B.J. Hickey, and C.H. Marrows, *Domain-wall pinning, nonadiabatic spin-transfer torque, and spin-current polarization in Permalloy wires doped with Vanadium*, Phys. Rev. B **81**, 020413 (2010).

[141] S. Lepadatu, A. Vanhaverbeke, D. Atkinson, R. Allenspach, and C.H. Marrows, *Dependence of domain-wall depinning threshold current on pinning profile*, Phys. Rev. Lett. **102**, 127203 (2009).

[142] T.A. Moore, M. Kläui, L. Heyne, P. Möhrke, D. Backes, J. Rhensius, U. Rüdiger, L.J. Heyderman, J.-U. Thiele, G. Woltersdorf, C.H. Back, A. Fraile Rodríguez, F. Nolting, T.O. Mentes, M. Á. Niño, A. Locatelli, A. Potenza, H. Marchetto, S. Cavill, and S.S. Dhesi, *Scaling of spin relaxation and angular momentum dissipation in Permalloy nanowires*, Phys. Rev. B **80**, 132403 (2009).

[143] G. Tatara, H. Kohno, and J. Shibata, *Microscopic approach to current-driven domain wall dynamics*, Phys. Rep. **468**, 213–301 (2008).

[144] A. Ghosh, J.F. Sierra, S. Aufrett, U. Ebels, and W.E. Bailey, *Dependence of nonlocal Gilbert damping on the ferromagnetic layer type in ferromagnet/Cu/Pt heterostructures*, Appl. Phys. Lett. **98**, 052508 (2011).

[145] C.W. Sandweg, Y. Kajiwara, A.C. Chumak, A.A. Serga, V.I. Vasyuchka, M.B. Jungfleisch, E. Saitoh, and B. Hillebrands, *Spin pumping by parametrically excited exchange magnons*, Phys. Rev. Lett. **106**, 216601 (2011).

[146] M. Hayashi, L. Thomas, Ya. B. Bazaliy, C. Rettner, R. Moriya, X. Jiang, and S.S.P. Parkin, *Influence of current on field-driven domain wall motion in permalloy nanowires from time resolved measurements of anisotropic magnetoresistance*, Phys. Rev. Lett. **96**, 197207 (2006).

[147] G. Meier, M. Bolte, R. Eiselt, B. Krüger, D.-H. Kim, and P. Fischer, *Direct imaging of stochastic domain-wall motion driven by nanosecond current pulses*, Phys. Rev. Lett. **98**, 187202 (2007).

[148] L. Thomas, M. Hayashi, X. Jiang, R. Moriya, C. Rettner, and S.S.P. Parkin, *Oscillatory dependence of current-driven magnetic domain wall motion on current pulse length*, Nature **443**, 197 (2006).

[149] L. Heyne, M. Kläui, D. Backes, T.A. Moore, S. Krzyk, U. Rüdiger, L.J. Heyderman, A.F. Rodríguez, F. Nolting, T.O. Mentes, M.Á. Niño, A. Locatelli, K. Kirsch, and R. Mattheis, *Relationship between nonadiabaticity and damping in permalloy studied by current induced spin structure transformations*, Phys. Rev. Lett. **100**, 066603 (2008).

[150] S. Lepadatu, J.S. Claydon, C.J. Kinane, T.R. Charlton, S. Langridge, A. Potenza, S.S. Dhesi, P.S. Keatley, R.J. Hicken, B.J. Hickey, and C.H. Marrows, *Domain-wall pinning, nonadiabatic spin-transfer torque, and spin-current polarization in permalloy wires doped with vanadium,* Phys. Rev. B **81**, 020413(R) (2010).

[151] B. Krüger, M. Najafi, S. Bohlens, R. Frömter, D.P.F. Möller, and D. Pfannkuche, *Proposal of a robust measurement scheme for the nonadiabatic spin torque using the displacement of magnetic vortices,* Phys. Rev. Lett. **104**, 077201 (2010).

[152] L. Heyne, J. Rhensius, D. Ilgaz, A. Bisig, U. Rüdiger, M. Kläui, L. Joly, F. Nolting, L.J. Heyderman, J.U. Thiele, and F. Kronas, *Direct determination of large spin-torque nonadiabaticity in vortex core dynamics,* Phys. rev. Lett. **105**, 187203 (2010).

[153] M. Eltschka, M. Wötzel, J. Rhensius, S. Krzyk, U. Nowak, M. Kläui, T. Kasama, R.E. Dunin-Borkowski, L.J. Heyderman, H.J. van Driel, and R.A. Duine, *Non-adiabatic spin torque investigated using thermally activated magnetic domain wall dynamics,* Phys. Rev. Lett. **105**, 056601 (2010).

[154] C. Burrowes, A.P. Mihai, D. Ravelosona, J.-V. Kim, C. Chappert, L. Vila, A. Marty, Y. Samson, F. Garcia-Sanchez, L.D. Buda-Prejbeanu, I. Tudosa, E.E. Fullerton and J.-P. Attané, *Non-adiabatic spin-torques in narrow magnetic domain walls,* Nature Physics **6**, 17 (2010).

[155] M. Yamanouchi, J. Ieda, F. Matsukura, S.E. Barnes, S. Maekawa, and H. Ohno, *Universality classes for domain wall motion in the ferromagnetic semiconductor,* Science **317**, 1726 (2007).

[156] S.A. Yang, G.S.D. Beach, C. Knutson, D. Xiao, Q. Niu, M. Tsoi, and J.L. Erskine, *Universal electromotive force induced by domain wall motion,* Phys. Rev. Lett. **102**, 067201 (2009).

9 Spin caloritronics

G. E. W. Bauer

9.1 Introduction

The coupling between spin and charge transport in condensed matter is studied in the lively field referred to as spintronics. Heat currents are coupled to both charge and spin currents [1, 2]. "Spin caloritronics" is the field combining thermoelectrics with spintronics and nanomagnetism, which recently enjoys renewed attention [3]. The term "caloritronics" (from *calor*, the Latin word for heat) has recently been introduced to describe the endeavor to control heat transport on micro- and nanometer scales. Alternative expressions such as "(mesoscopic) heattronics" or "caloric transport" have also been suggested. Specifically, spin caloritronics is concerned with new physics related to spin, charge, and entropy/energy transport in materials and nanoscale structures and devices. Examples are spin dependence of thermal conductance, the Seebeck and Peltier effects, heat current effects on spin-transfer torque, thermal spin, and anomalous Hall effects, etc. Heat and spin effects are also coupled by the dissipation and noise associated with magnetization dynamics.

The societal relevance of the topic is given by the imminent breakdown of Moore's law by the thermodynamic bottleneck: further decrease in feature size and transistor speed goes in parallel with intolerable levels of ohmic energy dissipation associated with the motion of electrons in conducting circuits. Thermoelectric effects in meso- [4] and nanoscopic [5] structures might help in managing the generated heat. Spin caloritronics is intimately related to possible solutions to these problems by making use of the electron spin degree of freedom.

Spin caloritronics is as old as spin electronics, starting in the late 1980s with M. Johnson and R.H. Silsbee's [1] visionary theoretical insights into the nonequilibrium thermodynamics of spin, charge and heat in metallic heterostructures with collinear magnetization configurations. Except for a few experimental studies on the thermoelectric properties of magnetic multilayers in the CIP (currents in the interface plane) configuration [6] in the wake of the discovery of giant magnetoresistance, the field remained dormant for many years. The Lausanne group started systematic experimental work on what we now call spin caloritronics in magnetic multilayer nanowires and further developed the theory [7].

Several new and partly unpublished discoveries in the field of spin caloritronics excite the community, such as the spin (wave) Seebeck effect in and signal transmission through magnetic insulators, the spin-dependent Seebeck effect in magnetic nanostructures, the magnonic thermal Hall effect, giant Peltier effect in constantan/gold nanopillars, and the thermal spin-transfer torque. After a brief introduction to the basics of how the spin affects classical thermoelectric phenomena, these topics will appear in the following sections.

9.2 Basic physics

We learn from textbooks that the electron–hole asymmetry at the Fermi energy in metals generates thermoelectric phenomena. A heat current $\dot{\mathbf{Q}}$ then drags charges with it, thereby generating a thermopower voltage or charge current $\mathbf{J}$ for open or closed circuit conditions, respectively. *Vice versa* a charge current is associated by a heat current, which can be used to heat or cool the reservoirs. In a diffusive bulk metal the relation between the local driving forces, i.e. the voltage gradient or electric field $\mathbf{E} = -\boldsymbol{\nabla}_{\mathbf{r}}V$ and temperature gradient $\boldsymbol{\nabla}_{\mathbf{r}}T$, reads

$$\begin{pmatrix} \mathbf{J} \\ \dot{\mathbf{Q}} \end{pmatrix} = \sigma \begin{pmatrix} 1 & S \\ \Pi & \kappa/\sigma \end{pmatrix} \begin{pmatrix} \boldsymbol{\nabla}_{\mathbf{r}}V \\ -\boldsymbol{\nabla}_{\mathbf{r}}T \end{pmatrix} \tag{9.1}$$

where σ is the electric conductivity, S the Seebeck coefficient, and κ the heat conductivity [8]. The Onsager–Kelvin relation between the Seebeck and Peltier coefficients $\Pi = ST$ is a consequence of Onsager reciprocity [9]. In the Sommerfeld approximation, valid when the conductivity as a function of energy varies linearly on the scale of the thermal energy $k_B T$ or, more precisely, when $\mathcal{L}_0 T^2 \left| \partial_\varepsilon^2 \sigma(\varepsilon) |_{\varepsilon_F} \right| \ll \sigma(\varepsilon_F)$,

$$S = -e\mathcal{L}_0 T \frac{\partial}{\partial \varepsilon} \ln \sigma(\varepsilon)|_{\varepsilon_F}, \tag{9.2}$$

where the Lorenz constant is $\mathcal{L}_0 = (\pi^2/3)(k_B/e)^2$ and $\sigma(\varepsilon)$ is the energy-dependent conductivity around the Fermi energy ε_F. In this regime the Wiedemann–Franz Law

$$\kappa = \sigma \mathcal{L}_0 T \tag{9.3}$$

holds. Thermoelectric phenomena at constrictions and interfaces are obtained by replacing the gradients by differences and the conductivities by conductances.

The spin dependence of the thermoelectric properties in isotropic and monodomain metallic ferromagnets can be expressed in the two-current model of majority and minority spins [1, 7, 12, 13]:

$$\begin{pmatrix} \mathbf{J}_c \\ \mathbf{J}_s \\ \dot{\mathbf{Q}} \end{pmatrix} = \sigma \begin{pmatrix} 1 & P & S \\ P & 1 & P'S \\ ST & P'ST & \mathcal{L}_0 T \end{pmatrix} \begin{pmatrix} \boldsymbol{\nabla}_{\mathbf{r}}\tilde{\mu}_c/e \\ \boldsymbol{\nabla}_{\mathbf{r}}\mu_s/2e \\ -\boldsymbol{\nabla}_{\mathbf{r}}T \end{pmatrix}, \tag{9.4}$$

where $\mathbf{J}_{c(s)} = \mathbf{J}^{(\uparrow)} \pm \mathbf{J}^{(\downarrow)}$ and $\dot{\mathbf{Q}} = \dot{\mathbf{Q}}^{(\uparrow)} + \dot{\mathbf{Q}}^{(\downarrow)}$ are the charge, spin, and heat currents, respectively. P and P' stand for the spin polarization of the conductivity and its energy derivative

$$P = \left. \frac{\sigma^{(\uparrow)} - \sigma^{(\uparrow)}}{\sigma^{(\uparrow)} + \sigma^{(\uparrow)}} \right|_{\varepsilon_F} \; ; \; P' = \left. \frac{\partial_\varepsilon \sigma^{(\uparrow)} - \partial_\varepsilon \sigma^{(\uparrow)}}{\partial_\varepsilon \sigma^{(\uparrow)} + \partial_\varepsilon \sigma^{(\uparrow)}} \right|_{\varepsilon_F} . \tag{9.5}$$

$\tilde{\mu}_c = \left(\mu^{(\uparrow)} + \mu^{(\downarrow)} \right) / 2$ is the charge electrochemical potential and $\mu_s = \mu^{(\uparrow)} - \mu^{(\downarrow)}$ is the difference between the chemical potentials of the two-spin species, i.e. the spin accumulation. The spin-dependent thermal conductivities obey the Wiedemann–Franz law $\kappa^{(\alpha)} \approx \mathcal{L}_0 T \sigma^{(\alpha)}$ when $S^{\uparrow(\downarrow)} \ll \sqrt{\mathcal{L}_0}$ and the total thermal conductivity $\kappa = \kappa^{(\uparrow)} + \kappa^{(\downarrow)} = \mathcal{L}_0 T \sigma$. In Eq. (9.4) the spin heat current $\dot{\mathbf{Q}}_s = \dot{\mathbf{Q}}^{(\uparrow)} - \dot{\mathbf{Q}}^{(\downarrow)}$ does not appear. This is a consequence of the implicit assumption that there is no spin temperature (gradient) $T_s = T^{(\uparrow)} - T^{(\downarrow)}$ due to effective interspin and electron–phonon scattering [12]. This approximation does not necessarily hold at the nanoscale and at low temperatures [14, 15]. Although initial experiments were inconclusive, a lateral spin-valve device has been proposed in which it should be possible to detect spin temperatures.

The above equations presume that the spin projections are good quantum numbers, which is not the case in the presence of noncollinear magnetizations or significant spin–orbit interactions. Both complications give rise to new physics in spintronics, such as the spin Hall effect and current-induced spin-transfer torques. Both have their spin caloritronic equivalents.

Lattice vibrations (phonons) provide a parallel channel for heat currents, as, in magnets, do spin waves (magnons). The study and control of spin waves is referred to as "magnonics" [17]. The coupling of different modes can be very important for thermoelectric phenomena, causing for instance the phonon-drag effect on the thermopower at lower temperatures. The heat current carried by magnons is a spin current and may affect the Seebeck coefficient [18]. In metallic ferromagnets the spin-wave heat current appears to be smaller than the thermoelectric heat current discussed above, but is the dominant mode of spin transport in magnetic insulators [19, 20]. The coupling between magnons and phonons has been recently demonstrated in the spin Seebeck effect (see Section 9.7 and Chapter 18).

9.3 Spin-dependent thermoelectric phenomena in metallic structures

A consequence of the basic physics sketched above is the existence of thermoelectric generalizations of the giant magnetoresistance (GMR), i.e. the modulation of the electric charge and heat currents by the spin configuration of magnetic multilayers, spin valves, and tunneling junctions as well as a family of thermal spin Hall effects.

9.3.1 *Magneto-Peltier and Seebeck effects*

The magneto-Peltier and magneto-Seebeck effects are caused by the spin dependence of the Seebeck/Peltier coefficients in ferromagnets [1, 7, 12]. The magneto-thermopower has been observed in multilayered magnetic nanowires [7]. A large

Peltier effect in constantan (CuNi alloy)/Au [21] has been associated with magnetism in the phase-separation magnetic phase [22].

A magneto-Seebeck effect in lateral spin valves has been demonstrated [23]. Here a temperature gradient is intentionally applied over an intermetallic interface. The spin-dependence of the Seebeck coefficient induce a spin-polarized current into the normal metal, in which Slachter *et al.* [23] detect the accompanying spin accumulation by an analyzing ferromagnetic contact. A corresponding spin-dependent Peltier effect has been reported very recently as well by the same group [76]. Spin-dependent thermopower has been predicted for molecular spin valves from first-principles theory [25]. A magneto-Seebeck effect in magnetic tunnel junctions has been observed [26, 27] and modelled by *ab initio* calculations [28] for MgO barriers. Large magnetothermolectric effects have been observed for Al_2O_3 tunnel junctions as well [29]. A spin-dependent Seebeck effect in Py–Si tunneling junctions has been observed by Le Breton *et al.* [24] by analyzing the magnetic field dephasing (Hanle effect) of a thermally injected spin accumulation. The thermoelectric figure of merit can possibly be improved by employing the conducting edge and surface states of topological insulators [30].

9.3.2   *Thermal Hall effects*

Thermal Hall effects exist in normal metals in the presence of external magnetic fields and can be classified into three groups [31]. The Nernst effect stands for the Hall voltage induced by a heat current. The Ettingshausen effect describes the heat current induced transverse to an applied charge current. The Hall heat current induced by a temperature gradient goes by the name of Righi-Leduc. The spin degree of freedom opens a family of spin caloritronic Hall effects in the absence of an external field which are not yet fully explored. We may add the label spin in order to describe effects in normal metals (spin Hall effect, spin Nernst effect, etc.). In ferromagnets we may distinguish the configuration in which the magnetization is normal to both currents (anomalous Hall effect, anomalous Nernst effect, etc.) from the configuration with in-plane magnetization (planar Hall effect, planar Nernst effect, etc.) as sketched in Fig. 9.1. Theoretical work has been carried out with emphasis on the intrinsic spin–orbit interaction [32–34].

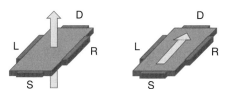

FIG. 9.1. A sketch of the configuration of anomalous (left figure) and planar (right figure) Hall effects in ferromagnets. S and D denote source and drain contacts and L and R left and right Hall contacts. The arrow denotes the magnetization direction.

Seki *et al.* [35] found experimental evidence for a thermal Hall effect in Au–FePt structures, which can be due either to an anomalous Nernst effect in FePt or a spin Nernst effect in Au. In GaMnAs the planar [36] and anomalous [37] Nernst effects have been observed, with intriguing temperature dependences. Slachter *et al.* [38] identified the anomalous Nernst effect and anisotropic magneto-heating in multiterminal permalloy–copper spin valves.

9.4   Thermal spin-transfer torques

A spin current is in general not conserved. In a metal, angular momentum can be dissipated to the lattice by spin-flip scattering. In the presence of a noncollinear magnetic texture, either in a heterostructure, such as a spin valve and tunnel junction, or a magnetization texture such as a domain wall or magnetic vortex, the magnetic condensate also absorbs a spin current, which by conservation of angular momentum leads to a torque on the magnetization that, if strong enough, can lead to coherent magnetization precessions and even magnetization reversal [39]. Just like a charge current, a heat current can exert a torque on the magnetization as well [11], which leads to purely thermally induced magnetization dynamics [40]. Such a torque can be measured under closed circuit conditions, in which part of the torque is simply exerted by the spin-dependent thermopower, and in an open circuit in which a charge current is suppressed [11].

9.4.1   *Spin valves*

The angular dependence of the thermal torque can be computed by circuit theory [11, 12]. Thermal spin-transfer torques have been detected in nanowire spin valves [41]. Slonczewski [48] studied the spin-transfer torque in spin valves in which the polarizer is a magnetic insulator that exerts a torque on a free magnetic layer in the presence of a temperature gradient. He concludes that the thermal torque can be more effective in switching magnetizations than a charge current-induced torque. Note that the physics of heat-current induced spin injection by magnetic insulators is identical to that of the longitudinal spin Seebeck effect as discussed briefly in Section 9.7.

9.4.2   *Magnetic tunnel junctions*

Large thermal torques have been predicted by first-principles calculations for magnetic tunnel junctions with thin barriers that compare favorable with those obtainable by an electric bias [46], but these have as yet not been confirmed experimentally.

9.4.3   *Textures*

Charge current-induced magnetization in magnetic textures has enjoyed a lot of attention in recent years. Domain wall motion can be understood easily in terms of angular momentum conservation in the adiabatic regime, in which the length-scale of the magnetization texture such as the domain wall width is much larger than the scattering mean free path or Fermi wavelength, as appropriate for most transition-metal ferromagnets. In spite of initial controversies, the importance of dissipation in the adiabatic regime [47] is now generally appreciated. In analogy to the Gilbert damping factor α the dissipation under an applied current is governed by a material parameter β_c that for itinerant magnetic materials is of the same order as α [49]. In the case of a heat-current induced domain wall motion, the adiabatic thermal spin transfer torque [11] is also associated with a dissipative β_T-factor that is independent of the charge current β_c [50, 51]. β_T has been explicitly calculated by Hals for GaMnAs [53]. Nonadiabatic corrections to the thermal spin transfer torque in fast-pitch ballistic domain walls have been calculated by first-principles [54]. Laser-induced domain-wall pinning might give clues for heat current effects on domain-wall motion [42].

In insulating ferromagnets, the domain wall can still be moved since part of the heat current is carried by spin waves, and therefore associated with angular momentum currents. In contrast to metals in which the angular momentum current can have either sign relative to the heat current direction, in insulators the magnetization current always flows against the heat current, which means that the adiabatic torque moves the domain wall to the hot region [43–45].

9.5   Magneto-heat resistance

The heat conductance of spin valves is expected to depend on the magnetic configuration, similar to the GMR, giving rise to a giant magneto-heat resistance [11] or a magneto-tunneling heat resistance. In contrast to the GMR, the magneto-heat resistance is very sensitive to inelastic (interspin and electron–phonon) scattering [14, 15].

Inelastic scattering leads to a breakdown of the Wiedemann–Franz law in spin valves. This is most easily demonstrated for half-metallic ferromagnetic contacts as sketched in Fig. 9.2 for a finite temperature bias over the sample. In the figure the distribution functions are sketched for the three spatial regions. Both spins form eigenstates in N, but in F only the majority spin exists. In Fig. 9.2(a) we suppose the absence of inelastic scattering between the spins, either by direct Coulomb interaction or indirect energy exchange via the phonons. When a strong interaction is switched on both spins in N will adopt the same temperature as sketched in Fig. 9.2(b). The temperature gradient on the right interface will induce a heat current, while a charge current is suppressed, clearly violating the Wiedemann–Franz law. A spin heat valve effect can therefore only exist when the interspin and spin-phonon interactions are sufficiently weak.

The heat conductance of tunnel junctions is expected to be less sensitive to inelastic scattering. A useful application for on-chip heat management could be a tunneling heat valve, i.e. a switchable heat sink as illustrated in Fig. 9.3.

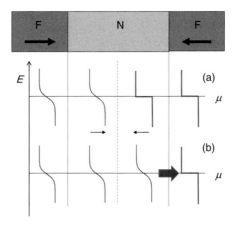

FIG. 9.2. A temperature difference over a spin valve with half-metallic contacts and an antiparallel configuration of the magnetic contacts. Plotted are the electron distribution functions in the ferromagnets and the normal metal spacer (μ is the chemical potential). In (a) the spins in the spacer are noninteracting; in (b) they are strongly interacting, thereby allowing a heat current to flow through the left interface.

FIG. 9.3. The dependence of the heat conductance of a magnetic tunnel junction or spin valve on the magnetic configuration can be used to control possible overheating of a substrate, such as a hot spot in an integrated circuit, when the necessity arises.

9.6 Spin caloritronic heat engines and motors

Onsager's reciprocal relations [9] reveal that seemingly unrelated phenomena can be expressions of identical microscopic correlations between thermodynamic variables of a given system [10]. The archetypal example is the Onsager–Kelvin identity of thermopower and Peltier cooling mentioned earlier. We have seen that spin and charge currents are coupled with each other and with the magnetization. Furthermore, mechanical and magnetic excitations are coupled by the Barnett and Einstein–de Haas effects [55, 56]. The thermoelectric response matrix including all these variables can be readily formulated for a simple model system consisting of a rotatable magnetic wire including a domain wall as sketched in Fig. 9.4. The linear response matrix then reads $\mathbf{J} = \hat{L}\mathbf{X}$, where the generalized currents $\mathbf{J}$ and forces $\mathbf{X}$

$$\mathbf{J} = \left(J_c, J_Q, \dot{\varphi}, \dot{r}_w \right)^T \tag{9.6}$$

$$\mathbf{X} = \left(-\Delta V, -\tfrac{\Delta T}{T}, \tau_{\text{ext}}^{\text{mech}}, -2AM_sH_{\text{ext}} \right)^T \tag{9.7}$$

are related by the response matrix

$$\hat{L} = \begin{pmatrix} L_{cc} & L_{cQ} & L_{c\varphi} & L_{cw} \\ L_{Qc} & L_{QQ} & L_{Q\varphi} & L_{Qw} \\ L_{\varphi c} & L_{\varphi Q} & L_{\varphi\varphi} & L_{\varphi w} \\ L_{wc} & L_{wQ} & L_{w\varphi} & L_{ww} \end{pmatrix}. \tag{9.8}$$

Onsager reciprocity implies that $L_{xy} = \pm L_{yx}$. The elements can be computed by scattering theory [51].

The matrix relation between generalized forces and currents implies a large functionality of magnetic materials. Each of the forces can give rise to all currents, where a temperature gradient is especially relevant here. The response coefficient L_{cQ} clearly represents the Seebeck effect, L_{QQ} the heat conductance, $L_{\varphi Q}$ a thermally driven (Brownian) motor, and L_{wQ} a heat current driven domain wall

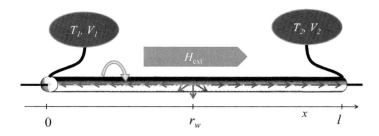

FIG. 9.4. Magnetic nanowire of length l in electrical and thermal contact with reservoirs. A domain wall is centered at position r_w. The wire is mounted such that it can rotate around the x-axis. A magnetic field and mechanical torque can be applied along x.

motion [50]. Onsager symmetry implies that $L_{wQ} = L_{Qw}$ and $L_{\varphi Q} = -L_{Q\varphi}$. For example, a Peltier effect can be expected by moving domain walls [50, 51] and mechanical rotations [51].

9.7 Spin Seebeck effect

The most spectacular development in recent in years in the field of spin caloritronics has been the discovery of the spin Seebeck effect, first in metals [60], and later in electrically insulating yttrium iron garnet (YIG) [61] and ferromagnetic semiconductors (GaMnAs) [62, 63]. The spin Seebeck effect stands for the electromotive force generated by a ferromagnet with a temperature bias over a strip of metal normal to the heat current. This effect is interpreted in terms of a thermally induced spin current injected into the normal metal that is transformed into a measurable voltage by the inverse spin Hall effect [64–66] in metals. A separate chapter of this book is devoted to the spin Seebeck effect, so the present section is kept brief.

It is important to point out the difference between the spin Seebeck effect and the magneto- or spin-dependent Seebeck effect measured by Slachter *et al.* [23] (see Section 9.3.1). Both are generated at an interface between a ferromagnet and a metal. In the magneto-Seebeck effect a temperature gradient is intentionally applied over an intermetallic interface, which is quite different from the spin Seebeck effect, and it can be explained by traditional spin caloritronics concepts [1]. On the other hand, in the spin Seebeck effect the ISHE contact is thermally floating and a standard thermoelectric explanation fails [67] (see, however, [68]).

There is consensus by now that the origin of the spin Seebeck effect is a net spin pumping current over the ferromagnet–metal interface induced by a nonequilibrium magnon distribution [69, 70]. Furthermore, the phonon–magnon drag has been found to be very important [71–73]. In magnetic insulators conventional thermoelectrics cannot be applied. A longitudinal configuration in which a temperature gradient is intentionally applied over the interface [74] can therefore be classified as a spin Seebeck effect. The Slachter experiments [23] might also be affected by the spin Seebeck effect, although the effect is probably overwhelmed by the spin-dependent thermoelectrics.

As mentioned in Section 9.4.1, the physics of the thermal torque induced by heat currents in spin valves with an insulator as polarizing magnet as proposed by Slonczewski [48] is identical to the longitudinal spin Seebeck effect [74], as explained theoretically by Xiao *et al.* [69]. The "loose" magnetic monolayer model hypothesized by Slonczewski appears to mimic the solution of the Landau–Lifshitz–Gilbert equation, which predicts a thin magnetically coherent layer that effectively contributes to the spin pumping. Slonczewski's claim that the heat-current-induced spin transfer torque through magnetic insulators should be large has been confirmed by first-principles calculations that predict that the spin-mixing conductance at the interface between YIG and silver is close to the intermetallic value [75]. This result is in stark contrast to the expectations from a Stoner model for the magnetic insulator [69], but can be explained by local magnetic moments at the interface [75].

From the discussion of the Onsager relations one might expect a spin Peltier effect as well, the observation of which has not yet been reported in public to date, however.

9.8 Conclusions

The field of spin caloritronics has gained momentum in recent years since experimental and theoretical groups have newly joined the community in the last few years. It should be obvious from the above summary that many effects predicted by theory have not yet been observed. The smallness of some effects are also a concern. If spin caloritronics is to become more than a scientific curiosity, the effects should be large enough to become useful. Therefore more materials research and device engineering, experimental and theoretical, is very welcome.

Acknowledgments

I am grateful for most pleasant collaboration on spin caloritronics with Arne Brataas, Xingtao Jia, Moosa Hatami, Tero Heikillä, Paul Kelly, Sadamichi Maekawa, Eiji Saitoh, Saburo Takahashi, Koki Takanashi, Yaroslav Tserkovnyak, Ken-ichi Uchida, Ke Xia, Jiang Xiao, and many others. This work was supported in part by the FOM Foundation, EU-ICT-7 "MACALO", and DFG Priority Programme 1538 "Spin-Caloric Transport".

References

[1] M. Johnson and R.H. Silsbee, Phys. Rev. B **35**, 4959 (1987); Phys. Rev. B **37,** 5326 (1988).

[2] M. Johnson, Solid State Commun. **150**, 543 (2010).

[3] G.E.W. Bauer, A.H. MacDonald, and S. Maekawa, Solid State Commun. **150**, 489 (2010).

[4] F. Giazotto, T.T. Heikkilä, A. Luukanen, A.M. Savin, and J.P. Pekola, Rev. Mod. Phys. **78**, 217 (2006).

[5] Y. Dubi and M. Di Ventra, Rev. Mod. Phys. **83**, 131 (2011).

[6] J. Shi, K. Pettit, E. Kita, S.S.P. Parkin, R. Nakatani, M.B. Salamon, Phys. Rev. B **54**, 15273 (1996) and references therein.

[7] L. Gravier, S. Serrano-Guisan, F. Reuse, and J.-Ph. Ansermet, Phys. Rev. B **73**, 024419 (2006); **73**, 052410 (2006).

[8] N.W. Ashcroft and N.D. Mermin, *Solid State Physics* (Saunders, Philadelphia, 1976).

[9] L. Onsager, Phys. Rev. **37**, 405 (1931).

[10] S.R. de Groot, *Thermodynamics of Irreversible Processes* (Interscience Publishers, New York, 1952).

[11] M. Hatami, G.E.W. Bauer, Q. Zhang, and P.J. Kelly, Phys. Rev. Lett. **99**, 066603 (2007).

[12] M. Hatami, G.E.W. Bauer, Q. Zhang, and P.J. Kelly, Phys. Rev. B **79**, 174426 (2009).

[13] Y. Takezoe, K. Hosono, A. Takeuchi, and G. Tatara, Phys. Rev. B **82**, 094451 (2010).

[14] T.T. Heikkilä, M. Hatami, and G.E.W. Bauer, Phys. Rev. B **81**, 100408 (2010).

[15] T.T. Heikkilä, M. Hatami, and G.E.W. Bauer, Solid State Commun. **150**, 475 (2010).

[16] A. Slachter, F.L. Bakker, and B.J. van Wees, Phys. Rev. B **84**, 020412 (2011).

[17] V.V. Kruglyak, S.O. Demokritov and D. Grundler, J. Phys. D: Appl. Phys. **43**, 26030 (2010); B. Lenk, H. Ulrichs, F. Garbs, and M. Münzenberg, Physics Rep. **507**, 107 (2011).

[18] A.A. Tulapurkar and Y. Suzuki, Solid State Commun. **150**, 489 (2010).

[19] C. Hess, Eur. Phys. J. Special Topics **151**, 73 (2007).

[20] F. Meier and D. Loss, Phys. Rev. Lett. **90**, 167204 (2003).

[21] A. Sugihara, M. Kodzuka, K. Yakushiji, H. Kubota, S. Yuasa, A. Yamamoto, K. Ando, K. Takanashi, T. Ohkubo, K. Hono, and A. Fukushima, Appl. Phys. Expr. **3**, 065204 (2010).

[22] N.D. Vu, K. Sato, and H. Katayama-Yoshida, Appl. Phys. Expr. **4**, 015203 (2011).

[23] A. Slachter, F.L. Bakker, J.P. Adam, and B.J. van Wees, Nature Physics **6**, 879 (2010).

[24] J.-Ch. Le Breton, S. Sharma, H. Saito, S. Yuasa, and R. Jansen, Nature **475**, 82 (2011).

[25] V.V. Maslyuk, S. Achilles, and I. Mertig, Solid State Commun. **150**, 500 (2010).

[26] N. Liebing, S. Serrano-Guisan, K. Rott, G. Reiss, J. Langer, B. Ocker, and H. W. Schumacher, Phys. Rev. Lett. **107**, 177201 (2011).

[27] M. Walter, J. Walowski, V. Zbarsky, M. Münzenberg, M. Schäfers, Daniel Ebke, G. Reiss, A. Thomas, P. Peretzki, M. Seibt, J.S. Moodera, M. Czerner, M. Bachmann, and Ch. Heiliger, Nature Materials **10**, 742 (2011).

[28] M. Czerner, M. Bachmann, and C. Heiliger, Phys. Rev. B **83**, 132405 (2011).

[29] W. Lin, M. Hehn, L. Chaput, B. Negulescu, S. Andrieu, F. Montaigne, S. Mangin, arXiv:1109.3421.

[30] O.A. Tretiakov, Ar. Abanov, S. Murakami, and J. Sinova, Appl. Phys. Lett. **97**, 073108 (2010).

[31] H.B. Callen, Phys. Rev. **73**, 1349 (1948).

[32] S. Onoda, N. Sugimoto, and N. Nagaosa, Phys. Rev. B **77**, 165103 (2008).

[33] Z. Ma, Solid State Commun. **150**, 510 (2010).

[34] X. Liu and X.C. Xie, Solid State Commun. **150**, 471 (2010).

[35] T. Seki, Y. Hasegawa, S. Mitani, S. Takahashi, H. Imamura, S. Maekawa, J. Nitta, K. Takanashi, Nature Materials. **7**, 125 (2008); T. Seki, I. Sugai, Y. Hasegawa, S. Mitani, K. Takanashi, Solid State Commun. **150**, 496 (2010).

[36] Y. Pu, E. Johnston-Halperin, D.D. Awschalom, and J. Shi, Phys. Rev. Lett. **97**, 036601 (2006).

[37] Y. Pu, D. Chiba, F. Matsukura, H. Ohno, and J. Shi, Phys. Rev. Lett. **101**, 117208 (2008).

[38] A. Slachter, F.L. Bakker, and B.J. van Wees, Phys. Rev. B **84**, 020412 (2011).

[39] For a review see: D.C. Ralph and M.D. Stiles, J. Magn. Magn. Mater. **320**, 1190 (2008).

[40] J.E. Wegrowe, Solid State Commun. **150**, 519 (2010).

[41] H. Yu, S. Granville, D.P. Yu, and J.-Ph. Ansermet, Phys. Rev. Lett. **104**, 146601 (2010).

[42] P. Möhrke, J. Rhensius, J.-U. Thiele, L.J. Heyderman, M. Kläui, Solid State Commun. **150**, 489 (2010).

[43] D. Hinzke and U. Nowak, Phys. Rev. Lett. **107**, 027205 (2011).

[44] A.A. Kovalev and Y. Tserkovnyak, Europhys. Lett. **97**, 67002 (2012).

[45] P. Yan, X.S. Wang, and X.R. Wang, Phys. Rev. Lett. **107**(17), 177207 (2011).

[46] X. Jia, K. Liu, K. Xia, and G.E.W. Bauer, Europhys. Lett. **96**, 17005 (2011).

[47] S. Zhang and Z. Li, Phys. Rev. Lett. **93**, 127204 (2004).

[48] J.C. Slonczewski, Phys. Rev. B **82**, 054403 (2010).

[49] For a review see: Y. Tserkovnyak, A. Brataas, and G.E.W. Bauer, J. Magn. Magn. Mater. **320**, 1282 (2008).

[50] A.A. Kovalev and Y. Tserkovnyak, Phys. Rev. B **80**, 100408 (2009).

[51] G.E.W. Bauer, S. Bretzel, A. Brataas, and Y. Tserkovnyak, Phys. Rev. B **81**, 024427 (2010).

[52] A.A. Kovalev and Y. Tserkovnyak, Solid State Commun. **150**, 500 (2010).

[53] K.M.D. Hals, A. Brataas, and G.E.W. Bauer, Solid State Commun. **150**, 461 (2010).

[54] Z. Yuan, S. Wang, and K. Xia, Solid State Commun. **150**, 548 (2010).

[55] S.J. Barnett, Phys. Rev. **6**, 239 (1915); Rev. Mod. Phys. **7**, 129 (1935).

[56] A. Einstein and W. J. de Haas, Deutsche Physikalische Gesellschaft, Verhandlungen **17**, 152 (1915).

[57] S. Serrano-Guisan, G. di Domenicantonio, M. Abid, J. P. Abid, M. Hillenkamp, L. Gravier, J.-P. Ansermet, and C. Félix, Nature Materials **5**, 730 (2006).

[58] O. Tsyplyatyev, O. Kashuba, and V.I. Fal'ko, Phys. Rev. B **74**, 132403 (2006).

[59] A. Brataas, G.E.W. Bauer, and P.J. Kelly, Phys. Rep. **427**, 157 (2006).

[60] K. Uchida, S. Takahashi, K. Harii, J. Ieda, W. Koshibae, K. Ando, S. Maekawa, and E. Saitoh, Nature **455**, 778 (2008); K. Uchida, T. Ota, K. Harii, S. Takahashi, S. Maekawa, Y. Fujikawa, E. Saitoh, Solid State Commun. **150**, 524 (2010).

[61] K. Uchida, J. Xiao, H. Adachi, J. Ohe, S. Takahashi, J. Ieda, T. Ota, Y. Kajiwara, H. Umezawa, H. Kawai, G.E.W. Bauer, S. Maekawa and E. Saitoh, Nature Materials **9**, 894 (2010).

[62] C.M. Jaworski, J. Yang, S. Mack, D.D. Awschalom, J.P. Heremans, and R.C. Myers, Nature Materials **9**, 898 (2010).

[63] S. Bosu, Y. Sakuraba, K. Uchida, K. Saito, T. Ota, E. Saitoh, and K. Takanashi, Phys. Rev. B **83**, 224401 (2011).

[64] E. Saitoh, M. Ueda, H. Miyajima, and G. Tatara, Appl. Phys. Lett. **88**, 182509 (2006).

[65] S.O. Valenzuela and M. Tinkham, Nature **442**, 176 (2006).

[66] T. Kimura, Y. Otani, T. Sato, S. Takahashi, and S. Maekawa, Phys. Rev. Lett. **98**, 156601 (2007).

[67] M. Hatami, G.E.W. Bauer, S. Takahashi, and S. Maekawa, Solid State Commun. **150**, 480 (2010).

[68] T.S. Nunner and F. von Oppen, Phys. Rev. B **84**, 020405 (2011).

[69] J. Xiao, G.E.W. Bauer, K. Uchida, E. Saitoh, and S. Maekawa, Phys. Rev. B **81**, 214418 (2010).

[70] H. Adachi, J. Ohe, S. Takahashi, and S. Maekawa, Phys. Rev. B **83**, 094410 (2011).

[71] H. Adachi, K. Uchida, E. Saitoh, J. Ohe, S. Takahashi, and S. Maekawa, Appl. Phys. Lett. **97**, 252506 (2010).

[72] C.M. Jaworski, J. Yang, S. Mack, D.D. Awschalom, R.C. Myers, and J.P. Heremans, Phys. Rev. Lett. **106**, 186601 (2011).

[73] K. Uchida, H. Adachi, T. An, T. Ota, M. Toda, B. Hillebrands, S. Maekawa, and E. Saitoh, Nature Materials **10**, 737 (2011).

[74] K. Uchida, T. Nonaka, T. Ota, and E. Saitoh, Appl. Phys. Lett. **97**, 172505 (2010).

[75] X. Jia, K. Liu, K. Xia, G.E.W. Bauer, Europhys. Lett. **96**, 17005 (2011).

[76] J. Flipse, F.L. Bakker, A. Slachter, F.K. Dejene, B.J. van Wees, Nature Nanotechnology **7**, 166 (2012).

10 Multiferroics

N. Nagaosa

10.1 Introduction

Recent developments in the physics of multiferroics are discussed from the viewpoint of the spin current and "emergent electromagnetism" for constrained systems. Starting from the Dirac equation, the projection of the wavefunctions onto the low-energy subspaces leads to a gauge structure analogous to electromagnetism. When the SU(2) spin space is preserved, it leads to a SU(2) non-abelian gauge field which is coupled to the spin current, corresponding to the spin–orbit interaction (SOI). When the wavefunctions are further projected onto one of the spin states assuming the magnetic ordering, the gauge field becomes a U(1) abelian gauge field similar to the electromagnetic field (emf). Therefore, there are three sources of U(1) gauge fields, i.e. (i) the Berry phase associated with the noncollinear spin structure, (ii) the spin-orbit interaction (SOI), and (iii) the usual emf. These three fields interact with each other, and lead to a variety of nontrivial phenomena in solids. In this chapter, we review multiferroic phenomena in noncollinear magnets from this viewpoint. Theories of multiferroic behavior of cycloidal helimagnets are discussed in terms of the spin current or vector spin chirality. Relativistic SOI leads to a coupling between the spin current and the electric polarization, and hence the ferroelectric and dielectric responses are a new and important probe for the spin states and their dynamical properties. Microscopic theories of the ground state polarization for various electronic configurations, collective modes including the electromagnon, and some predictions including photoinduced chirality switching are discussed with comparison to experimental results.

The current is one of the most important concepts in physics. It can carry physical quantities and information, and the conservation law for the charge and electric current is fundamental to all electromagnetic phenomena. The basic theory describing electrons coupled to an electromagnetic field (emf) is quantum electrodynamics (QED), where the Dirac relativistic electrons and their charge current are minimally coupled to the emf. Therefore, it appears that there is no chance of the spin current playing role in a electromagnetic phenomena. More explicitly, in natural units where $\hbar = c = 1$, the QED Lagrangian reads [1]

$$L = \bar{\psi}^{\dagger}[i\gamma^{\mu}\hat{D}_{\mu} - m]\psi \tag{10.1}$$

where ψ is the four-component spinor field operator, $\bar{\psi} = \psi^\dagger \gamma^0$, γ^μ are the Dirac matrices, and $\hat{D}_\mu = \partial_\mu - ieA_\mu$ is the gauge covariant derivative with $\mu = 0, 1, 2, 3$. The four-component charge current density is defined as

$$j^\mu = -\frac{\partial L}{\partial A_\mu} = -e\bar{\psi}\gamma^\mu\psi \tag{10.2}$$

whose zero-component is the charge density ρ, while the spatial components are the current density $\vec{j}$. From gauge invariance, the conservation law of the charge is derived through Noether's theorem as

$$\partial_\mu j^\mu = \frac{\partial \rho}{\partial t} + \nabla \cdot \vec{j} = 0. \tag{10.3}$$

By taking the variation, one can derive the Maxwell equation

$$\partial_\mu F^{\mu\nu} = j^\nu \tag{10.4}$$

and Dirac equation

$$[i\gamma^\mu \hat{D}_\mu - m]\psi = 0. \tag{10.5}$$

As is well known, the solutions to the Dirac equation are classified into two classes, i.e. the positive energy and negative energy states separated by twice the rest-mass energy of the electrons, $2mc^2$. Since the energy mc^2 is of the order of MeV, for low-energy phenomena typically of the order of $\sim eV$, the negative energy states are not relevant. Therefore, the nonrelativistic Schrödinger equation is usually used, which describes the dynamics of the two-component spinor wavefunctions for positive energy states. However, one needs to take into account one important aspect of "projection". Namely, the neglect of the negative energy states means the projection of the wavefunctions onto the positive energy states, i.e. sub-Hilbert space. Usually the subspace is not flat but curved, and associated geometrical structure is introduced. The derivation of the effective Lagrangian describing the low-energy physics is achieved by expansion with respect to $1/(mc^2)$, and the result reads [2, 3]

$$L = i\psi^\dagger D_0\psi + \psi^\dagger \frac{\vec{D}^2}{2m}\psi + \frac{1}{2m}\psi^\dagger\left[eq\tau^a\vec{A}\cdot\vec{A}^a + \frac{q^2}{4}\vec{A}^a\cdot\vec{A}^a\right]\psi \tag{10.6}$$

where ψ is now a two-component spinor, $D_0 = \partial_0 + ieA_0 + iqA_0^a\frac{\tau^a}{2}$, and $D_i = \partial_0 - ieA_i - iqA_i^a\frac{\tau^a}{2}$ ($i = 1, 2$) are the gauge covariant derivatives with q being the quantity proportional to the Bohr magneton [2, 3]. A_μ is the vector potential for the emf, while the SU(2) gauge potential is defined as $A_0^a = B_a$, $A_i^a = \epsilon_{ia\ell}E_\ell$. The former is coupled to the charge current, and the latter to the four-component spin current $j_0^a = \psi\sigma^a\psi$, $j_i^a = \frac{1}{2mi}[\psi^\dagger\sigma^a D_i\psi - D_i\psi^\dagger\sigma^a\psi]$. Note that the spin current is a tensor quantity with one suffix for the direction of the spin polarization while the other is for the direction of the flow. Note, however, an important difference between the emf and the SU(2) gauge field. The former has gauge symmetry, i.e. the freedom to choose the arbitrary gauge for the vector potential A_μ, while the "vector potential" A_μ^a for the latter is given by the physical field strength $\vec{B}$ and $\vec{E}$. Actually, the relation $\partial^\mu A_\mu^a = 0$ holds. Therefore, the SU(2) gauge symmetry is absent. This is the basic reason why the spin is not conserved in the presence of the relativistic SOI. (Note that SU(2) gauge theory is a nonlinear theory and the gauge field is "charged," and the sum of the spin current by the matter field and the gauge field is conserved in the non-abelian gauge theory as Yang and Mills first showed [1].) Instead, the spin current is "covariantly" conserved and satisfies [2, 3]

$$D_0 J_0^a + \vec{D} \cdot \vec{J}^a = 0. \tag{10.7}$$

This means that in the co-moving frame the spin is conserved while in the laboratory frame the spin source or sink appears when the electron forms a loop and comes back to the same position in space since the frame has changed. Zaanen *et al.* [3] studied the physical meaning of this conservation law by separating the spin current into two parts, i.e. the coherent part and the noncoherent part. The former is associated with order such as magnetism or superfluidity, and recovers its conservation law thanks to the single-valueness of the order parameter, and is irrelevant to spin accumulation. The noncoherent part, on the other hand, is associated with particle transport, and does not give any "soft" modes. The spin current discussed in this chapter corresponds to the former, associated with the noncollinear spin structure, while spin-current transport such as the spin Hall effect (SHE) [4] is due to the latter, as described briefly below.

Even though the usual conservation law for the spin current is absent, the coupling between A_μ^a and j_μ^a leads to several interesting phenomena. For example, it is suggested that the electric field drives the spin current perpendicular to it, i.e.

$$j_i^a \propto \epsilon_{ia\ell} E_\ell, \tag{10.8}$$

This is the simplest form of the spin Hall effect (SHE) [4] even though the band structure and disorder effect are important in discussing the SHE in real materials. On the other hand, the electric polarization $\vec{P}$ is given by the derivative of the Lagrangian with respect to $\vec{E}$, i.e.

$$P_i \propto \epsilon_{ia\ell} j_\ell^a, \tag{10.9}$$

which means that the spin current produces the ferroelectric moment.

When magnetic ordering occurs, the wavefunctions are further projected on to the spin component at each site. In the continuum approximation,

$$\psi_\sigma = z_\sigma f \tag{10.10}$$

with f being the spinless fermion corresponding to the charge degrees of freedom. This leads to the three "electromagnetic fields" in magnetic systems, i.e. (i) the Berry phase associated with the noncollinear spin structure, (ii) the spin-orbit interaction (SOI), and (iii) the usual Maxwell emf as described below. We call these U(1) gauge fields "emergent electromagnetism."

Putting Eq. (10.10) into Eq. (10.6), we obtain the effective Lagrangian for the f field as

$$L_{\text{eff.}} = f^{\dagger}\left[i\partial_0 + a_0^B + a_0^{SO} + A_0 + \frac{(\vec{\nabla} + i\vec{a}^B + i\vec{a}^{SO} + ie\vec{A})^2}{2m}\right]f \qquad (10.11)$$

where $a_\mu^B = i\langle z|\partial_\mu|z\rangle$ is the U(1) field originating from the Berry connection of the spin wavefunctions, and $a_\alpha^{SO} = A_\alpha^a\langle z|\tau^a\partial_\alpha|z\rangle$, $a_0^{SO} = A_0^a\langle z|\tau^a|z\rangle$ are the U(1) field coming from the SOI. These three U(1) gauge fields and their interplay describe a variety of novel phenomena in magnets as listed below.

(i) The spin chirality induced anomalous Hall effect where $\vec{b} = \nabla \times \vec{a}^B$ is produced by the noncoplanar spin and induces the Chern–Simon term $\propto \varepsilon_{\mu\nu\lambda}A_\mu\partial_\nu A_\lambda$ for the Maxwell emf [5].

(ii) The U(1) gauge field of SOI can lead to a fictitious magnetic field which cancels within the unit cell of the crystal, but gives rise to the distribution of the Berry curvature in momentum space, leading to the anomalous Hall effect. Especially near the band (anti)crossing structures, the Berry curvature is enhanced giving the dominant contribution to the Hall conductivity [5].

(iii) The Dzyaloshinskii–Moriya (DM) spin–orbit interaction leads to the SU(2) gauge field and hence $\vec{a}^{SO}$ in the CP^1 representation, which under a magnetic field produces the Skyrmion lattice structure with the effective magnetic field $\nabla \times \vec{a}^B$, which supports the topological Hall effect [6, 7]. This is an example where three gauge fields are entangled with each other.

(iv) The electromotive force due to spin. The time dependence of the spin Berry phase $\vec{b}^B = \nabla \times \vec{a}^B$ leads to the effective electric field $\vec{e}^B$ and hence the voltage drop $\vec{E}$ in metallic magnets [8–10].

These are just a few examples and many more are unexplored. In the following, we study the physical consequences of this emergent electromagnetism in insulating magnetic systems. In an insulator, there is no transport current [11]. However, the current and spin current have a non-dissipative nature, which leads to various interesting phenomena. Multiferroics is the most representative arena from this viewpoint, and is discussed in most of the sections below. In the last section, we also mention other possible systems for emergent electromagnetism.

10.2 Multiferroics—a generic consideration

The close relation between the electric and magnetic fields has the essence of electromagnetism described by the Maxwell equations. Namely, the electric and magnetic fields are two sides of the single field A_μ (vector potential). In solids, the charge and spin of electrons determine the electric and magnetic properties. More explicitly, the electromagnetic responses are described by the function $K_{\mu\nu}(q, \omega)$ ($\mu, \nu = 0, 1, 2, 3$, q: momentum; ω: frequency) which relates the current J_μ to the external electromagnetic field A_ν as $J_\mu = K_{\mu\nu} A_\nu$. The Onsager reciprocal theorem gives the constraint that

$$K_{\mu\nu}(q, \omega, B) = K_{\nu\mu}(-q, \omega, -B) \tag{10.12}$$

where B is the magnetic field representing time-reversal symmetry breaking and can be replaced by the magnetization M [12].

The SOI in the previous section is written in the case of a spherically symmetry potential as

$$H_{SO} = \lambda \vec{\ell} \cdot \vec{s} \tag{10.13}$$

with $\vec{\ell}$ being the orbital angular momentum $\vec{\ell} = \vec{r} \times \vec{p}$, and λ is the spin–orbit interaction strength, and is proportional to Z^4 with Z being the atomic number. Compared with free electrons in a vacuum, the strength of the relativistic SOI can be enhanced by a factor of $\sim 10^6$ which is the ratio of the rest mass of the electrons, mc^2, and the band gap. For $3d$ electrons in transition metal atoms, λ is typically of the order of $\sim$20–50 meV, while it becomes $\sim$ 0.5 eV for $5d$ electrons. The electron correlation energy, on the other hand, decreases from $3d$ to $5d$ since the wavefunction is more expanded for $5d$ electrons. In the cubic crystal field in transition metal oxides, the five-fold degeneracy of d orbitals is lifted due to the ligand field of oxygen atoms. As a result, three-fold degenerate t_{2g} orbitals (xy, yz, zx orbitals) with lower energy, and doubly degenerate e_g orbitals ($x^2 - y^2, 3z^2 - r^2$ orbitals) with higher energy are formed. The matrix elements of the orbital angular momentum $\vec{\ell}$ are zero within the e_g orbitals. On the other hand, they are nonzero among the t_{2g} orbitals and also between the e_g and t_{2g} orbitals. This SOI is the origin of the relativistic coupling between magnetism and electric polarization.

The linear magneto-electric (ME) effect is given by the formula [13, 14]

$$P = \alpha H$$

$$M = \alpha^t E \tag{10.14}$$

where α is the ME tensor and α^t is its transpose. This relation can be derived from the term $-\alpha_{ij} E_i H_j$ in the free energy F, $P_i = -\partial F/\partial E_i$, and $M_j = -\partial F/\partial H_j$. For this term to be present, the symmetries of time reversal T and space-inversion I need to be broken because P (M) is T-even and I-odd (T-odd and I-even), while the free energy should be even for both symmetries. The I-symmetry breaking in insulators is naturally accompanied by ferroelectricity, while the T-symmetry is associated with magnetism. Therefore, the coexistence of both orders, i.e. multiferroics, is most relevant to the giant ME effect [15].

However, the coexistence of ferroelectric and magnetic orders has been considered to be difficult, because magnetism requires partially filled d-orbitals while ferroelectricity was assumed to be driven by completely filled or empty d-orbitals or lone-pair electrons. Even if both orders coexist, usually they are almost decoupled from each other with separate transition temperatures.

This situation has changed since the discovery of multiferroic behavior in $RMnO_3(R = \text{Gd,Tb,Dy})$ [16]. In this material, the spontaneous electric polarization P_s is induced by the magnetic order and they are necessarily strongly coupled. In $RMnO_3$, there are two successive magnetic phase transitions, and P_s appears only below the second one [16–18]. This fact suggests that a particular type of magnetic order is responsible for ferroelectricity. From this viewpoint, Eq. (10.9) gives a clue, i.e. the spin current associated with the magnetic order induces the electric polarization. Based on this idea, the spin-current model of ferroelectricity has been theoretically developed as described in the next section.

10.3   Spin-current model of ferroelectricity

Let us start with a schematic explanation why the spin current is related to the electric polarization. We base our discussion on duality in electromagnetism. It is well known that a charge current produces a circulating magnetic field around it. Two slightly shifted opposite charges produce an electric dipole, and its motion produces a magnetic field, which is obtained by superimposing the two magnetic fields on the two charges, perpendicular to both the direction of the motion and the electric polarization. By duality, we can replace the charge by a magnetic charge (monopole) and the magnetic field by an electric field. Although there are no magnetic monopoles in nature, a magnetic dipole exists and its motion is nothing but the spin current. Therefore, the spin current is expected to produce the electric polarization as described in Eq. (10.9). Then the next question is how the spin current flows in magnets. As discussed in Section 10.1 the spin current is classified into two categories, i.e. coherent and non-coherent. We are interested in the equilibrium state, and hence only the coherent part is possible.

The key idea is that the quantum nature of the spin operator leads to a spin current for noncollinear spin structures. The commutation relation of the spin components

$$[S^\alpha, S^\beta] = i\hbar\varepsilon_{\alpha\beta\gamma}S^\gamma \qquad (10.15)$$

is translated into

$$[S^z, S^\pm] = \pm i\hbar S^\pm, \qquad (10.16)$$

where $S^\pm = S^x \pm iS^y$. Let us define the "phase" θ relative to the xy-component of the spin as $S^\pm \sim e^{\pm i\theta}$. Then the commutation relation (10.16) can be translated to

$$[S^z, \theta] = i\hbar. \qquad (10.17)$$

This is analogous to the relation between the particle number n and the phase φ of a bosonic field operator, and a magnetically ordered state, i.e. the fixed θ state, corresponds to a superfluid of spin current. Therefore, the spatial gradient of the phase $\nabla\theta$ leads to a super-spin-current. Combining this with Eq. (10.9), one concludes that the electric polarization $\vec{P}$ is given by

$$\vec{P} = \eta\vec{e}_{ij} \times (\vec{S}_i \times \vec{S}_j), \qquad (10.18)$$

where η is a coupling constant proportional to the SOI [19]. This is the spin–current mechanism of electric polarization.

To embody this schematic consideration, the cluster model of magnetic ions sandwiching an oxygen ion has been studied theoretically by taking into account the SOI when deriving the superexchange interaction [19]. Since this theory has already been reviewed in several articles [20], we only quote its final results. As mentioned above, the spin current flows between the two noncollinear spins $\vec{S}_i$ and $\vec{S}_j$, which produces an electric polarization $\vec{P}$ given by

$$\vec{P} \cong -\frac{4e}{9}\left(\frac{V}{\Delta}\right)^3 I\vec{e}_{12} \times (\vec{e}_1 \times \vec{e}_2). \qquad (10.19)$$

where $I = \langle p_x|z|d_{zx}\rangle$, $\vec{e}_{12}$ is a unit vector connecting the two magnetic ions, and Δ (V) is the energy difference (hybridization) between the p orbitals and the d orbitals. The SOI is implicitly included in this model by picking up one doublet after splitting by the SOI. Applying this result to various magnetic structures, one can easily predict the presence or absence, and the direction, of the polarization. This theory does not contradict the symmetry argument developed for magnets [12], but stresses the physical mechanism of the spin-current-induced polarization. One needs to be careful that this is not the only mechanism of the magnetic origin electric polarization as will be discussed below.

After the present theory was published, it was revealed that the magnetic structure is cycloidal and the above scenario has been established in RMnO$_3$ [21–24]. On the theory side, there are some other works related to this spin-current model. Mostovoy [25] wrote down the form of the free energy for the spin-current mechanism, and discussed furthermore the charge accumulation $\nabla \cdot \vec{P}$ due to the spin texture such as the vortex. Also a theory taking into account the atomic displacements has been developed [26]. A detailed group-theoretical analysis of multiferroics can be found in [27].

After this spin-current model succeeded in explaining multiferroic behavior in RMnO$_3$, extensive experimental studies have been done to look for other systems, and Ni$_3$V$_2$O$_8$ [28], Ba$_{0.5}$Sr$_{1.5}$Zn$_2$Fe$_{12}$O$_{22}$ [29], CoCr$_2$O$_4$ [30], MnWO$_4$ [31], CuFeO$_2$[32], LiCuVO$_4$ [33], and LiCu$_2$O$_2$ [34] were discovered to be multiferroics. Multiferroics is not a special phenomenon but is a rather universal phenomenon in insulating correlated electrons. These experimental findings urged systematic theoretical studies of the microscopic mechanisms of spin-related electric polarization.

For this purpose, we have considered the more general case of the cluster model taking into account the five d orbitals, and also other possible origins of the electric polarization [35, 36]. The perturbative approach in both V/Δ and λ/Δ is employed, where V and Δ represent the transfer integral and the charge transfer energy between the transition metal (TM) d and ligand (L) p orbitals. The SOI at the ligand oxygen site is also considered. Therefore, the electric polarization due to the SOI is proportional to λ in first-order perturbation, which is more realistic because $\lambda \sim 20$ meV is smaller than the energy denominators such as Δ, which are of the order of a fraction of an eV at least.

This analysis concludes that the polarization $\vec{P}_{\vec{r}+\frac{\vec{e}}{2}}$ appearing at the bond between the sites $\vec{r}$ and $\vec{r} + \vec{e}$ is given by

$$\vec{P}_{\vec{r}+\frac{\vec{e}}{2}} = P^{\mathrm{ms}}(\vec{m}_{\vec{r}} \cdot \vec{m}_{\vec{r}+\vec{e}})\vec{e} + P^{\mathrm{sp}}\vec{e} \times (\vec{m}_{\vec{r}} \times \vec{m}_{\vec{r}+\vec{e}})$$

$$+P^{\mathrm{orb}}\left[(\vec{e} \cdot \vec{m}_{\vec{r}})\vec{m}_{\vec{r}} - (\vec{e} \cdot \vec{m}_{\vec{r}+\vec{e}})\vec{m}_{\vec{r}+\vec{e}}\right], \qquad (10.20)$$

where $\vec{m}_{\vec{r}}$ is the spin direction at $\vec{r}$. The first term $P^{\mathrm{ms}} \propto (V/\Delta)^3$ is the polarization due to magnetostriction, which is nonzero when the inversion symmetry between $\vec{r}$ and $\vec{r} + \vec{e}$ is absent because the two intermediate states of doubly occupied d-orbitals becomes inequivalent. This term does not require the SOI, and hence is considered to be larger than the rest of the terms if it exists. The second term $P^{\mathrm{sp}} \propto (\lambda/\Delta)(V/\Delta)^3$ is due to the spin-current mechanism already discussed. The third term $P^{\mathrm{orb}} \sim \min(\lambda/V, 1)(V/\Delta)$ is nonzero for the partially filled t_{2g} orbitals and comes from the modification of the single-spin anisotropy due to the electric field [35, 36]. These three contributions appear differently depending on the wavevector of the polarization. Therefore, experiments with momentum resolution such as X-ray and neutron scattering can contribute to the identification of the microscopic mechanism of the electric polarization.

A recent development is the microscopic studies of the polarization $\vec{P}_s$ in multiferroic materials by first-principles band structure calculations [37, 38]. It is well known that the electric polarization is related to the Berry phase of the Bloch wavefunctions [39, 40], which enabled the estimation of $\vec{P}_s$ even for periodic boundary conditions. By applying this method to TbMnO$_3$ with LDA+U, two groups examined the origin of the polarization. The obtained conclusions are:

(i) the calculated polarization with atomic displacements is an order of magnitude larger than the purely electronic one without atomic displacements;

(ii) the direction of the polarization is in accordance with the prediction of the spin-current model [19], but its sign depends on the details of the electronic states; and

(iii) the atomic displacements are determined by various mechanisms and not only by the Dzyaloshinski–Moriya (DM) interaction. Therefore, the identification of the microscopic mechanisms requires a detailed analysis for each material, but it also turns out that the spin-current model captures the qualitative features of the polarization. On the other hand, the other approach to this problem is to consider the effective spin Hamiltonian which is consistent with the phase diagram and electric polarization value observed experimentally. In the next section, we pursue this direction for RMnO$_3$. Note here that the third term in Eq. (10.20) is thought to be the origin of the multiferroic polarization in delafossite compounds Cu(Fe,Al)O$_2$ [41].

10.4 Spin Hamiltonian for RMnO$_3$

Although the spin-current model has been successful in explaining the various features of the multiferroic behavior of RMnO$_3$ as described above, a quantitative understanding is desired as the next step. For that purpose, we have constructed a realistic spin Hamiltonian for RMnO$_3$ including the spin–phonon coupling, and reproduced the entire phase diagram in the plane of the temperature and Mn-O-Mn bond angle. This offers the basis of an electromagnon spectrum in the following section.

In RMnO$_3$, the nearest neighbor spin exchange interaction is rather small ($\sim$1 meV) compared with that in other perovskite compounds (e.g. $\sim$15 meV in LaTiO$_3$) because of the cancellation among various contributions from t_{2g} and e_g orbitals [42]. Therefore, the next-neighbor antiferromagnetic (AF) coupling becomes comparable to the nearest neighbor ferromagnetic (FM) coupling, which leads to various competing phases including multiferroic phases with nontrivial spin structures and the ferroelectric polarization $\vec{P}$ [15, 16]. Also the magnetic-field-induced $\vec{P}$ flops [16, 43], and colossal magnetocapacitance [43–45] has been experimentally observed.

The electronic configuration of Mn^{3+} is d^4 with three electrons in t_{2g} orbitals and one electron in e_g orbitals, whose spins are aligned parallel due to Hund's coupling forming the spin $S = 2$. We treat this spin as classical vectors on a cubic lattice [42, 46], and the spin Hamiltonian reads

$$\mathcal{H} = \sum_{\langle i,j \rangle} J_{ij} \vec{S}_i \cdot \vec{S}_j + D \sum_i S_{\zeta_i}^2$$
$$+ E \sum_i (-1)^{i_x + i_y} (S_{\xi_i}^2 - S_{\eta_i}^2)$$
$$+ \sum_{\langle i,j \rangle} \vec{d}_{ij} \cdot (\vec{S}_i \times \vec{S}_j) + K \sum_i (\delta_{i,i+\hat{x}}^2 + \delta_{i,i+\hat{y}}^2), \quad (10.21)$$

where i_x, i_y, i_z represent integer coordinates of the i-th Mn ion with respect to the cubic x, y and z axes.

Figure 10.1(a) schematically shows the location of each interaction. The first term in Eq. (10.21) is the spin-exchange interactions, while the second and third terms represent the single-ion anisotropies. For the local axes ξ_i, η_i and ζ_i

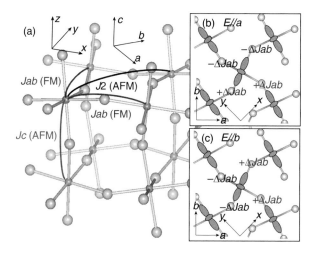

FIG. 10.1. (a) Superexchange interactions in RMnO$_3$. See Eq. (10.21) in the text. The small spheres are Mn atoms, while large spheres are O atoms. (b) Modulations of the in-plane nearest neighbor ferromagnetic exchanges under $\vec{E} \| a$. (c) Those under $\vec{E} \| b$. Here FM and AFM denote ferromagnetic and antiferromagnetic exchanges. (Reproduced from Ref. [46].)

attached to the MnO$_6$ octahedron, we use the structural data of DyMnO$_3$ [47]. The fourth term denotes the DM interaction with DM vectors $\vec{d}_{ij}$ being given by five DM parameters, α_{ab}, β_{ab}, γ_{ab}, α_c, and β_c [48]. The last term represents the elastic energy of the lattice with K being the elastic constant. Here $\delta_{i,j}$ is the shift of the oxygen ion between the i-th and j-th Mn ions normalized by the MnO bond length, which modulates the in-plane exchange coupling as $J_{ij} = J_{ab} + J'_{ab}\delta_{i,j}$ leading to the spin–phonon interaction ($J'_{ab} = \partial J_{ab}/\partial \delta$).

The values of J_{ab}, J_c, J_b, D, E, and the five DM parameters have been microscopically determined in Ref. [42] for several RMnO$_3$ compounds. Except for J_b, they are almost unchanged upon variation of the R-site in the vicinity of the multiferroic phases. We take $J_{ab} = -0.8, J_c = 1.25, D = 0.2, E = 0.25$, $(\alpha_{ab}, \beta_{ab}, \gamma_{ab}) = (0.1, 0.1, 0.14)$, and $(\alpha_c, \beta_c) = (0.42, 0.1)$ in energy unit of meV. We also found that very weak FM exchange J_a is required to realize the E phase, and adopt $J_a = -0.1$. The value of K is chosen to reproduce the experimental P in the E phase as described below (see Fig. 10.3 a), which mostly comes from the $(\vec{S} \cdot \vec{S})$-type contribution. We estimated $J'_{ab} = -2$ from the Δ_o-dependence of J_{ab} for several R species (see Fig. 10.1 c). $J'_{ab} = \partial J_{ab}/\partial \Delta_o = -2$. We choose J_b as a variable which increases (decreases) as r_R decreases (increases).

By the replica exchange Monte Carlo method [49] applied to Eq. (10.21), we obtain the phase diagram in the T-J_b plane given in Fig. 10.2, which is in good agreement with experiments. Four phases successively emerge at high T as J_b decreases; the A, ab spiral, bc spiral, and E phases. In the A (E) phase, the FM

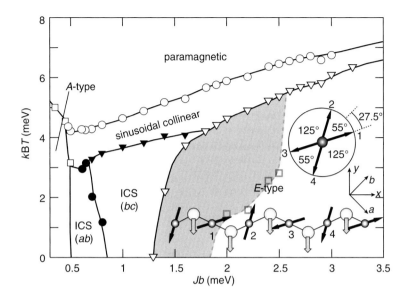

FIG. 10.2. Theoretical phase diagram of RMnO$_3$ in the plane of J_b and T. ICS denotes the incommensurate spiral phase. The E and ICS states coexist in the shaded area. The inset shows the spin configuration in the E phase. The ion shifts due to the electric polarization of $(\vec{S} \times \vec{S})$-type magnetostriction are shown by gray arrows. (Reproduced from Ref. [50].)

(up-up-down-down) Mn-spin layers stack antiferromagnetically, while in the ab (bc) spiral phase, the Mn spins rotate within the ab (bc) plane (P_{bnm} setting) to form transverse cycloids [21, 23]. As T increases, these four phases turn into the sinusoidal collinear phase. Here the magnetic structure is commensurate (C) with $q_b = 0.5$ in the E phase, whereas it is incommensurate (IC) in the ab and bc spiral phases. Note that the sinusoidal collinear state is also IC even above the E phase (e.g. $q_b = 0.458$ for $J_b = 2.4$), and the spin–phonon coupling is a source of the IC-C transition between them with lowering T.

As for the magnitude of the electric polarization in each phase, we consider both $\vec{P}_S$ due to $(\vec{S}_i \cdot \vec{S}_j)$-type magnetostriction, and $\vec{P}_{AS}$ of $(\vec{S}_i \times \vec{S}_j)$-type due to a spin current. The magnitude of the magnetostriction is estimated in the point-charge model [50].

In Fig. 10.3(a), we show the calculated P_S and P_{AS}, and their sum as $T \to 0$ as functions of J_b. There is a finite P_S in the ab spiral phase (e.g. $P_S \sim 500 \mu C/m^2$ for $J_b = 0.7$), while it is zero in the bc spiral phase. We also plot the experimentally measured P for the solid solutions $\mathrm{Eu}_{1-x}\mathrm{Y}_x\mathrm{MnO}_3$ and $\mathrm{Y}_{1-y}\mathrm{Lu}_y\mathrm{MnO}_3$ for comparison [51]. The calculated sum $P_S + P_{AS}$ reproduces well the experimental P. Here the only fitting parameter is the elastic constant $K \cong 500$, which is determined to reproduce the experimental P in the E phase. It turns out that

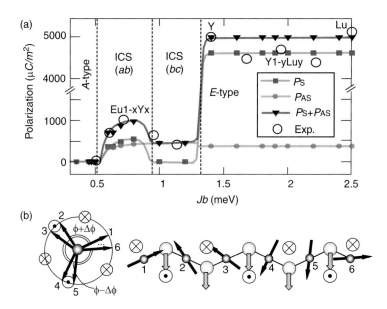

FIG. 10.3. (a) Polarizations P vs J_b at low temperature, i.e. the $(\vec{S} \cdot \vec{S})$ contribution P_S, the $(\vec{S} \times \vec{S})$ contribution P_{AS}, and experimentally measured P
in $Eu_{1-x}Y_xMnO_3$ and $Y_{1-y}Lu_yMnO_3$ [51]. The total polarization $P_S + P_{AS}$
reproduces well the experimental value. (b) Alternation of the spin directions
in the ab spiral state due to the staggered DM vectors where $\odot$ ($\otimes$) denotes
the positive (negative) c-component of the vector. Shifts of the oxygen ions by
$(\vec{S} \cdot \vec{S})$-type magnetostriction are shown by gray arrows. (Reproduced from
Ref. [50].)

P_S can be comparable to or even larger than P_{AS}, which explains why P is much
larger in the ab spiral phase than in the bc spiral phase. This is consistent with
the experiment in $DyMnO_3$.

In the E phase, we find that the spin structure is noncollinear with a spiral
modulation within the ab plane as shown in the inset of Fig. 10.2. Single-ion
anisotropy or the alternation of the in-plane easy magnetization axes due to
$d_{3x^2-r^2}/d_{3y^2-r^2}$-type orbital ordering is the origin of the cycloidal deformation.

With dominant up-up-down-down spin b-axis components, the oxygen ions
between nearly (anti)parallel Mn-spin pairs shift positively (negatively) to modulate the FM exchanges leading to the ferroelectric polarization (see the inset
of Fig. 10.2) [52, 53]. In Fig. 10.3(a), we indeed see a very large P_S ($\sim$4600
$\mu C/m^2$) and a finite P_{AS} due to this cycloidal deformation in the E phase in
agreement with the experimental observations [51]. As shown in Fig. 10.3 (a),
the nonmonotonic behavior of the spontaneous polarization as a function of J_b
is well reproduced by the two types of exchangestriction, i.e. $(\vec{S} \cdot \vec{S})$-type and
$(\vec{S} \times \vec{S})$-type.

10.5 Electromagnons in multiferroics

Up to now we have discussed the ground state properties or thermal equilibrium properties of the multiferroics. The next important direction is their dynamics and nonequilibrium properties. The small-amplitude fluctuations are the first issue to be studied, which is discussed in this section. The small-amplitude vibration around the ground state spin configuration is called a spin wave or magnon. Its dynamics is different for different ground states. For the ferromagnetic state with spontaneous magnetization along the z-axis, the commutation relation Eq. (10.15) leads to the commutation relationship

$$[M_x, M_y] = i\hbar M_z \cong i\hbar\langle M_z\rangle \qquad (10.22)$$

where the z-component of the uniform magnetization M_z is replaced by its expectation value, i.e. the spontaneous magnetization. This is justified by the long-range ordering where the quantum mechanical operator can be regarded as a classical variable when it happens to be the order parameter. Equation (10.22) means that M_x and M_y are canonical conjugate variables and constitute a harmonic oscillator with Hamitonian

$$H = D[(M_x)^2 + (M_y)^2] \qquad (10.23)$$

where D is the easy-axis spin anisotropy energy. The uniform magnetization has the meaning of the generators of the uniform spin rotations, which correspond to the Goldstone modes of the ordered state. In the ferromagnetic state given above, M_x and M_y acting on the ground state produce different (excited) states, while M_z does not change the ground state. (Note that when $D = 0$, rotations by M_x and M_y generate other possible ground states.) In the case of an antiferromagnet, on the other hand, the order parameter is the staggered magnetization $\vec{M}_s$. Suppose that $\vec{M}_s \parallel \hat{z}$; again M_x and M_y are the generators of the Goldstone modes. Therefore, in this case, there are two sets of canonical conjugate pairs, i.e. M_x, M_{sx} and M_y, M_{sy}, to constitute the harmonic oscillators. In the case of noncollinear magnets, all three components M_x, M_y, and M_z are the generators of the Goldstone modes [54]. Therefore, the number of Goldstone modes is determined by the pattern of the symmetry breaking.

In the case of multiferroics, the spinwave is coupled to the electric polarization and/or the atomic displacements, and hence is called an electromagnon [55, 56]. In the case of a cycloidal magnet, the fluctuation of the electric polarization $\vec{P}$ is coupled to the rotation of the spin plane along the direction of the spiral wavevector $\vec{q}$, leading to infrared absorption perpendicular to both $\vec{P}$ and $\vec{q}$.

Experimentally, Pimenov et al. [57] observed the peak of $\Im\varepsilon$ at around 20 cm^{-1} with a magnitude of 1–2 in GdMnO$_3$ and TbMnO$_3$. This 20 cm^{-1} is identified with ω_-, and the integration of $-\mathrm{Im}\varepsilon_{yy}(\omega)$ over ω gives $I_- \sim 12$ cm^{-1}. An interpretation of the experiments on the infrared absorption of RMnO$_3$ in teta-Hz region [57] in terms of these electromagnons was proposed [56], but the observed oscillator strength was a bit larger than the theoretical estimate. Also a neutron scattering experiment [58] reported the identification of one of the spin wave mode branches as the electromagnon. However, recent experiments have revealed that the oscillator strength grows and this discrepancy increases even more as the temperature is further lowered [59–61]. An even more serious puzzle is that the anisotropy of the optical absorption does not change even when the spiral plane changes from the bc to ac plane, while the electric polarization associated with the electromagnon should change direction.

Recently, this puzzle has been resolved [62]; it is shown that the conventional exchange-striction effect, i.e.

$$\vec{P} = \sum_{ij} \vec{\Pi}_{ij} \vec{S}_i \cdot \vec{S}_j \qquad (10.24)$$

contributes to the single magnon absorption. In RMnO$_3$, the vector $\vec{P}_{ij}$ in Eq. (10.24) is nonzero since the inversion symmetry is absent at the center of the Mn-O-Mn bond because of the orthorhombic lattice distortion and/or the staggered $3x^2 - r^2/3y^2 - r^2$ orbital ordering. This contribution cancels out in the ground state due to symmetry, but the dynamical fluctuations of $\vec{P}$ in Eq. (10.24) contribute to the optical absorption. In particular, when the ground state spin configuration is noncollinear, it gives a one-magnon absorption process, while it gives only two-magnon absorption in the collinear case. This is easily understood as

$$\vec{S}_i \cdot \vec{S}_j = (\langle \vec{S}_i \rangle + \delta \vec{S}_i) \cdot (\langle \vec{S}_j \rangle + \delta \vec{S}_j)$$
$$= \langle \vec{S}_i \rangle \cdot \langle \vec{S}_j \rangle + \langle \vec{S}_i \rangle \cdot \delta \vec{S}_j + \delta \vec{S}_i \cdot \langle \vec{S}_j \rangle + \delta \vec{S}_i \cdot \delta \vec{S}_j \qquad (10.25)$$

where the second and third terms correspond to the one-magnon process while the last term corresponds to the two-magnon process in the second line. Considering the fact that the fluctuation $\delta \vec{S}_i$ is perpendicular to $\langle \vec{S}_i \rangle$, the one-magnon contriution survives only when $\langle \vec{S}_i \rangle$ and $\langle \vec{S}_j \rangle$ are not collinear. However, in the experimental data analyzed for RMnO$_3$ in Ref. [62] the dominant absorption occurs at higher energy ($\sim 8~meV$) while the spectral shape depends rather sensitively on the material, and the lower energy peak around ~ 2–$3~meV$ is stronger than that at higher energy. This problem can be addressed only with the accurate spin Hamiltonian obtained in the previous section.

A clue to this issue is the proximity to collinear spin phases, i.e. the A-type and E-type spin phases. Near the phase boundary, the spin configuration is not a simple spiral but suffers from significant elliptical modulation and involves higher harmonics, which is sensitively enhanced by the tiny spin–phonon coupling or by the weak magnetic anisotropy. We employ the realistic spin model discussed in the last section. The only difference is that the phonon degrees of freedom are integrated out to result in the bi-quadratic interaction as given by

$$H_{\text{biq}} = -B_{\text{biq}} \sum_{\langle i,j \rangle}^{ab} (\vec{S}_i \cdot \vec{S}_j)^2, \tag{10.26}$$

which replaces the terms containing phonon coordinates in Eq. (10.21). We study the electromagnon optical spectra (OS) and the phonon diepsersion which can be detected by neutron scattering experiments. We perform the calculations using two sets of model parameters ((a) and (b)) corresponding to the ab plane spiral in $Eu_{1-x}Y_xMnO_3$ ($x = 0.45$) with $q_b \sim 0.3\pi$ and the bc plane spiral in $DyMnO_3$ with $q_b = 0.39\pi$, respectively.

We solve numerically the following Landau–Lifshitz–Gilbert equation by the Runge–Kutta method,

$$\frac{\partial \vec{S}_i}{\partial t} = -\vec{S}_i \times \vec{H}_i^{\text{eff}} + \frac{\alpha_G}{S} \vec{S}_i \times \frac{\partial \vec{S}_i}{\partial t}, \tag{10.27}$$

where $\alpha_G (= 0.1\text{–}0.2)$ is the dimensionless Gilbert damping coefficient. The coupling term $-\vec{E} \cdot \vec{P}$ between the electric field $\vec{E}$ and the polarization $\vec{P}$ given by Eq. (10.24) is taken into account following Ref. [62]. This coupling effectively

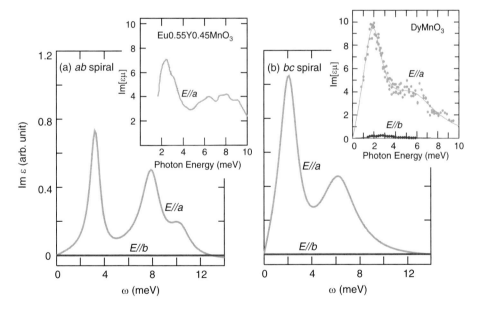

FIG. 10.4. Theoretical optical absorption spectra due to electromagnons for (a) an ab plane spiral state ($q_b = \pi/3$) and (b) a bc plane spiral state ($q_b = 2\pi/5$) with the parameter sets corresponding to $Eu_{0.55}Y_{0.45}MnO_3$ and $DyMnO_3$, respectively. The insets show the corresponding experimental results for each material. (Reproduced from Ref. [46].)

modulates the nearest neighbor ferromagnetic exchanges in the ab plane as shown in Fig. 10.1(b) (Fig. 10.1c) in Section 10.4.

We apply the delta-functional pulse of the electric field $\vec{E} \parallel a$ or $\vec{E} \parallel b$ at $t = 0$, and calculate the time evolution of $\vec{P}$. The electromagnon spectrum Im $\varepsilon(\omega)$ is obtained from the Fourier transformation of $\vec{P}(t)$, as shown in Fig. 10.4, with the parameter sets corresponding to $Eu_{0.55}Y_{0.45}MnO_3$ and $DyMnO_3$, respectively. Independent of the spiral-plane orientation, a large spectral weight emerges at low energy when $\vec{E} \parallel a$. No response to $\vec{E} \parallel b$ is observed for both cases in agreement with the experimental observations. The experimental results for each

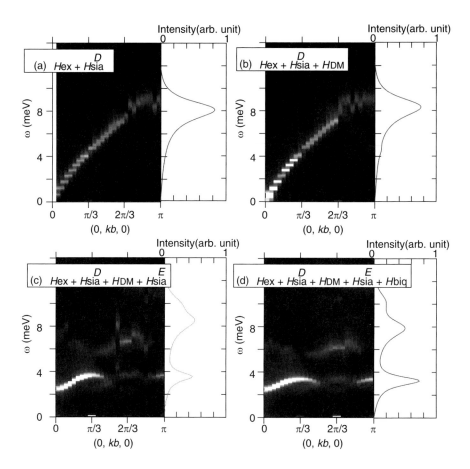

FIG. 10.5. Theoretical magnon dispersions and electromagnon spectra for various cases of interactions. From (a) to (d), the interactions are added one by one, and (d) is for the full Hamiltonian. The folding of the magnon dispersion due to the higher harmonics gives the oscillator strength to the lower energy region around 2 meV. (Reproduced from Ref. [46].)

material are shown in the insets, which are in good agreement with the theoretical calculations.

The two-peak structure is due to the folding of the magnon dispersion due to the higher harmonics of the ground state spin configuration. Namely, due to the E-term in Eq. (10.21) and the bi-quadratic term in Eq. (10.26), the spin rotation angle is not uniform, and the higher harmonic components lead to Umklapp scattering and hence the folding of the spin wave dispersion as shown in Fig. 10.5. Note that this higher harmonic, i.e. the deviation from the uniform rotation of the spin, is experimentally detected by the ratio $\eta = \sqrt{\hat{S}_a(\vec{q}_b)/\hat{S}_b(\vec{q}_b)}$ where $\hat{S}_i(\vec{q}_b)$ is the spin structure factor along the $i(= a, b)$-direction. This quantity is usually regarded as the "ellipticity," but it does not necessarily mean the modulation of the spin length. The deviation of η from unity originates also from the nonuniform rotation of the spins with fixed length as in the case of the present calculation. The above results suggest that as $1 - \eta$ increases, the oscillator strength of the lower energy peak increases, which seems to be the case experimentally also.

The concept of the electromagnon can be extended in many directions. One interesting direction is to consider the quantum and thermal fluctuation beyond the small-amplitude vibration. A Ginzburg–Landau theory has been developed to study this problem for thermal fluctuation, and the self-consistent mode-coupling approximation leads to a chiral spin liquid with a finite vector chirality above the magnetic transition temperature [63]. Recently, an experiment on Gd(hfac)$_3$NITEt has been reported which found the two-step phase transition and suggested the chiral spin liquid state [64]. This is a quasi-one-dimensional system and Gd has spin 7/2. Therefore, it can be regarded as a strongly fluctuating classical helimagnet, and offers an ideal arena to test the theory in Ref. [63].

10.6   Ultrafast switching of spin chirality by optical excitation

Since we have obtained an accurate spin Hamiltonian describing RMnO$_3$, it is possible to predict some new phenomena based on it. Here we study the nonlinear processes driven by the intense light pulse irradiation of picosecond order, which excites the electromagnons. As mentioned in the last section, there are basically two peaks in the optical absorption spectrum of an electromagnon at around 3 meV and 8 meV due to the mechanism of the exchange-striction. We solve the Landau–Lifshitz–Gilbert equation (10.27) with this strong pulse and trace the resulting change of the spin structure using the fourth-order Runge–Kutta method.

We assume that the electric field is along the a-axis, i.e. the direction of the polarization for the strongest absorption, as

$$E_a = -E_0 \sin \omega t \exp\left[-\frac{(t - t_0)^2}{2\sigma^2}\right] \tag{10.28}$$

where ω is taken to be 2.1 THz (8.2 meV), while the half-width of the pulse $2\sqrt{2\log 2}\sigma$ is taken to be 0.5 psec. Figure 10.6 shows the time evolution of the three components of the vector spin chirality defined as $\vec{C} = \frac{1}{2N}\sum_i[\vec{S}_i \times \vec{S}_{i+x} + \vec{S}_i \times \vec{S}_{i+y}]$. It is shown that a chirality switch occurs from $\vec{C} \parallel -a$ to $\vec{C} \parallel +a$ for $E_0 = 14$ MV/cm while a chirality flop occurs from $\vec{C} \parallel -a$ to $\vec{C} \parallel +c$ for $E_0 = 13$ MV/cm. The microscopic mechanism of this phenomenon is the combination of the change in the spin tilting angle due to the change in J_{ij} and the DM interaction, which modulates the relative energy of the various chirality states. The process, however, is highly dynamical by the intensive and nonlinear excitation of the electromagnon with large amplitude followed by the

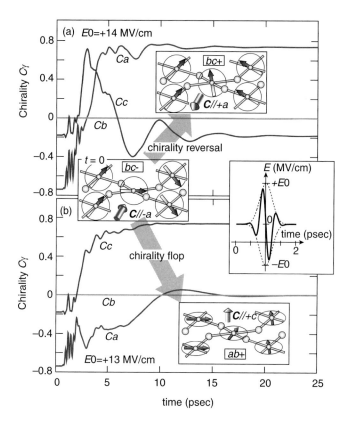

FIG. 10.6. Time evolution of the total vector spin chirality $\vec{C} = (C_a, C_b, C_c)$ after application of the pulse with (a) $E_0 = +14$ MV/cm and (b) $E_0 = +13$ MV/cm. (a) shows the chirality-reversal from bc_- ($C_a < 0$) to bc_+ ($C_a > 0$), while (b) shows the chirality flop from bc_- ($\vec{C} \parallel -\vec{a}$) to ab_+ ($\vec{C} \parallel +\vec{c}$). The insets show the spin states before or after applying the pulse, and the time profile of the applied pulse $E_a(t)$. (Reproduced from Ref. [65].)

inertial motion of the spins due to the mass generated by the SOI. The domain structure emerges during the phase change with the spins parallel to the b-axis acting as the nodes of the spin configuration. By the shape and intensity of the optical pulse, one can control the chirality switch processes. For example, one can reverse the direction of the rotation of the vector spin chirality $\vec{C}$ by reversing the sign of E_a. Usually, the chirality is determined by the structure of the molecules or crystals and is difficult to change. In the spin system, on the other hand, it can be switched in picoseconds as predicted theoretically above.

10.7   Quasi-one-dimensional quantum multiferroics

Quantum fluctuation in quasi-one-dimensional systems is also of great interest. In particular, there have appeared quasi-one-dimensional [34, 66], and quasi-two-dimensional [28] helimagnetic systems. Schwinger boson theory to treat this problem has been developed, and the length of the spin can be "soft" in the quantum spin case [67]. The two-step transition from paramagnetic to collinear, and from collinear to spiral states is interpreted in the following way. The collinear ordering is described by the spin density wave of the Schwinger bosons, while the spiral spin state appears once the bose condensation occurs. Therefore, the elliptic ratio is interpreted as the ratio of the classical condensed part and the quantum mechanical fluctuating part of the Schwinger bosons [67]. Experimentally, there has been no signature of the quantum fluctuation up to now, and further studies, both theoretically and experimentally are desired.

Quasi-one-dimensional multiferroics are not restricted to spiral magnets. The spin-Peierls systems in donor–acceptor mixed stack charge transfer compounds are genuine ferroelectrics when the interchain coupling is ferroic. Because the inversion symmetry between the donor and acceptor is absent, the polarization due to the exchange-striction reads [68]

$$\vec{P} = \sum_i \vec{\Pi} \vec{S}_i \cdot \vec{S}_{i+1}. \tag{10.29}$$

The idea is that the spinon (spin-$\frac{1}{2}$ object) is the most fundamental particle in the 1D antiferromagnetic Heisenberg model [69], which turns into the *electro-spinon* and governs the infrared activity [70]. By a Jordan–Wigner transformation, the spin operator can be represented by fermion operators. The XY-interaction corresponds to the transfer of fermions while the Ising model corresponds to the fermion–fermion interaction. Spin excitations are described as the particle–hole excitations of the fermions, and the Ising interaction leads to the attraction between the particle and hole. This interaction leads to the Tomonaga–Luttinger behavior for the undimerized gapless case. In the dimerized case, the gap opens, and the interaction gives the particle–hole attractive force and hence the bound states, i.e. exciton formation. These ideas can be formulated more rigorously as follows.

In the undimerized state, the low-energy asymptotic behavior of the optical spectrum $\sigma(\omega)$ can be analyzed by conformal field theory (CFT) with $c = 1$ [69]. $c = 1$ CFT is characterized by the exponent K, and the AF Heisenberg model corresponds to $K = 1$. This leads to the following conclusion: for $\omega \ll T$, $\sigma(\omega) \propto \omega^2 T^{K-3}$, while $\sigma(\omega) \propto \omega^{K-1}$ for $\omega \gg T$. It is noted that $\sigma(\omega)$ is asymptotically constant at the Heisenberg point ($K = 1$).

In the low-temperature dimerized case, one can map the model into the quantum sine-Gordon model, which can be solved by the Bethe ansatz and form factor expansion [71]. In the sine-Gordon model, the "exciton" is described as breathers, i.e. the bound states of a soliton and anti-soliton. The energy and the form factor for this breather are known, and one can predict the exact energy position and oscillator strength of this excitation. For more details, readers are referred to the original paper [70].

10.8   Summary and conclusions

In this chapter, we have reviewed theoretical studies of multiferroic helimagnets from the viewpoint of the spin current or the vector spin chirality. This introduces a new point of view to frustrated spin systems, i.e. the ferroelectric and dielectric responses associated with the vector spin chirality. The ground and excited states of the spin systems are characterized by the electric polarization, and it is now recognized that the charge degrees of freedom in Mott insulators are not silent at all even in the low-energy region, and rich physics is there. Here we discuss some of the important issues left for future studies, and perspectives.

From the viewpoint of spintronics, it is highly desirable to develop the spintronics without dissipation. The spin current in insulating magnets is an ideal laboratory to develop this idea. The magneto-electric effect in insulators is a promising direction for this purpose. The enhancement due to dynamical resonance is a possible direction of future research although the electromagnon is rather heavily damped experimentally. The theoretical analysis of this damping is still lacking, and is desired. As for the dynamical aspect, a recent preprint [72] has studied the most generic mechanism for the coupling between the electric field and the spins in multi-orbital systems by considering the dynamics of the spins in the intermediate states for the exchange interaction. Roughly speaking, this can be understood as $\vec{e} \cdot \vec{E}$ with $\vec{e}$ being the "electric field" associated with the time-dependent Berry connection of the spins. This is analogous to the spin-motive force in metallic ferromagnetic systems by which the domain wall or vortex motions produce the voltage drop [8–10].

The other direction is to pursue the physics of noncollinear spin structures, i.e. spin textures. The scalar spin chirality defined as $S = \vec{S}_i \cdot (\vec{S}_j \times \vec{S}_k)$ is another important physical quantity to characterize the noncollinear spin structure. This scalar spin chirality corresponds to the solid angle subtended by the three spins, and acts as the effective magnetic field for the conduction electrons coupled to these spins. Therefore, it is expected that the Hall effect, especially the anomalous Hall effect, occurs due to the spin chirality [5].

The DM interaction in noncentrosymmstric magnets often leads to spiral spin structures. A typical example is the MnSi with B20 structure. Recently, a neutron scattering experiment identified the mysterious A-phase in MnSi as the Skyrmion crystal state stabilized by the external magnetic field and thermal fluctuations [73–75] . Note that the conical spin structure is the most stable state in all the other regions of the phase diagram. However, when one reduces the thickness of the sample to smaller than the wavelength of the spiral, the conical state is not possible when the external magnetic field is perpendicular to the film. Actually, a Monte Carlo simulation of the 2D magnet with DM interaction concluded that the Skyrmion crystal state is stable in a much wider region of the phase diagram including the zero-temperature case[76]. Motivated by these expectations, a recent experiment using Lorentz microscopy succeeded in real-space observation of the Skyrmion crystal in a thin film of (Fe,Co)Si [6]. This finding offers an ideal arena to study the manipulation of spin textures by an electric current or an electric field, which will be an important issue in the future.

The topological nature of the multiferroic behavior is, even though implicit, the background of the discussion given above. From this viewpoint, the recently discovered topological insulators (TIs) offer an interesting possibility for the multiferroic phenomenon. First, the topological magneto-electric effect has been proposed using the three-dimensional TI, which is described by the following effective action

$$S_\theta = \left(\frac{\theta}{2\pi}\right)\left(\frac{\alpha}{2\pi}\right)\int d^3x dt \vec{E} \cdot \vec{B} \tag{10.30}$$

for the electromagnetic field [77]. ($\alpha \cong \frac{1}{137}$ is the fine structure constant.) This action is similar to the θ term discussed for QCD [1], and there is a periodicity with respect to $\theta \to \theta \pm 2\pi$. If the system is time-reversal symmetric, θ and $-\theta$ should be equivalent, which restricts the θ value to be 0 or π (or plus an integer times 2π). TI corresponds to $\theta = \pi$, while an ordinary insulator corresponds to $\theta = 0$. Since $\vec{E} \cdot \vec{B}$ can be written as the divergence of the Chern–Simons term, it does not affect the equations of motion in the system without the boundary, i.e. the system with periodic boundary conditions. When the boundary is there, which is usually the case for real materials, the surface current density and charge density are proportional to $\nabla\theta(r)$ which is localized at the surface of the sample:

$$\vec{j} \propto \nabla\theta \times \vec{E}$$
$$\rho \propto -\nabla\theta \cdot \vec{B} \tag{10.31}$$

and the integrated 2D current density or charge density is proportional to the discontinuity of θ inside and outside of the sample, i.e. $\Delta\theta = \pm\pi$. This is described by the 2D Chern–Simons term for the surface, which is obtained by integrating over the Dirac fermions with the mass m corresponding to the time-reversal symmetry breaking and assuming the Fermi energy being within the mass gap. Therefore, the value $\theta = \pm\pi$ is dictated by the topological property of the bulk states, but the choice of θ is determined by the surface. The magneto-electric (ME) effect derived from Eq. (10.30) gives rise to the bulk orbital magnetization by the surface current, and is different from the conventional ME effect in which the bulk magnetism is required. Also the quntization of the ME effect from the defnite value of $\theta = \pm\pi$ is a unique feature of this topological insulator. As argued in Ref. [77], the TI is related to the second Chern form in higher dimensions, i.e. (4+1)D. Interestingly, the spin-current mechanism of the polarization discussed above is also related to the second Chern form [78], which suggests a deep connection between the multiferroic behavior and the topological structure of the electronic states. This direction is worth exploring more in the future.

In summary, we have discussed recent developments in the research into multiferroic phenomena and materials from the viewpoint of the relativistic spin–orbit interaction as the gauge field. The concept of the spin current emerges as a consequence of the projection onto the positive energy subspce of the solutions to the Dirac equation, which results in the relativistic spin–orbit interaction as a gauge field. This idea leads to various interesting phenomena interpreted from the geometrical viewpoint, and multiferroics is one of the representative examples. These gauge structures offer an interesting direction of the future research in condensed matter physics.

Acknowledgments

The author thanks H. Katsura, M. Mochizuki, N. Furukawa, A.V. Balatsky, S. Onoda, H.J. Han, C. Jia, M. Sato, T. Furuta, K. Nomura and M. Mostovoy for collaboration, and Y. Tokura, T. Arima, N. Kida, M. Kawasaki, D.I. Khomskii, and A. Aharony for useful discussions. This work is supported by Priority Area Grants, Grant-in-Aids under the Grant numbers 19048015, 19048008, and 21244053, and NAREGI Nanoscience Project from the Ministry of Education, Culture, Sports, Science, and Technology, Japan, and also by Funding Program for World-Leading Innovative R and D on Science and Technology (FIRST Program).

References

[1] See for example M.E. Peshkin and D.V. Schroeder, *Introduction to Quantum Field Theory* (Addison-Wesley, New York, 1995).

[2] J. Froelich and U.M. Studer, Rev. Mod. Phys. **65**, 733 (1993).

[3] B.W.A. Leurs, Z. Nazario, D.I. Santiago, and J. Zaanen, Ann. Phys. **323**, 907 (2008).

[4] S. Murakami, N. Nagaosa, and S.C. Zhang, Science **301**, 1348 (2003); J. Sinova *et al.*, Phys. Rev. Lett. **92**, 126603 (2004).

[5] N. Nagaosa, J. Sinova, S. Onoda, A.H. MacDonald, and N.P. Ong, Rev. Mod. Phys. **82**, 1539 (2010).

[6] X.Z. Yu *et al.*, Nature **465**, 901 (2010).

[7] J.H. Han *et al.*, Phys. Rev. B **82**, 094429 (2010).

[8] L. Berger, Phys. Rev. B **33**, 1572 (1986).

[9] S.E. Barnes and S. Maekawa, Phys. Rev. Lett. **98**, 246601 (2007).

[10] S.A. Yang *et al.*, Phys. Rev. Lett. **102**, 067201 (2009).

[11] N. Nagaosa, Y. Tokura, Phys. Scr., **T146**, 014020 (2012).

[12] L.D. Landau, E.M. Lifshitz, and L.P. Pitaevskii, *Electrodynamics of Continuous Media* (Elsevier, Oxford, 2008).

[13] P. Curie, J. Phys. **3**, 393 (1894).

[14] M. Fiebig, J. Phys. D: Appl. Phys. **38**, R123 (2005).

[15] For recent reviews, Y. Tokura, Science **312**, 1481 (2006); S.-W. Cheong and M. Mostovoy, Nature Mater. **6**, 13 (2007); Y. Tokura, J. Magn. Magn. Mater. **310**, 1145 (2007).

[16] T. Kimura *et al.*, Nature **426**, 55 (2003).

[17] T. Kimura *et al.*, Phys. Rev. B **68**, 060403(R) (2003).

[18] T. Goto *et al.*, Phys. Rev. Lett. **92**, 257201 (2004).

[19] H. Katsura, N. Nagaosa, and A.V. Balatsky, Phys. Rev. Lett. **95**, 057205 (2005).

[20] N. Nagaosa, J. Phys. Cond.-Mat. **20**, 434207 (2008); N. Nagaosa, J. Phys. Soc. Jpn. **77**, 031010 (2008).

[21] M. Kenzelmann *et al.*, Phys. Rev. Lett. **95**, 087206 (2005).

[22] Y. Yamasaki, H. Sagayama, T. Goto, M. Matsuura, K. Hirota, T. Arima, and Y. Tokura, Phys. Rev. Lett. **98**, 147204 (2007)

[23] Y. Yamasaki *et al.*, Phys. Rev. Lett. **101**, 097204 (2008).

[24] Y. Tokura, Science **312**, 1481 (2006).

[25] M. Mostovoy, Phys. Rev. Lett. **96**, 067601 (2001).

[26] I.A. Sergienko and E. Dagotto, Phys. Rev. B **73**, 094434 (2006).

[27] A.B. Harris and G. Lawes, in *The Handbook of Magnetism and Advanced Magnetic Materials*, ed. H. Kronmuller and S. Parkin (John Wiley, New York 2006); A.B. Harris *et al.*, Phys. Rev. B **73**, 184433 (2006).

[28] G. Lawes, A.B. Harris, T. Kimura, N. Rogado, R.J. Cava, A. Aharony, O. Entin-Wohlman, T. Yildirim, M. Kenzelmann, C. Broholm, and A.P. Ramirez, Phys. Rev. Lett. **95**, 087205 (2005).

[29] T. Kimura, G. Lawes, and A.P. Ramirez, Phys. Rev. Lett. **94**, 137201 (2005).

[30] Y. Yamasaki, S. Miyasaka, Y. Kaneko, J.-P. He, T. Arima, and Y. Tokura, Phys. Rev. Lett. **96**, 207204 (2006).

[31] K. Taniguchi, N. Abe, T. Takenobu, Y. Iwasa, and T. Arima, Phys. Rev. Lett. **97**, 097203 (2006).

[32] T. Kimura, J.C. Lashley, and A.P. Ramirez, Phys. Rev. B **73**, 220401 (2006).

[33] Y. Naito, K. Sato, Y. Yasui, Y. Kobayashi, Y. Kobayashi, and M. Sato, J. Phys. Soc. Jpn. **76**, 023708 (2007).

[34] S. Park, Y.J. Choi, C.L. Zhang, and S-W. Cheong, Phys. Rev. Lett. **98**, 057601 (2007).

[35] C. Jia, S. Onoda, N. Nagaosa, and J.H. Han., Phys. Rev. B **74**, 224444 (2006).

[36] C. Jia, S. Onoda, N. Nagaosa, and J.H. Han, Phys. Rev. B **76**, 144424 (2007).

[37] H.J. Xiang, S.-H. Wei, M.-H. Whangbo, and J.L. Da Silva, Phys. Rev. Lett. **101**, 037209 (2008).

[38] A. Malashevich and D. Vanderbilt, Phys. Rev. Lett. **101**, 037210 (2008).

[39] R. Resta, Rev. Mod. Phys. **66**, 899 (1994).

[40] R.D. King-Smith and D. Vanderbilt, Phys. Rev. B **47**, 1651(1993).

[41] T. Arima, J. Phys. Soc. Jpn. **76**, 073702 (2007).

[42] M. Mochizuki, and N. Furukawa, J. Phys. Soc. Jpn. **78**, 053704 (2009); Phys. Rev. B **80**, 134416 (2009).

[43] T. Kimura *et al.*, Phys. Rev. B **71**, 224425 (2005).

[44] F. Kagawa, M. Mochizuki, Y. Onose, H. Murakawa, Y. Kaneko, N. Furukawa, and Y. Tokura, Phys. Rev. Lett. **102**, 057604 (2009).

[45] F. Schrettle, P. Lunkenheimer, J. Hemberger, V. Yu. Ivanov, A.A. Mukhin, A.M. Balbashov, and A. Loidl, Phys. Rev. Lett. **102**, 207208 (2009).

[46] M. Mochizuki, N. Furukawa, and N. Nagaosa, Phys. Rev. Lett. **104**, 177206 (2010).

[47] B. Dabrowski *et al.*, J. Sol. Stat. Chem. **178**, 629 (2005).

[48] I. Solovyev, N. Hamada, and K. Terakura, Phys. Rev. Lett. **76**, 4825 (1996).

[49] K. Hukushima and K. Nemoto, J. Phys. Soc. Jpn. **65**, 1604 (1996).

[50] M. Mochizuki, N. Furukawa, and N. Nagaosa, Phys. Rev. Lett. **105**, 037205 (2010).

[51] S. Ishiwata, Y. Kaneko, Y. Tokunaga, Y. Taguchi, T. Arima, and Y. Tokura, arXiv:0911.4190.

[52] I.A. Sergienko, C. Sen, and E. Dagotto, Phys. Rev. Lett. **97**, 227204 (2006).

[53] S. Picozzi, K. Yamauchi, B. Sanyal, I.A. Sergienko, and E. Dagotto, Phys. Rev. Lett. **99**, 227201 (2007).

[54] T. Nagamiya, in *Solid State Physics*, Vol. 20, ed. F. Seitz, D. Turnbull, and H. Ehrenreich (Academic Press, New York, 1967), p. 305.

[55] V.G. Baryakhtar and I.E. Chupis, Sov. Phys. Solid State **11**, 2628 (1970); G.A. Smolenskii and I.E. Chupis, Sov. Phys. Usp. **25**, 475 (1982).

[56] H. Katsura, A.V. Balatsky, and N. Nagaosa, Phys. Rev. Lett. **98**, 027203 (2007).

[57] A. Pimenov *et al.*, Nature Physics **2**, 97 (2006).

[58] D. Senff *et al.*, Phys. Rev. Lett. **98**, 137206 (2007).

[59] A. Pimenov *et al.*, Phys. Rev. B **74**, 100403(R) (2006).

[60] N. Kida *et al.*, Phys. Rev. B **78**, 104414 (2008).

[61] N. Kida, Y. Takahashi, J.S. Lee, R. Shimano, Y. Yamasaki, Y. Kaneko, S. Miyahara, N. Furukawa, T. Arima, and Y. Tokura, J. Opt. Soc. Am. B **26**, A35 (2009).

[62] R. Valdes Aguilar, M. Mostovoy, A.B. Sushkov, C.L. Zhang, Y.J. Choi, S.-W. Cheong, and H.D. Drew, Phys. Rev. Lett. **102**, 047203 (2009).

[63] S. Onoda and N. Nagaosa, Phys. Rev. Lett. **99**, 027206 (2007).

[64] F. Cinti *et al.*, Phys. Rev. Lett. **100**, 057203 (2008).

[65] M. Mochizuki and N. Nagaosa, Phys. Rev. Lett. **105**, 147202 (2010).

[66] Y. Naito *et al.*, cond-mat/0611659.

[67] H. Katsura, S. Onoda, J.H. Han, and N. Nagaosa, Phys. Rev. Lett. **101**, 187207 (2008).

[68] Y. Tanabe, T. Moriya, and S. Sugano, Phys. Rev. Lett. **15**, 1023 (1965).

[69] A.O. Gogolin, A.A. Nersesyan, and A.M. Tsvelik, *Bosonization and Strongly Correlated Systems* (Cambridge University Press, Cambridge, 1998).

[70] H. Katsura, M. Sato, T. Furuta, and N. Nagaosa, Phys. Rev. Lett. **103**, 177402 (2009).

[71] F.A. Smirnov, *Form Factors in Completely Integrable Models of Quantum Field Theory* (World Scientific, Singapore, 1992).

[72] M. Mostovoy, K. Nomura, and N. Nagaosa, Phys. Rev. Lett. **106**, 047204 (2011).

[73] C. Pfleiderer *et al.*, Nature **427**, 227 (2004).

[74] U.K. Rosler, A.N. Bogdanov, and C. Pfleiderer, Nature **442**, 797 (2006).

[75] S. Muhlbauer *et al.*, Science **323**, 915 (2009).

[76] S.D. Yi, S. Onoda, N. Nagaosa, and J.H. Han, Phys. Rev. B **80**, 054416 (2009).

[77] X.L. Qi, T.L. Hughes, and S.C. Zhang, Phys. Rev. B **78**, 195424 (2008); X.L. Qi, and S.C. Zhang, Rev. Mod. Phys. **83**, 1057 (2011).

[78] D. Xiao, J. Shi, D.P. Clougherty, and Q. Niu, Phys. Rev. Lett. **102**, 087602 (2009).

Part II Spin Hall effect

11 Introduction

S. O. Valenzuela

11.1 Historical background

Spin Hall effects are a group of phenomena that result from spin–orbit interaction, which links orbital motion to spin direction and acts as a spin-dependent magnetic field. In its simplest form, an electrical current gives rise to a transverse spin current that induces spin accumulation at the boundaries of the sample, the direction of the spins being opposite at opposing boundaries. It can be intuitively understood by analogy with the Magnus effect where a spinning ball in a fluid deviates from its straight path in a direction that depends on the sense of rotation. Spin Hall effects can be associated to a variety of spin–orbit mechanisms, which can have intrinsic or extrinsic origin, and depend on the sample geometry, impurity band structure and carrier density but do not require a magnetic field or any kind of magnetic order to occur.

The phenomenon was predicted by Dyakonov and Perel in 1971 [1, 2]. This prediction was scarcely noticed until 1999, when Hirsch rediscovered it and introduced the term "spin Hall effect" [3]. The effect is indeed analogous to the normal Hall effect, where charges of opposite sign accumulate at the boundaries of the sample due to the Lorenz force in a magnetic field (Fig. 11.1). In order to predict its existence Hirsch simply argued that the presence of the familiar anomalous Hall effect in ferromagnetic metals [4], known since 1880, was experimental proof that electrons carrying a spin are subject to a transverse force when they are moving. Possible mechanisms for this force include the side-jump and Mott-skew scattering by impurities and phonons. Because of the magnetic order in the ferromagnet, an electric current will be spin polarized. Thus the transverse force results in charge accumulation perpendicular to the current flow direction, and therefore to the anomalous Hall effect. Following this argument, in a paramagnet, or in the same ferromagnet above its Curie temperature, the same scattering mechanisms that induce the anomalous Hall effect should scatter electrons with spin up and spin down preferentially in opposing directions. Given that there is an equal number of spin-up and spin-down electrons there is no charge accumulation but spin accumulation occurs.

This simple reasoning motivated a vast body of theoretical, and later on, experimental research. Two- and three-dimensional electron systems with spin–orbit interaction were studied theoretically in order to clarify the possible competing mechanisms involved [5]. Experimentalists have been able to demonstrate and quantitatively study the spin Hall effect and its inverse (i.e. the generation

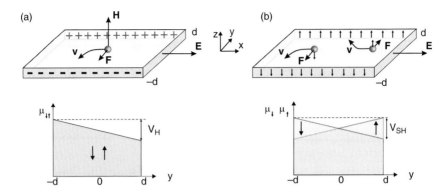

FIG. 11.1. (a) In the Hall effect, the presence of a magnetic field **B** and the associated Lorentz force **F** generate charge accumulation. The electrochemical potential for spin-up and spin-down carriers, assumed to be electron-like, are the same. A voltage difference V_H between the two edges is built up. (b) In the spin Hall effect, spin–orbit coupling causes spin accumulation. The electrochemical potential for spin up and spin down carriers is different. A voltage difference V_{SH} between the two edges is built up for each spin orientation but with opposite sign.

of a transverse charge current by a spin current) in a variety of systems, which include semiconductors like GaAs and metals like Al, and Pt [6–10] (see Chapters 14, 15 and 16). The effect is very robust and has been observed at room temperature. In the bulk of the aforementioned materials, early experiments pointed to a spin Hall effect of extrinsic origin, that is, due to side-jump and Mott-skew scattering mechanisms, as the phenomenology of early theories of the anomalous Hall effect would suggest. As scattering mechanisms are better understood and samples that engineer them fabricated, larger effects are obtained and it is becoming clear that the extrinsic spin Hall effect can play an important role in spintronic applications.

The possibility of an intrinsic spin Hall effect was also put forward [11, 12]. Here the mechanism depends only on the electronic structure of the material with scattering playing a minor role. The intrinsic effect would be relatively large and potentially allow control of spin currents with electric fields, which could flow without dissipation. Two model Hamiltonians were originally considered, a p-doped three-dimensional system (spin-3/2 valence band in GaAs [11]) and a two-dimensional electron gas with Rashba-type coupling [12]. Experiments in two-dimensional layers of p-GaAs have shown results which are consistent with these predictions [7] (Chapters 13 and 16).

In the intrinsic spin Hall effect, the origin of the spin current can be traced to a topological phase collected by the carriers as they move through momentum space. This is the result of an effective gauge field due to spin–orbit coupling that acts as a spin-dependent magnetic field and leads to an anomalous velocity as in the quantum Hall effect. This effective magnetic field acts differently on the two spin orientations leading to a net spin current. The possibility of a dissipationless character of the spin current is a consequence of time reversal. Charge currents, which have units of charge times velocity, are odd under time reversal, while electric fields are even. The coupling between them (i.e. the conductivity) has to be odd and therefore dissipative. However, spin currents have dimensions of angular momentum times velocity, therefore they are even under time reversal just like the electric fields. The coupling between them is therefore even and not necessarily dissipative.

Similar ideas were applied to insulators with time-reversal symmetry, leading to the concepts of a spin Hall insulator [13], and the quantum spin Hall effect in two-dimensional systems [14–16]. The quantum spin Hall phase is a topological phase in the sense that certain fundamental properties are insensitive to small changes in material parameters [15]. For these fundamental properties to change there should be a phase transition. The phase was later generalized to three dimensions [17–19] and is usually known as a "topological insulator" [18] (Chapter 17). It is characterized by an insulating bulk and gapless states localized in the system boundaries when placed in vacuum or in contact with an ordinary insulator. These metallic boundaries originate from topological invariants and cannot change as long as the bulk remains insulating and time reversal symmetry is not broken. It is remarkable that topological insulators can be understood in the framework of the band theory of solids and more than 50 compounds have already been predicted.

In a simple two-dimensional (2D) picture, topological insulators can be understood as two copies of the integer quantum Hall effect (Fig. 11.2). The quantum Hall effect occurs in semiconductors at low temperatures when a magnetic field is applied. There the electrons only travel at the edge in one direction, therefore they cannot scatter back when they encounter an impurity and their motion is nondissipative (Fig. 11.2a). In the idealized quantum spin Hall effect, or 2D topological insulator, spin-up and spin-down electrons are independent and are in oppositely directed quantum Hall states. Spin-up electrons are in an integer quantum Hall effect induced by an effective magnetic field pointing up, while spin-down electrons are in an equivalent state induced by an effective magnetic field pointing down (Fig. 11.2b). Because the magnetic fields are in opposite direction, the direction of motion of spin-up and spin-down electrons at the edge is also opposite. A system with such edge states is said to be in a quantum spin Hall state because edge currents carry spin instead of charge. As opposed to the quantum Hall effect, there are both forward and backward movers. However, in a backscattering event the change of direction should be accompanied by change in spin orientation, which in this system cannot occur unless time reversal is broken. Backscattering by nonmagnetic impurities is thus forbidden (Fig. 11.2c).

In the simplest three dimensional case, the surface state (or interface with an ordinary insulator) can be described by two-dimensional massless Dirac fermions with a dispersion forming a Dirac cone with the crossing point located at the

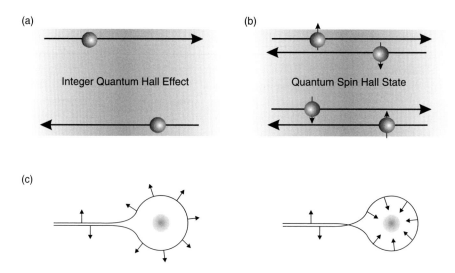

F<small>IG</small>. 11.2. (a) Edge states of the integer quantum Hall effect with one propagating mode. Electrons propagate in one direction determined by the orientation
of the magnetic field. There are no back movers and therefore the state is
robust and goes around an impurity without scattering. (b) In an idealized
two-dimensional topological insulator or quantum spin Hall state, spin-up
and spin-down electrons move in opposite directions. They are equivalent
to two independent quantum Hall effects with opposite magnetic fields. (c)
Backscattering by a nonmagnetic impurity in the quantum spin Hall effect.
This is possible in principle because there are backward and forward movers.
The left and right graphs show two possible paths. In the left path spin
rotates by π, while in the right path, it rotates by $-\pi$. Overall a geometrical
(Berry) phase factor of -1 associated with the total rotation of 2π of the
spin leads to destructive interference and suppression of backscattering. The
states are robust against backscattering as long as time reversal symmetry is
not broken. Adapted from Ref. [20].

time-reversal invariant momentum $\mathbf{k} = 0$ and a spin arrangement as shown in
Fig. 11.3. The degeneracy at $\mathbf{k} = 0$ and the surface metallic states are protected
by time inversion symmetry and electrons traveling on such a surface state
are weakly sensitive to localization and their spins have opposite orientation
at momenta $\mathbf{k}$ and $-\mathbf{k}$. The spin arrangement contributes a Berry phase of
π to the wavefunction protecting the surface states against backscattering by
nonmagnetic impurities. Nothing prevents carriers from scattering in all other
directions (Fig. 11.3b), however, the reduction of backscattering compared with
ordinary metals has major consequences for electron localization.

   Experiments have demonstrated the presence of boundary states in topological insulators. A series of experiments in HgTe quantum wells that studied charge

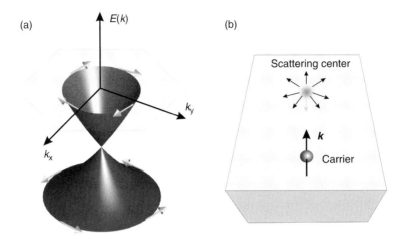

FIG. 11.3. (a) Energy and momentum dependence of the local density of states for an idealized three dimensional topological insulator. (b) In a normal metal a point-like scattering center can scatter a carrier in any direction. In a topological insulator backscattering is suppressed. In a backscattering process, the carrier motion is reversed (say from $\mathbf{k}$ to $-\mathbf{k}$). If the scattering center is nonmagnetic, the spin remains unaffected. However, the only spin state available at $-\mathbf{k}$ is opposite to the one of the incident carrier and backscattering cannot occur (see also Fig. 11.2c).

transport detected the edge states [21, 22], while experiments using angle-resolved photoemission spectroscopy (ARPES) in Bi_xSb_{1-x} alloys, and Bi_2Se_3 and Bi_2Te_3 crystals mapped the unusual surface bands [23, 24].

All of these phenomena are fascinating and the physics is extremely rich. The rest of this chapter will be dedicated to introducing basic concepts and describe early experimental observations of the spin Hall effects. The chapters that follow will provide a thorough description of these concepts and of recent experimental progress.

11.2 Spin–orbit interaction

The spin–orbit interaction (SOI) is a relativistic effect in which the magnetic moment of a moving particle in an electric field couples to an effective magnetic field. In vacuum, the Dirac equation can be reduced to the Pauli equation which contains the spin–orbit interaction as a correction with the following form,

$$H_{\text{SO}} = -\eta_{\text{so}} \sigma \cdot [\mathbf{k} \times \nabla V_{\text{vac}}(\mathbf{r})], \qquad (11.1)$$

where $\eta_{\text{so}} = (\hbar/2mc)^2 \approx 3.7 \times 10^{-6} \text{Å}^2$, $\mathbf{k} = \mathbf{p}/\hbar$, $V_{\text{vac}}(\mathbf{r})$ is the potential acting on the electron with momentum $\mathbf{p}$, σ is the vector of the Pauli matrices, and m and c are the free electron mass and the velocity of light.

In practice the previous equation is a starting point to define an effective spin–orbit Hamiltonian. For instance, in a solid the potential acting on the electron can be split into two components: a periodic one related to the lattice $V_L(\mathbf{r})$, and a nonperiodic one, $V(\mathbf{r})$, which reflects the influence of impurities, boundaries, and external applied fields. In analogy to atoms, where the effective interaction is proportional to the dot product between spin and angular momentum $\sim \mathbf{S} \cdot \mathbf{L}$, the periodic potential leads to the appearance of an effective interaction of the form,

$$H_{SO,int} = -\frac{1}{2}\sigma \cdot \mathbf{B}(\mathbf{k}), \qquad (11.2)$$

where $\mathbf{B}(\mathbf{k})$ is an effective $\mathbf{k}$-dependent magnetic field for the electron band considered, which depends on $V_L(\mathbf{r})$, and $\mathbf{k}$ is now the crystal wavevector. This spin–orbit contribution arises even in the absence of impurities and is usually referred to as *intrinsic*. When considering the nonperiodic component, the coupling has a similar form to that in vacuum, Eq. (11.1),

$$H_{SO,ext} = -\bar{\eta}_{so}\sigma \cdot [\mathbf{k} \times \nabla V(\mathbf{r})], \qquad (11.3)$$

where $\bar{\eta}_{so}$ can be orders of magnitude larger than η_{so} because of the interaction of electrons with the nuclei at velocities that are nearly relativistic. This spin–orbit contribution is usually referred to as *extrinsic*. Together with the intrinsic contribution, they give place to the spin Hall effects.

As a consequence of the spin–orbit interaction the velocity and coordinate operators become spin dependent. When an electron scatters with an impurity (Eq. 11.3), the scattering cross-section depends on the spin state and results in different scattering angles for spin-up and spin-down electrons, as represented in Fig. 11.4(a). This effect is known as Mott-skew scattering and has been recognized as a source for the spin Hall effects in early predictions [1–3]. Additionally, for impurity scattering with momentum transfer $\delta\mathbf{k}$, a lateral displacement of the electron $\delta\mathbf{r} = -\bar{\eta}_{so}[\delta\mathbf{k} \times \sigma]$ occurs, which is known as the side-jump mechanism (Fig. 11.4b) (see also Chapter 12).

The spin Hall effect can also arise from the intrinsic spin–orbit coupling in the band structure, Eq. (11.2). The basic mechanism depends on the effect of the $\mathbf{k}$-dependent magnetic field $\mathbf{B}(\mathbf{k})$. When an electric field is applied, the charge carriers in the material are accelerated. As the carrier momentum k changes, so does the effective spin–orbit field $\mathbf{B}(\mathbf{k})$ (see Fig. 11.4b and further explanation below).

The effect appears in the conduction band of asymmetric quantum wells and in the spin-3/2 valence band of GaAs described by the Luttinger model [11, 12]. The first case is described by the Rashba Hamiltonian [25],

$$H_{SO,R} = \alpha(\mathbf{k} \times \sigma) \cdot \hat{\mathbf{z}}, \qquad (11.4)$$

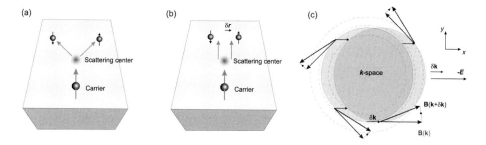

FIG. 11.4. (a) Schematic picture of the skew-scattering mechanism. An electron with spin up (down) scatters preferably with a positive (negative) angle. (b) Schematic picture of the side-jump mechanism. The path of electrons is shifted to the left (right) side of the scattering center for spin down (up) states. Schematic picture of intrinsic spin–orbit generated spin-currents. An electric field in the $-x$ direction displaces the Fermi distribution by $\delta\mathbf{k}$. Carriers experience a torque that tilts them according to their spins. The tilting is opposite for opposite momenta and it generates a spin current in the y-direction.

which corresponds to $\mathbf{B}(\mathbf{k}) = 2\alpha\hat{\mathbf{z}} \times \mathbf{k}$. The coupling parameter α depends on the well-confining potential and on an external field that may be applied by gates. There, the effective magnetic field is perpendicular to the momenta and leads to Rashba splitting for the two spin orientations (Fig. 11.5). Due to its simplicity, this model has attracted great attention and can be used to visualize the mechanism of the intrinsic spin Hall effect described above, as illustrated in

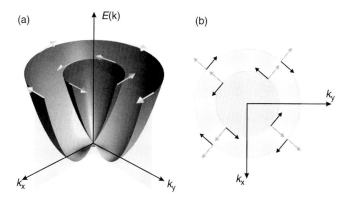

FIG. 11.5. (a) Dispersion relation and spin splitting induced by Rashba-type spin–orbit interaction in a two-dimensional electron system. (b) The effective magnetic field causes the spin (black arrows) to align perpendicular to the momenta (gray arrows).

Ref. [12] and reproduced in Fig. 11.4(c). The electric field along the x-direction displaces the Fermi surface. The electric field changes $\mathbf{k}$ and forces the electrons out of alignment with $\mathbf{B}(\mathbf{k})$. Therefore, while moving in momentum space, electrons experience an effective torque which tilts the spins up for $k_y > 0$ and down for $k_y < 0$, because the spins tilt in opposing directions on opposite sides of the Fermi surface, it creates a spin current in the y-direction (see also Chapters 13 and 16).

Because the spin–orbit interaction in solids is influenced by the nuclei, it has been reasoned that large effects should be observed in heavy elements or when heavy impurities are present. Such a simple argument is supported by experiments, which find that the magnitude of the effect is largest in materials such as Pt or Au or in light materials such as Cu with heavy impurities such as Ir or Bi. See Chapters 12 and 14 for a detailed description and comparison.

Surface and edge states in topological insulators are also the result of a large coupling between orbital and spin motion. However, contrary to the spin Hall effect, where large effects are expected in heavy elements, not all heavy compounds turn out to be topological insulators. A topological insulator is an insulator that always has a boundary that is metallic when placed in contact with an ordinary insulator or vacuum. The boundary states originate from topological invariants. Systems that have an energy gap separating the ground states from excited states can be topologically classified, and topological and ordinary insulators have different topologies: all time-reverse invariant insulators classified by a Z_2 order parameter fall into two distinct classes (see Chapter 17). Any smooth change of the Hamiltonian would not close the gap and therefore does not change the topology of the insulator, in analogy to the classic example of a coffee cup transforming into a doughnut. However, because the invariants have to change at an interface between ordinary and topological insulators, such an interface cannot remain insulating.

The first experimental observation of a topological insulator was realized in HgTe/CdTe quantum wells where an inverted electronic gap occurs because of spin–orbit interaction [21]. CdTe has a band ordering similar to GaAs with an s-like conduction band and p-like valence band. HgTe, on the other hand, has inverted bands, where the p levels are above the s levels. In a CdTe/HgTe/CdTe structure with thin HgTe the behavior is similar to that of CdTe but if the thickness of HgTe is increased, a critical value d_c should be reached where the gap closes and the bands become inverted. At this point a quantum phase transition from an ordinary to a topological insulator with protected edge states occurs [16].

11.3 The family of spin Hall effects

From the previous discussion it is clear that spin Hall effects form a large family of phenomena, which exist in the absence of magnetic fields and have the spin–orbit coupling and spin currents as a common link between them. They can be extrinsic or intrinsic, depending on the origin of the spin–orbit interaction. They are observed in insulators, in metals, or in semiconductors. From symmetry considerations it has been reasoned that, if one can generate spin currents from charge currents, the opposite should be possible and for each *direct* spin Hall effect, an *inverse* spin Hall effect is in order [1, 3]. Sometimes the experimental techniques that are used sense an indirect consequence of the spin Hall effect or a different probe to isolate it is used. In those situations, and in order to avoid confusion with other experiments, a different name for the effect is coined. In this section, we briefly summarize the main effects discussed above and the terminology found in the literature.

The oldest and most widely known effect is the anomalous Hall effect in ferromagnets [4]. Here both a transverse Hall voltage and a transverse spin accumulation are present. This is due to the combination of asymmetric spin scattering and the presence of spin polarization in the current that is inherent to ferromagnets. In the spin Hall effect, such spin polarization does not exist and therefore only the spin accumulation remains. In the inverse spin Hall effect (or spin-current induced Hall effect [26]), a spin-polarized current is applied and therefore a transverse charge current and associated Hall voltage are induced. Normally this term is used when the spin current is "pure", that is, not accompanied by a charge current. More recently, when both charge currents and spin currents are present but the spin current originates from spin injection (optical or electrical), the term spin-injection Hall effect has been used [27] (see Chapter 16).

For each of the above effects there is a related topological state, which is of intrinsic origin. In particular, the equivalent to the spin Hall effect in two dimensions is known as the quantum spin Hall effect, quantum spin Hall insulator, or just two-dimensional topological insulator. It is also referred to as the quantum version of the spin Hall effect, following the comparison between the quantum Hall and conventional Hall effects. Although time-reversal symmetry is essential in a quantum spin Hall state, it has been predicted that a related state should exist where time-reversal symmetry is broken [28]. This state is known as the quantum anomalous Hall state in analogy to the classical equivalent and it would be obtained by magnetically doping the quantum spin Hall state. Its main signature is to have only one spin-up (or spin-down) edge state. In three-dimensional systems, the predominant term is just a topological insulator, although the surface states have been refereed to as a Kramer's metal. In a doped system with magnetic impurities a ferromagnetic phase has also been discussed.

11.4 Experimental observation

Experimental evidence of the spin Hall effect in semiconductors came from spatially resolved electron spin-polarization measurements near the edges of n-type GaAs channels using Kerr rotation microscopy (Fig. 11.6) and by the polarization of the recombination radiation of holes in a light-emitting diode (LED) structure (Fig. 11.7).

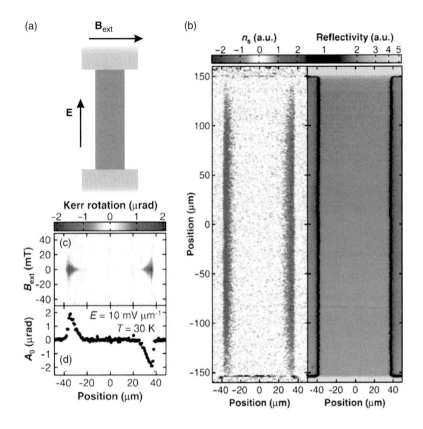

FIG. 11.6. (a) Schematics of the GaAs sample and the experimental geometry in
Kerr rotation detection of the spin Hall effect. (b) Two-dimensional images of
spin density n_s (left) and reflectivity (right) for the unstrained GaAs sample
measured at $T = 30$ K and $E = 10$ mV μm^{-1}. (c) Kerr rotation as a function
of x and $\mathbf{B}_{ext}$ for $E = 10$ mV μm^{-1}. (d) Spatial dependence of peak Kerr
rotation A_0 across the channel. Adapted from Ref. [6].

The first experiments [6] were performed on n-GaAs samples grown by
molecular beam epitaxy, on (001) semi-insulating GaAs substrates. They were
doped with Si with $n = 3 \times 10^{16}$ cm^{-3} in order to obtain long spin lifetimes.
Static Kerr rotation measurements were achieved with a pulsed Ti:sapphire
laser tuned to the absorption edge of the semiconductor with normal incidence
to the sample. In this technique, the laser beam is linearly polarized and the
polarization axis of the reflected beam is determined. The rotation angle is
proportional to the net magnetization along the beam direction. Figure 11.6(a)
shows a schematic representation of the experimental geometry. An electric
field was applied along the channel while a magnetic field B could be applied
perpendicular to it in the sample plane. Figure 11.6(b) shows a two-dimensional

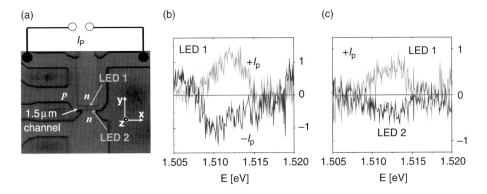

FIG. 11.7. Spin Hall experiment in a two-dimensional hole gas. (a) Scanning electron microscopy image of the device. The top (LED 1) or bottom (LED 2) n contacts are used to measure the electroluminiscence at opposite edges of the hole gas p channel. The current I_p is along the channel. (b) Polarization along the z-axis measured with active LED 1 for two opposite I_p current orientations. (c) Polarization along the z-axis measured with fixed I_p and for biased LED 1 or LED 2. Adapted from Ref. [7].

scan of the sample, which demonstrates the existence of spin accumulation close to the edges. The polarization has opposite signs at the two edges and decreases rapidly with the distance from the edge as expected from the spin Hall effect. This is clearly seen in the one-dimensional profile in Fig. 11.6(c). The magnitude of the polarization reached about 0.1 %. Further experiments demonstrated the effect of spin precession under the influence of an applied B.

Measurements were repeated in strained n-InGaAs channels but no significant crystal orientation dependence was observed, which indicates that the spin Hall effect observed in these experiments is of extrinsic origin. This is consistent with the order of magnitude of the effect and a spin Hall conductivity of about 1 (Ωm^{-1}) as expected from modeling based on scattering by screened and short-range impurities [5, 30]. Follow-up experiments demonstrated that the observed spin accumulation is due to a transverse bulk electron spin current, which can drive spin polarization tens of microns into a region in which there is minimal electric field [31]. More recently, time-resolved measurement of the dynamics of spin accumulation generated by the extrinsic spin Hall effect was also studied in n-GaAs using pumped time-resolved Kerr rotation [32]. Researchers succeeded in imaging the spin accumulation, precession, and decay dynamics. Additional experiments using the same methods investigated the spin Hall effect in a two-dimensional electron gas in (110) AlGaAs quantum wells [33] and in bulk ZnSe [34], at room temperature. All of these experiments were in close agreement with extrinsic theory.

The experiments performed in a two-dimensional hole gas in p-type GaAs (Fig. 11.7) were ascribed to the intrinsic mechanism [7, 35]. The device comprised coplanar $p - n$ junction light-emitting diodes that were fabricated in (Al,Ga)A/GaAs heterostructures grown by molecular beam epitaxy. The detection of spin polarization at the sides of a p channel, while a current I_p was applied, was performed by measuring the circular polarization of emitted light due to recombination near p–n junctions. By using two LEDs in the opposite sides of the channel, it was possible to compare the polarization of the light in the two sides and the behavior under current reversal (see also Chapter 16). The intrinsic character of the effect was further tested in Ref. [36] where polarizations in the order of 1% were observed. The signal was independent of the channel width as expected from the theory of the spin Hall effect.

The electrical detection of the spin Hall effect was elusive because the transverse spin currents do not lead to a measurable voltage. The first experiments aimed at an electrical measurement in metals therefore focused on the detection of the inverse spin Hall effect [8–10]. They used spin pumping (Fig. 11.8a) and nonlocal lateral spin injection and detection techniques (Fig. 11.8b). Both methods rely on the injection of a spin-polarized current. In nonlocal devices this current is provided by a ferromagnetic injector. A current from the injector is applied into the paramagnetic metal, which creates a pure spin current towards

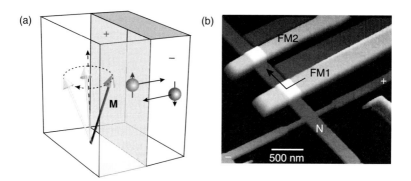

FIG. 11.8. (a) Schematic picture of spin pumping. The precession of the magnetization **M** in the ferromagnet (left) injects a spin current in the normal metal (right). Due to the inverse spin Hall effect charge is accumulated in the lateral walls. See Ref. [8]. (b) Nonlocal spin device for the detection of the inverse spin Hall effect. A Hall cross of a normal metal N (Al in this case) is contacted with two ferromagnetic electrodes widths (FM1 and FM2). A current I is injected out of FM1 into the N film and away from the Hall cross. A spin Hall voltage is measured between the two Hall probes. The second ferromagnetic electrode in this device is for control measurements. Adapted from Ref. [9].

a remote Hall cross. As the spin current flows in the metal the inverse spin Hall effect converts it into a measurable voltage (see Chapters 12 and 14). The spin-pumping method operates by ferromagnetic resonance (see Chapter 15). In this case, the magnetization precession results in the emission of a spin current into a paramagnetic metal in contact with the ferromagnet and again a voltage signal develops in the metal due to the inverse spin Hall effect. These experiments are related to pioneer work in semiconductors where the spin injection is obtained by optical orientation [37] and to the detection of the inverse spin Hall effect with optically generated spin currents in intrinsic GaAs using a two-color optical technique with orthogonally polarized laser pulses [38].

The above electrical methods are extremely useful from a practical point of view. Recent advances permitted the study of spin Hall effects in a variety of metals and even semiconductors [39, 40] and to engineer the spin Hall angle by the addition of impurities [41]. They opened the door for the discovery of novel phenomena such as the spin Seebeck effect [42] (see Chapter 18) and may find applications in spintronics and thermoelectricity, for example for memory technologies via spin-torque switching originating from the spin current generation due to the Rashba effect [43] or the spin Hall effect [44] (see Chapter 24) or for energy harvesting.

Finally, the first signatures of the quantum spin Hall state in HgTe were obtained from measurements of the electrical conductance [21] (Fig. 11.9a). A quantized conductance of $2e^2/h$ was observed and associated to the pairs of states in the edges. This and subsequent experiments, which established the nonlocality of the edge states, can be understood within the Landauer–Buttiker formalism [22].

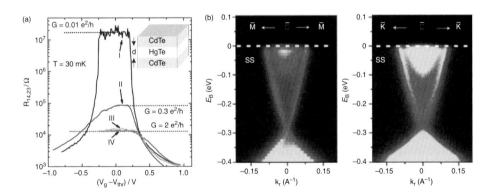

FIG. 11.9.   (a) Conductance as a function of gate voltage that tunes the Fermi energy through the bulk gap. Sample I ($d < d_c$) is insulating. Sample II and IV ($d < d_c$) show quantized transport associated to the edge states. Inset: HgTe/CdTe quantum well structure. Adapted from Ref. [21]. (b) ARPES data for the dispersion of the surface states of Bi_2Se_3 along directions Γ–M (left) and Γ–K in the Brillouin zone. Adapted from Ref. [24].

The experiments addressed the transport of HgTe/(Hg,Cd)Te quantum wells as a function of thickness, gate voltage, and external magnetic field. Only thick samples $d > d_c$ showed quantization behavior while thin samples $d < d_c$ were insulating. Application of a small magnetic field perpendicular to the quantum well in the thick films resulted in a transition to insulating behavior, which agrees with theoretical expectations. The edge states are robust under time-reversal symmetry. A magnet breaks this symmetry and turns on a gap on the otherwise degenerate edge states [16].

Three-dimensional topological insulators were first observed by ARPES [23, 24]. In ARPES experiments, a high-energy photon ejects an electron from the crystal. The analysis of the momentum of this emitted electron provides information on the surface and bulk electronic structure. The surface states were first observed in $Bi_x Sb_{1-x}$ alloys, and then in $Bi_2 Se_3$ and $Bi_2 Te_3$. $Bi_2 Se_3$ and $Bi_2 Te_3$ show topological behavior with the simplest surface state allowed. With a band gap in excess of 0.1 eV they conserve the topological behavior up to higher temperatures than $Bi_x Sb_{1-x}$ alloys, which have a gap of just 0.01 eV. Figure 11.9(b) shows the measured surface state of $Bi_2 Se_3$, which is similar to the idealization shown in Fig. 11.3. Additionally, interference patterns near defects or steps on the surface show that electrons are never completely reflected, as observed with scanning tunneling microscope measurements. This property also protects the surface states from Anderson localization. Further experimental results are reviewed in Chapter 17 and Refs. [45, and 46].

Acknowledgments

We acknowledge support from the Spanish Ministry of Science and Innovation, MICINN (MAT2010-18065).

References

[1] Dyakonov, M. I. and Perel, V. I. (1971). Possibility of orienting electron spins with current. *JETP Lett.*, **13**, 467.

[2] D'yakonov, M. I. and Perel', V. I. (1971). Current induced spin orientation of electrons in semiconductors. *Phys. Lett. A*, **35**, 459.

[3] Hirsch, J. E. (1999). Spin Hall effect. *Phys. Rev. Lett.*, **83**, 1834.

[4] Nagaosa, N., Sinova, J., Onoda, S., MacDonald, A. H., and Ong, N. P. (2010). Anomalous Hall effect. *Rev. Mod. Phys.*, **82**, 1539.

[5] Engel, H. A., Rashba, E. I., and Halperin, B. I. (2007). Theory of spin Hall effects in semiconductors, in *Handbook of Magnetism and Advanced Magnetic Materials*, H. Kronmüller and S. Parkin (eds.). John Wiley, New York.

[6] Kato, Y. K., Myers, R. C., Gossard, A. C., and Awschalom, D. D. (2004). Observation of the spin Hall effect in semiconductors. *Science*, **306**, 1910.

[7] Wunderlich, J., Kaestner, B., Sinova, J., and Jungwirth, T. (2005). Experimental observation of the spin-Hall effect in a two-dimensional spin-orbit coupled semiconductor system. *Phys. Rev. Lett.*, **94**, 047204.

[8] Saitoh, E., Ueda, M., Miyajima, H., and Tatara, G. (2006). Conversion of spin current into charge current at room temperature: Inverse spin-Hall effect. *Appl. Phys. Lett.*, **88**, 182509.

[9] Valenzuela, S. O. and Tinkham, M. (2006). Direct electronic measurement of the spin Hall effect. *Nature*, **442**, 176.

[10] Kimura, T., Otani, Y., Sato, T., Takahashi, S., and Maekawa, S. (2007). Room-temperature reversible spin Hall effect. *Phys. Rev. Lett.*, **98**, 156601.

[11] Murakami, S., Nagaosa, N., and Zhang, S.-C. (2003). Dissipationless quantum spin current at room temperature. *Science*, **301**, 1348.

[12] Sinova, J., Culcer, D., Niu, Q., Sinitsyn, N. A., Jungwirth, T. and MacDonald, A. H. (2004). Universal intrinsic spin Hall effect. *Phys. Rev. Lett.*, **92**, 16603.

[13] Murakami, S., Nagaosa, N., and Zhang, S.-C. (2004). Spin Hall insulator. *Phys. Rev. Lett.*, **93**, 156804.

[14] Kane, C. L., and Mele, C. J. (2005). Quantum spin Hall effect in graphene. *Phys. Rev. Lett.*, **95**, 226801.

[15] Kane, C. L., and Mele, C. J. (2005). Z2 Topological order and the quantum spin Hall effect. *Phys. Rev. Lett.*, **95**, 146802.

[16] Bernevig, B. A., Hughes, T. L., and Zhang, S.-C. (2006). Quantum spin Hall effect and topological phase transition in HgTe quantum wells. *Science*, **314**, 1757.

[17] Fu, L., Kane, C. L., and Mele, C. J. (2007). Topological insulators in three dimensions. *Phys. Rev. Lett.*, **98**, 106803.

[18] Moore, J. E. and Balents, L. (2007). Topological invariants of time-reversal-invariant band structures. *Phys. Rev. Lett.*, **75**, 121306(R).

[19] Roy, R. (2009). Topological phases and the quantum spin Hall effect in three dimensions. *Phys. Rev. B*, **75**, 195322.

[20] Qi, X.-L. and Zhang, S.-C. (2010). The quantum spin Hall effect and topological insulators. *Phys. Today*, **63**, 133.

[21] König, M., Wiedmann, S., Brüne, C., Roth, A., Buhmann, H., Molenkamp, L. W., Qi, X.-L., and Zhang, S.-C. (2007). Quantum spin Hall state in HgTe quantum wells. *Science*, **318**, 766.

[22] Roth, A., Brüne, C., Buhmann, H., Molenkamp, L. W., Maciejko, J., Qi, X.-L., and Zhang, S.-C. (2007). Nonlocal edge state transport in the quantum spin Hall state. *Science*, **325**, 294.

[23] Hsieh, D., Qian, D., Wray, L., Xia, Y., Hor, Y. S., Cava, R. J., and Hasan, M. Z. (2008). A topological Dirac insulator in a quantum spin Hall phase. *Nature*, **452**, 970.

[24] Xia, Y., Qian, D., Hiseh, D., Wray, L., Pal. A., Lin, H., Bansil, A., Grauer, D., Hor, Y. S., Cava, R. J., and Hasan, M. Z. (2009). Observation of a large-gap topological-insulator class with a single Dirac cone on the surface. *Nature Phys.*, **5**, 398.

[25] Bychkov, Y. A. and Rashba, E. I. (1984). Properties of a 2D electron gas with lifted spectral degeneracy. *JETP Lett.*, **39**, 78.

[26] Valenzuela, S. O. and Tinkham, M. (2007). Nonlocal electronic spin detection, spin accumulation and the spin Hall effect. *J. Appl. Phys.*, **101**, 09B103.

[27] Wunderlich, J., Irvine, A. C., Sinova, J., Park, B. G., Zârbo, L. P., Xu, X. L., Kaestner, B., Novák, V., and Jungwirth, T. (2009). Spin-injection Hall effect in a planar photovoltaic cell. *Nature Phys.*, **5**, 675.

[28] Fu, L., Kane, C. L., and Mele, C. J. (2007). Quantum anomalous Hall affect in $Hg_{1-y}Mn_y Te$ quantum wells. *Phys. Rev. Lett.*, **101**, 146802.

[29] Engel, H. A., Rashba, E. I., and Halperin, B. I. (2005). Theory of spin Hall conductivity in n-doped GaAs. *Phys. Rev. Lett.*, **95**, 166605.

[30] Tse, W.-K. and Das Sarma, S. (2006). Spin Hall effect in doped semiconductor structures. *Phys. Rev. Lett.*, **96**, 56601.

[31] Sih, V., Lau, W. H., Myers, R. C., Horowitz, V. R., Gossard, A. C., and Awschalom, D. D. (2006). Generating spin currents in semiconductors with the spin hall effect. *Phys. Rev. Lett.*, **97**, 096605.

[32] Stern, N. P., Steuerman, D. W., Mack, S., Gossard, A. C., and Awschalom, D. D. (2008). Time-resolved dynamics of the spin Hall effect. *Nature Phys.*, **4**, 843.

[33] Sih, V., Myers, R. C., Kato, Y. K., Lau, W. H., Gossard, A. C., and Awschalom, D. D. (2005). Spatial imaging of the spin Hall effect and current-induced polarization in two-dimensional electron gases. *Nature Phys.*, **1**, 31.

[34] Stern, N. P., Ghosh, S., Xiang, G., M. Zhu, M., Samarth, N., and Awschalom, D. D. (2006). Current-induced polarization and the spin Hall effect at room temperature. *Phys. Rev. Lett.*, **97**, 126603.

[35] Schliemann, J. and Loss, D. (2005). Spin-Hall transport of heavy holes in III-V semiconductor quantum wells. *Phys. Rev. B*, **71**, 085308.

[36] Nomura, K., Wunderlich, J., Sinova, J., Kaestner, B., MacDonald, A., and Jungwirth, T. (2005). Edge spin accumulation in semiconductor two-dimensional hole gases. *Phys. Rev. B*, **72**, 245330.

[37] Bakun, A. A., Zakharchenya, B. P., Rogachev, A. A., Tkachuk, M. N., and Fleisher, V. G. (1984). Detection of a surface photocurrent due to electron optical orientation in a semiconductor. *Sov. Phys. JETP Lett.*, **40**, 1293.

[38] Zhao, H., Loren, E. J., van Driel, H. M., and Smirl, A. L. (2006). Coherence control of Hall charge and spin currents *Phys. Rev. Lett.*, **96**, 24660.

[39] Garlid, E. S., Hu, Q. O., Chan, M. K., Palmstrøm, C. J., and Crowell, P. A. (2010). Electrical measurement of the direct spin Hall effect in Fe/In$_x$Ga$_{1-x}$As heterostructures. *Phys. Rev. Lett.*, **105**, 156602.

[40] Brüne, C., Roth, A., Novik, E. G., König, M., Buhmann, H., Hankiewicz, E. M., Hanke, W., Sinova, J., and Molenkamp, L. W. (2010). Evidence for the ballistic intrinsic spin Hall effect in HgTe nanostructures. *Nature Phys.*, **6**, 448.

[41] Niimi, Y., Morota, M., Wei, D. H., Deranlot, C., Basletic, M., Hamzic, A., Fert, A., and Otani, Y. (2011). Extrinsic spin Hall effect induced by iridium impurities in copper. *Phys. Rev. Lett.*, **106**, 126601.

[42] Uchida, K., Takahashi, S., Harii, K., Ieda, J., Koshibae, W., Ando, K., Maekawa, S., and Saitoh, E. (2008). Observation of the spin Seebeck effect. *Nature*, **455**, 778.

[43] Miron, I. M., Garello, K., Gaudin, G., Zermatten, P.-J., Costache, M. V., Auffret, S., Bandiera, S., Rodmacq, B., Schuhl, A., and Gambardella, P. (2011). Perpendicular switching of a single ferromagnetic layer induced by in-plane current injection. *Nature*, **476**, 189.

[44] Liu, L. Q., Moriyama, T., Ralph, D. C., and Buhrman, R. A. (2011). Spin torque ferromagnetic resonance induced by the Spin Hall effect. *Phys. Rev. Lett.*, **106**, 036601.

[45] Hasan, M. Z. and Kane, C. L. (2010). Topological insulators. *Rev. Mod. Phys.*, **82**, 3045.

[46] Qi, X.-L. and Zhang, S.-C. (2011). Topological insulators and superconductors. *Rev. Mod. Phys.*, **83**, 1057.

12 Spin Hall effect

S. Maekawa and S. Takahashi

12.1 Introduction

The anomalous Hall effect (AHE) originates from the spin–orbit interaction between the spin and orbital motion of electrons in metals or semiconductors [1, 2]. Conduction electrons are scattered by local potentials created by impurities or defects in a crystal. The spin–orbit interaction in local potentials causes a spin-asymmetric scattering of conduction electrons. In ferromagnetic materials, the electrical current is carried by up-spin (majority) and down-spin (minority) electrons, in which the flow of up-spin electrons is slightly deflected in a transverse direction while that of down-spin electrons is deflected in the opposite direction, resulting in an electron flow in the direction perpendicular to both the applied electric field and the magnetization directions. Since up-spin and down-spin electrons are strongly imbalanced in ferromagnets, both spin and charge currents are generated in the transverse direction by AHE, the latter of which are observed as the electrical Hall voltage.

Spin injection from a ferromagnet (F) to a nonmagnetic conductor (N) in nanostructured devices [5–14] provides a new opportunity for investigating AHE in nonmagnetic conductors, the so-called "spin Hall effect" (SHE) [15–24]. A pure spin current is created in N via nonlocal spin injection, in which up-spin and down-spin electrons flow in equal amounts but in opposite directions without charge current, and therefore the up-spin and down-spin currents flowing in opposite directions are deflected in the same direction to induce a charge current in the transverse direction, resulting in a charge accumulation on the edges of N. Inversely, when an unpolarized charge current flows in N as a result of an applied electric field, the up-spin and down-spin currents flowing in the same direction are deflected in the opposite direction to induce a pure spin current in the transverse direction, resulting in a spin accumulation with the spin diffusion length from the edges of N. As a consequence, the spin (charge) degrees of freedom are converted to charge (spin) degrees of freedom because of spin–orbit scattering in nonmagnetic conductors. Recently, SHE has been observed using nonlocal spin injection in metal-based nanostructured devices [7, 9–12]. In addition to these extrinsic SHEs, intrinsic SHEs have been intensively studied in metals and semiconductors which do not require impurities or defects [25–30].

In this chapter, we consider the effect of spin–orbit scattering on spin and charge transport in nonmagnetic metals (N) such as Cu, Al, and Ag, and discuss SHE by taking into account the side-jump (SJ) and skew-scattering (SS) mechanisms [31, 32], and derive formulas for the SHE induced by spin–orbit scattering in nonmagnetic metals.

12.2   Spin Hall effect due to side jump and skew scattering in diffusive metals

The spin–orbit interaction in the presence of nonmagnetic impurities in a metal is derived as follows [34]. The impurity potential $u(\mathbf{r})$ gives rise to an additional electric field $\mathbf{E} = -(1/e)\nabla u(\mathbf{r})$. When an electron passes through the field with velocity $\hat{\mathbf{p}}/m = (\hbar/i)\nabla/m$, the electron feels an effective magnetic field $\mathbf{B}_{\text{eff}} = -(1/mc)\hat{\mathbf{p}} \times \mathbf{E}$, which leads to the spin-orbit coupling

$$u_{\text{so}}(\mathbf{r}) = \mu_{\text{B}}\boldsymbol{\sigma} \cdot \mathbf{B}_{\text{eff}} = \eta_{\text{so}}\boldsymbol{\sigma} \cdot [\nabla u(\mathbf{r}) \times \nabla/i], \tag{12.1}$$

where $\boldsymbol{\sigma}$ is the Pauli spin operator, μ_{B} is the Bohr magneton, and η_{so} is the spin–orbit coupling parameter. Though the value of $\eta_{\text{so}} = (\hbar/2mc)^2$ in the free-electron model is too small to account for SHE as well as AHE observed in experiments, the value of η_{so} in real metals may be enhanced by several orders of magnitude for Bloch electrons [1]. In the following, η_{so} is treated as a phenomenological (renormalized) parameter. The total impurity potential $U(\mathbf{r})$ is the sum of the ordinary impurity potential and the spin–orbit potential: $U(\mathbf{r}) = u(\mathbf{r}) + u_{\text{so}}(\mathbf{r})$.

The one-electron Hamiltonian H in the presence of the impurity potential $U(\mathbf{r})$ is given by

$$H = \sum_{\mathbf{k},\sigma} \xi_{\mathbf{k}} a^{\dagger}_{\mathbf{k}\sigma} a_{\mathbf{k}\sigma} + \sum_{\mathbf{k},\mathbf{k}'} \sum_{\sigma,\sigma'} \langle \mathbf{k}'\sigma'|U|\mathbf{k}\sigma\rangle a^{\dagger}_{\mathbf{k}'\sigma'} a_{\mathbf{k}\sigma}. \tag{12.2}$$

Here, the first term is the kinetic energy of conduction electrons with one-electron energy $\xi_{\mathbf{k}} = (\hbar k)^2/2m - \varepsilon_{\text{F}}$ measured from the Fermi level ε_{F}, and the second term describes the scattering of conduction electrons between different momentum and spin states with the scattering amplitude

$$\langle \mathbf{k}'\sigma'|U|\mathbf{k}\sigma\rangle = u_{\mathbf{k}'\mathbf{k}}\delta_{\sigma'\sigma} + i\eta_{\text{so}}u_{\mathbf{k}'\mathbf{k}}\left[\boldsymbol{\sigma}_{\sigma'\sigma} \cdot (\mathbf{k}' \times \mathbf{k})\right], \tag{12.3}$$

where $u_{\mathbf{k}'\mathbf{k}} = \langle \mathbf{k}'|u|\mathbf{k}\rangle$ and the first and second terms are the the matrix elements of ordinary and spin–orbit potentials, respectively. For a short-range impurity potential, $u(\mathbf{r}) \approx u_{\text{imp}} \sum_i \delta(\mathbf{r} - \mathbf{r}_i)$, at position $\mathbf{r}_i$ and $u_{\mathbf{k}'\mathbf{k}} \approx (u_{\text{imp}}/V) \sum_i e^{i(\mathbf{k}-\mathbf{k}')\cdot\mathbf{r}_i}$, where V is the volume.

The velocity $\mathbf{v}_{\mathbf{k}}^{\sigma}$ of an electron in the presence of the spin–orbit potential is calculated by taking the matrix element $\mathbf{v}_{\mathbf{k}}^{\sigma} = \langle \mathbf{k}^+\sigma|\hat{\mathbf{v}}|\mathbf{k}^+\sigma\rangle$ of the velocity operator

$$\hat{\mathbf{v}} = d\mathbf{r}/dt = (i\hbar)^{-1}[\mathbf{r}, H] = \hat{\mathbf{p}}/m + (\eta_{\text{so}}/\hbar)[\boldsymbol{\sigma} \times \nabla u(\mathbf{r})] \tag{12.4}$$

between the scattering state $|\mathbf{k}^+\sigma\rangle = |\mathbf{k}\sigma\rangle + \sum_{\mathbf{k}'} u_{\mathbf{k}'\mathbf{k}}(\xi_{\mathbf{k}} - \xi_{\mathbf{k}'} + i\delta)^{-1}|\mathbf{k}'\sigma\rangle$ within the Born approximation, and becomes

$$\mathbf{v}_{\mathbf{k}}^\sigma = \mathbf{v}_{\mathbf{k}} + \boldsymbol{\omega}_{\mathbf{k}}^\sigma, \qquad \boldsymbol{\omega}_{\mathbf{k}}^\sigma = \alpha_{\mathrm{SH}}^{\mathrm{SJ}}\left(\boldsymbol{\sigma}_{\sigma\sigma} \times \mathbf{v}_{\mathbf{k}}\right), \tag{12.5}$$

where $\mathbf{v}_{\mathbf{k}} = \hbar\mathbf{k}/m$ is the usual velocity, $\boldsymbol{\omega}_{\mathbf{k}}^\sigma$ is the anomalous velocity, $\boldsymbol{\sigma}_{\sigma\sigma} = \sigma\hat{\mathbf{z}}$ is the polarization vector, and $\alpha_{\mathrm{SH}}^{\mathrm{SJ}}$ is the spin Hall angle due to side jump

$$\alpha_{\mathrm{SH}}^{\mathrm{SJ}} = \frac{\hbar\bar{\eta}_{\mathrm{so}}}{2\varepsilon_{\mathrm{F}}\tau_{\mathrm{tr}}^0} = \frac{\bar{\eta}_{\mathrm{so}}}{k_{\mathrm{F}}l}, \tag{12.6}$$

where $\tau_{\mathrm{tr}}^0 = [(2\pi/\hbar)n_{\mathrm{imp}}N(0)u_{\mathrm{imp}}^2]^{-1}$ is the scattering time due to impurities, n_{imp} is the impurity concentration, $\bar{\eta}_{\mathrm{so}} = k_{\mathrm{F}}^2\eta_{\mathrm{so}}$ is the dimensionless spin–orbit coupling parameter, k_{F} is the Fermi momentum, and l is the mean-free path.

Introducing the current operator $\hat{\mathbf{J}}_\sigma = e\sum_{\mathbf{k}}(\mathbf{v}_{\mathbf{k}} + \boldsymbol{\omega}_{\mathbf{k}}^\sigma) a_{\mathbf{k}\sigma}^\dagger a_{\mathbf{k}\sigma}$ for the spin channel σ, the total charge current $\mathbf{J}_q = \mathbf{J}_\uparrow + \mathbf{J}_\downarrow$ and the total spin current $\mathbf{J}_s = \mathbf{J}_\uparrow - \mathbf{J}_\downarrow$ are expressed as

$$\mathbf{J}_q = \mathbf{J}_q' + \alpha_{\mathrm{SH}}^{\mathrm{SJ}}\left(\hat{\mathbf{z}} \times \mathbf{J}_s'\right), \qquad \mathbf{J}_s = \mathbf{J}_s' + \alpha_{\mathrm{SH}}^{\mathrm{SJ}}\left(\hat{\mathbf{z}} \times \mathbf{J}_q'\right), \tag{12.7}$$

where

$$\mathbf{J}_q' = e\sum_{\mathbf{k}}\mathbf{v}_{\mathbf{k}}\left(f_{\mathbf{k}\uparrow} + f_{\mathbf{k}\downarrow}\right), \qquad \mathbf{J}_s' = e\sum_{\mathbf{k}}\mathbf{v}_{\mathbf{k}}\left(f_{\mathbf{k}\uparrow} - f_{\mathbf{k}\downarrow}\right) \tag{12.8}$$

are the charge and spin currents for electrons with velocity $\mathbf{v}_{\mathbf{k}}$ and the distribution function $f_{\mathbf{k}\sigma} = \langle a_{\mathbf{k}\sigma}^\dagger a_{\mathbf{k}\sigma}\rangle$ of momentum $\mathbf{k}$ and spin σ. The second terms in Eqs. (12.7) are the charge and spin currents due to side jump. In addition to the side-jump contribution, there is the skew-scattering contribution, which originates from the modification of $f_{\mathbf{k}\sigma}$ in $\mathbf{J}_q'$ and $\mathbf{J}_s'$ due to the asymmetric scattering by the spin–orbit interaction.

The distribution function $f_{\mathbf{k}\sigma}$ is calculated based on the Boltzmann transport equation in the steady state,

$$\mathbf{v}_{\mathbf{k}} \cdot \nabla f_{\mathbf{k}\sigma} + \frac{e\mathbf{E}}{\hbar} \cdot \nabla_{\mathbf{k}} f_{\mathbf{k}\sigma} = \left(\frac{\partial f_{\mathbf{k}\sigma}}{\partial t}\right)_{scatt}, \tag{12.9}$$

where $\mathbf{E}$ is the external electric field and the collision term due to impurity scattering is written as [36]

$$\left(\frac{\partial f_{\mathbf{k}\sigma}}{\partial t}\right)_{scatt} = \sum_{\mathbf{k}'\sigma'}\left[P_{\mathbf{k}\mathbf{k}'}^{\sigma\sigma'} f_{\mathbf{k}'\sigma'} - P_{\mathbf{k}'\mathbf{k}}^{\sigma'\sigma} f_{\mathbf{k}\sigma}\right], \tag{12.10}$$

where the first and second terms in the bracket represent the scattering-in $(\mathbf{k}'\sigma' \to \mathbf{k}\sigma)$ and the scattering-out $(\mathbf{k}\sigma \to \mathbf{k}'\sigma')$, respectively,

$$P_{\mathbf{k}'\mathbf{k}}^{\sigma'\sigma} = (2\pi/\hbar)n_{\mathrm{imp}}|\langle\mathbf{k}'\sigma'|\hat{T}|\mathbf{k}\sigma\rangle|^2\delta(\xi_{\mathbf{k}} - \xi_{\mathbf{k}'})$$

is the scattering probability from state $|\mathbf{k}\sigma\rangle$ to state $|\mathbf{k}'\sigma'\rangle$, and $\hat{T}$ is the scattering matrix, whose matrix elements are calculated up to the second-order Born approximation as

$$\langle \mathbf{k}'\sigma'|\hat{T}|\mathbf{k}\sigma\rangle = \left[u_{\mathbf{k}'\mathbf{k}} + \sum_{\mathbf{k}''} \frac{u_{\mathbf{k}'\mathbf{k}''}u_{\mathbf{k}''\mathbf{k}}}{\xi_{\mathbf{k}} - \xi_{\mathbf{k}''} + i\delta} \right] \delta_{\sigma'\sigma} + i\eta_{\mathrm{so}} u_{\mathbf{k}'\mathbf{k}}(\mathbf{k}' \times \mathbf{k}) \cdot \boldsymbol{\sigma}_{\sigma'\sigma}. \quad (12.11)$$

After averaging over impurity positions, we find that the scattering probability has the symmetric (non-skew) contribution

$$P_{\mathbf{k}'\mathbf{k}}^{\sigma'\sigma\,(1)} = \frac{2\pi}{\hbar} \frac{n_{\mathrm{imp}}}{V} u_{\mathrm{imp}}^2 \left(\delta_{\sigma\sigma'} + \eta_{\mathrm{so}}^2 \left| (\mathbf{k}' \times \mathbf{k}) \cdot \boldsymbol{\sigma}_{\sigma\sigma'} \right|^2 \right) \delta(\xi_{\mathbf{k}'} - \xi_{\mathbf{k}}), \quad (12.12)$$

and the asymmetric (skew) contribution

$$P_{\mathbf{k}'\mathbf{k}}^{\sigma'\sigma\,(2)} = -\frac{(2\pi)^2}{\hbar} \eta_{\mathrm{so}} \frac{n_{\mathrm{imp}}}{V} u_{\mathrm{imp}}^3 N(0) \delta_{\sigma\sigma'} \left[(\mathbf{k}' \times \mathbf{k}) \cdot \boldsymbol{\sigma}_{\sigma\sigma'} \right] \delta(\xi_{\mathbf{k}'} - \xi_{\mathbf{k}}). \quad (12.13)$$

The skew contribution arises from third order with respect to the scattering potential, i.e. first order in the asymmetric potential $u_{\mathrm{so}}(\mathbf{r})$ and second order in the symmetric potential $u(\mathbf{r})$.

In solving the Boltzmann equation, it is convenient to separate $f_{\mathbf{k}\sigma}$ into three parts

$$f_{\mathbf{k}\sigma} = f_{k\sigma}^0 + g_{\mathbf{k}\sigma}^{(1)} + g_{\mathbf{k}\sigma}^{(2)}, \quad (12.14)$$

where $f_{k\sigma}^0$ is a nondirectional distribution function, and $g_{\mathbf{k}\sigma}^{(1)}$ and $g_{\mathbf{k}\sigma}^{(2)}$ are the directional distribution functions which vanish by averaging with respect to the solid angle $\Omega_{\mathbf{k}}$ of $\mathbf{k}$, i.e. $\int g_{\mathbf{k}\sigma}^{(i)} d\Omega_{\mathbf{k}} = 0$, and are related to the symmetric and asymmetric contributions, respectively.

We first consider spin transport in the absence of skew scattering, in which case the Boltzmann equation becomes [17, 37]

$$\boldsymbol{v}_{\mathbf{k}} \cdot \nabla f_{\mathbf{k}\sigma} + \frac{e\mathbf{E}}{\hbar} \cdot \nabla_{\mathbf{k}} f_{\mathbf{k}\sigma} = -\frac{g_{\mathbf{k}\sigma}^{(1)}}{\tau_{\mathrm{tr}}} - \frac{f_{k\sigma}^0 - f_{k-\sigma}^0}{\tau_{\mathrm{sf}}(\theta)}, \quad (12.15)$$

where $\tau_{\mathrm{tr}}^{-1} = \sum_{\mathbf{k}'\sigma'} P_{\mathbf{k}\mathbf{k}'}^{\sigma\sigma'\,(1)} = (1/\tau_{\mathrm{tr}}^0)\left(1 + 2\bar{\eta}_{\mathrm{so}}^2/3\right)$ is the transport relaxation time, $\tau_{\mathrm{sf}}^{-1}(\theta) = \sum_{\mathbf{k}'} P_{\mathbf{k}\mathbf{k}'}^{\uparrow\downarrow\,(1)} = (\bar{\eta}_{\mathrm{so}}^2/3\tau_{\mathrm{tr}}^0)\left(1 + \cos^2\theta\right)$ is the spin-flip relaxation time, and θ is the angle between $\mathbf{k}$ and the z axis. In Eq. (12.15), the first term in the right-hand side describes the momentum relaxation due to impurity scattering and the second term the spin relaxation due to spin-flip scattering. Since $\tau_{\mathrm{tr}} \ll \tau_{\mathrm{sf}}$, momentum relaxation occurs first, followed by slow spin relaxation.

The distribution function $f_{\mathbf{k}\sigma}^0$ describes the spin accumulation by the shift in the chemical potential ε_F^σ from the equilibrium one ε_F and is expanded as

$$f_{\mathbf{k}\sigma}^0 \approx f_0(\xi_\mathbf{k}) + \left(-\frac{\partial f_0}{\partial \xi_\mathbf{k}}\right)(\varepsilon_F^\sigma - \varepsilon_F), \tag{12.16}$$

where $f_0(\xi_\mathbf{k})$ is the Fermi distribution function. Replacing $f_{\mathbf{k}\sigma}$ in Eq. (12.15) with $f_{\mathbf{k}\sigma}^0$ and disregarding the term of order of $\tau_{\mathrm{tr}}/\tau_{\mathrm{sf}}$, we obtain

$$g_{\mathbf{k}\sigma}^{(1)} \approx -\tau_{\mathrm{tr}}\left(-\frac{\partial f_0}{\partial \xi_\mathbf{k}}\right)\boldsymbol{v}_\mathbf{k} \cdot \nabla \mu_N^\sigma, \tag{12.17}$$

where $\mu_N^\sigma = \varepsilon_F^\sigma + e\phi$ is the electrochemical potential (ECP) and ϕ is the electric potential ($\mathbf{E} = -\nabla\phi$).

Substituting Eqs. (12.16) and (12.17) into the Boltzmann equation (12.15) and summing over $\mathbf{k}$, one obtains the spin diffusion equation

$$\nabla^2(\mu_N^\uparrow - \mu_N^\downarrow) = \frac{1}{\lambda_N^2}(\mu_N^\uparrow - \mu_N^\downarrow), \tag{12.18}$$

where $\lambda_N = \sqrt{D\tau_S}$ is the spin diffusion length, $D = (1/3)\tau_{\mathrm{tr}}v_F^2$ is the diffusion constant, v_F is the Fermi velocity, $\tau_S = \tau_{\mathrm{sf}}/2$ is the spin relaxation time, and τ_{sf} is the spin-flip relaxation time defined by $\tau_{\mathrm{sf}}^{-1} = \langle \tau_{\mathrm{sf}}^{-1}(\theta)\rangle_{\mathrm{av}}$. The ratio of the transport relaxation time to the spin-flip relaxation time is related to the spin-orbit coupling parameter:

$$\tau_{\mathrm{tr}}/\tau_{\mathrm{sf}} \approx (4/9)\bar{\eta}_{\mathrm{so}}^2. \tag{12.19}$$

The asymmetric part of the distribution function $g_{\mathbf{k}\sigma}^{(2)}$ due to skew scattering is determined by the asymmetric terms of the Boltzmann equation $\sum_{\mathbf{k}'\sigma'}[-P_{\mathbf{k}'\mathbf{k}}^{\sigma'\sigma(1)}g_{\mathbf{k}\sigma}^{(2)} + P_{\mathbf{k}'\mathbf{k}}^{\sigma'\sigma(2)}g_{\mathbf{k}'\sigma'}^{(1)}] = 0$, which, together with Eqs. (12.12), (12.13), and (12.17), yields

$$g_{\mathbf{k}\sigma}^{(2)} = \alpha_{\mathrm{SH}}^{\mathrm{SS}}\tau_{\mathrm{tr}}\left(-\frac{\partial f_0}{\partial \xi_\mathbf{k}}\right)(\boldsymbol{\sigma}_{\sigma\sigma} \times \boldsymbol{v}_\mathbf{k}) \cdot \nabla \mu_N^\sigma(\mathbf{r}), \tag{12.20}$$

where $\alpha_{\mathrm{SH}}^{\mathrm{SS}}$ is the spin Hall angle due to skew scattering

$$\alpha_{\mathrm{SH}}^{\mathrm{SS}} = -(2\pi/3)\bar{\eta}_{\mathrm{so}}N(0)u_{\mathrm{imp}}. \tag{12.21}$$

Therefore the distribution function $f_{\mathbf{k}\sigma}$ becomes

$$\begin{aligned} f_{\mathbf{k}\sigma} \approx {} & f_0(\xi_\mathbf{k}) + \left(-\frac{\partial f_0}{\partial \xi_\mathbf{k}}\right)(\varepsilon_F^\sigma - \varepsilon_F) \\ & - \tau_{\mathrm{tr}}\left(-\frac{\partial f_0}{\partial \xi_\mathbf{k}}\right)[\boldsymbol{v}_\mathbf{k} + \alpha_{\mathrm{SH}}^{\mathrm{SS}}\boldsymbol{\sigma}_{\sigma\sigma} \times \boldsymbol{v}_\mathbf{k}] \cdot \nabla \mu_N^\sigma(\mathbf{r}). \end{aligned} \tag{12.22}$$

12.3 Spin and charge currents

Using the distribution function $f_{\mathbf{k}\sigma}$ obtained above, the equations in Eq. (12.8) are calculated as $\mathbf{J}'_q = \mathbf{j}_q + \alpha_{\mathrm{SH}}^{\mathrm{SS}}(\hat{\mathbf{z}} \times \mathbf{j}_s)$ and $\mathbf{J}'_s = \mathbf{j}_s + \alpha_{\mathrm{SH}}^{\mathrm{SS}}(\hat{\mathbf{z}} \times \mathbf{j}_q)$, where the first terms are the longitudinal charge and spin currents:

$$\mathbf{j}_q = \sigma_{\mathrm{N}}\mathbf{E}, \qquad \mathbf{j}_s = -(\sigma_{\mathrm{N}}/2e)\nabla\delta\mu_{\mathrm{N}}, \tag{12.23}$$

with the electrical conductivity $\sigma_{\mathrm{N}} = 2e^2 N(0)D$ and the chemical potential splitting $\delta\mu_{\mathrm{N}} = (\mu_{\mathrm{N}}^{\uparrow} - \mu_{\mathrm{N}}^{\downarrow})$, and the second terms are the Hall charge and spin currents caused by skew scattering, respectively. Therefore, the total charge and spin currents in Eq. (12.7) are written as

$$\mathbf{J}_q = \mathbf{j}_q + \alpha_{\mathrm{SH}}(\hat{\mathbf{z}} \times \mathbf{j}_s), \tag{12.24}$$

$$\mathbf{J}_s = \mathbf{j}_s + \alpha_{\mathrm{SH}}(\hat{\mathbf{z}} \times \mathbf{j}_q), \tag{12.25}$$

where $\alpha_{\mathrm{SH}} = \alpha_{\mathrm{SH}}^{\mathrm{SJ}} + \alpha_{\mathrm{SH}}^{\mathrm{SS}}$. Equations (12.24) and (12.25) indicate that the spin current $\mathbf{j}_s$ induces the transverse charge current $\mathbf{j}_q^{\mathrm{SH}} = \alpha_{\mathrm{SH}}(\hat{\mathbf{z}} \times \mathbf{j}_s)$, and the charge current $\mathbf{j}_q$ induces the transverse spin current $\mathbf{j}_s^{\mathrm{SH}} = \alpha_{\mathrm{SH}}(\hat{\mathbf{z}} \times \mathbf{j}_q)$.

Figures 12.1 (a) and (b) show the SHE and ISHE in a stripe film. In the case of SHE, the external current $\mathbf{j}_q$ in the x-direction is converted to the spin current $\mathbf{j}_s^{\mathrm{SH}}$ in the y-direction via SHE and accumulates spin near the edges of the stripe, which in turn generates the counter spin current $\mathbf{j}_s$ so as to satisfy the boundary condition that the spin current vanishes $\mathbf{J}_s(\pm w/2) = 0$ at the edges. The resulting spin splitting due to spin accumulation is given by [17]

$$\delta\mu_{\mathrm{N}} = 2\alpha_{\mathrm{SH}}e\rho_{\mathrm{N}}\lambda_{\mathrm{N}}j_q \frac{\sinh(y/\lambda_{\mathrm{N}})}{\cosh(w/2\lambda_{\mathrm{N}})}, \tag{12.26}$$

where w is the stripe width, indicating the spin accumulation is built up within the spin diffusion length from the stripe edges. In the case of ISHE, the spin current $\mathbf{j}_s$ in the x-direction is converted to the charge current $\mathbf{j}_q^{\mathrm{SH}}$ in the

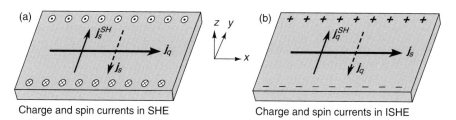

Charge and spin currents in SHE   Charge and spin currents in ISHE

FIG. 12.1. (a) Spin Hall effect (SHE). An applied charge current j_q induces spin currents and spin accumulation in the transverse y-direction. (b) Inverse spin Hall effect (ISHE). An injected spin current j_s induces charge currents and charge accumulation in the transverse y-direction.

y-direction via SHE and accumulates surface charge on the sample edges, by which the transverse electric field builds up to flow the counter charge current $\mathbf{j}_q$ so as to make the total charge current vanish $\mathbf{J}_q = 0$. The resulting Hall voltage is given later [see Eq. (12.33)].

The spin Hall conductivity is given by the sum of the side-jump and skew-scattering contributions: $\sigma_{\mathrm{SH}} = \sigma_{\mathrm{SH}}^{\mathrm{SJ}} + \sigma_{\mathrm{SH}}^{\mathrm{SS}}$. The SJ conductivity

$$\sigma_{\mathrm{SH}}^{\mathrm{SJ}} = \alpha_{\mathrm{SH}}^{\mathrm{SJ}}\sigma_{\mathrm{N}} = (e^2/\hbar)\eta_{\mathrm{so}}n_e \tag{12.27}$$

(n_e is the carrier density) is independent of the impurity concentration. By contrast, the SS conductivity

$$\sigma_{\mathrm{SH}}^{\mathrm{SS}} = \alpha_{\mathrm{SH}}^{\mathrm{SS}}\sigma_{\mathrm{N}} = -(2\pi/3)\bar{\eta}_{\mathrm{so}}[N(0)u_{\mathrm{imp}}]\sigma_{\mathrm{N}} \tag{12.28}$$

depends on the strength, sign, and distribution of impurity potentials. When the impurities have a narrow distribution of potentials with definite sign, as in doped impurities, the SS contribution is dominant for SHE, whereas when the impurity potentials are distributed with positive and negative contributions and their average over the impurity distribution vanishes ($\langle u_{\mathrm{imp}} \rangle \approx 0$), then the SJ contribution is dominant. The spin Hall resistivity $\rho_{\mathrm{SH}} \approx \sigma_{\mathrm{SH}}/\sigma_{\mathrm{N}}^2$ has linear and quadratic terms in ρ_{N} representing the contributions from side jump and skew scatterings, respectively:

$$\rho_{\mathrm{SH}} \approx a_{\mathrm{SS}}\rho_{\mathrm{N}} + b_{\mathrm{SJ}}\rho_{\mathrm{N}}^2, \tag{12.29}$$

where $a_{\mathrm{SS}} = -(2\pi/3)\bar{\eta}_{\mathrm{so}}N(0)u_{\mathrm{imp}}$ and $b_{\mathrm{SJ}} = (2/3\pi)\bar{\eta}_{\mathrm{so}}(e^2/h)k_{\mathrm{F}}$.

12.4 Spin–orbit coupling

The electrical resistivity and the spin diffusion length are key parameters for spin and charge transport [48]. By writing the product of the resistivity ρ_{N} and the spin diffusion length λ_{N}, we obtain $\rho_{\mathrm{N}}\lambda_{\mathrm{N}} = (\sqrt{3}\pi/2k_{\mathrm{F}}^2)(h/e^2)(\tau_{\mathrm{sf}}/2\tau_{\mathrm{tr}})^{1/2}$, where k_{F} is the Fermi momentum, $h/e^2 \approx 25.8\,\mathrm{k}\Omega$, and $(\tau_{\mathrm{tr}}/\tau_{\mathrm{sf}}) \approx (4/9)\bar{\eta}_{\mathrm{so}}^2$. Thus, we have a simple relation between $\bar{\eta}_{\mathrm{so}}$ and $\rho_{\mathrm{N}}\lambda_{\mathrm{N}}$ [22, 49]:

$$\bar{\eta}_{\mathrm{so}} \approx \frac{3\sqrt{3}\pi}{4\sqrt{2}}\frac{1}{k_{\mathrm{F}}^2}\frac{h}{e^2}\frac{1}{\rho_{\mathrm{N}}\lambda_{\mathrm{N}}}, \tag{12.30}$$

which implies that the spin–orbit coupling parameter $\bar{\eta}_{\mathrm{so}}$ is readily obtained by measuring ρ_{N} and λ_{N}, providing a useful method of evaluating the spin–orbit coupling in nonmagnetic metals. Table 12.1 shows the experimental data of ρ_{N} and λ_{N} for various metals and the values of the spin–orbit coupling parameter $\bar{\eta}_{\mathrm{so}}$ estimated from Eq. (12.30). It is noteworthy that $\bar{\eta}_{\mathrm{so}}$ is small for Al (light metal), large for Pt (heavy metal), and intermediate for Cu, Ag, and Au. In the case of Al, the spin–orbit coupling parameters of different samples are very close to each other, despite the scattered values of λ_{N} and ρ_{N} in those samples.

Table 12.1 Spin–orbit coupling parameter $\bar{\eta}_{\text{so}} = k_{\text{F}}^2 \eta_{\text{so}}$ for Al, Mg, Cu, Ag, Au, and Pt. Here, the Fermi momenta, $k_{\text{F}} = 1.75 \times 10^8$ cm^{-1} (Al), 1.36×10^8 cm^{-1} (Mg, Cu), 1.20×10^8 cm^{-1} (Ag), and 1.21×10^8 cm^{-1} (Au) are taken [38], and 1×10^8 cm^{-1} (Pt) is assumed.

	λ_{N} (nm)	ρ_{N} ($\mu\Omega$ cm)	$\tau_{\text{sf}}/\tau_{\text{tr}}$	$\bar{\eta}_{\text{so}}$	Ref.
Al (4.2K)	650	5.90	5.6×10^4	0.0063	[6]
Al (4.2K)	455	9.53	7.2×10^4	0.0056	[7]
Al (4.2K)	705	5.88	6.5×10^4	0.0059	[7]
Al (4.2K)	850	4.00	4.4×10^4	0.0072	[39]
Mg (10K)	720	4.00	5.8×10^3	0.014	[40]
Cu (4.2K)	1000	1.43	2.8×10^3	0.028	[5]
Cu (50K)	1300	0.76	1.1×10^3	0.045	[41]
Cu (4.2K)	546	3.44	4.9×10^3	0.021	[42]
Ag (4.2K)	162	4.00	3.7×10^2	0.079	[43]
Ag (4.2K)	195	3.50	4.1×10^2	0.075	[43]
Ag (10K)	920	1.22	1.7×10^3	0.046	[45]
Ag (77K)	3000	1.10	9.2×10^3	0.016	[44]
Au (4.2K)	168	4.00	3.8×10^2	0.077	[46]
Au (10K)	63	1.36	6.2	0.27	[47]
Pt (5K)	14	12.4	13	0.42	[10]

The values of $\bar{\eta}_{\text{so}}$ extracted from the spin injection method are several orders of magnitude larger than the value of $\bar{\eta}_{\text{so}} = (\hbar k_{\text{F}}/2mc)^2$ in the free-electron model.

In the side jump, with the aid of the relation $(l/\lambda_{\text{N}}) = (6\tau_{\text{tr}}/\tau_{\text{sf}})^{1/2} = (8/3)^{1/2}\bar{\eta}_{\text{so}}$, the spin Hall angle (12.6) is rewritten in the form

$$\alpha_{\text{SH}}^{\text{SJ}} = (3/8)^{1/2}\frac{1}{k_{\text{F}}\lambda_{\text{N}}} \tag{12.31}$$

which depends only on the Fermi momentum and the spin diffusion length, enabling us to estimate the spin Hall angle and conductivity directly from the measured values of λ_{N} and σ_{N}. Using the data in Table 12.1, one can obtain the magnitude of $\alpha_{\text{SH}}^{\text{SJ}}$ and $\sigma_{\text{SH}}^{\text{SJ}}$. For Al, $\alpha_{\text{SH}}^{\text{SJ}} \sim (0.7\text{–}1.0) \times 10^{-4}$ and $\sigma_{\text{SH}}^{\text{SJ}} = (11\text{–}13) (\Omega \text{ cm})^{-1}$. For Pt, $\alpha_{\text{SH}}^{\text{SJ}} = 6 \times 10^{-3}$ and $\sigma_{\text{SH}}^{\text{SJ}} = 495 (\Omega \text{ cm})^{-1}$, which are much larger than those of Al, since Pt is a heavy metal element with large $\bar{\eta}_{\text{so}}$ and short λ_{N}.

12.5   Nonlocal spin Hall effect

In nonlocal spin injection devices, a pure spin current is created in a nonmagnetic metal [5, 6]. It is fundamentally important to verify the spin current flowing in a nonmagnetic metal. A simple and direct proof for the existence of the spin current is made by using a nonlocal spin Hall device shown in Fig. 12.2 [22, 49, 51].

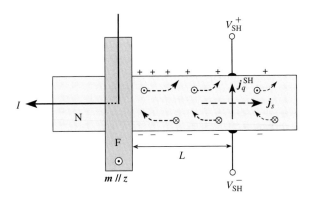

FIG. 12.2. Nonlocal spin Hall device with a ferromagnet (F) and a normal
conductor (N). The magnetization of F points in the perpendicular direction
to the plane. The injected spin current generates the spin Hall voltage
$V_{\mathrm{SH}} = V_{\mathrm{SH}}^+ - V_{\mathrm{SH}}^-$ in the transverse direction at distance L.

In this device, the magnetization of the ferromagnet (F) is in the z-direction
perpendicular to the plane. Spin injection is made by applying the current I
from F to the left end of N, while the Hall voltage (V_{SH}) is measured by the
Hall bar at distance L, where the pure spin current $\mathbf{j}_s = (j_s, 0, 0)$ flows in the
x-direction. Thus it follows from Eqs. (12.24) that $\mathbf{J}_s = \mathbf{j}_s$ and

$$\mathbf{J}_q = \sigma_{\mathrm{N}}\mathbf{E} + \alpha_{\mathrm{SH}}\left(\hat{\mathbf{z}} \times \mathbf{j}_s\right), \qquad (12.32)$$

where the second term is the Hall current induced by the spin current. In an open
circuit condition in the transverse direction, the ohmic current builds up in the
transverse direction as opposed to the Hall current such that the y-component
of $\mathbf{J}_q$ in Eq. (12.32) vanishes, resulting in the relation between the Hall electric
field E_y and the spin current j_s, $E_y = -\alpha_{\mathrm{SH}}\rho_{\mathrm{N}}j_s$, which is integrated over the
width w_{N} of N to yield the Hall voltage

$$V_{\mathrm{SH}} = \alpha_{\mathrm{SH}}w_{\mathrm{N}}\rho_{\mathrm{N}}j_s, \qquad (12.33)$$

indicating that the induced Hall voltage is proportional to the spin current. The
spin current at $x = L$ is given by

$$j_s \approx \frac{1}{2}P_{\mathrm{eff}}(I/A_{\mathrm{N}})e^{-L/\lambda_{\mathrm{N}}}, \qquad (12.34)$$

where P_{eff} is the effective spin polarization which takes the tunnel spin polar-
ization P_{T} for a tunnel junction and $P_{\mathrm{eff}} = [p_{\mathrm{F}}/(1 - p_{\mathrm{F}}^2)](R_{\mathrm{F}}/R_{\mathrm{N}})$ for a metallic
contact junction, where p_{F} is the current spin polarization of F and R_{F} and R_{N}
are the spin-accumulation resistances of the F and N electrodes, respectively [13].
Therefore, the nonlocal Hall resistance $R_{\mathrm{SH}} = V_{\mathrm{SH}}/I$ becomes [22, 49, 51]

$$R_{\mathrm{SH}} = \frac{1}{2} P_{\mathrm{eff}} \alpha_{\mathrm{SH}} \frac{\rho_{\mathrm{N}}}{d_{\mathrm{N}}} e^{-L/\lambda_{\mathrm{N}}}. \tag{12.35}$$

For typical values of device parameters ($P_{\mathrm{eff}} \sim 0.3$, $d_{\mathrm{N}} \sim 10\,\mathrm{nm}$, and $\rho_{\mathrm{N}} \sim 5\,\mu\Omega\,\mathrm{cm}$), and $\alpha_{\mathrm{SH}} \sim 0.1\text{--}0.0001$ for $\bar{\eta}_{\mathrm{so}} = 0.5\text{--}0.005$ (Table 12.1), $k_{\mathrm{F}}l \sim 100$, and $u_{\mathrm{imp}}N(0) \sim 0.1\text{--}0.01$, the expected value of ΔR_{SH} at $L = \lambda_{\mathrm{N}}/2$ is of the order of $0.05\text{--}5\,\mathrm{m\Omega}$, indicating that SHE is measurable using nonlocal Hall devices. Using the finite element method in three dimensions, a more realistic and quantitative calculation is possible to investigate the spin Hall effect and reveal the spatial distribution of spin and charge currents in a nonlocal device by taking into account the device structure and geometry [52].

Recently, the spin Hall effect was observed by using nonlocal spin injection devices: CoFe/I/Al (I = Al_2O_3) under high magnetic fields perpendicular to the device plane by Valenzuela and Tinkham [7, 53], Py/Cu/Pt using strong spin absorption by Pt [9, 10] in FePt/Au using a perpendicularly magnetized FePt [11], and by using ferromagnetic resonance (FMR) in Py/Pt bilayer films [8, 56, 57].

In the CoFe/I/Al devices, the measured spin Hall conductivity is $\sigma_{\mathrm{SH}} = 34\,\Omega^{-1}\,\mathrm{cm}^{-1}$ and $27\,\Omega^{-1}\,\mathrm{cm}^{-1}$ for the Al thickness of 12 nm and 25 nm, respectively, and the spin Hall angle is $\alpha_{\mathrm{SH}} = (1\text{--}3) \times 10^{-4}$ [7]. In the Py/Cu/Pt devices, $\sigma_{\mathrm{SH}} = 350\,\Omega^{-1}\,\mathrm{cm}^{-1}$ and $\alpha_{\mathrm{SH}} = 4 \times 10^{-3}$ [10]. These values of Al and Pt are comparable to those of the side-jump contribution expected from the formula of Eq. (12.31). By contrast, the measured spin Hall angle $\alpha_{\mathrm{SH}} = 0.1$ in the FePt device [11] is much larger than that of the side-jump contribution expected from Eq. (12.6), indicating that the skew scattering dominates the SHE.

It has been pointed out that a large spin Hall effect is caused by the skew-scattering mechanism in a nonmagnetic metal with magnetic impurities [54, 55]. Recently, based on a first-principles band structure calculation [58] and a quantum Monte Carlo simulation [59] for Fe impurities in a Au host metal, a novel type of Kondo effect due to strong electron correlation at iron impurities tremendously enhances the spin–orbit interaction of the order of the hybridization energy ($\sim$ eV), leading to a large spin Hall angle comparable to that observed in experiments of the giant spin Hall effect [11].

12.6 Anomalous Hall effect (AHE) in a ferromagnet

The anomalous Hall effect (AHE) in ferromagnetic metals has a long history and has been frequently discussed numerously based on various theoretical models and calculation techniques [1, 2, 25, 31, 32, 60, 61]. Here we briefly discuss the anomalous Hall effect in ferromagnets based on a Stoner model of ferromagnets, in which the up-spin and down-spin bands are split by the exchange energy. This simple model enables us to calculate the side-jump and skew-scattering contributions by a straightforward extension of the spin Hall effect in nonmagnetic conductors to the case of the exchange splitting bands. Since up-spin (majority) and down-spin (minority) electrons are imbalanced in ferromagnets, the flows of the majority and minority electrons by an applied electric field (E_x) are deflected in the opposite directions by spin–orbit scattering to induce a charge current in the transverse direction (j_{qy}), which is measured as the anomalous Hall conductivity $\sigma_{\mathrm{AH}} = j_{qy}/E_x$ of ferromagnets.

The skew-scattering contribution to the Hall conductivity is calculated as

$$\sigma_{\mathrm{AH}}^{\mathrm{SS}} = -(2\pi/3)\left[\left(\frac{n_\uparrow - n_\downarrow}{n_\uparrow + n_\downarrow}\right) + \left(\frac{\sigma_\uparrow - \sigma_\downarrow}{\sigma_\uparrow + \sigma_\downarrow}\right)\right]\bar{\eta}_{\mathrm{so}}u_{\mathrm{imp}}N_{\mathrm{eff}}(0)\sigma_{xx}, \quad (12.36)$$

where $N_{\mathrm{eff}}(0) = (m/4\pi^2\hbar^2)(k_{\mathrm{F}}^{\uparrow 3} + k_{\mathrm{F}}^{\downarrow 3})/\bar{k}_{\mathrm{F}}^2$ is the effective density of states and $\bar{k}_{\mathrm{F}} = (k_{\mathrm{F}}^\uparrow + k_{\mathrm{F}}^\downarrow)/2$. It is interesting to note that the skew-scattering contribution depends on the spin polarizations of the electron density and electrical conductivity.

The side-jump contribution to σ_{xy} is

$$\sigma_{\mathrm{AH}}^{\mathrm{SJ}} = \frac{e^2}{\hbar}\eta_{\mathrm{so}}n_e\left(\frac{n_\uparrow - n_\downarrow}{n_\uparrow + n_\downarrow}\right), \quad (12.37)$$

where $n_e = n_\uparrow + n_\downarrow$. Since the magnetization M_z is given by $M_z = \mu_{\mathrm{B}}(n_\uparrow - n_\downarrow)$, the side-jump contribution is proportional to the magnetization $\sigma_{\mathrm{AH}}^{\mathrm{SJ}} \propto M_z$. On the other hand, the skew-scattering contribution depends not only on the magnetization but also the asymmetry in the conductivities of up-spin and down-spin electrons, so that the Hall resistivity of skew scattering is no longer scaled to the magnetization.

12.7 Summary

In this chapter, we briefly discussed the basic aspects of the spin Hall effect in diffusive metallic conductors based on the semiclassical Boltzmann transport theory. The spin Hall effect makes it possible to interconvert the spin and charge current owing to the spin-dependent asymmetric scattering of conduction electrons by the spin–orbit interaction in nonmagnetic conductors. The electrical current in nonmagnetic conductors creates a spin current in the transverse direction by the SHE and accumulated spin near the sample edge, which provides a spin source without use of magnetic materials. In an inverse way, a "pure" spin current in nonmagnetic conductors generates charge current in the transverse direction by the ISHE, which is detected as an electric signal. The recent observations of SHE and ISHE have demonstrated the interconversion between charge and spin currents, and gives information for the spin Hall angle due to the skew-scattering and side-jump contributions. In addition, the spin diffusion–length and the electrical conductivity enable us to estimate the strength of the spin–orbit coupling in each specific sample of nonmagnetic conductors. Further experimental and theoretical studies of the mechanism of SHE with large spin Hall angle will open a new avenue in research into spintronics.

References

[1] Chien, C. L. and Westgate, C. R. (eds.) (1980). *The Hall Effect and its Applications*. Plenum, New York.

[2] Nagaosa, N., Sinova, J., Onoda, S., MacDonald, A. H., and Ong, N. P. (2010). Anomalous Hall effect. *Rev. Mod. Phys.*, **82**, 1539.

[3] Maekawa, S. (ed.) (2006). *Concepts in Spin Electronics*. Oxford University Press.

[4] Žutić, I., Fabian, J., and Das Sarma, S. (2004). Spintronics: fundamentals and applications. *Rev. Mod. Phys.*, **76**, 323.

[5] Jedema, F. J., Filip A. T., and van Wees, B. J. (2001). Electrical spin injection and accumulation at room temperature in an all-metal mesoscopic spin valve. *Nature*, **410**, 345.

[6] Jedema, F. J., Heersche, H. B., Filip, A. T., Baselmans, J. J. A., and van Wees, B. J. (2002). Electrical detection of spin precession in a metallic mesoscopic spin valve. *Nature*, **416**, 713.

[7] Valenzuela, S. O. and Tinkham, M. (2006). Direct electronic measurement of the spin Hall effect. *Nature*, **442**, 176.

[8] Saitoh, E., Ueda, M., Miyajima, H., and Tatara, G. (2006). Conversion of spin current into charge current at room temperature: Inverse spin-Hall effect. *Appl. Phys. Lett.*, **88**, 182509.

[9] Kimura, T., Otani, Y., Sato, T., Takahashi, S., and Maekawa, S. (2007). Room-temperature reversible spin Hall effect. *Phys. Rev. Lett.*, **98**, 156601.

[10] Vila, L., Kimura, T. and Otani, Y. (2007). Room-temperature reversible spin Hall effect. *Phys. Rev. Lett.*, **99**, 226604.

[11] Seki, T., Hasegawa, Y., Mitani, S., Takahashi, S., Imamura, H., Maekawa, S., Nitta, J., and Takanashi, K. (2008). Giant spin Hall effect in perpendicularly spin-polarized FePt/Au devices. *Nature Mater.*, **7**, 125.

[12] Niimi, Y., Morota, M., Wei, D.-H., Deranlot, C., Basletic, M., Hamzic, A., Fert, A., and Otani, Y. (2011). *Phys. Rev. Lett.*, **106**, 126601.

[13] Takahashi, S. and Maekawa, S. (2003). Spin injection and detection in magnetic nanostructures. *Phys. Rev. B*, **67**, 052409.

[14] Takahashi, S. and Maekawa, S. (2008). Spin current in metals and superconductors. *J. Phys. Soc. Jpn.*, **77**, 031009.

[15] D'yakonov, M. I. and Perel', V. I. (1971). Current induced spin orientation of electrons in semiconductors. *Phys. Lett. A*, **35**, 459.

[16] Hirsch, J. E. (1999). Spin Hall effect. *Phys. Rev. Lett.*, **83**, 1834.

[17] Zhang, S. (2001). Spin Hall effect in the presence of spin diffusion. *Phys. Rev. Lett.*, **85**, 393.

[18] Takahashi, S. and Maekawa, S. (2002). Hall effect induced by a spin-polarized current in superconductors. *Phys. Rev. Lett.*, **88**, 116601.

[19] Shchelushkin, R. V. and Brataas, A. (2005). Spin Hall effects in diffusive normal metals. *Phys. Rev. B*, **71**, 045123.

[20] Tse, W.-K. and Das Sarma, S. (2006). Spin Hall effect in doped semiconductor structures. *Phys. Rev. Lett.*, **96**, 56601.

[21] Engel, H. A., Rashba, E. I., and Halperin, B. I. (2005). Theory of spin Hall conductivity in *n*-doped GaAs. *Phys. Rev. Lett.*, **95**, 166605.

[22] Takahashi, S., Imamura, H. and Maekawa, S. (2006). Chapter 8 in *Concepts in Spin Electronics*, S. Maekawa, (ed.). Oxford University Press.

[23] Sinitsyn, N. A. (2008). Semiclassical theories of the anomalous Hall effect. *J. Phys.: Condens. Matter.*, **20**, 023201.

[24] Dyakonov, M. I. and Khaetskii, A. V. (2008). Spin Hall effect, in *Spin Physics in Semiconductors*. Dyakonov, M. I. and Mikhail, I. (eds.), Springer Series in Solid-State Sciences, Springer, Berlin.

[25] Karplus, R. and Luttinger, J. M. (1954). Hall effect in ferromagnetics. *Phys. Rev.*, **95**, 1154.

[26] Murakami, S., Nagaosa, N., and Zhang, S.-C. (2003). Dissipationless quantum spin current at room temperature. *Science*, **301**, 1348.

[27] Sinova, J., Culcer, D., Niu, Q., Sinitsyn, N. A., Jungwirth, T. and MacDonald, A. H. (2002). Universal intrinsic spin Hall effect. *Phys. Rev. Lett.*, **92**, 126603.

[28] Inoue, J., Bauer, G. E. W., and Molenkamp, L. W. (2004). Suppression of the persistent spin Hall current by defect scattering. *Phys. Rev. B*, **70**, 041303.

[29] Kato, Y. K., Myers, R. C., Gossard, A. C., and Awschalom, D. D. (2004). Observation of the spin Hall effect in semiconductors. *Science*, **306**, 1910.

[30] Wunderlich, J., Kaestner, B., Sinova, J., and Jungwirth, T. (2005). Experimental observation of the spin-Hall effect in a two-dimensional spin-orbit coupled semiconductor system. *Phys. Rev. Lett.*, **94**, 047204.

[31] Smit, J. (1958). The spontaneous Hall effect in ferromagnetics II. *Physica*, **24**, 39.

[32] Berger, L. (1970). Side-jump mechanism for the Hall effect of ferromagnets. *Phys. Rev. B*, **2**, 4559.

[33] Crépieux, A. and Bruno, P. (2001). Theory of the anomalous Hall effect from the Kubo formula and the Dirac equation. *Phys. Rev. B*, **64**, 14416.

[34] Sakurai, J. J. and Tuan, S. F. (1985). *Modern Quantum Mechanics*, Addison-Wesley.

[35] Lyo, S. K. and Holstein, T. (1972). Side-jump mechanism for ferromagnetic Hall effect. *Phys. Rev. Lett.*, **29**, 423.

[36] Kohn, W. and Luttinger, J. M. (1957). Quantum theory of electrical transport phenomena. *Phys. Rev.*, **108**, 590.

[37] Ansermet, J.-P. (1998). Perpendicular transport of spin-polarized electrons through magnetic nanostructures. *J. Phys.: Condens. Matter*, **10**, 6027.

[38] Ashcroft, N. W. and Mermin, D. (1976). *Solid State Physics*, Saunders College.

[39] Urech, M., Korenivski, V., Poli, N., and Haviland, D. B. (2006). Direct demonstration of decoupling of spin and charge currents in nanostructures. *Nano Lett.*, **6**, 871.

[40] Idzuchi, H., Fukuma, Y., Wang, L., and Otani, Y. (2010). Spin diffusion characteristics in magnesium nanowires. *Appl. Phys. Exp.*, **3**, 063002.

[41] Kimura, T., Sato, T., and Otani, Y. (2008). Temperature evolution of spin relaxation in a NiFe/Cu Lateral spin valve. *Phys. Rev. Lett.*, **100**, 066602.

[42] Garzon, S., Žutić, I., and Webb, R. A. (2005). Temperature-dependent asymmetry of the nonlocal spin-injection resistance: Evidence for spin nonconserving interface scattering. *Phys. Rev. Lett.*, **94**, 176601.

[43] Godfrey, R. and Johnson, M. (2006). Spin injection in mesoscopic silver wires: Experimental test of resistance mismatch. *Phys. Rev. Lett.*, **96**, 136601.

[44] Kimura, T. and Otani, Y. (2007). Large spin accumulation in a permalloy-silver lateral spin valve. *Phys. Rev. Lett.*, **99**, 196604.

[45] Fukuma, Y., Wang, L., Idzuchi, H., Takahashi, S., Maekawa, S., and Otani, Y. (2011). Giant enhancement of spin accumulation and long-distance spin precession in metallic lateral spin valves. *Nature Mater.*, **10**, 527.

[46] Ku, J. H., Chang, J., Kim, H., and Eom, J. (2006). Effective spin injection in Au film from Permalloy. *Appl. Phys. Lett.*, **88**, 172510.

[47] Ji, Y., Hoffmann, A., Jiang, J. S., and Bader, S. D. (2004). Spin injection, diffusion, and detection in lateral spin-valves. *Appl. Phys. Lett.*, **85**, 6218.

[48] Bass, J. and Pratt Jr., W. P. (2007). Spin-diffusion lengths in metals and alloys, and spin-flipping at metal/metal interfaces: an experimentalist's critical review. *J. Phys. Condens. Matter.*, **19**, 183201.

[49] Takahashi, S. and Maekawa, S. (2006). Spin injection and transport in magnetic nanostructures. *Physica C*, **437-438**, 309.

[50] Takahashi, S. and Maekawa, S. (2008). Spin current, spin accumulation and spin Hall effect. *Sci. Technol. Adv. Mater.*, **9**, 014105.

[51] Takahashi, S. and Maekawa, S. (2004). *Spin Electronics – Basic and Forefront*, K. Inomata, ed., CMC Publishing, Tokyo, p. 28.

[52] Sugano, R., Ichimura, M., Takahashi, S., and Maekawa, S. (2008). Three dimensional simulations of spin Hall effect in magnetic nanostructures. *J. Appl. Phys.*, **103**, 07A715.

[53] Valenzuela, S. O. and Tinkham, M. (2007). Nonlocal electronic spin detection, spin accumulation and the spin Hall effect. *J. Appl. Phys.*, **101**, 09B103.

[54] Fert, A. and Levy, P. M. (2011). Spin Hall effect induced by resonant scattering on impurities in metals. *Phys. Rev. Lett.*, **106**, 157208.

[55] Gradhand, M., Fedorov, D. V., Zahn, P., and Mertig, I. (2010). Extrinsic spin Hall effect from first principles. *Phys. Rev. Lett.*, **104**, 186403.

[56] Ando, K., Takahashi, S., Harii, K., Sasage, K., Ieda, J., Maekawa, S., and Saitoh, E. (2008). Electric manipulation of spin relaxation using the spin Hall effect. *Phys. Rev. Lett.*, **101**, 036601.

[57] Liu, L., Moriyama, T., Ralph, D. C. and Buhrman, R. A. (2011). Spin-torque ferromagnetic resonance induced by the spin Hall effect. *Phys. Rev. Lett.*, **106**, 036601.

[58] Guo, G.-Y., Maekawa, S., and Nagaosa, N. (2009). Enhanced spin Hall effect by resonant skew scattering in the orbital-dependent Kondo effect. *Phys. Rev. Lett.*, **102**, 036401.

[59] Gu, B., Gan, J.-Y., Bulut, N., Ziman, T., Guo, G.-Y., Nagaosa, N., and Maekawa, S. (2010). Quantum renormalization of the spin Hall effect. *Phys. Rev. Lett.*, **105**, 086401.

[60] Kondo, J. (1964). Anomalous Hall effect and magnetoresistance of ferromagnetic metal. *Prog. Theor. Phys.*, **27**, 772.

[61] Fert, A. (1972). Skew scattering in alloys with cerium impurities. *J. Phys. F*, **3**, 2126.

13 Spin generation and manipulation based on spin–orbit interaction in semiconductors

J. Nitta

Motivated by the tremendous commercial success of spintronics in metallic systems, the electron spin degree of freedom has also become the center of interest in semiconductor spintronics [1]. Exploitation of the spin degree of freedom for the conduction carriers provides a key strategy for finding new functionalities. In particular the electrostatic control of the spin degree of freedom is an advantageous technology over metal-based spintronics. However, carriers in semiconductors are not spin-polarized, and generation of spin–polarized carriers is crucial for semiconductor spintronics. The key phase of this chapter is spin–orbit interaction which gives rise to an effective magnetic field. In particular the Rashba spin–orbit interaction (SOI) [2] is important since the strength is controlled by the gate voltage on top of the semiconductor's is two-dimensional electron gas (2DEG). By utilizing the effective magnetic field induced by the SOI, spin generation and manipulation are possible in electrostatic ways. In this chapter, we will discuss the origin of spin–orbit interaction in semiconductors and the electrical manipulation of spins by a gate electric field.

13.1 Origin of spin–orbit interaction (SOI) in semiconductors

Spin–orbit interaction (SOI) plays important roles for the generation of spin current and for the electrical manipulation of the electron spins. The essence of SOI is that the moving electrons in an electric field feel an effective magnetic field even without any external magnetic field. This effective magnetic field can be used for generation and manipulation of spins. In the III-V compound semiconductor heterostructure, the main contributions of the SOI are the Dresselhaus SOI caused by bulk inversion asymmetry (BIA) [3] and the Rashba SOI caused by structural inversion asymmetry (SIA) [2]. The internal electric field of the Dresselhaus SOI originates from the microscopic Coulomb potential gradient of the atomic core region, the strength of which is generally difficult to modulate. The Dresselhaus SOI is considered to be a material-constant parameter. On the other hand, the internal electric field of the Rashba SOI originates from both the *microscopic* Coulomb potential induced by the atomic core and the *macroscopic* potential gradient caused by the heterointerface and band bending in the semiconductor heterostructure [4]. Although the microscopic electric field is a material-constant parameter, as is the Dresselhaus SOI, the macroscopic electric field can be modulated by applying an external gate bias voltage on top of the two-dimensional electron gas (2DEG). This enables us to electrically control the effective magnetic field [5, 6]. It should be noted that the SOI effect in solids is much enhanced in contrast to that in vacuum. This is because the electric field near the atomic core is large and the electron wavefunction varies rapidly in space.

In vacuum, the SOI is described by the Thomas term in the Pauli equation,

$$H_{SO} = -\frac{1}{2m_0c^2}\mu_B\sigma \cdot (\vec{p} \times \nabla V_0) = -\mu_B\sigma \cdot \left(\frac{\vec{p} \times \vec{E}}{2m_0c^2}\right). \qquad (13.1)$$

Here μ_B, σ and V_0 are the Bohr magneton, the Pauli spin matrix, and the scalar potential, respectively. In analogy to the Zeeman Hamiltonian, $H_Z = (1/2)g_0\mu_B\sigma \cdot \vec{B}$, the effective magnetic field of the SOI is

$$B_{eff} = \frac{\vec{p} \times \vec{E}}{2m_0c^2}. \qquad (13.2)$$

It is clear that the effective magnetic field is induced perpendicular to both the electron momentum and the electric field. In relativistic quantum theory, $2m_0c^2$ is the energy gap between an electron and a positron, which is a negative-energy particle with the negative mass predicted by Dirac. The energy scale of $2m_0c^2$ is $\sim 1\,MeV$; thus, the SOI is negligible for a particle with nonrelativistic momentum in a vacuum. On the other hand, in crystalline solids, the energy band gap is reduced to $\sim 1\,eV$ in typical semiconductors. Since the Dirac gap is replaced by the energy band gap according to $k\cdot p$ perturbation theory [4], it results in an enhancement of about six orders of the SOI in semiconductors.

The electron wavefunction in semiconductors is characterized by both a Bloch function and an envelope function. In the Bloch part, the electron wave is rapidly modulated by the atomic core potential, while the electron wave in the envelope part is gradually modulated by the periodic crystalline structure. The origin of the enhancement of the SOI in semiconductors is due to the Bloch part, where both the electron momentum and the electric field are enlarged by the fast oscillation of the electron wave and the strong confinement near the atomic core, respectively [4].

In the III-V semiconductor heterostructures, the spin degeneracy is lifted due to the SIA of the confining potential of the 2DEG quantum well (QW). The schematic band profile of the QW is shown in Fig. 13.1. When the confinement potential of the QW is symmetric as shown in Fig. 13.1(a), the Rashba SOI caused by the macroscopic electric field in the QW is zero. By applying an external gate bias on top of the 2DEG, the potential profile can be modulated, and the asymmetric QW potential causes a finite Rashba SOI that lifts the spin degeneracy as shown in Fig. 13.1(b). The advantage of the Rashba SOI is that the strength of the SOI can be controlled by the gate voltage [5–7]. The Hamiltonian of the Rashba SOI in a 2DEG is given by

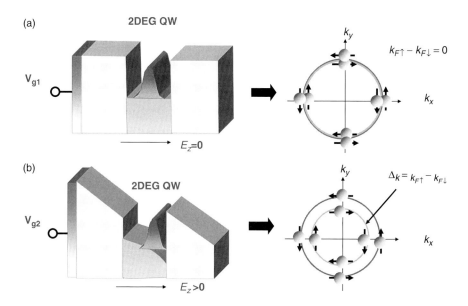

FIG. 13.1. Electrical control of the Rashba SOI. The band profile of the QW can be tuned by an external gate bias voltage. When the QW potential is symmetric, the Rashba SOI caused by an electric field in the QW is zero and spin states are degenerate. An asymmetric potential profile in the QW by tuning the gate voltage lifts the spin degeneracy since the electric field in the QW is finite. The spin configuration at the Fermi energy is shown in the bottom right figure. The Fermi momentum difference $\Delta k = k_{F\uparrow} - k_{F\downarrow}$ between spin up and down is proportional to the Rashba SOI parameter α.

$$H_R = \alpha \left(k_x \sigma_y - k_y \sigma_x \right), \tag{13.3}$$

where α is the Rashba SOI parameter, $\hbar$ is Planck's constant, $\sigma_i (i = x, y)$ are the x and y-components of the Pauli spin matrix, and the x- and y-axes are parallel to the 2DEG plane. The Rashba SOI parameter α depends on the band parameters and the electric field in the QW.

According to $k \cdot p$ perturbation theory [4, 7], the Rashba SOI parameter α is given by the following equation

$$\alpha = \frac{\hbar^2 E_p}{6 m_0} \langle \psi(z)| \frac{d}{dz} \left(\frac{1}{E_F - E_{\Gamma_7}(z)} - \frac{1}{E_F - E_{\Gamma_8}(z)} \right) |\psi(z)\rangle \tag{13.4}$$

where $\Psi(z)$ is the wavefunction for the confined electron, E_p is the inter-band matrix element, E_F is the Fermi energy in the conduction band, and $E_{\Gamma_7}(z)$ and $E_{\Gamma_8}(z)$ are the positions of the band edge energies for Γ_7 (spin split off band) and Γ_8 (the highest valence band) bands, respectively. The contribution to Eq. (13.4) can be split into two parts: (i) the field part, which is related to the electric field in the QW and (ii) the interface part, which is related to band discontinuities at the heterointerface.

For many years, there has been intense discussion about the Rashba SOI. It was thought that the Rashba SOI should be very small because the average electric field for the bound state is zero, i.e. $\langle E \rangle = 0$ in order to satisfy the condition that there is no force acting on a bound state. In fact this controversy is resolved by Eq. (13.4). It is clear that the Rashba spin splitting in the conduction band originates from the electric field in the valence band [4, 7]. Equation (13.4) also shows that the strength of the Rashba SOI can be controlled by the gate electric field on top of the 2DEG [5–7].

The spin splitting energy at the Fermi energy, $\Delta = 2\alpha k_F$, is calculated from Eq. (13.3), where k_F is the Fermi wavenumber. By comparing the Zeeman energy, the effective magnetic field is given by $B_{eff} = 2\alpha k_F/\mu_B$, and the momentum difference between spin-up and spin-down at the Fermi energy is described by $k_{F\uparrow} - k_{F\downarrow} = 2\alpha m^*/\hbar^2$, where m^* is the effective mass of electron. The spin precession angle $\Delta\theta$ is given by

$$\Delta\theta = \frac{2\alpha m^*}{\hbar^2} L \qquad (13.5)$$

when an electron spin travels in a length of L. The field effect spin transistor [8] was proposed with an assumption of the gate controlled Rashba SOI.

In III-V compound semiconductors, the electric field due to the ionized atoms in the crystal can be the origin of SOI in Eq. (13.1). This is the so-called Dresselhaus SOI. The derivation of the Dresselhaus SOI is obtained from $k \cdot p$ perturbation theory based on a Hamiltonian with a 14×14 matrix and is too complex. Here, the linear Dresselhaus SOI is given by the following equation.

$$H_D = \beta \left(k_x \sigma_x - k_y \sigma_y \right). \qquad (13.6)$$

It should be noted that the Dresselhaus cubic term is negligible when the QW confinement $k_z \approx \pi/d_{QW}$ is larger than the Fermi wavenumber. Here, d_{QW} is the thickness of the QW. The Dresselhaus SOI parameter is given by $\beta = \gamma \langle k_z^2 \rangle$. Here, γ is the bulk Dresselhaus SOI parameter, which is a material constant.

13.2 Gate controlled Rashba SOI

One of the ways to obtain the Rashba SOI parameter α is to measure the beating pattern in the Shubnikov–de Haas (SdH) oscillations [5–7]. The origin of the beating in the SdH oscillations comes from the spin-split two-Fermi circle in momentum space as shown in Fig. 13.1(a). Note that one should be careful about the origin of the beating when we have a second subband in a QW since the magneto intersubband scattering (MIS) also makes a beating pattern in the SdH oscillations [9]. In the present QW as shown below, the Fermi energy obtained from the carrier density range is much below the second subband energy level, and we can exclude the beating pattern due to MIS. The oscillations plotted as the longitudinal resistance R vs $1/B$ have a characteristic frequency f which is proportional to the carrier density, $n_e = f/\Phi_0$ (where f has units of tesla, and $\Phi_0 = h/2e$). If there is SOI the two spin directions have slightly different densities with two slightly different frequencies, visible as a beating pattern in the SdH oscillations. Thus, we can calculate α with the formula [6]

$$|\alpha| = \frac{\Delta n_e \hbar^2}{m^*}\sqrt{\frac{\pi}{2(n_e - \Delta n_e)}} = \frac{f_1 - f_2}{2m^*}\sqrt{\frac{e\hbar^3}{f_2}},\tag{13.7}$$

where f_1 and f_2 are the high and low frequencies, respectively, and Δn_e is the difference between spin concentrations, respectively.

The 2DEG channel in the present experiment is formed in an InP/InGaAs(10 nm)/InAlAs heterostructure. A Hall bar sample was made by standard lift-off techniques. A 50 nm SiO_2 insulator was deposited by electron cyclotron resonance (ECR) sputtering, and on top of that was an Au gate, used to control the carrier density and the SOI parameter α. We measured the SdH oscillations in a wide range of carrier densities.

Figure 13.2 shows the gate voltage dependence of SdH oscillations at 0.3 K. The magnetic field was applied perpendicular to the 2DEG. Beating patterns are observed in the SdH oscillations because of the existence of two closely spaced SdH frequency components with similar amplitudes. These observed beating patterns are attributed to the spin splitting since the second subband is not occupied in the QW, therefore we can rule out MIS. When the gate voltage is decreased, the node position shifts to a higher magnetic field. Although the beating pattern was not very pronounced, the FFT spectra showed clear double

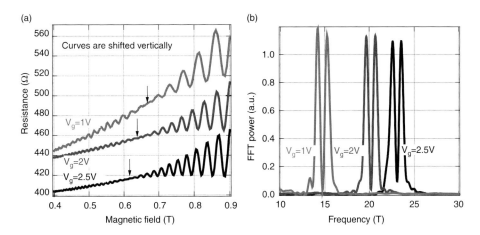

FIG. 13.2. (a) Gate voltage dependence of the beating pattern appearing in the Shubunikov–de Haas (SdH) oscillations. Curves are vertically shifted for clarity. (b) Fast Fourier transform (FFT) spectra of the SdH oscillations.

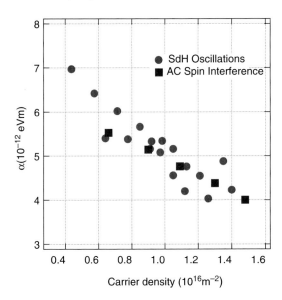

FIG. 13.3. The SOI parameter (α) dependence of the carrier density (n_e). Black
    dots and square symbols are obtained from the SdH oscillations and from the
    zero crossing points of the AC spin interference, respectively.

peaks as shown in Fig. 13.2(b), therefore, the spin–orbit interaction parameters
α were obtained from Eq. (13.7) using the double peaks in the FFT spectra.
The carrier density (n_e) dependence of the spin–orbit interaction α is plotted
as black dots in Fig. 13.3. The spin–orbit interaction parameter decreases with
increasing n_e since the potential profile of the QW becomes more symmetric.
This result shows that the SOI parameter α can be controlled by the gate
voltage. It has also been confirmed that the spin–orbit interaction can be controlled by the design of asymmetry in QWs from weak anti-localization analysis
[10, 11].

13.3 Spin relaxation and its suppression

Spins of electrons in solids are not conserved quantities in contrast to charges of
electrons. The information of spins is lost and leads to so-called spin relaxation for
several reasons. In epitaxially grown III-V compound QWs with the Rashba and
Dresselhaus SOIs, the D'yakonov–Perel (DP) mechanism is generally dominant
for spin relaxation [12]. The Rashba SOI causes an effective magnetic field B_{eff}
which is pointing perpendicular to the momentum direction in the 2DEG plane.
The spin of electrons is precessing around the effective field. In a diffusive 2DEG,
the momentum direction of the electron changes frequently, and hence so does
the direction of B_{eff}. Due to the random change in B_{eff}, the spin orientation

is randomized and the spin loses its memory of its initial spin direction. DP spin relaxation is caused not only by the Rashba SOI but also by the Dresselhaus SOI. Generally speaking, DP spin relaxation is expected if the system has a momentum-dependent effective field B_{eff} due to the SOI spin splitting energy Δ.

From the above DP spin relaxation picture, we can estimate the spin relaxation time. The spin is initially precessing around a certain effective magnetic field direction with a typical frequency of $\omega = \Delta/\hbar$ and during a typical scattering time τ. After a scattering event, the direction of B_{eff} is randomly changed, and the spin starts to precess around the new B_{eff} direction. Hence, after a certain number of scattering events there is no correlataion anymore between the initial and final spin states. The precise time-scale on which the spin loses its memory depends on the typical spin precession angle between scattering events $\delta\phi = \omega\tau = \Delta\tau/\hbar$. For $\delta\phi \ll 1$, the precession angle is small between succeeding scattering events, so that the spin vector experiences a slow angle diffusion. During a time interval t, the number of random steps is t/τ. For uncorrelated steps in the precession angle we have to sum the squared precession angles $\delta\phi^2 = (\Delta\tau/\hbar)^2$, and the total squared precession angle after time t is $(\Delta\tau/\hbar)^2 t/\tau$. The spin relaxation time τ_s can be defined as the time at which the total precession angle becomes of the order of unity. Hence the spin relaxation time is given by $\tau_s \approx \hbar^2/\Delta^2\tau$.

Gate controlled SOI which gives rise to an effective magnetic field provides an electrical way to manipulate spins. On the other hand, a momentum-dependent effective magnetic field due to the SOI randomizes spin orientations after several momentum scattering events. The SOI is a double-edged sword because it can be used for spin manipulation, however, at the same time it causes spin relaxation. Therefore, it is very crucial to suppress the spin relaxation while keeping the strength and the controllability of SOI. One of the ways to suppress spin relaxation is to confine electrons to moving one-dimensionally by narrow wire structures whose width is of the same scale as the bulk spin diffusion length $L_{SO} = \hbar^2/\alpha m^*$ due to the Rashba SOI. This suppression of spin relaxation due to the lateral confinement effect has been theoretically investigated [13, 14] and has been experimentally demonstrated with optical methods [15] and weak anti-localization analysis [16, 17].

Most effective way to suppress the DP spin relaxation is to utilize the so-call persistent spin helix (PSH) condition [18, 19], where the Rashba SOI strength α is equal to the linear Dresselhaus SOI β. In this PSH condition, conservation of spin polarization is preserved even after scattering events. This conservation is predicted to be robust against all forms of spin-independent scattering, including the electron–electron interaction, but is broken by spin-dependent scattering and the cubic Dresselhaus term. Recently, the PSH in semiconductor quantum wells was confirmed by optical transient spin-grating spectroscopy by Koralek et al. [20]. They found enhancement of the spin lifetime by two orders magnitude near the exact PSH point. This experimental demonstration is a breakthrough toward minimizing and controlling spin relaxation.

13.4 Spin Hall effect based on Rashba and Dresselhaus spin–orbit interaction

In contrast to metal systems, a band structure affected by the spin–orbit interaction in semiconductor systems leads to an intrinsic spin Hall effect even in the absence of scattering events. The intrinsic spin Hall effect in valence-band holes was first predicted by Murakami et al. [21]. A universal intrinsic spin Hall effect in a ballistic Rashba 2DEG system was calculated by J. Sinova et al. [22]. The universal spin Hall conductivity is given by e/8π. It is important to note that the spin Hall conductivity vanishes in a diffusive Rashba 2DEG system [23]. The spin Hall effect in heavy holes in a confined QW was studied by Schielmann and Loss [24]. In contrast to the Rashba 2DEG system, the intrinsic spin Hall effect in a hole system does not vanish even in the diffusive case.

An intuitive picture of the spin Hall effect in the Rashba 2DEG system is provided by the spin transverse force [25]. Using the Heisenberg equation of motion and the commutation relation of the Pauli spin matrix, the second derivative of the position operator gives the following spin transverse force on a moving electron [25]

$$m^* \frac{\partial^2 \vec{r}}{\partial t^2} = -\frac{2m^* \alpha^2 \sigma_z}{\hbar^3} \vec{p} \times \hat{z}. \tag{13.8}$$

This spin transverse force tends to form a σ_z spin current perpendicular to the momentum direction.

Experimental observation of the spin Hall effect in bulk GaAs and strained InGaAs was demonstrated with the use of Kerr rotation microscopy [26]. Without applying any external magnetic field, out-of-plane spin-polarized carriers with opposite sign were detected at the two edges of the sample. The amplitude of the spin polarization was very weak, and the mechanism originated from the extrinsic spin Hall effect. The spin Hall effect in a GaAs 2D hole system was observed by Wunderlich et al. [27]. The out-of-plane spin polarized carriers were detected by light-emitting diodes, and the sign of the spin polarization was switched by the electric field direction. The authors claimed that the intrinsic spin Hall effect is responsible for their system.

The experimentally observed spin polarization due to the spin Hall effect was very small. So the Stern–Gerlach spin filter was proposed (Fig. 13.4) by using the spatial gradient of the Rashba SOI [28]. The spatial gradient of the effective magnetic field due to the Rashba SOI causes Stern–Gerlach type spin separation. Almost 100% spin polarization can be realized even without applying any external magnetic fields and without attaching ferromagnetic contacts. In this case, the spin-polarized orientation is not out-of-plane but in-plane. The spin polarization persists even in the presence of randomness.

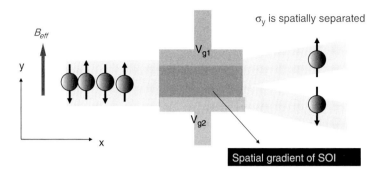

FIG. 13.4. Schematic structure of the proposed Stern–Gerlach spin filter. The spatial gradient of the Rashba SOI is induced by two gate electrodes.

13.5 Aharonov–Casher spin interference; theory

A rotation operator for spin $1/2$ produces a minus sign under 2π rotation [29]. Neutron spin interference experiments performed by two groups have verified this extraordinary prediction of quantum mechanics [30, 31]. In solids, an electron spin interference experiment in an n-GaAs interference loop has been conducted using optical pump and probe methods [32]. A local magnetic field due to dynamic nuclear spin polarization caused spin precession of the wavepacket in one of the interference paths. In the above spin interference experiments, spin precession was controlled by a local magnetic field.

An electron acquires a phase around the magnetic flux due to the vector potential leading to the Aharonov–Bohm (AB) effect in an interference loop [33, 34]. From the viewpoint of inherent symmetries between the magnetic field and electric field in the Maxwell equations, Aharonov and Casher have predicted that a magnetic moment acquires a phase around a charge flux line [35]. It should be noted that the original Aharonov–Casher (AC) effect was proposed for charge-neutral particles since the electric field modifies the trajectory of charged particles in the same sense as the original AB effect was predicted in the situation where magnetic flux should not exist in an electron path. It is pointed out that the AC phase shift can be derived from spin–orbit interaction (SOI) [36]. A. G. Aronov and Y. B. Lyanda-Geller have derived a spin–orbit Berry phase in conducting rings with SOI [37]. T. Qian and Z. Su have obtained the AC phase, which is the sum of the spin–orbit Berry phase and the spin dynamical phase in a one-dimensional ring with SOI [38]. The major difference between the AC effect and the AB effect is that the AC effect is not observable if the electric field is not in the paths although the AB effect can take place even if there is no magnetic field in the electron paths. Cimmino *et al.* [39] managed to perform the AC interference experiment in a neutron (having spin $1/2$ but no charge) beam loop using a voltage of 45 kV to create the electric field. However, the modified precession angle of the neutron spin was only 2.2 mrad since the SOI is not strong in vacuum.

Mathur and Stone have theoretically shown [40] that the effects of SOI in disordered conductors are manifestations of the AC effect in the same sense as the effects of weak magnetic fields are manifestations of the AB effect. They have proposed the electronic AC effect in a mesoscopic interference loop made of GaAs 2DEG with the Dresselhaus SOI. It is emphasized that a 1000-fold improvement in this experiment can be expected in the electronic AC effect since the SOI in semiconductors is much enhanced compared with that in vacuum. AC spin interference was not reported by utilizing the Dresselhaus SOI since the strength is not controlled by an electrostatic way.

Electrostatic manipulation of spins is crucial for spintronics. An AC spin-interference device was proposed on the basis of the gate controlled Rashba SOI [41]. The schematic structure of the spin interference device is shown in Fig. 13.5. The spin interference can be expected in an AB ring with the Rashba SOI because the spins of electrons precess in opposite directions between clockwise and counterclockwise travelling directions in the ring. The relative difference in spin precession angle at the interference point causes the phase difference in the spin wavefunctions. The gate electrode, which covers the whole area of the AB ring, controls the Rashba SOI, and therefore, the interference. The advantage of this proposed spin-interference device is that conductance modulation is not washed out even in the presence of multiple modes.

The total Hamiltonian of a one-dimensional ring with the Rashba SOI in cylindrical coordinates reads [42]

$$H(\phi) = \frac{\hbar^2}{2m^*r^2}\left(-i\frac{\partial}{\partial\phi} + \frac{\phi}{\phi_0}\right)^2 + \frac{\alpha}{r}\left(\cos\phi\sigma_x + \sin\phi\sigma_y\right)\left(-i\frac{\partial}{\partial\phi} + \frac{\phi}{\phi_0}\right)$$
$$-i\frac{\alpha}{2r}\left(\cos\phi\sigma_y - \sin\phi\sigma_x\right) + \frac{\hbar\omega_B}{2}\sigma_z \qquad (13.9)$$

Where the an external magnetic field B_z is applied in the z-direction which is perpendicular to the ring plane, and the magnetic flux through the ring is given by $\Phi = B_z\pi r^2$ with ring radius r, and $\Phi_0 = h/e$ is the flux quantum. The polar

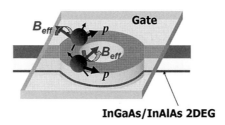

FIG. 13.5. Schematic structure of the proposed spin interference device. The Rashba SOI is tunable by a gate voltage.

angle is given by ϕ and $\omega_B = 2\mu_B B_z/\hbar$ is the Larmor frequency. In an isolated ring, the wavefunction is given by the following form

$$\psi = \frac{1}{\sqrt{2\pi}} \begin{pmatrix} C_n^+ e^{in\phi} \\ C_n^- e^{in\phi} \end{pmatrix} \qquad (13.10)$$

where C_n^+ and C_n^- are the coefficients of the spin-up and spin-down eigenstates, respectively. When the Zeeman term is negligible, the energy eigenvalues can be written as [5, 43, 44]

$$E_{n,s} = \hbar\omega_0 \left[n + \frac{\phi}{\phi_0} - \frac{\phi_{AC}^s}{2\pi} \right]^2 \qquad (13.11)$$

when $\omega_0 = \hbar/2m^*r^2$, n is an integer, and the AC phase is Φ_{AC}^s. The AC phase is given by [5, 43, 44]

$$\Phi_{AC}^s = -\pi \left[1 + s\sqrt{\left(\frac{2rm^*\alpha}{\hbar^2}\right)^2 + 1} \right], \qquad s = \pm. \qquad (13.12)$$

This AC phase can be viewed as an effective spin-dependent magnetic flux through the ring which modulates the conductance of the ring. Here $s = \pm$ corresponds to spin-up and spin-down along the effective magnetic field. From the above calculation, the conductance when electrons travel halfway around the ring at $B_z = 0$ is written as [5, 43, 44]

$$G = \frac{e^2}{h} \left[1 - \cos\left\{ \pi\sqrt{1 + \left(\frac{2m\alpha r}{\hbar^2}\right)^2} \right\} \right] \qquad (13.13)$$

$$= \frac{e^2}{h} \left[1 + \cos\left\{ 2\pi r \frac{m\alpha}{\hbar^2} \sin\theta - \pi\left(1 - \cos\theta\right) \right\} \right].$$

The acquired AC phase can be written as the sum of two phases, $2\pi r m^*/\hbar^2$ and $\pi(1 - \cos\theta)$, as shown in the last expression in Eq. (13.13). The former term is sometimes called the dynamical part of the AC phase, because of its dependence on the distance traveled by the electrons. The latter term is a geometrical phase since it only depends on the solid angle θ, and not on spatial parameters. Such geometrical phases were discovered by Berry from the basic laws of quantum mechanics [45], and received considerable attention. Berry showed that the wavefunction obtains a nontrivial phase when the parameter in the Hamiltonian is changed in a cyclic and adiabatic way.

From the above expression it follows that the conductance of the ring depends crucially on the Rashba SOI strength α. This equation shows that the conductance of the ring oscillates as a function of α. This SOI dependence is very similar to the conductance of the spin-FET proposed by Datta and Das, in which they need ferromagnetic electrodes for spin injection and detection [8]. This proposed spin interferometer works without ferromagnetic electrodes. It is worth pointing out that the above described spin interference effect is expected to be robust since the acquired phase difference does not depend on the Fermi energy. Furthermore, the independence of the phase difference with respect to the Fermi energy also implies that the conductance modulation will be present in a multi-mode ring if the different radial modes do not mix in the interference process; the conductance modulation of a multi-mode ring is still given by Eq. (13.13).

13.6   Aharonov–Casher spin interference; experiment

The resistance of a mesoscopic ring is affected by several quantum interference effects. The well-known AB effect results in a resistance oscillation with a magnetic flux period of h/e. The AB effect is sample-specific and very sensitive to the Fermi wavelength, therefore, the interference pattern is rapidly changed by the gate voltage. In order to detect the AC effect we used another quantum interference phenomenon, the Al'tshuler–Aronov–Spivak (AAS) effect [46]. The AAS effect is an AB effect of time-reversal symmetric paths, where the two wavefunction parts go all around back to the origin on identical paths, but in opposite directions. In this situation any phase which is due to the path geometry will be identical and will not affect the interference. This also means that it is independent of the Fermi energy E_F (and consequently the carrier density n_e). However, the AAS effect is sensitive to the spin phase when the SOI plays a role. If there is magnetic flux inside the paths the resistance will oscillate with the period of $h/2e$. When the flux is increased the resistance oscillates with the period $h/2e$, but the AAS oscillation amplitude decays after a several periods because of averaging between different paths in the ring, with different areas. If there is SOI in the ring, the electron spin will start precessing around the effective magnetic field and change the interference at the entry point. Note that the effective magnetic field due to the SOI is much stronger than the external magnetic field to pick up AAS oscillations. The precession axes for the two parts of the wavefunction are opposite and therefore the relative precession angle is twice the angle of each part. If the relative precession angle is π the spins of the two parts are opposite and cannot interfere, and the AAS oscillations disappear. If the relative angle is 2π the two parts will have the same spin but opposite signs because of the $1/2$ spin quantum laws (a 4π rotation is required to return to the original wavefunction), effectively changing the phase of the AAS oscillations by π, which we interpret as a negative amplitude.

By using arrays rather than single rings we get a stronger spin signal and we average out some of the universal conductance fluctuations (UCF) and sample-specific AB oscillations [47]. Complex gate voltage dependence has been reported in an Aharonov–Bohm type AC experiment in a single ring fabricated from HgTe/HgCdTe QWs [48]. Therefore, a detailed analysis is necessary to compare with the AC theory.

The ring arrays were etched out in an electron cyclotron resonance (ECR) dry-etching process from an InP/InGaAs/InAlAs based 2DEG, the same as used for the SdH measurements as shown in Fig. 13.2. The electron mobility was 7–11 m^2/Vs, depending on the carrier density and the effective electron mass m^* was $0.050m_0$ as determined from the temperature dependence of SdH oscillation amplitudes. Figure 13.6 shows an example of the ring array which consists of 4×4 rings of 1.0 μm radius. Note that the actually measured sample was a 5×5 ring array. The rings were covered with a 50 nm thick SiO$_2$ insulator layer, and an Au gate electrode, used to control the carrier density and the SOI parameter α. In the present sample, we design the array with a small number of rings in order to escape the gate tunneling leakage problem. The advantage of using a small number of rings rather than a large array is that the gate tunneling leakage is much smaller and we can use a relatively high gate voltage [49]. This makes it possible to see several oscillations of AC interference. Earlier experiments on square loop arrays with very large number of loops showed convincing spin interference results, but only up to one interference period [50].

The experiment was carried out in a ^{3}He cryostat at the base temperature which varied between 220 mK and 270 mK. The sample was put in the core of a superconducting magnet with the field B perpendicular to the 2DEG plane. We measured the resistance R of the ring array simultaneously with the Hall resistance R_H of the Hall bar close to the rings, while stepping the magnetic field and the gate voltage V_G. Close to the arrays and in the same current path and under the same gate was a Hall bar, 5 μm wide and 20 μm long, used to measure the carrier density. We calculated the carrier density n_e from the slope of the R_H vs. B ($n_e^{-1} = e \, dR_H/dB$) graph, and the carrier concentration is linearly increased with the gate voltage V_G.

In order to reduce noise and UCF effects we averaged ten resistance versus magnetic field (R vs. B) curves with slightly different gate voltages. This averaging preserves the AAS oscillations but the averaging of M curves reduces

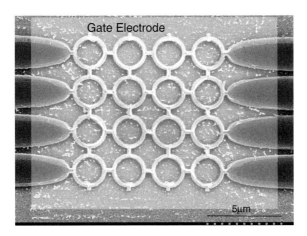

FIG. 13.6. An SEM image of an array of rings with 1 μm radius. The whole area is covered by the gate electrode.

the AB amplitude roughly as $M^{-1/2}$. We took the FFT spectrum (using an exact Blackman window) of this average and got a spectrum with two peaks, corresponding to the AB oscillations and the AAS oscillations at twice the frequency. We integrated the area of the AAS peak to get the amplitude and determined the sign by analyzing the phase of the central part of the filtered R vs. B data.

In Fig. 13.7, we display the $h/2e$ magnetoresistance oscillations due to the AAS effect at five different gate voltages. The oscillations in the top and bottom curves are reversed compared to the middle one because of the AC effect. The second and fourth ones have almost no oscillations; the spin precession rotates the spins of the two wavefunction parts in opposite directions. Figure 13.8 shows the gray scale plot after digital band-pass filtering of the AAS oscillations which are visible as vertical stripes in the figure. We can clearly see the oscillations switching phase as we increase the gate voltage. We then plotted the amplitude against the gate voltage as shown in Fig. 13.9. The AAS amplitude oscillates as a function of the gate voltage which changes the SOI parameter α. As we discuss below using Eqs. (13.5) and (13.13) the amplitude crosses zero, inverting the AAS oscillations. Each period represents one extra 2π spin precession of an electron moving around a ring.

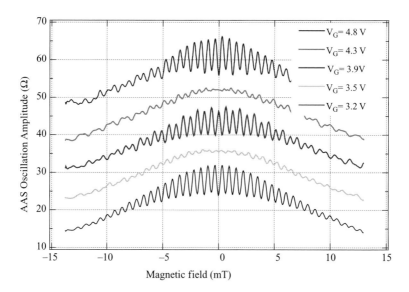

FIG. 13.7. Magnetoresistance oscillations with period of $h/2e$ due to the AAS effect at five different gate voltages. The curves are shifted vertically for clarity. The oscillations in the top and bottom curves are reversed compared to the middle one because of the Aharonov–Casher effect. The oscillation amplitudes for the second and fourth curves are suppressed. These gate voltage dependent AAS oscillations are due to the AC effect.

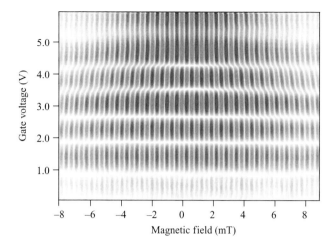

FIG. 13.8. The $h/2e$ oscillations are plotted against the gate voltage. The oscillations reverse phase at several gate voltages due to the Aharonov–Casher effect.

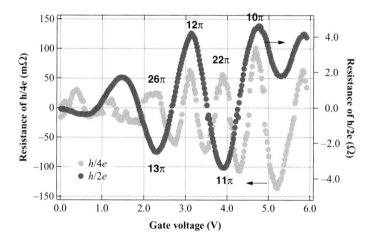

FIG. 13.9. The time-reversal Aharonov–Casher oscillations of the first and second harmonics. The second harmonic corresponds to two turns around the ring before interfering. The period of the second harmonic is half the first harmonic period as expected. The precession angles θ correspond to the argument of the cosine in Eq. (13.14).

In the FFT spectra there is also a small peak at $h/4e$. This is due to the wavefunction parts going twice around the ring before interfering. If we do the same analysis on this peak we get an oscillating amplitude with half the period compared to the $h/2e$ amplitude. This is expected because the distance is doubled and therefore the precession angle is also doubled. Both the $h/2e$ and h/4e oscillation amplitudes increase with increasing gate voltage V_G. This is because the phase coherence length of the ring becomes longer with increasing diffusion constant which depends on the carrier density.

The precession angle θ of an electron moving along a straight narrow channel is given by Eq. (13.5). The modulation of the $h/2e$ oscillation amplitude can be expressed as a function of α,

$$\frac{\delta R_\alpha}{\delta R_{\alpha=0}} = \cos\left\{2\pi\sqrt{1 + \left(\frac{2m^*\alpha}{\hbar^2}r\right)^2}\right\} \qquad (13.14)$$

whereδR_α and $\delta R_{\alpha=0}$ are the $h/2e$ amplitudes with and without SOI, respectively. In relating the result of the spin interference experiment to the spin precession angle, the argument of the cosine in Eq. (13.13) reduces to the spin precession angle θ in the limit of strong SOI or large ring radius because the distance traveled around the ring is $2\pi r$.

As shown in Fig. 13.3, the SOI strength α obtained from the beating pattern of the SdH oscillations shows a carrier density dependence. The gate voltage sensitivity $\Delta\alpha/\Delta V_G$ is about 0.51×10^{-12} eVm/V in the present heterostructure. From the gate voltage dependence of SOI, α, we estimate the spin precession angles at several different gate voltages. The estimated spin precession angles at the peak and dips of the AC oscillations are shown in Fig. 13.9. These precession angles correspond to the argument of the cosine in Eq. (13.14). It is found that the spin precession angle is controlled over the range of 4π by the gate electric field. We could observe more than a 20π spin precession angle. The squares in Fig. 13.3 are α values obtained from the consecutive zero-crossing points of the AC spin interference experiment, which are consistent with the SdH measurement result.

This clear demonstration of AC interference controlled by the gate electric field can be attributed to the fact that the SOI is much enhanced in semiconductor heterostructures compared to the SOI in vacuum. The AC effect is of fundamental importance for quantum interference phenomena and quantum interactions.

References

[1] D. D. Awschalom and M. E. Flatte, Nature Phys. **3**, 153 (2007).
[2] E. I. Rashba, Sov. Phys. Solid State **2**, 1109 (1960); Y. A. Bychkov and E. I. Rashba, J. Phys. **C17**, 6039 (1984).
[3] G. Dresselhaus, Phys. Rev. **100**, 580, (1955).

[4] R. Winkler, *Spin–Orbit Coupling Effects in Two-Dimensional Electron and Hole Systems*, Springer-Verlag, Berlin, 2003.

[5] J. Nitta, T. Akazaki, H. Takayanagi, and T. Enoki, Phys. Rev. Lett. **78**, 1335 (1997).

[6] G. Engels, J. Lange, Th. Schäpers, and H. Lüth, Phys. Rev. B **55,** R1958 (1997).

[7] Th. Schäpers, G. Engels, J. Lamge, Th. Klocke, M. Hollfelder, and H. Lüth, J. Appl. Phys. **83**, 4324 (1998).

[8] S. Datta and B. Das, Appl. Phys. Lett. **56**, 665 (1990).

[9] A. C. H. Rowe, J. Nehls, and R. A. Stradling, Phys. Rev. B **63**, 201307 (2001).

[10] T. Koga, J. Nitta, T. Akazaki, and H. Takayanagi, Phys. Rev. Lett. **89**, 046801 (2002).

[11] Y. Lin, T. Koga, and J. Nitta, Phys. Rev. B **71**, 045328-1 (2005).

[12] M. I. Dyakonov and V. I. Perel', Sov. Phys. JETP **33**, 1053 (1971).

[13] A. G. Mal'shukov and K. A. Chao, Phys. Rev. B **61**, R2413 (2000).

[14] S. Kettemann, Phys. Rev. Lett. **98**, 176808 (2007).

[15] A. W. Holleitner, V. Sih, R. C. Myers, A. C. Gossard, and D. D. Awschalom, Phys. Rev. Lett. **97**, 036805 (2006).

[16] Th. Schäpers, V. A. Guzenko, M. G. Pala, U. Zülich, M. Governale, J. Knobbe, and H. Hardtdegen, Phys. Rev. B **74**, 081301(R) (2006).

[17] Y. Kunihashi, M. Kohda, and J. Nitta, Phys. Rev. Lett. **102**, 226601 (2009).

[18] J. Schliemann, J. Carlos Egues, and D. Loss, Phys. Rev. Lett. **90**, 146801 (2003).

[19] B. A. Bernevig, J. Orenstein, and S.-C. Zhang, Phys. Rev. Lett. **97**, 236601 (2006).

[20] J. D. Koralek, C. P. Weber, J. Orenstein, B. A. Bernevig, S.-C. Zhang, S. Mack, and D. D. Awschalom, Nature, **458**, 610 (2009).

[21] S. Murakami, N. Nagaosa, and S. C. Zhang, Science **301**, 1348 (2003).

[22] J. Sinova, D. Culcer, Q. Niu, N. A. Sinitsyn, T. Jungwirth, and A. H. MacDonald, Phys. Rev. Lett. **92**, 126603 (2004).

[23] J. Inoue, G. E. Bauer, and L. W. Molenkamp, Phys. Rev. B **70**, 041303(R) (2004).

[24] J. Schiemann and D. Loss, Phys. Rev. B **71**, 085308 (2005).

[25] J. Li, L. Hu, and S.-Q Shen, Phys. Rev. B **71**, 241305 (R) (2005).

[26] Y. K. Kato, R. C. Myers, A. C. Gossard, and D. D. Awschalom, Science **306**, 1910 (2004).

[27] J. Wunderlich, B. Kaestner, J. Sinova, and T. Jungwirth, Phys. Rev. Lett. **94**, 047204 (2005).

[28] J. Ohe, M. Yamamoto, T. Ohtsuki, and J. Nitta, Phys. Rev. B **72**, 041308(R) (2005).

[29] J. J. Sakurai: *Modern Quantum Mechanics*, Benjamin/Cummings, New York, 1985.

[30] S. A. Werner, R. Colella, A. W. Overhauser, and C. F. Eagen, Phys. Rev. Lett. **35**, 1053 (1975).

[31] H. Rauch, A. Zeilinger, G. Badurek, A. Wilfing, W. Bauspiess, and U. Bonse, Phys. Lett. **54A**, 425 (1975).

[32] Y. K. Kato, R. C. Myer, A. C. Gossard, and D. D. Awschalom, Appl. Phys. Lett. **86**, 162107 (2005).

[33] Y. Aharonov and D. Bohm, Phys. Rev. **115**, 485 (1959).

[34] A. Tonomura, N. Osakabe, T. Matsuda, T. Kawasaki, and J. Endo, Phys. Rev. Lett. **56**, 792 (1986).

[35] Y. Aharonov and A. Casher, Phys. Rev. Lett. **53**, 2964 (1984).

[36] A. Balatsky and B. Al'tshuler, Phys. Rev. Lett. **70**, 1678 (1993).

[37] A. G. Aronov and Y. B. Lyanda-Geller, Phys. Rev. Lett. **70**, 343 (1993).

[38] T.-Z. Qian and Z.-B. Su, Phys. Rev. Lett. **72**, 2311 (1994).

[39] A. Cimmino, G. I. Opat, A. G. Klein, H. Kaiser, S. A. Werner, M. Arif, and R. Clothier, Phys. Rev. Lett. **63**, 380 (1989).

[40] H. Mathur and A. D. Stone, Phys. Rev. Lett. **68**, 2964 (1992).

[41] J. Nitta, F. E. Meijer, and H. Takayanagi, Appl. Phys. Lett. **75**, 695 (1999).

[42] F. E. Meijer, A. F. Morpurgo, and T. M. Klapwijk, Phys. Rev. B **66**, 033107 (2002).

[43] X. F. Wang and P. Vasilopoulos, Phys. Rev. B **72**, 165336 (2005).

[44] D. Frustaglia and K. Richter, Phys. Rev. B **69**, 235310 (2004).

[45] M. V. Berry, Proc. R. Soc. London A **392**, 45 (1984).

[46] B. L. Al'tshuler, A. G. Aronov, and B. Z. Spivak, JETP Lett. **33**, 94 (1981).

[47] C. P. Umbach, C. Van Haesendonck, R. B. Laibowitz, S. Washburn, and R. A. Webb, Phys. Rev. Lett. **56**, 386 (1986).

[48] M. König, A. Tschetschetkin, E. M. Hankiewiccz, J. Sinova, V. Hock, V. Daumer, M. Schaefer, C. R. Becker, H. Buhmann, and L. W. Molenkamp, Phys. Rev. Lett. **96**, 076804 (2006).

[49] T. Bergsten, T. Kobayashi, Y. Sekine, and J. Nitta, Phys. Rev. Lett. **97**, 196803 (2006).

[50] T. Koga, J. Nitta, and M. van Veenhuisen, Phys. Rev. B **70**, R161302 (2004); M. J. van Veenhuizen, T. Koga, and J. Nitta, Phys. Rev. B **73**, 235315 (2006); T. Koga, Y. Sekine, and J. Nitta, Phys. Rev. B **74**, 041302(R) (2006).

14 Experimental observation of the spin Hall effect using electronic nonlocal detection

S. O. Valenzuela and T. Kimura

14.1 Observation of the spin Hall effect

Owing to its technological implications and its many subtleties, the experimental observation of the spin Hall effect (SHE) has received a great deal of attention and has been accompanied by an extensive theoretical debate [1, 2]. The SHE has been described as a source of spin-polarized electrons for electronic applications without the need of ferromagnets or optical injection. Because spin accumulation does not produce an obvious measurable electrical signal, electronic detection of the SHE proved to be elusive and was preceded by optical demonstrations [3, 4]. Several experimental schemes for the electronic detection of the SHE had been originally proposed [5–8], including the use of ferromagnetic electrodes to determine the spin accumulation at the edges of the sample. However, the difficulty of sample fabrication and the presence of spin-related phenomena such as anisotropic magnetoresistance or the anomalous Hall effect in the ferromagnetic electrodes could mask or even mimic the SHE signal in those sample layouts. The first successful experiments, which took these effects into account, were reported in 2006 and 2007 [9–12]. They used nonlocal lateral spin-injection structures, which can be shaped easily into multi-terminal devices with output signals that are only determined by the spin degree of freedom. In the following, we first describe briefly the basic aspects of spin injection, transport, and detection in conventional lateral structures [13–16], which are commonly used as a reference in spin Hall experiments [16]. We then describe the experimental advantages for nonlocal spin Hall detection which is the main focus of the rest of the chapter. We place particular emphasis on device fabrication and the different device layouts that have been designed for SHE detection. We also review the experimental values of the spin Hall angles for specific materials and discuss the origin of the SHE for each of them.

14.2 Nonlocal spin injection and detection

In 1985 Johnson and Silsbee first reported [13] the injection and detection of nonequilibrium spins using a device that consisted of a nonmagnetic metal N with two ferromagnetic (F1, F2) electrodes attached (Fig. 14.1(a)). In this device, spin-polarized electrons are injected from F1 into N by applying a current I from F1 that results in spin accumulation in N. The population of, say, spin-up

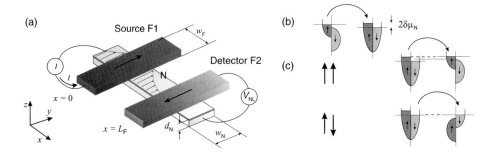

FIG. 14.1. (a) Nonlocal spin detection and spin accumulation. (a) Schematic
    illustration of the device layout. An injected current I on the source (F1)
    generates spin accumulation in the normal metal (N) which is quantified
    by the detector (F2) voltage V_{NL}. The sign of V_{NL} is determined by the
    relative magnetization orientations of F1 and F2. The current is injected away
    from F2. Electron spins diffuse isotropically from the injection point. (b)
    Schematic representation of the spin splitting in the electrochemical potential
    induced by spin injection. The splitting decays over characteristic lengths λ_N
    over the N side. (c) Detector behavior for an idealized Stoner ferromagnet
    with a full spin subband. The electrochemical potential in F2 equilibrates
    with the N spin-up electrochemical potential for the parallel magnetization
    orientation (top) and with the spin-down electrochemical potential for the
    antiparallel magnetization orientation (bottom) resulting in a voltage V_{SH}
    between C and D.

electrons in N increases by shifting the electrochemical potential by $\delta\mu_N$, while
the population of spin-down electrons decreases by a similar shift of $-\delta\mu_N$.
Overall, this corresponds to a spin-accumulation splitting of $2\delta\mu_N$ (Fig. 14.1(b)).
The spin accumulation diffuses away from the injection point and reaches the F2
detector, which measures its local magnitude. As first suggested by Silsbee [17],
the spin accumulation in N can be probed by the voltage V_{NL}, which is induced at
F2. The magnitude of V_{NL} is associated to $\delta\mu_N$, while its sign is determined by the
relative magnetization orientation of F1 and F2. If we consider the pedagogical
case where both F1 and F2 are half-metallic, when the magnetization of F2 is
parallel to that of F1, the electrochemical potential of F2 coincides with the spin-
up electrochemical potential of N, and when the magnetization is antiparallel, it
coincides with the spin-down electrochemical potential (Fig. 14.1(c)).

Because the current is applied to the left on N, there is no charge current
towards the right, where the detector F2 lies. Therefore, the detection is implemented nonlocally, where no charge current circulates by the detection point, and
thus V_{NL} is sensitive to the spin degree of freedom only. Accordingly, nonlocal
measurements eliminate the presence of spurious effects such as anisotropic
magnetoresistance or the Hall effect that could mask subtle signals related to spin

injection. Therefore, nonlocal devices usually exhibit a small output background allowing sensitive spin-detection experiments.

Nonlocal spin injection and detection was originally performed in a non-magnetic strip (bulk aluminum, Al) with two ferromagnetic (FM, permalloy) electrodes attached and spin transport was reported over lengths of several μm. More recent demonstrations in thin-film devices using advanced nanolithography techniques [18–25] increased the interest in nonlocal structures. A number of research groups used different geometries, materials, and interfaces between the ferromagnetic electrodes and the nonmagnetic metal. In the last few years, the nonlocal detection technique has been successfully utilized in diverse systems comprising one [19–21, 23–31] and zero-dimensional [32] metallic structures, superconductors [22, 33], graphene [34, 35], and a variety of semiconductors [36–38], using both transparent and tunneling interfaces. These devices shed light on spin transport in many materials, demonstrated electrical detection of spin precession [9, 19, 32], the study of the spin polarization of tunneling electrons as a function of the bias voltage [25], the ferromagnet–nonmagnetic metal interface properties [21, 27, 39], and the implementation of magnetization reversal of a nanoscale ferromagnetic particle with pure spin currents [40].

A common characteristic of these conventional nonlocal devices is that detection is sensitive to the local spin accumulation [13, 14], whereas the bulk spin current is determined only indirectly. Below, we describe electrical detection of spin currents and the spin Hall effect using the nonlocal geometry described in Section 12.5. The detection technique is based on a spin-current induced Hall effect, which is the reciprocal of the spin Hall effect [5, 41, 42] or inverse spin Hall effect (ISHE). By using a ferromagnetic electrode, a spin-polarized current is injected in a nonmagnetic strip, while measuring the laterally induced voltage that results from the conversion of the injected spin current into charge imbalance owing to the spin–orbit coupling in the nonmagnetic strip (Fig. 14.2).

These ideas were first introduced in Refs. [9, 11, 43]. As described below, a similar device was later used to measure both the SHE and the ISHE (see Fig. 12.1). Note that, according to the Onsager symmetry relations, the SHE and the ISHE are mathematically equivalent [5, 8, 44]. This is schematically shown in Fig. 14.2. The spin-polarized current I_{AB} between contacts A and B induces a voltage $V_{CD} = R_{AB,CD}(\mathbf{M})I_{AB}$ between contacts C and D. As explained in Section 12.5, the coefficient $R_{AB,CD}(\mathbf{M})$ is a function of the nonmagnetic metal properties, the orientation of the magnetization $\mathbf{M}$ of the ferromagnetic electrode, and the degree of polarization of the electrons transmitted through the interface. Alternatively, if a current I_{CD} between C and D is applied, spin accumulation builds up underneath the ferromagnet, owing to the SHE, and this results in a voltage $V_{AB} = R_{CD,AB}(\mathbf{M})I_{CD}$ between A and B with $R_{CD,AB}$ proportional to the SHE coefficient of the nonmagnetic metal (Fig. 14.2). Therefore, V_{AB} is a direct consequence of the SHE. According to the Onsager symmetry relations, the measurements of both experiments are equivalent with

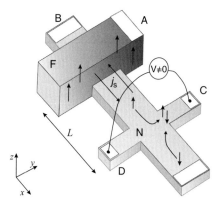

FIG. 14.2. Spin-current induced Hall effect or inverse spin Hall effect (ISHE). Schematic representation of an actual device where the pure spin current is generated by spin injection through a ferromagnet with out-of-plane magnetization. Due to spin–orbit interaction a transverse charge current and an associated voltage are induced.

$R_{AB,CD}(\mathbf{M}) = R_{CD,AB}(-\mathbf{M})$ [8, 44], a relationship that has been proved experimentally.

14.3 The electronic spin Hall experiments

Nonlocal spin Hall devices are prepared either with single-step (shadow) or multiple-step electron-beam lithography processing. For example, the device shown in Fig. 14.3 is fabricated with a two-angle shadow-mask evaporation technique to produce tunnel barriers *in situ* between the ferromagnet and the nonmagnetic metal. It was first fabricated to demonstrate the ISHE in Al. The shadow mask is made on a Si/SiO_2 substrate with a methyl-methacrylate (MMA)–polymethyl-methacrylate (PMMA) bilayer, using the fact that the base resist (MMA) has a sensitivity that is ∼5 times larger than the top resist (PMMA) [9, 11, 45]. This way, suspended masks with controlled undercut can be fabricated by selective electron-beam exposure. A nonmagnetic cross of N (Al), is first deposited at normal incidence onto the Si/SiO_2 substrate using electron beam evaporation. The voltage leads are much narrower than the main channel in order not to affect the spin diffusion in the latter. Next, insulating barriers for tunneling injection are generated. In the present device Al is oxidized in pure oxygen (150 mtorr for 40 min) to grow a thin layer of Al_2O_3. After the vacuum recovers, two ferromagnetic electrodes, F1 and F2, of different widths are deposited at an angle of 45–50°, measured from the normal to the substrate surface. N is deposited through all of the mask features but no image of the Hall cross appears when depositing F1 and F2. The axis of rotation (indicated by a dashed line in Fig. 14.3) is selected such that the ferromagnet deposits

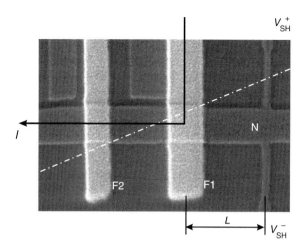

FIG. 14.3. Scanning electron microscope picture of a nonlocal spin Hall device (CoFe/Al). F1 and F2 appear brighter than the N cross. The diagonal dashed line represents the axis of rotation for shadow evaporation.

on the wall of the top resist and it is removed by lift-off, except for the lines that define the ferromagnet electrodes, and for some features that are far away from the Hall cross and that are not relevant for the experiment (see [9, 11] for details). The ferromagnetic electrodes in the final device form tunnel junctions where they overlap with the N strip. For the ferromagnetic electrodes, Co, Fe, or alloys are commonly used because they provide a large polarization when combined with Al_2O_3 as a tunneling barrier [21, 46]. The tunnel barrier is also relevant to generate a uniformly distributed injection current. The difference in the ferromagnetic electrode widths is necessary to obtain different coercive fields.

The device layout in Fig. 14.3 is more sophisticated than that represented in Fig. 14.2, where only F1 is required. The second electrode (F2), together with F1 and the N strip, form a spin-injection/detection device (Fig. 14.1) for the purpose of calibration. Calibration procedures are necessary to demonstrate consistency with standard nonlocal methods, and are common in nonlocal SHE experiments. Explicitly, this device can be utilized to measure the spin accumulation in the nonmagnetic metal and then determine its associated spin relaxation length λ_N, the spin polarization of the injected electrons P, and the magnetization orientation of the ferromagnetic electrodes θ in the presence of an external magnetic field (perpendicular to the substrate). For this purpose, batches of samples are commonly used where the distance between the two ferromagnets, L_F, is modified. The distance of F1 relative to the Hall cross, L, is also modified in order to test the consistency of the spin relaxation results.

Measurements are usually performed using standard lock-in techniques in a set-up where the angle of the substrate relative to an external magnetic field can be readily controlled. In Ref. [9, 11] samples with different N thickness d_N (12 and 25 nm) were fabricated in order to study the spin-current induced Hall signal in devices with different λ_N.

Device characterization   As mentioned above, F1 and F2 are used to obtain P, λ_N, and θ at $B_\perp \neq 0$. In this case, the current I is applied from F1 towards the Hall cross and the voltage is measured between F2 and the end of the wide nonmagnetic arm that is opposite to the Hall cross. Both P and λ_N are obtained by measuring the spin transresistance $\Delta R_{NL} = \Delta V/I$ as a function of L_F, where ΔV is the difference in the output voltage between parallel and antiparallel magnetization configurations of the ferromagnetic electrodes at zero magnetic field. P and λ_N are obtained by fitting to [13, 14, 19, 47]:

$$R_{NL} = P_{\text{eff}}^2 \frac{\lambda_N}{\sigma_N A} e^{-L_F/\lambda_N}. \tag{14.1}$$

where $\sigma_N = (\rho_N)^{-1}$ is the nonmagnetic metal conductivity and A its cross sectional area.

Spin injection in the nonmagnetic film occurs with a defined spin direction given by the magnetization orientation of the ferromagnetic electrode. Consequently, in the SHE experiments V_{SH} is expected to vary when a magnetic field perpendicular to the substrate, $B_\perp$, is applied and the magnetization $\mathbf{M}$ of the electrode is tilted an angle θ out of the substrate plane. For arbitrary spin orientation, Eq. (12.35) can be generalized by adding a factor $\sin\theta$:

$$R_{SH} = \frac{1}{2} P_{\text{eff}} \frac{\alpha_{SH}}{\sigma_N d_N} e^{-L/\lambda_N} \sin\theta. \tag{14.2}$$

As long as $\mathbf{M}$ is parallel to $\mathbf{B}$ or $\mathbf{B}$ is perpendicular to the Hall-cross plane, the output signal is not affected by spin precession as the component of the spins perpendicular to the substrate is not modified by this effect. The tilting angle can be obtained from spin precession measurements under a variable magnetic field using F1 and F2 and the original spin injection and detection technique [13, 14]. Examples of these measurements are shown in Fig. 14.4(a) for two different Al samples with $L_F = 1$ and 2 μm, showing consistent results. At $B_\perp = 0$, $\theta = 0$ due to the shape anisotropy of the ferromagnetic electrodes. When $B_\perp$ is applied, the magnetization follows the Stoner–Wohlfarth model [48] with a saturation field $B_\perp^{\text{sat}}$ of about 1.55 T for which $\sin\theta$ approaches one, and the magnetization aligns with the field. The magnetization can also be rotated in-plane, which usually requires smaller applied magnetic fields [10, 12]. Moreover, recent experiments used FePt ferromagnetic contacts where the magnetization is naturally pointing out of plane and thus the magnetic field can be eliminated altogether [49]. The latter samples were fabricated by a multiple step lithography process where first a FePt ferromagnetic layer was epitaxially grown on a MgO substrate and patterned to form the spin injector using electron beam lithography with a negative resist and ion etching. Subsequently, an N layer (Au in this case) was deposited on the sample surface, which was also patterned into the shape of the Hall cross by electron beam lithography and ion etching.

Experimental results and interpretation Having determined P, λ_N, and θ, the ISHE can be studied using the measurement configuration shown in Fig. 14.3 by injecting current from F1 away from the Hall cross and measuring the induced lateral voltage V_{SH}. Figure 14.4(b) shows typical R_{SH} measurements as a function of $B_\perp$ for an Al sample with $L = 590$ nm (circles). A positive $B_\perp$ is pointing out of the page in the $\hat{z}$-direction. $B_\perp$ is swept between -3.5 and 3.5 T, enough to saturate the magnetization of the F1 along the field [Fig. 14.4(a)]. A linear response is observed around $B_\perp = 0$, followed by a saturation on the scale of $B_\perp^{sat}$, both for positive and negative $B_\perp$. The saturation in R_{SH} for $|B_\perp| > B_\perp^{sat}$ demonstrates that the device output is related to the magnetization orientation of the ferromagnetic electrode and the spin Hall effect, an idea that is reinforced by comparing $R_{SH}(B_\perp)$ with the magnetization component perpendicular to the substrate, which as discussed above is proportional to $\sin\theta(B_\perp)$ [the lines in Fig. 14.4(b)].

Figure 14.5(a) shows the overall change of R_{SH}, ΔR_{SH}, at magnetic field values beyond saturation. Different points correspond to different orientations of the magnetic field relative to the substrate that determine the magnetization orientation at saturation and the angle θ. The measurements shown were

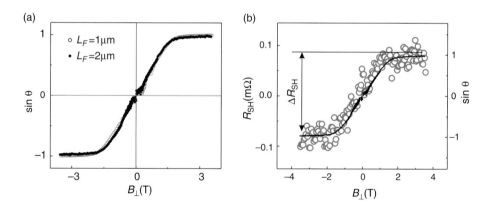

FIG. 14.4. (a) Experimentally determined $\sin\theta$ from spin precession measurements using the standard nonlocal technique (see Fig. 14.1): $L_F = 1$ and 2 μm, for open and full circles, respectively. (b) Spin Hall resistance R_{SH} versus the perpendicular field $B_\perp$, $L = 590$ nm. For comparison, the measured $\sin\theta$ in (a) is also shown. ΔR_{SH} is the overall change of R_{SH} between negative and positive magnetic fields at magnetization saturation.

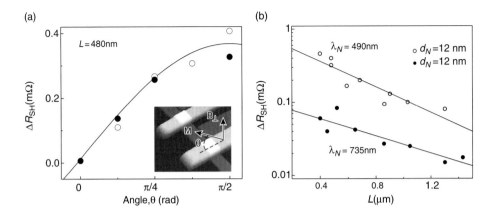

FIG. 14.5. (a) $\Delta R_{\rm SH}$ versus the orientation of the magnetic field, which determines θ at saturation. Results for two samples with $L = 480$ nm are shown.
The line is a fit to $\sim \sin \theta$. The inset shows the magnetization orientation.
For the measurements in Fig. 14.4 the magnetic field is perpendicular to the
substrate. In the current experiment $\mathbf{B} \parallel \mathbf{M}$. (b) $\Delta R_{\rm SH}$ versus L. Lines are
best fits to Eq. (12.39) from which $\lambda_{\rm N}$ and $\sigma_{\rm SH}$ are obtained.

performed in two samples with $L = 480$ nm and $d_{\rm N} = 12$ nm. The line shows a
fit to $\sim \sin \theta$, which closely follows the experimental results.

Figure 14.5(b) shows $\Delta R_{\rm SH}$ as a function of L in a semilogarithmic plot for the
magnetic field perpendicular to the substrate. Consistent with Eq. (14.2), $\Delta R_{\rm SH}$
decreases exponentially as a function of L. By fitting the data to Eq. (14.2), $\lambda_{\rm N}$
and $\sigma_{\rm SH}$ can be obtained and compared with the results obtained independently
with the reference devices described previously.

Direct inspection of Eqs. (14.1), and (14.2) shows that $R_{\rm SH}$ differs by a factor
$R_{\rm SH}/R_{\rm NL} \sim \alpha_{\rm SH}/P_{\rm eff}$ when compared with $R_{\rm NL}$ of spin accumulation devices
with tunnel barriers (Eq. 14.1). The spin Hall angle $\alpha_{\rm SH}$ for different materials
is in the range 0.0001–0.1 [16], indicating that $R_{\rm SH}$ can vary significantly when
using different materials but it could be as large as $R_{\rm NL}$ for spin accumulation
devices with tunnel barriers ($P \sim 0.1$). There is, however, a fundamental dis-
tinction in the origin of $R_{\rm SH}$ and $R_{\rm NL}$ in spite of the similarities of Eqs. (14.1),
and (14.2). The voltage output of the SHE device is directly proportional to
the spin current $j_{\rm s}$ [Eq. (12.33)]. In contrast, nonlocal spin accumulation devices
are sensitive to the spin accumulation but are not explicitly affected by the
spin flow. The spin accumulation and SHE based detection techniques are thus
complementary and the magnitudes of their respective device outputs are not
directly comparable. It is possible to envision situations where, although the
local spin accumulation is zero, i.e. $\delta\mu_{\rm N} = (\mu_{\rm N}^\uparrow - \mu_{\rm N}^\downarrow) = 0$, there exists a local
spin current, i.e. $j_s = -(\sigma_{\rm N}/2e)\nabla \mid \delta\mu_{\rm N} \mid\neq 0$, or *vice versa*.

The Onsager relation $R_{AB,CD}(\mathbf{M}) = R_{CD,AB}(-\mathbf{M})$ has been experimentally verified using a second device layout (Fig. 14.6). This device structure is similar to that in Fig. 14.2, but enables us to access the electrical detection of SHEs in materials with short λ_N less than 10 nanometers. Here, the transverse arm consisting of the material with the large spin–orbit coupling metal N2 acts as either a spin current source for the SHE or a spin current absorber for the ISHE. The longitudinal arm, on the other hand, is made of a metal N1 with a long spin diffusion length that fulfils the purpose of transporting spin information between the ferromagnet electrode (F) and N2.

The way the measurements are performed is sketched in Fig. 14.6(b). To study the ISHE, a charge current is injected from F into N1 that induces a spin current towards N2 [Fig. 14.6(b), left]. When the distance between F and the cross is smaller than the spin diffusion length of N1, the spin current is preferentially absorbed into the transverse arm N2 because of the strong spin relaxation of N2. (see Section 21.1). The injected spin current into N2 vanishes in a short distance from the N1/N2 interface because of the short spin diffusion length of N2 and generates a voltage via the ISHE as in Fig. 14.2. To study the SHE, the bias configuration is modified as shown in Fig. 14.6(b) (right). Here, N2 acts as a spin-current source, which induces a spin accumulation in N1 that is detected with the ferromagnetic electrode. Examples of measurements where N1 = Cu and N2 = Pt are presented in Fig. 14.7 [12]. As before, consistency checks with conventional nonlocal devices (Fig. 14.1) are carried out in order to determine λ_N and the degree of spin absorption in N2 [12].

Using the above techniques, the spin Hall angle α_{SH} and spin Hall conductivities σ_{SH} were determined in a large variety of materials, which are listed

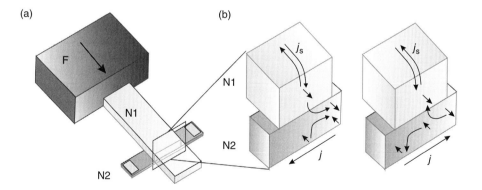

FIG. 14.6. (a) Schematic illustration of a nonlocal device to measure the direct and inverse spin Hall effect in materials (N2) with short λ_N. (b) Transformation from spin to transverse charge current (left) and from spin to charge current (right).

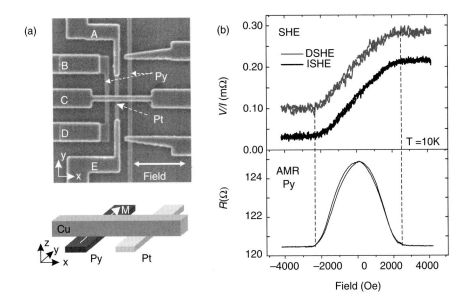

FIG. 14.7. (a) Scanning electron microscope image of a typical device for SHE
measurements as illustrated in Fig. 14.6. (b) Direct and inverse spin Hall
effect (SHE and ISHE) recorded at 10 K in a Pt device with thickness 20
nm (top). Anisotropic magnetoresistance (AMR) from the injector consisting
of a Permalloy (Py) wire is measured under the same conditions. SHE
measurement corresponds to V_{BC}/I_{AE}, and ISHE to V_{EA}/I_{BC}; with V the
voltage, I the applied current; and A, B, C and E are the contact leads as
denoted in (a) (see also Fig. 14.2).

in Table 14.1. For completeness, we have also included in Table 3.2 the α_{SH}
obtained via ferromagnetic resonance techniques (see Chapter 15).

The SHE in Al is well explained via extrinsic [side-jump, Eq. (12.31)]
mechanisms. The predicted σ_{SH}, when considering δ-like scattering centers, is
$\bar{\eta}_{\text{so}}\hbar e^2 N_0/3m$ [6, 7], where $\hbar$ is Planck's constant divided by 2π, $N_0 = 2.4 \times 10^{28}$
states/eV m^3 is the density of states of Al at the Fermi energy [51], $\bar{\eta}_{\text{so}} \sim$
0.006–0.008 is the dimensionless spin–orbit coupling constant of Al [52], and
e and m are the charge and mass of the electron. Without free parameters,
we obtain $\sigma_{\text{SH}} \sim 1$–$1.4 \times 10^3$ $(\Omega$ m$)^{-1}$ and $\alpha_{\text{SH}} = \sigma_{\text{SH}}/\sigma_{\text{N}} \sim 0.4$–$1.4 \times 10^{-4}$, in
reasonable agreement with the experimental results. The difference between the
spin Hall angles for the two thin-film thicknesses reported can be attributed
to the larger influence of the surface in spin scattering events for the thinner
films [21].

The most frequently studied material using electrical methods has been Pt.
For Pt, the reported α_{SH} varies considerably and is in the range between 0.004
and 0.02 at room temperature. The lower end of this range, however, could be

Table 14.1 Experimental spin Hall angles α_{SH} for Al, Au, CuIr, Mo, Pd, Pt, and Ta. CuIr parameters aggregate the results for Ir concentrations in the range between 1% and 12%. References marked with the symbol (†) are based on spin-pumping and spin-torque methods. The values marked with (*) are not measured but assumed from literature in the corresponding references. To calculate the spin–orbit coupling parameter $\bar{\eta}_{so} = k_F^2 \eta_{so}$, the Fermi momenta, $k_F = 1.75 \times 10^8$ cm^{-1} (Al), 1.21×10^8 cm^{-1} (Au), and 1.18×10^8 cm^{-1} (Nb), are taken [50], and 1×10^8 cm^{-1} (Mo, Pd, Ta, Pt) is assumed.

	λ_N (nm)	σ_N $(\Omega\text{cm})^{-1}$	$\bar{\eta}_{so}$	α_{SH} (%)	Ref.
Al (4.2K)	455 ± 15	1.05×10^5	0.0079	0.032 ± 0.006	[9, 11]
Al (4.2K)	705 ± 30	1.70×10^5	0.0083	0.016 ± 0.004	[9, 11]
Au (295K)	86 ± 10	3.70×10^5	0.3	11.3	[49]
Au (295K)	$35 \pm 3^*$	2.52×10^5	0.52	0.35 ± 0.03	[56]†
CuIr (10K)	$5 - 30$			2.1 ± 0.6	[61]
Mo (10K)	10	3.03×10^4	0.32	-0.20	[69]
Mo (10K)	10	6.67×10^3	0.07	-0.075	[69]
Mo (10K)	8.6 ± 1.3	2.8×10^4	0.34	$-(0.8 \pm 0.18)$	[53]
Mo (295K)	$35 \pm 3^*$	4.66×10^4	0.14	$-(0.05 \pm 0.01)$	[56]†
Nb (10K)	5.9 ± 0.3	1.1×10^4	0.14	$-(0.87 \pm 0.20)$	[53]
Pd (295K)	9^*	1.97×10^4	0.23	1.0	[68]†
Pd (10K)	13 ± 2	2.2×10^4	0.18	1.2 ± 0.4	[53]
Pd (295K)	$15 \pm 4^*$	4.0×10^4	0.28	0.64 ± 0.10	[56]†
Pt (295K)		6.41×10^4	0.74	0.37	[10]
Pt (5K)	14	8.0×10^4	0.61	0.44	[12]
Pt (295K)	10	5.56×10^4	0.58	0.9	[12]
Pt (10K)	11 ± 2	8.1×10^4	0.77	2.1 ± 0.5	[53]
Pt (295K)	7^*	6.4×10^4	0.97	8.0	[55]†
Pt (295K)	$3 - 6$	5.0×10^4	0.88-1.75	$7.6^{+8.4}_{-2.0}$	[57]†
Pt (295K)	$10 \pm 2^*$	2.4×10^4	0.25	1.3 ± 0.2	[56]†
Ta (10K)	2.7 ± 0.4	3.0×10^3	0.17	$-(0.37 \pm 0.11)$	[53]

compromised due to underestimations that stem from the wrong assumption about complete spin current absorption into the Pt wire [10] and from improper boundary conditions at the Pt interface with N (Cu in this case) [12]. The more refined analysis [53] probably gives the best estimate of $\alpha_{SH} \sim 0.021$, which is considerably larger than those reported in the first experiments. In addition, the results gathered with ferromagnetic resonance techniques [54–57] are within a factor of 4 of the latter value with $0.013 < \alpha_{SH} < 0.076$. The discrepancy between these results can be related to the assumed λ_N, which spreads over 3 and 10 nm; a better agreement would be obtained using the largest estimates for λ_N.

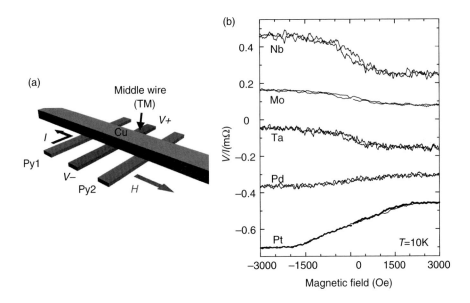

FIG. 14.8. (a) Schematic illustration of the probe configuration for the inverse
    SHE measurement. (b) Inverse SHE signals measured at 10 K for various
    transition metal wire insertions.

The origin of the spin Hall effect in $4d$ and $5d$ transition metals is still a matter
of debate. Early measurements in Pt indicated that the side-jump mechanism
was dominant [12]. As expected in the side-jump origin of the SHE, the spin Hall
resistivity was found to be proportional to the resistivity of Pt squared [12] and
the values for σ_{SH} and α_{SH} were comparable to those obtained from Eq. (12.31).
However, a recent report in various $4d$ and $5d$ transition metals (Mo, Nb, Pd, Pt,
and Ta) shows that the sign of the spin Hall conductivity changes systematically
depending on the number of d electrons, as shown in Fig. 14.8.[53] This is in
agreement with calculations based on the intrinsic properties of the materials
[58, 59], namely the degeneracy of d orbits, and together with the experimental
results suggest an intrinsic origin of the SHE.

The spin Hall angle can be enhanced by introducing impurities in a host metal
[60]. In this way a large $\alpha_{SH} \sim 0.02$ has been obtained in CuIr throughout an Ir
concentration range between 1% and 12% [61]. Similarly, the large discrepancy
between the results obtained with Au (Table 14.1) would result from a strong
enhancement of the spin-orbit interaction due to Fe impurities in Au [62]
(Section 12.5).

The spin Hall effect has also been observed experimentally in semiconductors
combining optical and electrical techniques (see also Chapter 16). Electrical
currents in n-GaAs layers induced a spin Hall effect which was detected optically,
at low temperatures (30 K), using Kerr microscopy [3]. The spin Hall angles

are similar to those in Al, $\alpha_{SH} \sim 2 \times 10^{-4}$. Because the results showed little dependence on crystal orientation, it was concluded that the origin of the effect was extrinsic. The data can indeed be well described by extrinsic models based on scattering by impurities [63, 64]. Subsequent experiments in ZnSe at room temperature [65] present effects of similar magnitude, which are also in agreement with extrinsic modeling. In another experiment performed in two-dimensional layers of p-GaAs, the spin accumulation due to the spin Hall effect at the edge of the sample is revealed by detecting the polarization of the recombination radiation of holes [4]. The magnitude of the spin accumulation is larger, and was ascribed to the intrinsic mechanism as supported by theoretical results [66, 67]. More recently, fully electrical measurements have also been used in semiconductors such as GaAs [70] and HgTe [71].

Acknowledgments

We acknowledge support from the Spanish Ministry of Science and Innovation, MICINN (MAT2010-18065), the US NSF, the US ONR, JST-CREST, and NEDO.

References

[1] Sinova, J., Murakami, S., Shen, S.-Q., and Choi, M.-S. (2006). Spin-Hall effect: Back to the beginning at a higher level. *Solid State Comm.*, **138**, 214.

[2] Engel, H. A., Rashba, E. I., and Halperin, B. I. (2007). Theory of spin Hall effects in semiconductors, in *Handbook of Magnetism and Advanced Magnetic Materials*, H. Kronmüller and S. Parkin (eds.). John Wiley, New York.

[3] Kato, Y. K., Myers, R. C., Gossard, A. C., and Awschalom, D. D. (2004). Observation of the spin Hall effect in semiconductors. *Science*, **306**, 1910.

[4] Wunderlich, J., Kaestner, B., Sinova, J., and Jungwirth, T. (2005). Experimental observation of the spin-Hall effect in a two-dimensional spin-orbit coupled semiconductor system. *Phys. Rev. Lett.*, **94**, 047204.

[5] Hirsch, J. E. (1999). Spin Hall effect. *Phys. Rev. Lett.*, **83**, 1834.

[6] Zhang, S. (2000). Spin Hall effect in the presence of spin diffusion. *Phys. Rev. Lett.*, **85**, 393.

[7] Shchelushkin, R. V. and Brataas, A. (2005). Spin Hall effects in diffusive normal metals. *Phys. Rev. B*, **71**, 045123.

[8] Hankiewicz, E. M., Li, J., Jungwirth, T., Niu, Q., Shen, S-Q., and Sinova, J. (2005). *Phys. Rev. B*, **72**, 155305.

[9] Valenzuela, S. O. and Tinkham, M. (2006). Direct electronic measurement of the spin Hall effect. *Nature*, **442**, 176.

[10] Kimura, T., Otani, Y., Sato, T., Takahashi, S., and Maekawa, S. (2007). Room-temperature reversible spin Hall effect. *Phys. Rev. Lett.*, **98**, 156601.

[11] Valenzuela, S. O. and Tinkham, M. (2007). Nonlocal electronic spin detection, spin accumulation and the spin Hall effect. *J. Appl. Phys.*, **101**, 09B103.

[12] Vila, L., Kimura, T. and Otani, Y. (2007). Room-temperature reversible spin Hall effect. *Phys. Rev. Lett.*, **99**, 226604.

[13] Johnson, M. and Silsbee, R. H. (1985). Interfacial charge–spin coupling: injection and detection of spin magnetization in metals. *Phys. Rev. Lett.*, **55**, 1790.

[14] Johnson, M. and Silsbee, R. H. (1988). Coupling of electronic charge and spin at a ferromagnetic–paramagnetic metal interface. *Phys. Rev. B*, **37**, 5312.

[15] Johnson, M. and Silsbee, R. H. (1988). Spin injection experiment. *Phys. Rev. B*, **37**, 5326.

[16] Valenzuela, S. O. (2009). Nonlocal electronic spin detection, spin accumulation and the spin Hall effect. *Int. J. Mod. Phys. B* **23**, 2413.

[17] Silsbee, R. H. (1980). Novel method for the study of spin transport in conductors. *Bull. Magn. Reson.* **2**, 284.

[18] Jedema, F. J., Filip, A. T., and van Wees, B. J. (2001). Electrical spin injection and accumulation at room temperature in an all-metal mesoscopic spin valve. *Nature*, **410**, 345.

[19] Jedema, F. J., Heersche, H. B., Filip, A. T., Baselmans, J. J. A., and van Wees, B. J. (2002). Electrical detection of spin precession in a metallic mesoscopic spin valve. *Nature*, **416**, 713.

[20] Kimura, T., Hamrle, J., Otani,Y., Tsukagoshi, K., and Aoyagi, Y. (2004). Spin-dependent boundary resistance in the lateral spin valve structure. *Appl. Phys. Lett.*, **85**, 3501.

[21] Valenzuela, S. O. and Tinkham, M. (2004). Spin polarized tunneling in room temperature spin valves. *Appl. Phys. Lett.* **85**, 5914.

[22] Beckmann, D., Weber, H.B. and Löhneysen, H. v. (2004). Evidence for crossed Andreev reflection in superconductor-ferromagnet hybrid structures *Phys. Rev. Lett.*, **93**, 197003.

[23] Ji, Y., Hoffmann, A., Jiang, J. S., and Bader, S. D. (2004). Spin injection, diffusion, and detection in lateral spin-valves. *Appl. Phys. Lett.*, **85**, 6218.

[24] Garzon, S., Žutić, I. and Webb, R. A. (2005). Temperature-dependent asymmetry of the nonlocal spin-injection resistance: Evidence for spin nonconserving interface scattering. *Phys. Rev. Lett.*, **94**, 176601.

[25] Valenzuela, S. O., Monsma, D. J., Marcus, C. M., Narayanamurti, V., and Tinkham, M. (2005). Spin polarized tunneling at finite bias *Phys. Rev. Lett.* **94**, 196601.

[26] Ji, Y., Hoffmann, A., Pearson, J. E. and Bader, S. D. (2006). Enhanced spin injection polarization in Co/Cu/Co nonlocal lateral spin valves. *Appl. Phys. Lett.*, **88**, 052509.

[27] Godfrey, R. and Johnson, M. (2006). Spin injection in mesoscopic silver wires: Experimental test of resistance mismatch. *Phys. Rev. Lett.*, **96**, 136601.

[28] Ku, J. H., Chang, J., Kim, H., and Eom, J. (2006). Effective spin injection in Au film from Permalloy, *Appl. Phys. Lett.*, **88**, 172510.

[29] Kimura, T. and Otani, Y. (2007). Large spin accumulation in a permalloy–silver lateral spin valve. *Phys. Rev. Lett.*, **99**, 196604.

[30] Kimura, T., Sato, T., and Otani, Y. (2008). Temperature evolution of spin relaxation in a NiFe/Cu Lateral spin valve. *Phys. Rev. Lett.*, **100**, 066602.

[31] Idzuchi, H., Fukuma, Y., Wang, L., and Otani, Y. (2010). Spin diffusion characteristics in magnesium nanowires. *Appl. Phys. Exp.*, **3**, 063002.

[32] Zaffalon, M. and van Wees, B. J. (2003). Zero-dimensional spin accumulation and spin dynamics in a mesoscopic metal island. *Phys. Rev. Lett.*, **91**, 186601.

[33] Urech, M., Johansson, J., Poli, N., Korenivski, V., and Haviland, D. B. (2006). Enhanced spin accumulation in superconductors *J. Appl. Phys.*, **99**, 08M513.

[34] Tombros, N., Jozsa, C., Popinciuc, M., Jonkman, H. T., and B.J. van Wees (2007). Electronic spin transport and spin precession in single graphene layers at room temperature. *Nature*, **448**, 571.

[35] Han, W., Pi, K., McCreary, K. M., Li, J. Y., Wong, J. I., Swartz, A. G., and Kawakami, R. K. (2010). Tunneling spin injection into single layer graphene. *Phys. Rev. Lett.*, **105**, 167202.

[36] Lou, X., Adelmann, C., Crooker, S. A., Garlid, E. S., Zhang, J., Madhukar Reddy, K. S., Flexner, S. D., Palmstrøm, C. J., and Crowell, P. A. (2007). Electrical detection of spin transport in lateral ferromagnetsemiconductor devices. *Nature Phys.*, **3**, 197.

[37] van't Erve, O. M. J., Hanbicki, A. T., Holub, M., Li, C. H., Awo-Affouda, C., Thompson, P. E., and Jonker, B. T. (2007). Electrical injection and detection of spin-polarized carriers in silicon in a lateral transport geometry. *Appl. Phys. Lett.*, **91**, 212109.

[38] Salis, G., Fuhrer, A., and Alvarado, S. F. (2008). Signatures of dynamically polarized nuclear spins in all-electrical lateral spin transport devices. *Phys. Rev. B*, **80**, 115332.

[39] Fukuma, Y., Wang, L., Idzuchi, and Otani, Y. (2010). Enhanced spin accumulation obtained by inserting low-resistance MgO interface in metallic lateral spin valves. *Appl. Phys. Lett.*, **97**, 012507.

[40] Kimura, T., Otani, Y., and Hamrle, J. D. (2006). Switching magnetization of a nanoscale ferromagnetic particle using nonlocal spin injection. *Phys. Rev. Lett.*, **96**, 037201.

[41] Dyakonov, M. I. and Perel, V. I. (1971). Possibility of orienting electron spins with current. *JETP Lett.*, **13**, 467.

[42] D'yakonov, M. I. and Perel', V. I. (1971). Current induced spin orientation of electrons in semiconductors. *Phys. Lett. A*, **35**, 459.

[43] Takahashi, S., Imamura, H. and Maekawa, S. (2006). Chapter 8 in *Concepts in Spin Electronics*. S. Maekawa (ed.), Oxford University Press.

[44] Adagideli, I., Bauer, G. E., and Halperin, B. I. (2006). Detection of current-induced spins by ferromagnetic contacts. *Phys. Rev. Lett.*, **97**, 256601.

[45] Costache, M. V. and Valenzuela, S. O. (2005). Experimental spin ratchet. *Science*, **330**, 1645.

[46] Monsma, D. J. and Parkin, S. S. P. (2000). Spin polarization of tunneling current from ferromagnet/Al_2O_3 interfaces using copper-doped aluminum superconducting films. *Appl. Phys. Lett.*, **77**, 720.

[47] Takahashi, S. and Maekawa, S. (2003). Spin injection and detection in magnetic nanostructures. *Phys. Rev. B*, **67**, 052409.

[48] O'Handley, R. C. (2000). *Modern Magnetic Materials*. John Wiley, New York.

[49] Seki, T., Hasegawa, Y., Mitani, S., Takahashi, S., Imamura, H., Maekawa, S., Nitta, J., and Takanashi, K. (2008). Giant spin Hall effect in perpendicularly spin-polarized FePt/Au devices. *Nature Mater.*, **7**, 125.

[50] Ashcroft, N. W. and Mermin, D. (1976). *Solid State Physics*, Saunders College.

[51] Papaconstantopoulos, D. A. (1986). *Handbook of the Band Structure of Elemental Solids*. Plenum, New York.

[52] Shchelushkin, R. V. and Brataas, A. (2005). Spin Hall effect, Hall effect, and spin precession in diffusive normal metals *Phys. Rev. B*, **72**, 073110.

[53] Morota, M., Niimi, Y., Ohnishi, K., Wei, D. H., Tanaka, T., Kontani, H., Kimura, T., and Otani, Y. (2011). Indication of intrinsic spin Hall effect in $4d$ and $5d$ transition metals. *Phys. Rev. B*, **83**, 174405.

[54] Saitoh, E., Ueda, M., Miyajima, H., and Tatara, G. (2006). Conversion of spin current into charge current at room temperature: Inverse spin-Hall effect. *Appl. Phys. Lett.*, **88**, 182509.

[55] Ando, K., Takahashi, S., Harii, K., Sasage, K., Ieda, J., Maekawa, S., and Saitoh, E. (2008). Electric manipulation of spin relaxation using the spin Hall effect. *Phys. Rev. Lett.*, **101**, 036601.

[56] Mosendz, O., Vlaminck, V., Pearson, J. E., Fradin, F. Y., Bauer, G. E. W., Bader, S. D., and Hoffmann, A. (2010). Detection and quantification of inverse spin Hall effect from spin pumping in permalloy/normal metal bilayers. *Phys. Rev. B*, **82**, 214403.

[57] Liu, L., Moriyama, T., Ralph, D. C., and Buhrman, R. A. (2011). Spin-torque ferromagnetic resonance induced by the spin Hall effect. *Phys. Rev. Lett.*, **106**, 036601.

[58] Guo, G.-Y., Murakami, S., Chen, T.-W., and Nagaosa, N. (2008). Intrinsic spin Hall effect in platinum: First-principles calculations. *Phys. Rev. Lett.*, **100**, 096401.

[59] Kontani, H., Tanaka, T., Hirashima, D. S., Yamada, K., and Inoue, J. (2009). Giant orbital Hall effect in transition metals: Origin of large spin and anomalous Hall effects. *Phys. Rev. Lett.*, **102**, 016601.

[60] Fert, A., Friederich, A., and Hamzic, A. (1981). Hall effect in dilute magnetic alloys. *J. Magn. Magn. Mater.*, **24**, 231.

[61] Niimi, Y., Morota, M., Wei, D. H., Deranlot, C., Basletic, M., Hamzic, A., Fert, A., and Otani, Y. (2011). Extrinsic spin Hall effect induced by iridium impurities in copper. *Phys. Rev. Lett.*, **106**, 126601.

[62] Guo, G.-Y., Maekawa, S., and Nagaosa, N. (2009). Enhanced spin Hall effect by resonant skew scattering in the orbital-dependent Kondo effect. *Phys. Rev. Lett.*, **102**, 036401.

[63] Engel, H. A., Rashba, E. I. and Halperin, B. I. (2005). Theory of spin Hall conductivity in n-doped GaAs. *Phys. Rev. Lett.*, **95**, 166605.

[64] Tse, W.-K. and Das Sarma, S. (2006). Spin Hall effect in doped semiconductor structures. *Phys. Rev. Lett.*, **96**, 56601.

[65] Stern, N. P., Ghosh, S., Xiang, G., M. Zhu, M., Samarth, N., and Awschalom, D. D. (2006). Current-induced polarization and the spin Hall effect at room temperature. *Phys. Rev. Lett.*, **97**, 126603.

[66] Schliemann, J. and Loss, D. (2005). Spin-Hall transport of heavy holes in III-V semiconductor quantum wells. *Phys. Rev. B*, **71**, 085308.

[67] Nomura, K., Wunderlich, J., Sinova, J., Kaestner, B., MacDonald, A., and Jungwirth, T. (2005). Edge spin accumulation in semiconductor two-dimensional hole gases. *Phys. Rev. B*, **72**, 245330.

[68] Ando, K. and Saitoh, E. (2010). Inverse spin Hall effect in palladium at room temperature. *J. Appl. Phys.*, **108**, 113925.

[69] Morota, M., Ohnishi, K., Kimura, T., and Otani, Y. (2009). Spin Hall effect in molybdenum wires. *J. Appl. Phys.*, **105**, 07C712.

[70] Garlid, E. S., Hu, Q. O., Chan, M. K., Palmstrøm, C. J. and Crowell, P. A. (2010). Electrical measurement of the direct spin Hall effect in $Fe/In_xGa_{1-x}As$ heterostructures. *Phys. Rev. Lett.*, **105**, 156602.

[71] Brüne, C., Roth, A., Novik, E. G., König, M., Buhmann, H., Hankiewicz, E. M., Hanke, W., Sinova, J., and Molenkamp, L. W. (2010). Evidence for the ballistic intrinsic spin Hall effect in HgTe nanostructures. *Nature Phys.*, **6**, 448.

15 Experimental observation of the spin Hall effect using spin dynamics

E. Saitoh and K. Ando

15.1 Inverse spin Hall effect induced by spin pumping

In one of the first reports on the inverse spin Hall effect (ISHE), the spin-pumping effect was used for spin-current generation [1]. Spin pumping is the generation of spin currents as a result of magnetization $M(t)$ precession [2, 3]; in a ferromagnetic/paramagnetic bilayer system, a conduction-electron spin current is pumped out of the ferromagnetic layer into the paramagnetic conduction layer in a ferromagnetic resonance (FMR) condition as shown in Fig. 15.1(b). In the standard model of spin pumping [3], the dc component of the generated spin current density j_s is expressed as

$$j_s = \frac{\omega}{2\pi} \int_0^{2\pi/\omega} \frac{\hbar}{4\pi} g_r^{\uparrow\downarrow} \frac{1}{M_s^2} \left[M(t) \times \frac{dM(t)}{dt} \right]_z dt, \qquad (15.1)$$

where ω, $\hbar$, $g_r^{\uparrow\downarrow}$, and M_s are the angular frequency of the magnetization precession, the Dirac constant, the spin mixing conductance [3], and the saturation magnetization. Here, $[M(t) \times dM(t)/dt]_z$ denotes the z-component of $M(t) \times dM(t)/dt$. The z-axis is directed along the magnetization-precession axis.

In the following, we describe an experiment on the ISHE induced by spin pumping [1, 4–6]. The sample is a $Ni_{81}Fe_{19}/Pt$ bilayer film comprising a 10-nm-thick ferromagnetic $Ni_{81}Fe_{19}$ layer and a 10-nm-thick paramagnetic Pt layer as shown in Fig. 15.1(a). The surface of the $Ni_{81}Fe_{19}$ layer is of a 1.0 mm × 1.2 mm rectangular shape. The Pt layer was sputtered on a thermally oxidized Si substrate and then the $Ni_{81}Fe_{19}$ layer was evaporated on the Pt layer in a high vacuum. Two electrodes are attached to both ends of the Pt layer.

For the measurement, the sample system is placed near the center of a TE_{011} microwave cavity at which the magnetic-field component of the microwave mode is maximized while the electric-field component is minimized. During the measurement, a microwave mode with frequency $f = 9.44$ GHz exists in the cavity and an external magnetic field H along the film plane is applied perpendicular to the direction across the electrodes, as illustrated in Fig. 15.1(a). Since the magnetocrystalline anisotropy in $Ni_{81}Fe_{19}$ is negligibly small, the magnetization in the $Ni_{81}Fe_{19}$ layer is uniformly aligned along the magnetic field direction. When H and f fulfill the FMR condition, a pure spin current with a spin

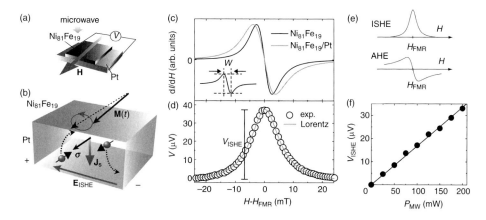

FIG. 15.1. (a) A schematic illustration of the $Ni_{81}Fe_{19}$/Pt film used in the present study. H is the external magnetic field. (b) A schematic illustration of the spin pumping and the inverse spin Hall effect in the $Ni_{81}Fe_{19}$/Pt film. $M(t)$ is the magnetization in the $Ni_{81}Fe_{19}$ layer. E_{ISHE}, J_s, and σ denote the electromotive force due to the inverse spin Hall effect, the spatial direction of the spin current, and the spin-polarization vector of the spin current, respectively. (c) Field (H) dependence of the FMR signals $dI(H)/dH$ for the $Ni_{81}Fe_{19}$/Pt film and the $Ni_{81}Fe_{19}$ film. Here, I denotes the microwave absorption intensity. H_{FMR} is the resonance field. The inset shows the definition of the spectral width W in the present study. (d) H dependence of the electric-potential difference V for the $Ni_{81}Fe_{19}$/Pt film under 200 mW microwave excitation. The open circles are the experimental data. The curve shows the fitting result using a Lorentz function for the V data. (e) The spectral shape of the electromotive force due to the inverse spin Hall effect (ISHE) and the anomalous Hall effect (AHE). (f) The microwave power, P_{MW}, dependence of the electromotive force, V_{ISHE}, for the $Ni_{81}Fe_{19}$/Pt film. V_{ISHE} is estimated as the peak height of the resonance shape in the V spectrum as shown in (d).

polarization σ parallel to the magnetization-precession axis in the $Ni_{81}Fe_{19}$ layer is injected into the Pt layer by spin pumping [see Fig. 15.1(b)] [1]. This injected spin current is converted into an electric voltage due to the strong ISHE in the Pt layer as shown in Fig. 15.1(b). By measuring the electric voltage, we can detect the ISHE induced by spin pumping. We measured the FMR signal and the electric potential difference V between the electrodes attached to the end of the Pt layer. All the measurements were performed at room temperature.

Figure 15.1(c) shows the FMR spectra $dI(H)/dH$ measured for the $Ni_{81}Fe_{19}$/Pt film and a $Ni_{81}Fe_{19}$ film where the Pt layer is missing. Here, I denotes the microwave absorption intensity. The spectral width W [see the inset to Fig. 15.1(c)] for the $Ni_{81}Fe_{19}$ film is clearly enhanced by attaching the Pt layer. This result shows that the magnetization-precession relaxation is enhanced by attaching the Pt layer, since the spectral width W is proportional to the Gilbert damping constant α [7]. This spectral width enhancement demonstrates the emission of a spin current from the magnetization precession induced by the spin pumping; since a spin current carries spin angular momentum, the spin-current emission deprives the magnetization of its spin angular momentum and thus gives rise to additional magnetization-precession relaxation, or enhances α.

Figure 15.1(d) shows the dc electromotive force signal V measured for the $Ni_{81}Fe_{19}/Pt$ film under 200 mW microwave excitation. In the V spectra, an electromotive force signal appears around the resonance field H_{FMR}. Notable is that the spectral shape of this electromotive force is well reproduced using a Lorentz function, as expected for the ISHE induced by spin pumping [8].

The symmetric Lorentz shape of the electromotive force signal shows that extrinsic electromagnetic effects are eliminated in this measurement; the electromotive force observed here is due to the ISHE induced by the spin pumping [8]. The in-plane component of the microwave electric field may induce a rectified electromotive force via the anomalous Hall effect (AHE) in cooperation with FMR [1]. The electromotive force due to the ISHE and AHE can be distinguished in terms of their spectral shapes [8]. Since the magnitude of the electromotive force due to the ISHE induced by spin pumping, $V_{ISHE}(H)$, is proportional to the microwave absorption intensity, $V_{ISHE}(H)$ is maximized at the FMR condition. In contrast, the sign of the electromotive force due to the AHE, $V_{AHE}(H)$, is reversed across the ferromagnetic resonance field, since the magnetization-precession phase shifts by π at the resonance. Therefore, the electromotive force due to the ISHE and AHE are of the Lorentz shape and the dispersion shape, respectively, as shown in Fig. 15.1(e).

Figure 15.1(f) shows the microwave power, P_{MW}, dependence of the voltage, V_{ISHE}, where V_{ISHE} is estimated as the peak height of the resonance shape in the V spectra as shown in Fig. 15.1(d). Figure 15.1(f) shows that V_{ISHE} increases linearly with the microwave power, which is consistent with the prediction of a direct-current-spin-pumping model. Equation (15.1) shows that the dc component of a spin current generated by spin pumping is proportional to the projection of $\boldsymbol{M}(t) \times d\boldsymbol{M}(t)/dt$ onto the magnetization-precession axis. This projection is proportional to the square of the magnetization-precession amplitude. In this case, therefore, the induced spin current or the electromotive force due to the ISHE is proportional to the square of the magnetization-precession amplitude or the microwave power, P_{MW}.

To further buttress the above result, we show that the ISHE voltage appears also in an $Y_3Fe_4GaO_{12}/Pt$ film, in which the metallic $Ni_{81}Fe_{19}$ layer is replaced by an insulator magnet $Y_3Fe_4GaO_{12}$ [9]. This result strongly supports the view that the ISHE induced by spin pumping is responsible for the electromotive force observed in ferromagnetic/paramagnetic bilayer systems.

Figure 15.2(a) shows a schematic illustration of an $Y_3Fe_4GaO_{12}$/Pt bilayer film. Here, $Y_3Fe_4GaO_{12}$ is a ferrimagnetic insulator. A polycrystalline 100-nm-thick $Y_3Fe_4GaO_{12}$ film was grown on a 0.5-mm-thick $Gd_3Ga_5O_{12}$ (111) single-crystal substrate by metal organic decomposition. Then, a 10-nm-thick Pt layer was sputtered on the $Y_3Fe_4GaO_{12}$ layer. Immediately before the sputtering, the surface of the $Y_3Fe_4GaO_{12}$ film was cleaned by Ar-ion bombardment in a vacuum. The surface of the $Y_3Fe_4GaO_{12}$ layer is of a 1.0 mm × 4.0 mm rectangular shape. Two electrodes are attached to both ends of the Pt layer as shown in Fig. 15.2(a).

Figure 15.2(b) shows the microwave absorption signal $dI(H)/dH$ and the electric-potential difference V measured for the $Y_3Fe_4GaO_{12}$/Pt film when $\phi_H = 0$ at $P_{MW} = 200$ mW. Here, the in-plane magnetic field angle ϕ_H is defined in Fig. 15.2(a). In the V spectrum, an electromotive force signal appears at the resonance field. This indicates that the electromotive force is induced in the Pt layer concomitant with FMR in the $Y_3Fe_4GaO_{12}$ layer. This electromotive force is found to disappear in a $Y_3Fe_4GaO_{12}$/Cu film [see Fig. 15.2(b)], where the Pt layer is replaced by a Cu layer in which the spin–orbit interaction is very

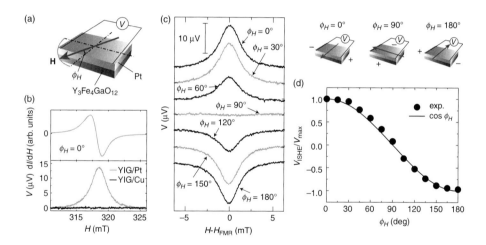

FIG. 15.2. (a) A schematic illustration of the $Y_3Fe_4GaO_{12}$/Pt film. $\boldsymbol{H}$ is the external magnetic field. ϕ_H is the in-plane magnetic field angle. (b) Field (H) dependence of the microwave absorption signal $dI(H)/dH$ and the electric-potential difference V for the $Y_3Fe_4GaO_{12}$/Pt (YIG/Pt) film and the $Y_3Fe_4GaO_{12}$/Cu (YIG/Cu) film under 200 mW microwave excitation. (c) The in-plane magnetic field angle, ϕ_H, dependence of V for the $Y_3Fe_4GaO_{12}$/Pt film. (d) The in-plane magnetic-field-angle, ϕ_H, dependence of the ISHE signal measured for the $Y_3Fe_4GaO_{12}$/Pt film. V_{ISHE}/V_{max} is the normalized spectral intensity. The filled circles are the experimental data. The solid curve shows $\cos \phi_H$.

weak [10], indicating that the spin-orbit interaction in the Pt layer is responsible for the voltage generation.

In Fig. 15.2(d), the normalized ISHE signal $V_{\mathrm{ISHE}}/V_{\mathrm{max}}$ is plotted as a function of the in-plane magnetic field angle ϕ_H. With increasing magnetic field angle ϕ_H from $\phi_H = 0$, V_{ISHE} decreases monotonically and changes its sign when $90° < \phi_H < 180°$. Notably, this variation is well reproduced using $\cos \phi_H$, being consistent with the model of the ISHE: $\boldsymbol{E}_{\mathrm{ISHE}} \propto \boldsymbol{J}_s \times \boldsymbol{\sigma}$, where $\boldsymbol{E}_{\mathrm{ISHE}}$, $\boldsymbol{J}_s$, and $\boldsymbol{\sigma}$ denote the electromotive force due to the inverse spin Hall effect, the spatial direction of the spin current, and the spin-polarization vector of the spin current, respectively [1, 11]. Since the spin polarization $\boldsymbol{\sigma}$ of the dc component of a spin current generated by the spin pumping is directed along the magnetization-precession axis, or the external magnetic field direction, the ISHE model predicts $V_{\mathrm{ISHE}} \propto |\,\boldsymbol{J}_s \times \boldsymbol{\sigma}\,|_x \propto \cos \phi_H$. Here, $|\,\boldsymbol{J}_s \times \boldsymbol{\sigma}\,|_x$ denotes the x-component of $\boldsymbol{J}_s \times \boldsymbol{\sigma}$. The x-axis is parallel to the direction across the electrodes. These results indicate that the electromotive force observed in the $Y_3Fe_4GaO_{12}/Pt$ film is attributed to the ISHE induced by the spin pumping due to the finite mixing conductance of the conduction electrons in the Pt layer [9]. The appearance of the electromotive force in the $Y_3Fe_4GaO_{12}/Pt$ film is direct evidence that the electromotive force observed in ferromagnetic/paramagnetic bilayer films is due to the ISHE induced by spin pumping; electromagnetic artifacts are irrelevant in the measurement, since $Y_3Fe_4GaO_{12}$ is an insulator.

15.2   Spin-Hall-effect induced modulation of magnetization dynamics

In the above experiment, we found that, in the $Ni_{81}Fe_{19}$ film, the spectral width W, or Gilbert damping constant α, is enhanced by attaching the Pt layer as shown in Fig. 15.1(c). This enhancement is due to the emission and the absorption of a spin current induced by spin pumping. Now, consider the inverse of the above electric current generation due to the ISHE induced by spin pumping; what happens when an electric current is injected in the Pt layer attached to the $Ni_{81}Fe_{19}$ layer? One may expect from reciprocity that spin relaxation α, namely, the width of the FMR spectra, may be modulated via the SHE in the Pt layer, enabling manipulation of the magnetization-precession relaxation of the ferromagnetic film in an electric manner. In the following, we demonstrate that this is actually the fact. In a $Ni_{81}Fe_{19}/Pt$ film, spin relaxation is manipulated by an electric current due to the macroscopic spin transfer induced by the SHE. The model calculation based on the standard Valet–Fert model and the spin torque well reproduces the experimental results [10].

In this Pt layer, the SHE converts an electric current $\boldsymbol{J}_c$ into a pure spin current, $\boldsymbol{J}_s$, which propagates into the $Ni_{81}Fe_{19}$ layer through the $Ni_{81}Fe_{19}/Pt$ interface. The spin polarization of this spin current is directed along $\boldsymbol{J}_c \times \boldsymbol{n}$ [see Fig. 15.3(a)], where $\boldsymbol{n}$ represents the normal vector of the interface. During the measurement, a microwave mode with frequency of $f = 9.44$ GHz exists in the cavity, and an external magnetic field $\boldsymbol{H}$ is applied along the film plane at the angle θ to the direction across the electrodes. The FMR spectra were measured by applying $\boldsymbol{J}_c$ through electrodes attached to both ends of the Pt layer. All the measurements were performed at room temperature.

In Fig. 15.3(b), we plot $\Delta W \equiv W^*(J_c) - W^*(-J_c)$: the asymmetric component of the spectral width W^* with respect to J_c, which enables us to eliminate heating effects from the FMR spectra. Here, $W^*(J_c) \equiv W(J_c)/W(0)$. ΔW at $\theta = 0$ [see the inset to Fig. 15.3(b)] is almost zero for all the J_c values as shown in the inset to Fig. 15.3(b). In contrast, when $\theta = 90°$, ΔW clearly increases with J_c in the $Ni_{81}Fe_{19}/Pt$ film, demonstrating that the spin relaxation α is manipulated electrically.

The $\boldsymbol{J}_c$-induced modulation of the spin relaxation observed in the $Ni_{81}Fe_{19}/Pt$ film cannot be attributed to magnetic-field effects; the $\boldsymbol{J}_c$ dependence of ΔW for a $Ni_{81}Fe_{19}/Cu$ film and a plain $Ni_{81}Fe_{19}$ film show no ΔW modulation even when an electric current is applied at $\theta = 90°$ as shown in Fig. 15.3(b), indicating that the observed $\boldsymbol{J}_c$-induced modulation of spin relaxation in the $Ni_{81}Fe_{19}/Pt$ film is not attributed to a possible small flow of electric currents in the $Ni_{81}Fe_{19}$ layer or the inhomogeneity of the magnetic field (Oersted field) induced by the electric current.

The observed modulation of the spin relaxation is interpreted in terms of the macroscopic spin transfer induced by the strong SHE in the Pt layer. Notable is

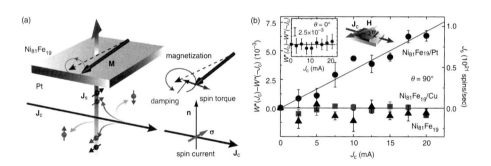

FIG. 15.3. (a) A schematic illustration of the spin Hall and the spin-torque effects. $\boldsymbol{H}$ and $\boldsymbol{J}_c$ represent the external magnetic field and the applied electric current, respectively. $\boldsymbol{M}$, $\boldsymbol{J}_s$, and $\boldsymbol{\sigma}$ denote the magnetization in the $Ni_{81}Fe_{19}$ layer, the flow direction of the spin current, and the spin polarization vector of the spin current, respectively. (b) The J_c dependence of $\Delta W \equiv W^*(J_c) - W^*(-J_c)$ for the $Ni_{81}Fe_{19}/Pt$ bilayer film, the $Ni_{81}Fe_{19}/Cu$ bilayer film, and the simple $Ni_{81}Fe_{19}$ film at $\theta = 90°$. Here, $W^*(J_c) \equiv W(J_c)/W(0)$. The inset shows ΔW for the $Ni_{81}Fe_{19}/Pt$ bilayer film at $\theta = 0$. J_s is the spin-current amplitude injected into the $Ni_{81}Fe_{19}$ layer in the $Ni_{81}Fe_{19}/Pt$ film estimated using Eq. (15.3).

that, when $\theta = 90°$, at which the $\boldsymbol{J}_c$-induced FMR modulation is observed, the external magnetic field is along the spin-polarization direction of the spin current $\boldsymbol{J}_c \times \boldsymbol{n}$ [see Fig. 15.3(a)] generated from the SHE. This spin current is injected into the whole $Ni_{81}Fe_{19}$ layer. In this situation, the spin torque acting on the magnetization in the $Ni_{81}Fe_{19}$ layer draws the magnetization toward ($J_c > 0$) or away from ($J_c < 0$) the external magnetic field direction. Since this torque is parallel or antiparallel to the Gilbert damping torque, it modulates the relaxation of the magnetization precession in the whole $Ni_{81}Fe_{19}$ layer.

To describe this effect quantitatively, we performed a model calculation based on the standard Valet–Fert model [12], in which the SHE parameterized by the spin-Hall angle θ_{SHE} (the ratio of the spin-Hall conductivity to the electrical conductivity) is considered. The spin relaxation coefficient α is calculated using the generalized LLG equation for local spins in the $Ni_{81}Fe_{19}$ layer [13]:

$$\frac{d\boldsymbol{M}}{dt} = -\gamma\boldsymbol{M} \times \boldsymbol{H}_{\text{eff}} + \frac{\alpha}{M_s}\boldsymbol{M} \times \frac{d\boldsymbol{M}}{dt} - \frac{\gamma J_s}{M_s^2 V_F}\boldsymbol{M} \times (\boldsymbol{M} \times \boldsymbol{\sigma}), \quad (15.2)$$

where J_s and V_F are the spin current injected into the ferromagnet and the volume of the ferromagnet, respectively. By ignoring the second-order contribution of the precession amplitude, we obtained α as $\alpha = \alpha_0 + \Delta\alpha_{SHE}$ at $\theta = 90°$, where α_0 is independent of J_c and

$$\Delta\alpha_{SHE} = \left(\frac{\gamma}{2\pi f M_s V_F}\right) J_s. \quad (15.3)$$

The asymmetric component of the relaxation coefficient is given by $\Delta\alpha \equiv \alpha(J_c) - \alpha(-J_c) = 2\Delta\alpha_{SHE}$ owing to the cancellation of α_0, and therefore $\Delta\alpha_{SHE}$ is directly related to the W modulation, since $\alpha = (\sqrt{3}\gamma/4\pi f)W$ [7]. When $\theta = 0$, $\Delta\alpha_{SHE}$ vanishes because the spin torque due to the SHE is canceled out during the precession motion. These J_c-dependent features are consistent with the experimental results shown in Fig. 15.3(b).

The calculated W modulation, i.e. $W(J_c) - W(-J_c) = (4\pi f/\sqrt{3}\gamma)2\Delta\alpha_{SHE} = (4/\sqrt{3}M_s V_F)J_s$, is directly proportional to the amplitude of the spin current injected into the $Ni_{81}Fe_{19}$ layer and the proportionality coefficient comprises macroscopic parameters only. This situation allows us to know the spin-current amplitude by monitoring W without assuming any microscopic material parameters. This modulation can thus be used as a spin-current meter. In Fig. 15.3(b), we show the amount of the spin current injected into the $Ni_{81}Fe_{19}$ layer estimated from the W modulation and Eq. (15.3).

References

[1] Saitoh, E., Ueda, M., Miyajima, H., and Tatara, G. (2006). *Appl. Phys. Lett.*, **88**, 182509.

[2] Mizukami, S., Ando, Y., and Miyazaki, T. (2002). *Phys. Rev. B*, **66**, 104413.

[3] Tserkovnyak, Y., Brataas, A., and Bauer, G. E. W. (2002). *Phys. Rev. Lett.*, **88**, 117601.

[4] Ando, K., Yoshino, T., and Saitoh, E. (2009). *Appl. Phys. Lett.*, **94**, 152509.

[5] Ando, K., Takahashi, S., Ieda, J., Kajiwara, Y., Nakayama, H., Yoshino, T., Harii, K., Fujikawa, Y., Matsuo, M., Maekawa, S., and Saitoh, E. (2011). *J. Appl. Phys.*, **109**(10), 103913.

[6] Yoshino, T., Ando, K., Harii, K., Nakayama, H., Kajiwara, Y., and Saitoh, E. (2011). *Appl. Phys. Lett.*, **98**, 132503.

[7] Morrish, A. H. (1980). *The Physical Principles of Magnetism*. Robert E. Krieger, New York.

[8] Inoue, H. Y., Harii, K., Ando, K., Sasage, K., and Saitoh, E. (2007). *J. Appl. Phys.*, **102**, 083915.

[9] Kajiwara, Y., Harii, K., Takahashi, S., Ohe, J., Uchida, K., Mizuguchi, M., Umezawa, H., Kawai, H., Ando, K., Takanashi, K., Maekawa, S., and Saitoh, E. (2010). *Nature (London)*, **464**, 262–266.

[10] Ando, K., Takahashi, S., Harii, K., Sasage, K., Ieda, J., Maekawa, S., and Saitoh, E. (2008). *Phys. Rev. Lett.*, **101**, 036601.

[11] Azevedo, A., Vilela-Leão, L. H., Rodríguez-Suárez, R. L., Lacerda Santos, A. F., and Rezende, S. M. (2011). *Phys. Rev. B*, **83**, 144402.

[12] Valet, T. and Fert, A. (1993). *Phys. Rev. B*, **48**, 7099–7113.

[13] Slonczewski, J. C. (1996). *J. Magn. Magn. Mater.*, **159**, L1–L7.

16 Spin-injection Hall effect

J. Wunderlich, L. P. Zârbo, J. Sinova, and T. Jungwirth

16.1 Spin-dependent Hall effects

In this chapter we will discuss the spin injection Hall effect (SiHE), yet another member of the spin-dependent Hall effects and closely related to the anomalous Hall effect (AHE), the spin Hall effect (SHE), and the inverse spin Hall effect (iSHE) as illustrated in Fig. 16.1. The microscopic origins responsible for the appearance of spin-dependent Hall effects are due to the spin–orbit (SO) coupling-related asymmetrical deflections of spin carriers. Depending on the relative strength of the SO coupling compared to the energy-level broadening of the quasi-particle states due to disorder scattering, scattering-related extrinsic mechanisms [1], or intrinsic band structure related deflection [2, 3], dominate the spin-dependent Hall response.

The AHE, recently reviewed by Nagaosa *et al.* [4], is the component of the Hall effect that exists in magnetic materials because electrical currents in ferromagnets are spin-polarized (Fig. 16.1a). The charge accumulation at sample boundaries transverse to the spin-polarized current can be detected by a voltage measurement.

Unlike the AHE, the SHE (Fig. 16.1b) occurs in nonmagnetic materials and is generated by an unpolarized charge current. Here, the asymmetrical deflection of spin carriers results in the accumulation of spins at sample boundaries transverse to the unpolarized current. The SHE was first detected by optical techniques [5, 6] and was recently measured electrically using nonlocal spin valves [7].

Both the iSHE (Fig. 16.1c) and the SiHE (Fig. 16.1d) require spin injection into a nonmagnetic system. The iSHE generated by a diffusive spin current electrically injected into aluminum was first measured by Valenzuela and Tinkham [8]. Similar to the AHE, a spin-polarized charge current flows in the case of the SiHE and the SO coupling generates the spin-dependent Hall signal. However, the SO-coupling can also affect the spin dynamics during electron propagation yielding a variation of the Hall response along the electron propagation channel. Precessing spin polarization was observed with the SiHE when spins were optically generated in the depletion layer of a quasi-lateral p-n junction [10] and injected into a two-dimensional electron gas (2DEG) [11, 13]. In the 2DEG, the SO coupling can be modulated electrically with electrostatic gates allowing us to control locally the spin polarziation. This concept of spin manipulation via

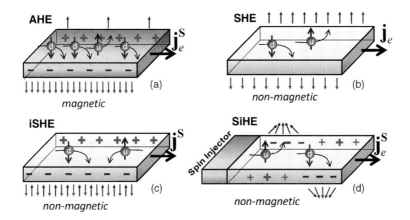

FIG. 16.1. Schematics of spin-dependent Hall effects: (a) anomalous Hall effect (AHE); (b) spin Hall effect (SHE); (c) inverse spin Hall effect (iSHE); and (d) spin-injection Hall effect (SiHE). A common property is their existence at zero external magnetic field (i.e. no ordinary Hall effect present) originating from the SO coupling. The variants of the spin-dependent Hall effects differ in the nature of the currents involved and the method of detection. In the case of the iSHE and SiHE, the Hall response can be combined with precessing spins due to, e.g. the single-particle transport analogue of the PSH state [23].

SO coupling in an electrically tunable semiconductor layer, originally proposed by Datta and Das, [12], was tested using the iSHE [13].

In the following we will describe the experiments in Refs. [11] and [13] and their theoretical analysis. Besides the spin-dependent Hall effects, we will review a concept which is the single-particle transport analogue of the persistent spin helix (PSH) [15, 22, 24], and which allows for long spin-relaxation lengths even in systems with strong SO coupling where otherwise fast spin relaxation is caused by the Dyakonov–Perel mechanism [16]. We will show that spin-relaxation lengths of several micrometers can be achieved without satisfying the PSH conditions but by restricting the spin propagation to quasi-one-dimensional transport. Finally, we will show the utility of the spin-dependent Hall effect combined with coherent spin precession in a microelectronic device geometry where a spin transistor with electrical detection directly along the gated semiconductor channel is realized and can be used to demonstrate a spin AND logic function in a semiconductor channel with two gates.

16.2 The spin-injection Hall effect experiment

In the experiment in Ref. [11] a polarized current is injected optically through a lateral p-n junction [14, 17] and its polarization is detected by transverse electrical SiHE signals along the semiconducting channel.

The lateral p-n junction devices are fabricated in GaAlAs/GaAs heterostructures grown by molecular beam epitaxy and using modulation donor (Si) and acceptor (Be) doping in the (Ga,Al)As barrier materials. The specifications of the GaAlAs/GaAs heterostructures can be found in Ref. [11]. A schematic of the wafer and numerical simulations of conduction and valence band profiles are shown in Figs. 16.2(a) and 2(b), respectively. The heterostructure is p-type in the as-grown wafer. Here, the band bending leads to the formation of a partly depleted rectangular quantum well in the conduction band at the lower interface and to an occupied triangular quantum well near the upper interface, forming a two-dimensional hole gas (2DHG). The lateral p-n junction is created by selectively removing the acceptor layer from the top of the wafer. The band bending in the etched part leads to a populated 2DEG in the conduction band well. The mobility of the 2DEG in the devices used for the SiHE experiments is only $\sim 3 \times 10^3$ cm^2/Vs because of the close proximity of the ionized donors to the quantum well.

In darkness, the lateral p-n junction has a strongly rectifying I–V characteristic and no current is flowing at zero or reverse bias. Shining light with sub-band-gap photon energy, about 100 meV lower than the band gap of GaAs, excites only 2D-hole to 3D-electron transitions near the etch step. The other parts of the wafer remain optically inactive at this wavelength. Owing to the optical selection rules, the out-of-plane spin polarization of the optically generated electrons and holes is determined by the sense and the degree of the circular polarization of the vertically incident light. A highly spin-polarized photocurrent of up to 100% can be generated when carriers from only one hole subband are excited, as illustrated in Fig. 16.2(c). The electric field in the depletion layer accelerates the photo-generated electrons vertically towards the 2DEG and counter-propagating electron and hole currents are flowing through the 2DEG and 2DHG, respectively, as illustrated in Fig 16.2(d).

In Figs. 16.2(e) and (f) we show measurements on a device type used in Ref. [11]. It has a depleted 2DEG on the unetched p-side of the sample which is optically inactive. The coplanar p-n junction acts as a self-focusing optical injection area ($\sim$ 100 nm around the selective etch step) increasing the resolution of the spin-injection point beyond the size of the focused light-spot diameter of about 1-2 μm. The transverse Hall signal measured in a Hall bar when a photo-generated electron current of about 1 μA is flowing through a 1 μm wide 2DEG channel is shown in Fig. 16.2(e). The Hall bar is 4 μm apart from the injection point. The measured signal increases linearly with the degree of circular light polarization and changes sign at opposite chirality. To verify that the detected transverse voltage is the response to the spin part of a spin-polarized electron current, the Hall signal is continuously recorded as the bias voltage at the lateral p-n junction is changing polarity. At negative (reverse) bias, the electron current is generated optically with a spin polarization depending on the sense and the degree of the circular polarization of the laser light. At forward bias, unpolarized electrons are flowing through the 2DEG channel and recombine with holes in

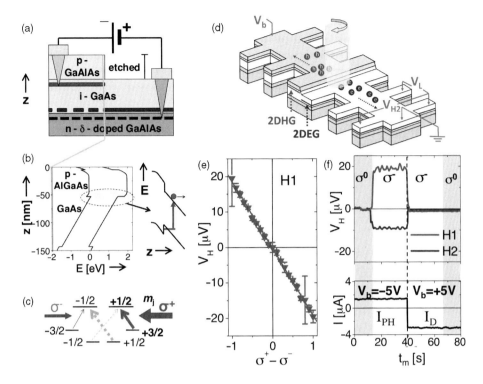

FIG. 16.2. (a) Schematics of the reverse-biased lateral p-n junction containing
a 2DHG in the as-grown part of the wafer and a populated 2DEG in the
region, where the p-GaAlAs layer is removed. (b) Calculated conduction
and valence band profile at a lateral position close to the etch step: the
populated triangular quantum well in the valence band at $z = -50$ nm
can be excited optically with sub-band-gap light generating free electrons
in the conduction band which get accelerated towards the 2DEG quantum
well at $z = -140$ nm. (c) Schematics of optical transitions from a 2DHG
quantum well, where the degeneration between heavy- and light hole states
is lifted. (d) Schematics of a Hall bar device used in the experiments: at
zero or reverse bias, photoexcited electrons and holes are counter-propagating
along the 2DEG and 2DHG, respectively. (e) Hall voltage V_H measured as
a function of the degree of circular polarization of the light used to excite
the photocurrent I_{PH}. Light intensity, bias voltage, and the corresponding
photocurrent remained constant during the experiment. (f) Hall voltages and
electrical current I (which corresponds to the photocurrent I_{PH} at zero
and reverse bias, or to the recombination current I_D at forward bias) in
dependence of bias polarity for unpolarized light σ^0 and circularly polarized
light σ^-, measured simultaneously at two Hall bars H1 and H2, 4 μm and
6 μm away from the injection point. (Data presented in (e, f) are taken from
Ref. [11].)

the p-n junction. As it becomes apparent from Fig. 16.2(f), a nonzero Hall signal
is only detected when photo-generated spin-polarized electrons are propagating
through the Hall bar. Furthermore, simultaneous Hall measurements on two
adjacent Hall bars labeled as H1 and H2, 2 μm apart from each other, show
signals of opposite signs. This is due to spin precession caused by the SO coupling
when spins propagate through the 2DEG channel.

In Fig. 16.3, measurements on a device type used in Ref. [13] are shown.
Here, the 2DEG is not completely depleted on the unetched p-side of this sample
and the optically active region on the p-side extends over a several μm range
from the etch step into the unetched p-type side of the epilayer. A focused laser
beam of ~1–2 μm spot diameter at the lateral p-n junction or near the junction
on the p-side of the epilayer was used to define the injection point. By shifting the
focused laser spot the position of the spin-injection point is smoothly changed
with respect to the detection Hall crosses. This results in damped oscillatory Hall
resistance, $R_H = V_H/I_{PH}$, measured at each of the two successive Hall crosses
labeled as H1 and H2, placed 6 and 8 μm from the lateral p-n junction. The
oscillations at each Hall cross and the phase shift between signals at the two
Hall crosses are consistent with a micron-scale spin precession period and with
a spin-diffusion length which extends over more than one precession period.

The experiments in Fig. 16.3 are performed in two distinct electrical measurement configurations. In Fig. 16.3(a), data obtained with the source and drain
electrodes at the far ends of the p- and n-type sides of the lateral junction are
shown, respectively. In this geometry, spin-polarized electrical currents reach
the detection Hall crosses, similar to the experiments performed in Ref. [11].
In Fig. 16.3(b) the electrical current is drained 4 μm before the first detection
Hall cross H1. In this case only pure spin-current [7–9] reaches crosses H1 and
H2 and generate transverse Hall signals. This demonstrates that in the 2DEG
microchannel Hall effect detection of injected spin-polarized electrical currents,
as well as pure spin currents, can be realized.

The possibility of observing and utilizing spin precession of an ensemble of
electrons in the diffusive regime is demonstrated by numerical Monte Carlo
simulations [40] shown in Fig. 16.3(c). The numerically obtained spin-precession
period is well described by an analytical formula derived from the dynamics of the
spin-density matrix [40], $L_{SO} = \pi\hbar^2/m(|\alpha| + |\beta|)$; $m = 0.067m_e$ is the electron
effective mass in GaAs and α and β are the Rashba and Dresselhaus SO-coupling
parameters, respectively (see the following section). There are two regimes in
which spin precession can be observed in the diffusive transport regime. In one
regime the width of the channel is not relevant and a spin-diffusion length larger
than the precession length occurs as a result of the single-particle transport
analogue of the PSH state [22, 23] realized at 2DEG Rashba and Dresselhaus
spin–orbit fields of equal or similar strengths, $\alpha \approx -\beta$ for the bar orientation
used in Refs. [11, 13]. We will discuss this scenario in the beginning of the
following theory section. When the two spin–orbit fields are not tuned to similar
strengths, the spin-diffusion length is approximately given by $\sim L_{SO}^2/w$ and spin

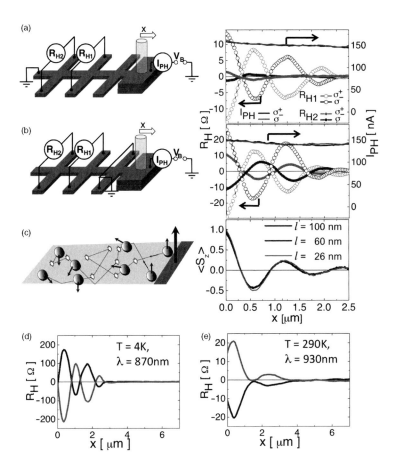

FIG. 16.3. (a) Schematics of the measurement set-up with an optically injected
spin-polarized electrical current propagating through the Hall bar and corresponding experimental Hall effect signals at crosses H1 and H2. The Hall
resistances, $R_H = V_H/I_{PH}$, for the two opposite helicities of the incident light
are plotted as a function of the focused ($\sim$1–2 μm) light spot position, i.e.
of the position of the injection point. Increasing x corresponds to shifting
the spot further away from the Hall detectors. (The focused laser beam is
indicated by the grayish cylinder in the schematics.) The optical current
I_{PH} is independent of the helicity of the incident light and varies only
weakly with the light spot position. (b) Same as (a) for a measurement
geometry in which electrical current is closed before the first detecting Hall
cross H1. (c) Schematics of the diffusive transport of injected spin-polarized
electrons and Monte Carlo simulations (explained in the next section) of the
out-of-plane component of the spin of injected electrons averaged over the
1 μm bar cross-section assuming a Rashba field $\alpha = 5.5$ meVÅ, Dresselhaus
field $\beta = -24$ meVÅ, and different values of the mean-free-path l. (d),(e)
Measurements of the Hall signal at the first Hall cross in the n-channel placed
2 μm from the coplanar p-n junction as a function of the laser spot position
at 4 K (d) and at room temperature (e). (Data presented in the figure are
taken from Ref. [13].)

precession is observable only when the width w of the channel is comparable to or smaller than the precession length [40–42]. As shown in Fig. 16.3(c), several precessions are readily observable in this quasi-1D geometry even in the diffusive regime and for $\alpha \neq -\beta$. As also demonstrated in Fig. 16.3(c), the spin-precession and spin-diffusion lengths are independent in this regime, of the mean-free-path, i.e. of the mobility of the 2DEG channel. In Figs. 16.3(d), (e), laser-spot shift measurements at 4 K and at room temperature are compared. The finite spin Hall signals measured at high temperatures indicate the independence of the spin-diffusion length on the mobility in the diffusive regime. Note that the precession length is temperature dependent which is attributed to the effective temperature dependence of the confining potential due to Fermi broadening and to the corresponding variation of the effective internal spin–orbit field.

16.3   Theory discussion

The theoretical approach is based on the observation that the micrometer length-scale governing the spatial dependence of the nonequilibrium spin polarization in the experiments of Refs. [11, 13] is much larger than the ~ 100 nm mean-free-path in the 2DEG which governs the transport coefficients. This allows us to first calculate the steady-state spin-polarization profile along the channel and then consider the iSHE or SiHE as a response to the local out-of-plane component of the polarization.

The calculations start from the electronic structure of GaAs whose conduction band near the Γ-point is formed dominantly by Ga s orbitals. This implies that spin–orbit coupling originates from the mixing of the valence-band p orbitals and from the broken inversion symmetry in the zincblende lattice. In the presence of an electric potential $V(\mathbf{r})$ the corresponding 3D spin–orbit coupling Hamiltonian reads

$$H_{3D-SO} = [\lambda^* \boldsymbol{\sigma} \cdot (\mathbf{k} \times \nabla V(\mathbf{r}))] + \left[\mathcal{B} k_x (k_y^2 - k_z^2)\sigma_x + \text{cyclic permutations} \right],$$

$$(16.1)$$

where $\boldsymbol{\sigma}$ are the Pauli spin matrices, $\mathbf{k}$ is the momentum of the electron, $\mathcal{B} \approx 10$ eV Å^3 , and $\lambda^* = 5.3$ Å^2 for GaAs [18, 19]. Equation (16.1) together with the 2DEG confinement yield an effective 2D Rashba and Dresselhaus SO-coupled Hamiltonian [20–22],

$$H_{2DEG} = \frac{\hbar^2 k^2}{2m} + \alpha(k_y \sigma_x - k_x \sigma_y) + \beta(k_x \sigma_x - k_y \sigma_y) + V_{\text{dis}}(\mathbf{r})$$

$$+ \lambda^* \boldsymbol{\sigma} \cdot (\mathbf{k} \times \nabla V_{\text{dis}}(\mathbf{r})),$$

$$(16.2)$$

where $\beta = -\mathcal{B}\langle k_z^2 \rangle \approx -0.02$ eV Å and $\alpha = e\lambda^* E_z \approx 0.01 - 0.03$ eV Å for the strength of the confining electric field, $eE_z \approx 2$–5×10^{-3} eV/Å pointing along the [001] direction; V_{dis} is the disorder potential. The strength of the confinement is obtained from a self-consistent Schrödinger–Poisson simulation of the conduction band profile of the GaAs/GaAlAs heterostructure [6]. Typically the strength of the Rashba SO term α can be tuned whereas the strength of the Dresselhaus SO term β is a material-dependent parameter fixed by the choice of growth direction and, to a smaller extent, the degree of confinement.

16.3.1 *Nonequilibrium polarization dynamics along the [1\bar{1}0] channel*

The realization of the original Datta–Dass device concept in a purely Rashba SO-coupled system has been unsuccessful until recently due to spin coherence issues; i.e. at the required length-scales in which transport is diffusive, no oscillating persistent precession states are present [12]. However, in a 2DEG where both Rashba and Dresselhaus SO-coupling have similar strengths a long lived precessing excitation of the system has been shown to exist along a particular direction [22, 24]. When α and β are equal in magnitude the component of the spin along the [110] direction for $\alpha = -\beta$ or along the [1\bar{1}0] direction for $\alpha = \beta$ is a conserved quantity [20], as well as a precessing spin wave, the PSH, of wavelength $\lambda_{\text{spin-helix}} = \pi\hbar^2/(2m\alpha)$ in the direction perpendicular to the conserved spin component [22, 23]. This PSH state has been observed through optical transient spin-grating experiments [15, 24].

Intuitively, one can visualize the PSH by considering electron-spin precession around a $\mathbf{k}$-dependent effective *internal* magnetic field consisting of Rashba and Dresselhaus SO fields, as shown in Figs. 16.4(a), (b). In the lowest order, their magnitudes increase linearly with k. In the particular case of $\alpha = -\beta$, the effective SO field is oriented along the [110] direction for all $\mathbf{k}$-vectors and its magnitude depends only on the [1\bar{1}0]-component of $\mathbf{k}$ (Fig. 16.3c). Let us consider that electron spins are injected at the point $\mathbf{r}_0$ with *up*-polarization (along the [001]-direction) and that the spins are detected at the point $\mathbf{r}_1$ displaced from $\mathbf{r}_0$ by a finite amount along the [1\bar{1}0] direction. At the point $\mathbf{r}_1$, all electron spins have precessed by exactly the same angle, independent of the particular path each individual electron took and of the number of scattering events each electron experienced along its path. Therefore, spins of an ensemble of spin-polarized electrons will not dephase but precess along the [1\bar{1}0] direction in such a way that they are all polarized along the same direction at a given point in position space.

For $\alpha = (+/-)\beta$, the Rashba–Dresselhaus Hamiltonian exhibits U(1) symmetry which means that an in-plane spin state parallel to this SO field direction is infinitely long lived. This state will be dephased if the cubic Dresselhaus term is present in the system [25–27]. Randomness in the SO coupling induced by remote impurities would cause additional spin relaxation [28]. Nevertheless, infinite spin lifetimes are theoretically still possible in SO-coupled 2DEGs if the spatially varying SO field can be described as a pure gauge and, thus, removed by a gauge transformation [29]. Furthermore, it was shown [23] that the many-electron system whose individual particles are described by the above U(1) symmetric single-particle Hamiltonian displays a SU(2) symmetry which is robust against both spin-independent disorder and electron–electron interactions. Owing to this symmetry, a collective spin state excited at a certain wavevector would have an infinite lifetime.

Let us now consider the strong scattering regime consistent with the structures used in the experiments in Refs. [11, 13]. In this case, αk_F and $\beta k_F \sim 0.5$ meV are much smaller than the disorder scattering rate $\hbar/\tau \sim 5$ meV, so that the system obeys a set of spin-charge diffusion equations [22] for arbitrary ratios of α and β:

$$\partial_t n = D\nabla^2 n + B_1 \partial_{x_+} S_{x_-} - B_2 \partial_{x_-} S_{x_+},$$

$$\partial_t S_{x_+} = D\nabla^2 \partial_t S_{x_+} - B_2 \partial_{x_-} n - C_1 \partial_{x_+} S_z - T_1 S_{x_+},$$

$$\partial_t S_{x_-} = D\nabla^2 \partial_t S_{x_-} + B_1 \partial_{x_+} n - C_2 \partial_{x_-} S_z - T_2 S_{x_-},$$

$$\partial_t S_z = D\nabla^2 \partial_t S_z + C_2 \partial_{x_-} S_{x_-} + C_1 \partial_{x_+} S_{x_+} - (T_1 + T_2) S_z,$$

where x_+ and x_- correspond to the [110] and [1$\bar{1}$0] directions, $B_{1/2} = 2(\alpha \mp \beta)^2(\alpha \pm \beta)k_F^2 \tau^2$, $T_{1/2} = \frac{2}{m}(\alpha \pm \beta)^2 \frac{k_F^2 \tau}{\hbar^2}$, $D = v_F^2 \tau/2$, and $C_{1/2}^2 = 4DT_{1/2}$. For the present device, where $\alpha \approx -\beta$, the 2DEG channel is patterned along the [1$\bar{1}$0] direction which corresponds to the direction of the PSH propagation. Within this direction the dynamics of S_{x_-} and S_z couple through the diffusion equations above. Seeking steady-state solutions of the form $\exp[qx_-]$ yields the transcendental equation $(Dq^2 + T_2)(Dq^2 + T_1 + T_2) - C_2^2 q^2 = 0$ which can be reduced to $q^4 + (\tilde{Q}_1^2 - 2\tilde{Q}_2^2)q^2 + \tilde{Q}_1^2 \tilde{Q}_2^2 + \tilde{Q}_2^4 = 0$, where $\tilde{Q}_{1/2} \equiv \sqrt{T_{1/2}/D} = 2m|\alpha \pm \beta|/\hbar^2$. This yields a physical solution for $q = |q|\exp[i\theta]$ of

$$|q| = \left(\tilde{Q}_1^2 \tilde{Q}_2^2 + \tilde{Q}_2^4\right)^{1/4} \quad \text{and} \quad \theta = \frac{1}{2}\arctan\left(\frac{\sqrt{2\tilde{Q}_1^2 \tilde{Q}_2^2 - \tilde{Q}_1^4/4}}{\tilde{Q}_2^2 - \tilde{Q}_1^2/2}\right). \tag{16.3}$$

The resulting damped spin precession of the out-of-plane polarization component for the parameter range of the device, where we have set $\beta = -0.02$ eV Å and varied α from -0.5β to -1.5β, is shown in Fig. 16.3(d).

These results are in agreement with Monte Carlo calculations on similar systems (modeling a InAs 2DEG) but with higher applied biases [30]. In the Monte Carlo calculations longer decaying lengths were observed at higher biases. However, the experiments in Refs. [11, 13] are well within the linear regime with very low driving fields; this results in shorter decay length-scales of the oscillations as compared to Ref. [30].

We note that Monte Carlo simulations including temperature broadening of the quasi-particle states confirm the validity of the above analytical results up to the high temperatures used in the experiment. The theoretical results show good agreement with the steady-state variations (changes in the length-scale of 1–2 μm) in the out-of-plane nonequilibrium polarization of the experimental system, observed indirectly through the Hall signals.

16.3.2 *Hall effect*

The Hall effect signal can be understood within the theory of the anomalous Hall effect. The contributions to the anomalous Hall effect in SO-coupled systems with nonzero polarization can be classified into two types: the first type arises from the SO-coupled quasi-particles interacting with the spin-independent disorder and the electric field, and the second type arises from the non-SO-coupled part of the quasi-particles scattering from the SO-coupled disorder potential (last term in Eq. 16.2). The contributions of the first type have been recently studied in 2DEG ferromagnetic systems with Rashba SO coupling [31–34]. These have shown that, within the regime applicable to the present devices, the anomalous Hall effect contribution due to the intrinsic and side-jump mechanisms vanish even in the presence of moderate disorder. In addition, the skew-scattering contribution from this type of contribution is also small (with respect to the contribution shown below) and furthermore is not linear in the polarization [31]. Hence we can disregard the contributions of the first type in the SiHE experiments.

This is not surprising since in the devices in Refs. [11, 13] $\alpha k_F, \beta k_F \ll \hbar/\tau$, and hence these contributions arising from the SO coupling of the bands are not expected to dominate. Instead the observed signal originates from contributions of the second type, i.e. from interactions with the SO-coupled part of the disorder [35, 36]. Within this type one contribution is due to the anisotropic scattering, the extrinsic skew scattering, and is obtained within the second Born approximation treatment of the collision integral in semiclassical linear transport theory [35, 36]:

$$|\sigma_{xy}|^{skew} = \frac{2\pi e^2 \lambda^*}{\hbar^2} V_{dis} \tau n(n_\uparrow - n_\downarrow) , \qquad (16.4)$$

where $n = n_\uparrow + n_\downarrow$ is the density of photoexcited carriers with polarization $p_z = (n_\uparrow - n_\downarrow)/(n_\uparrow + n_\downarrow)$. Using the relation for the mobility $\mu = e\tau/m$ and the relation between n_i, V_{dis}, and τ, $\hbar/\tau = n_i V_{dis}^2 m/\hbar^2$, the extrinsic skew-scattering contribution to the Hall angle, $\alpha_H \equiv \rho_{xy}/\rho_{xx} \approx \sigma_{xy}/\sigma_{xx}$, can be written as

$$\alpha_H^{skew} = 2\pi\lambda^* \sqrt{\frac{e}{\hbar n_i \mu}} \, n \, p_z (x_{[1\bar{1}0]})$$

$$= 2.44 \times 10^{-4} \frac{\lambda^*[\text{Å}^2](n_\uparrow - n_\downarrow)[10^{11}\text{cm}^{-2}]}{\sqrt{\mu[10^3\text{cm}^2/\text{Vs}]n_i[10^{11}\text{cm}^{-2}]}}$$

$$\sim 1.1 \times 10^{-3} p_z, \qquad (16.5)$$

where we have used $n = 2 \times 10^{11}$ cm^{-2}, p_z is the polarization, $\mu = 3 \times 10^3$ cm^2/Vs, and $n_i = 2 \times 10^{11}$cm^{-2}; the last estimate is introduced to give a lower bound to the Hall angle contribution within this model. In addition to this contribution, there also exists a side-jump scattering contribution from this SO-coupled disorder term given by

$$\alpha_{\mathrm{H}}^{\mathrm{s-j}} = \frac{2e^2\lambda^*}{\hbar\sigma_{\mathrm{xx}}}(n_\uparrow - n_\downarrow) \tag{16.6}$$

$$= 3.0 \times 10^{-4}\frac{\lambda^*[\text{Å}^2]}{\mu[10^3 \text{ cm}^2/\text{Vs}]}p_z \sim 5.3 \times 10^{-4}p_z. \tag{16.7}$$

As expected this is an order of magnitude lower than the skew-scattering contribution within this system. We can then combine the above result with the results from the previous section to predict, in this diffusive regime, the resulting theoretical α_H along the [1$\bar{1}$0] direction for the relevant range of Rashba and Dresselhaus parameters corresponding to the experimental structure [11]. This is shown in Fig. 16.4(d). We have assumed a donor impurity density n_i of the order of the equilibrium density $n_{2\mathrm{DEG}}$ =2.5×10^{11} cm^{-2} of the 2DEG in dark, which is an upper bound for the strength of the impurity scattering in the modulation-doped heterostructure and, therefore, a lower bound for the Hall angle. For the mobility of the injected electrons in the 2DEG channel one can consider the experimental value determined from ordinary Hall measurements without illumination, $\mu = 3 \times 10^3$ cm^2/Vs. The density of photoexcited carriers of $n \approx 2 \times 10^{11}$ cm^{-2} was obtained from the measured longitudinal resistance between successive Hall probes under illumination assuming constant mobility.

16.3.3 *Spin diffusion and spin precession in narrow 2DEG bars*

Long spin-diffusion lengths in 2DEGs can be achieved not only when the PSH state conditions are approximately satisfied but also when the device geometry restricts the spin transport to quasi-one-dimensional propagation. As we will discuss in this section, the relevant confinement length is not related to the scattering mean-free path of the electrons but quasi-one-dimensional spin transport takes place in 2DEG channels whose width is smaller than the length of a full spin precession which is inversely proportional to the strength of the internal spin–orbit fields. One can understand this intuitively by considering the SO fields in the 2DEG as momentum-dependent magnetic fields that couple to the electronic magnetic moment. If the channel is not one-dimensional, the **k**-states which are not collinear to the channel orientation cause decoherence of the spin polarization along the channel. However, if the SO fields are weak enough so that the corresponding spin-precession length is large compared to the channel width, the decoherence is reduced. Roughly speaking, the electron spin of electrons moving towards the channel edges have not precessed sufficiently to cause decoherence before they scatter back from the boundaries.

A more quantitative analysis of the channel width dependence of the spin dynamics of an ensemble of spin-polarized electrons injected in a diffusive microchannel with linear Rashba and Dresselhaus SO coupling is shown by numerical ensemble Monte Carlo (EMC) calculations where electron orbital degrees of freedom are described by classical momentum and position and the spin degree of freedom by a quantum-mechanical spin-density matrix,

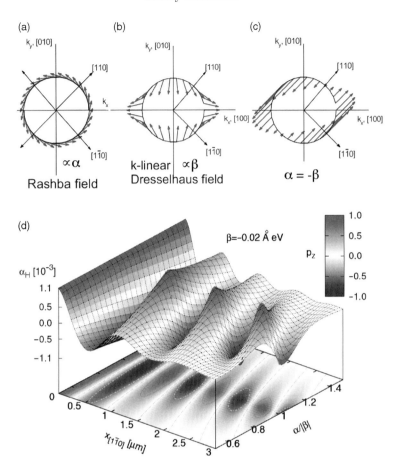

FIG. 16.4. Schematics of the (a) Rashba, (b) Dresselhaus, and (c) combined effective SO field for $\alpha = -\beta$. (d) Microscopic theory of the SiHE assuming spin–orbit coupled band-structure parameters of the experimental 2DEG system. The calculated spin-precession and spin-coherence lengths and the magnitude of the Hall angles are consistent with experiment. The gray-scale-coded surface shows the proportionality between the Hall angle and the out-of-plane component of the spin polarization. (Data presented in (d) are taken from Ref. [11].)

Fig. 16.5(a) [40]. In the diffusive regime the momentum and position of electrons can be treated as classical variables. We emphasize that the direct correspondence between the suppressed spin relaxation in the single-particle transport problem and the collective PSH state is valid only in this diffusive regime. Here the group velocity of an electron in the Rashba–Dresselhaus 2DEG can be approximated by its momentum divided by the mass. The spin-precession angle of such a particle

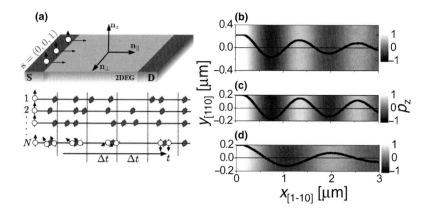

FIG. 16.5. (a) Schematic depiction of the EMC method as described in detail by Zârbo *et al.* [40]: Spin-*up* polarized electrons are injected from the source electrode in a Rashba and Dresselhaus SO-coupled channel. The time evolution of each particle belonging to the ensemble is sampled at equal intervals t called subhistories. The particle spin precesses in the SO field during the free flight time but is unaffected by collisions. (b)–(d) EMC calculation results of the spin-density distribution $S_z(\mathbf{r})$ in $[1\bar{1}0]$-oriented 2DEG channels with fixed Dresselhaus SO coupling $\beta = -2.0 \times 10^{-12}$ eVm for (b) 800 nm, (c) 400 nm wide channels with $\alpha = -0.5\beta$, and for a 400 nm wide channel with $\alpha = 0$ (d). The gray-scale plots show the 2D distribution of $S_z(\mathbf{r})$ in the channels; the bold black line corresponds to the $y_{[110]}$-averaged spin density as a function of $x_{[1\bar{1}0]}$. (Data presented in (b)–(d) are taken from Ref. [40].)

depends then only on the distance traveled along the direction perpendicular to the SO field and the resulting spin-density pattern of an ensemble of injected electron spins coincides with the spin-density pattern of the PSH spin wave [13]. In the opposite limit of strong SO coupling and weak disorder, the velocity and momentum are not simply proportional to each other and the direct link is lost between the one-particle and collective physics. This is because the expression for the velocity of SO-coupled electrons contains terms proportional to the SO coupling strength. For example, the velocity along the $[1\bar{1}0]$ direction of Rashba and Dresselhaus SO-coupled electrons in the PSH regime ($\alpha = -\beta$) is $\mathbf{v}_{[1\bar{1}0]} = \hbar \mathbf{k}_{[1\bar{1}0]}/m \pm 2\beta/\hbar$ for states with spins along $[\bar{1}\bar{1}0]$ and along $[110]$, respectively.

In Fig. 16.5 (b)–(d), we show EMC calculations of the spin-density distribution $S_z(\mathbf{r})$ in $[1\bar{1}0]$-oriented 2DEG channels as described in Ref. [40] with the chosen SO parameter matching the experimental observations. In case that the PSH conditions are not satisfied, e.g. for $\alpha = -0.5\beta$, the shorter spin relaxation length in the 800 nm bar compared to the narrower 400 nm bar is due to the more efficient randomization of the spin orientations in the wider bar. Even in the extreme case of the absence of the Rashba SO field ($\alpha = 0$), the $\sim 1~\mu$m spin relaxation length makes a full spin-precession possible along the narrow 400 nm 2DEG bar.

16.4 Spin Hall effect transistor

The theoretical proposal of electrical manipulation and detection of electron spin in a semiconductor channel is more than 20 years old [12]. However, its recent experimental realization [44] turned out to be unexpectedly difficult because of the fundamental physical problems related to the resistance mismatch [45] between ferromagnetic contacts for spin-injection/detection and the requirement that the electron dwell time in the semiconductor channel must be shorter than the spin lifetime [46].

In this paragraph, we demonstrate the applicability of the iSHE in a new type of a spin transistor, the SHE transitor. Similar to the Datta–Das proposal of the spin transistor, the active semiconductor channel in the SHE transistor is a two-dimensional electron gas (2DEG) in which the SO-coupling induced spin precession is controlled by external gate electrodes. The gates are realized by the p-type surface layer areas of the heterostructure which were locally masked and remained unetched during the fabrication of the n-channel Hall bar [17]. The detection is provided by transverse iSHE voltages measured along the 2DEG Hall bar. For spin injection the optical method is utilized. This way all three components of the spin transistor are realized within an all-semiconductor structure. The optical injection method is less scalable than electrical injection from ferromagnetic contacts but, on the other hand, it does not require any magnetic elements or external magnetic fields for the operation of the device. Because of the nondestructive nature of the iSHE detection, one semiconductor channel can accommodate multiple gates and Hall cross detectors and is therefore directly suitable for realizing spin logic operations.

The conventional field-effect transistor functionality in the 2DEG channel achieved by the p-layer top gate is demonstrated in Fig. 16.6(a) where the gate voltage dependence of the channel current and mobility underneath the gate are shown. At zero gate voltage, only a small residual channel current consistent with the partial depletion of the 2DEG in the unetched part of the heterostructure is detected. By applying forward or reverse voltages of an amplitude less than 1 V the 2DEG channel is opened or closed, respectively, at negligible gate-channel leakage current. Within the range of measurable signals, a gate-voltage-induced change of the channel current by five orders of magnitude is detected while the mobility changes by two orders of magnitude. The main effect of the gate voltage on the channel current is therefore via direct charge depletion/accumulation of the 2DEG but mobility changes are also significant. With increasing reverse gate voltage the mobility decreases because the 2DEG is shifted closer to the

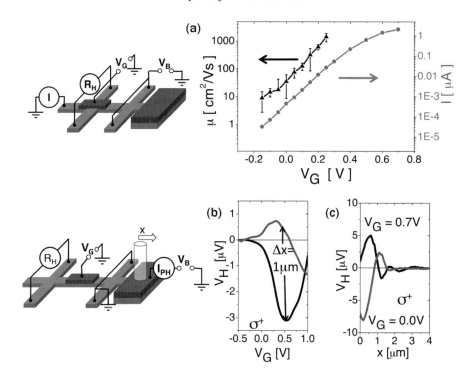

FIG. 16.6. Spin Hall effect transistor. (a) Schematics of the measurement set-up corresponding to the conventional field-effect transistor and experimental dependence of the electrical current (circle) through the channel and mobility (triangle) underneath the gate on the gate voltage. (b) Schematics of the set-up of the spin Hall transistor and experimental Hall signals as a function of the gate voltage at a Hall cross placed behind the gate electrode for two light spot positions with a relative shift of 1 μm and the gray curve corresponding to the spot shifted further away from the detection Hall cross. (c) Hall signals as a function of the spot positions for two different gate voltages, $V_G = 0$ (gray) and $V_G = 0.7$V (black). The thin gray and black curves in (b) and (c) correspond to the respective photocurrents. (Data presented in (a)–(c) are taken from Ref. [13].)

ionized donors on the other side of the GaAlAs/GaAs heterojunction and because screening of the donor impurity potential by the 2DEG decreases with depletion.

The sensitivity of the measured Hall signal at the cross placed behind the gate on the voltage applied to the gate electrode is shown in Fig. 16.6(b). In order to exclude any potential gate voltage dependence of spin-injection conditions in the device the experiments are performed with the electrical current drained before the gated part of the channel. The data show two regimes of operation of the SHE transistor. At large reverse voltages the Hall signals disappear as the diffusion of spin-polarized electrons from the injection region towards the detecting Hall cross is blocked by the repulsive potential of the intervening gate electrode. Upon opening the gate, the Hall signal first increases, in analogy to the operation of the conventional field-effect transistor. While the optically generated current I_{PH} is kept constant, the electrical current in the manipulation and detection parts of the transistor channel remains zero at all gate voltages. The onset of the output transverse electrical signal upon opening the gate is a result of a pure spin current. The initial increase of the detected output signal upon opening the gate shown in Fig. 16.6(b) is followed by a nonmonotonic gate voltage dependence of the Hall voltage. This is in striking contrast to the monotonic increase of the normal electrical current in the channel observed in the conventional field-effect-transistor measurement in Fig. 16.6(a). Apart from blocking the spin current at large reverse gate voltages, the intermediate gate electric fields modify spin precession of the injected electrons and therefore the local spin polarization at the detecting Hall cross when the channel is open. This is the spin manipulation regime analogous to the original Datta–Das proposal. The presence of this regime in the device is further demonstrated by comparing two measurements shown in Fig. 16.6(b), one where the laser spot is aligned close to the lateral p-n junction on the p-side (black, bold line) and the other one with the spot shifted by approximately 1 μm in the direction away from the detecting Hall crosses (gray, bold line). The reverse voltage at which the Hall signals disappear is the same in the two measurements. For gate voltages at which the channel is open, the signals are shifted with respect to each other in the two measurements, have opposite sign at certain gate voltages, and the overall magnitude of the signal is larger for smaller separation between injection and detection points, all confirming the spin-precession origin of the observed effect. Further evidence for the gate-voltage dependent variation of the spin precession underneath the gate electrode is shown in Fig. 16.6(c), where the phase-shifted Hall signal variations as a function of the spot position for two different gate voltages, $V_G = 0$ (gray, bold line) and $V_G = 0.7$V (black, bold line) are compared. The photocurrents (thin gray and black lines) drop down to zero when exceeding a light-spot position of 2 μm away from the etch-step in the p-region.

One of the important attributes of this nondestructive spin-detection method integrated, together with the electrical spin manipulation, along the semiconductor channel is the possibility to fabricate devices with a series of Hall cross detectors and also with a series of gates. In Fig. 16.7 the feasibility of this concept and of the ensuing logic functionality on a SHE transistor structure with two gates, the first placed before cross H1 and second before H2, is demonstrated. A scanning electron micrograph of the device is shown in Fig. 16.7(a). The measured data plotted in Fig. 16.7(b) demonstrate that Hall cross H1 responds strongly to the electric field on the first gate, with similar gate voltage characteristics as observed in the single-gate device in Fig. 16.6. As expected for the relative positions of the injection point, of Hall cross H1, and of the two gates in the device, the dependence of the signal at cross H1 on the second gate

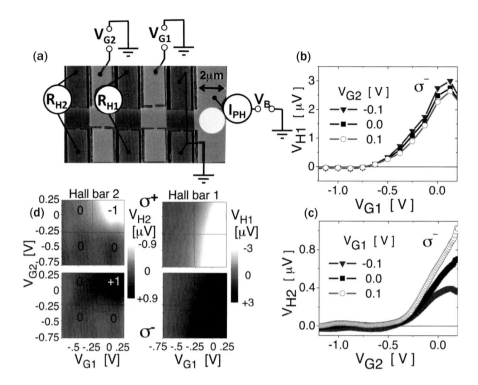

FIG. 16.7. (a) Scanning electron micrograph image and schematics of the device
with two detecting Hall crosses H1 and H2 and one gate placed before cross
H1 and the second gate placed behind cross H1 and before cross H2. Gates
and p-side of the lateral p-n junction are highlighted in gray. The focused
laser beam is indicated by the light-grayish spot. (b) Hall signals at cross H1
measured as a function of the first gate voltage. These gating characteristics
are similar to the single-gate device in Fig. 16.6(b) and have much weaker
dependence on the second gate voltage. (c) Hall signals at cross H2 measured
as a function of the second gate voltage. The curves show strong dependence
on the voltages on both gates. (d) Demonstration of the spin AND logic
function by operating both gates (input signals) and measuring the response
at Hall cross H2 (output signal). Measured data at cross H1 are also shown
for completeness. (Figure is taken from Ref. [13].)

is much weaker. On the other hand, Hall cross H2 responds strongly to both
gates (Fig. 16.7c). Before the spin can reach the detecting Hall cross H2 it is
manipulated by two external parameters. This is analogous to the measurement
in Fig. 16.6(b) in which the position of the injection point played the role of the
second parameter.

In Fig. 16.7(d), a simple AND logic functionality is demonstrated by operating
both gates and by measuring the Hall electrical signal at cross H2. Intermediate

gate voltages on both gates represent the input value 1 and give the largest electrical signal at H2 (positive for σ^- helicity of the incident light), representing the output value 1. By applying to any of the two gates a large reverse (negative) gate voltage, representing input 0, the electrical signal at H2 disappears, i.e. the output is 0. Note that additional information is contained in the polarization dependence of the detected Hall signals, as illustrated in Fig. 16.6(d).

The strength of the confining electric field of the 2DEG underneath the gate changes by up to a factor of ~ 2 in the range of applied gate voltages in the experiments. It implies comparably large changes in the strength of the internal SO fields in the 2DEG channel. The dependence on the SO field strength confirmed by MC simulations [40] (and the independence on the momentum of injected electrons) implies also comparably large changes of the spin-precession length. These estimates corroborate the observed spin manipulation in the SHE transistor by external electric fields applied to the gates.

16.5   Prospectives of spin-injection Hall effect

The spin-injection Hall effect in nonmagnetic semiconductor structures with the ability to control the SO coupling offer unprecedented possibilities to study coupled spin-charge dynamics without intervening spin and charge propagation. SiHE and iSHE enable us to detect locally the polarization of spin- or spin-polarized current with high spatial resolution limited only by the nano-fabrication capabilities, which is 1–2 orders of magnitude higher than the resolution of current magneto-optical scanning probes. Moreover, studying the spin-dependent Hall effects both in the diffusive regime [11, 13] and in the strong spin-orbit coupling, weak disorder regime [32] will enable us to analyze the different microscopic origins of the spin-dependent Hall effects.

From the application perspective, spin-injection Hall effect devices can be directly implemented as light-polarization sensors, so-called polarimeters, which convert the degree of light polarization into a directly proportional electrical signal. Moreover, in the last section we have shown that SiHE and iSHE can be implemented in a spin transistor type of device. An important next step towards the practical implementation of such devices is the replacement of optical spin injection by other solid sate means of spin injection. These lightless devices utilizing the spin-injection Hall effect can be fabricated in a broad range of materials including indirect-gap Si/Ge semiconductors [39]. Since the magnitude of the spin-injection Hall effect scales linearly with the spin–orbit coupling strength we expect $\sim 100\times$ weaker signals in Si/Ge 2DEGs as compared to the measurements in GaAs/AlGaAs in Refs. [11, 13], which is still readily detectable.

References

[1] M. I. Dyakonov and V. I. Perel, Phys. Lett. A **35**, 459 (1971).

[2] S. Murakami, N. Nagaosa, and S.-C. Zhang, Science **301**, 1348 (2003).

[3] J. Sinova *et al.*, Phys. Rev. Lett. **92**, 126603 (2004).

[4] N. Nagaosa, J. Sinova, S. Onoda, A. H. MacDonald, and N. P. Ong, Rev. Mod. Phy. **82**, 1539 (2010).

[5] Y. K. Kato, R. C. Myers, A. C. Gossard, and D. D. Awschalom, Science **306**, 1910 (2004).

[6] J. Wunderlich, B. Kaestner, J. Sinova, and T. Jungwirth, Phys. Rev. Lett. **94**, 047204 (2005); preprint at arXiv:cond-mat/ 0410295v3 (2004).

[7] E. S. Garlid, Q. O. Hu, M. K. Chan, C. J. Palmstrom, P. A. Crowell, arXiv:cond-mat/ 1006.1163 (2010).

[8] S. Valenzuela and M. Tinkham, Nature **442**, 176 (2006).

[9] C. Bruene, *et al.*, Nature Phys. **6**, 448 (2010).

[10] I. Zutic, J. Fabian, and S. Das Sarma, Phys. Rev. B **64**, 121201 (2001).

[11] J. Wunderlich, A. C. Irvine, J. Sinova, B. G. Park, X. L. Xu, B. Kaestner, V. Novak, and T. Jungwirth, Nature Phys. **5**, 675 (2009).

[12] S. Datta and B. Das, Appl. Phys. Lett. **56**, 665 (1990).

[13] J. Wunderlich, B.G. Park, A.C. Irvine, L. P. Zarbo, E. Rozkotova, P. Nemec, V. Novak, J. Sinova, and T. Jungwirth, Science **330**, 1801 (2010).

[14] B. Kaestner, D. G. Hasko, and D. A. Williams, D. A. Design of quasi-lateral pn junction for optical spin-detection in low-dimensional systems. Preprint at http://arxiv.org/abs/cond-mat/0411130 (2004).

[15] J. D. Koralek, *et al.*, Nature **458**, 610 (2009).

[16] M. I. Dyakonov and V. I. Perel, in *Optical Orientation*, ed. by F. Meier and B.P. Zakharchenya (North Holland, Amsterdam, 1984).

[17] B. Kaestner, J. Wunderlich, and T. J. B. M. Jamssen, J. Mod. Opt. **54**, 431 (2007).

[18] W. Knap, C. Skierbiszewski, A. Zduniak, E. Litwin-Staszewska, D. Bertho, F. Kobbi, J. L. Robert, G. E. Pikus, F. G. Pikus, S. V. Iordanskii, V. Mosser, K. Zekentes, and Y. B. Lyanda-Geller, Phys. Rev. B **53** (1996).

[19] R. Winkler, *Spin-Orbit Coupling Effects in Two-Dimensional Electron and Hole Systems.* Spinger-Verlag, New York (2003).

[20] J. Schliemann, J. C. Egues, and D. Loss, Phys. Rev. Lett. **90**, 146801 (2003).

[21] X. Cartoixa, D. Z.-Y. Ting, and Y.-C. Chang, Appl. Phys. Lett. **83**, 1462 (2003).

[22] B. A. Bernevig, T. L. Hughes, and S.-C. Zhang, Science **314**, 1757 (2006).

[23] B. A. Bernevig, J. Orenstein, and S.-C. Zhang, Phys. Rev. Lett. **97**, 236601 (2006).

[24] C. P. Weber, J. Orenstein, B. A. Bernevig, S.-C. Zhang, J. Stephens, and D. D. Awschalom, Phys. Rev. Lett. **98**, 076604 (2007).

[25] T. D. Stanescu and V. Galitski, Phys. Rev. B **75**, 125307 (2007).

[26] J. L. Cheng and M. W. Wu, J. Appl. Phys. **99**, 083704 (2006).

[27] J. L. Cheng, M. W. Wu, and I. C. da Cunha Lima, Phys. Rev. B **75**, 205328 (2007).

[28] E. Y. Sherman and J. Sinova, Phys. Rev. B **72**, 075318 (2005).

[29] I. Tokatly and E. Sherman, Ann. Phys. **325**, 1104 (2010).

[30] M. Ohno and K. Yoh, Phys. Rev. B **77**, 045323 (2008).

[31] A. A. Kovalev, K. Vyborny, and J. Sinova, Phys. Rev. B **78**, 41305 (2008).

[32] M. Borunda, T. Nunner, T. Luck, N. Sinitsyn, C. Timm, J. Wunderlich, T. Jungwirth, A. H. MacDonald, and J. Sinova, Phys. Rev. Lett. **99**, 066604 (2007).

[33] T. S. Nunner, N. A. Sinitsyn, M. F. Borunda, V. K. Dugaev, A. A. Kovalev, A. Abanov, C. Timm, T. Jungwirth, J.-i. Inoue, A. H. MacDonald, and J. Sinova, Phys. Rev. B **76**, 235312 (2007).

[34] S. Onoda, N. Sugimoto, and N. Nagaosa, Phys. Rev. B **77**, 165103 (2008).

[35] P. Nozieres and C. Lewiner, J. Phys. (Paris) **34**, 901 (1973).

[36] A. Crépieux and P. Bruno, Phys. Rev. B **64**, 014416 (2001).

[37] I. Zutic, J. Fabian, and S. Das Sarma, Rev. Mod. Phys. **76**, 323 (2004).

[38] I. Zutic, Nature Phys. **5**, 630 (2009).

[39] I. Zutic, J. Fabian, and S. C. Erwin, Phys. Rev. Lett. **97**, 026602 (2006).

[40] L. P. Zârbo, J. Sinova, I. Knezevic, J. Wunderlich, and T. Jungwirth, Phys. Rev. B **82**, 205320 (2010).

[41] A. A. Kiselev and K. W. Kim, Phys. Rev. B **61**, 13115 (2000).

[42] S. Kettemann, Phys. Rev. Lett. **98**, 176808 (2007).

[43] P. R. Hammar and M. Johnson, Phys. Rev. Lett. **88**, 066806 (2002).

[44] H. C. Koo, *et al.*, Science **325**, 1515 (2009).

[45] G. Schmidt and L. W. Molenkamp, Semicond. Sci. Technol. **17**, 310 (2002).

[46] A. Fert and H. Jaffrs, Phys. Rev. B **64**, 184420 (2001).

[47] I. Zutic, J. Fabian, and S. C. Erwin, Phys. Rev. Lett. **97**, 026602 (2006).

17 Quantum spin Hall effect and topological insulators

S. Murakami and T. Yokoyama

17.1 Quantum spin Hall systems

17.1.1 *Introduction*

Topological insulators (quantum spin Hall systems) are a new quantum state of matter theoretically proposed in 2005 [7, 42, 43], and have been experimentally observed in various methods later. Topological insulators can be realized in both two dimensions (2D) and in three dimensions (3D), and they are nonmagnetic insulators in the bulk, but have gapless edge states (2D) or surface states (3D) (see Fig. 17.1). These edge/surface states carry pure spin current and they are sometimes called helical. The novel property for these edge/surface states is that they originate from bulk topological order, and are robust against nonmagnetic disorder [7, 43, 119, 121].

We first explain how topological insulators are related to other spin transport phenomena. The spin–orbit coupling in solids arises from relativistic effects. The electrons are moving around the nuclei with a speed close to the speed of light. This spin-orbit coupling gives rise to spin-dependent orbital motions of the electrons, which enable us to manipulate spins by purely electric means. One of the methods to manipulate spins electrically is the spin Hall effect [64, 100], schematically shown in Fig. 17.2. In this effect, the electric field applied to the system induces a transverse spin current. After the theoretical predictions of the intrinsic spin Hall effect due to the band structure, the spin Hall effect has been observed in various semiconductors and metals with various methods.

The quantum spin Hall effect is the "quantum" version of the spin Hall effect, in a similar sense to the quantum Hall effect compared with the Hall effect. The (charge) Hall effect occurs in a system in a magnetic field. The Hall effect has a novel and interesting variant, called the quantum Hall effect. The quantum Hall effect is realized in a two-dimensional electron gas in a strong magnetic field. In this case the electrons form Landau levels, and the Fermi energy is between the Landau levels. Thus there are no bulk states at the Fermi energy, and the bulk is insulating. Nevertheless, at the Fermi energy there are some other states which are localized on the edge of the system. In integer quantum Hall states, these edge states are chiral, i.e. they go along the whole edge only in one direction, but not in the other. These states are responsible for the quantized Hall effect. The number of chiral edge states N is a topological number, and it is independent of

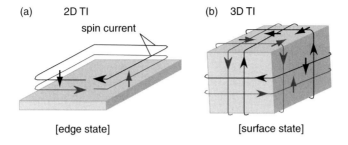

FIG. 17.1. Schematics for (a) 2D topological insulator and (b) 3D topological insulator.

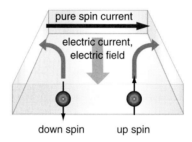

FIG. 17.2. Schematic of the spin Hall effect.

the details of the edge. This number N is called the Chern number. The Chern number gives the quantized Hall conductivity $\sigma_{xy} = Ne^2/h$.

The quantum Hall effect can be regarded as an insulator version of the Hall effect, because the bulk is insulating. In a similar sense, we can consider the insulator analog of the spin Hall effect: the quantum spin Hall effect [7, 42, 43]. Systems showing the quantum spin Hall effect are called topological insulators. Topological insulators can be schematically expressed as the lower panel of Fig. 17.3. This is a superposition of the two subsystems; the electrons with up-spins are under the $+B$ magnetic field and form $\sigma_{xy} = e^2/h$ quantum Hall states (Chern number $N = 1$), and electrons with down-spins are under the $-B$ magnetic fields and form $\sigma_{xy} = -e^2/h$ quantum Hall states (Chern number $N = -1$). To realize this system as a whole, we have to apply the magnetic field which is opposite for the up- and down-spins. This cannot be the usual magnetic field, but can be realized by the spin–orbit coupling inherent in solids. The resulting edge states consist of counterpropagating states with opposite spins.

These edge states constructed in this way are eigenstates of s^z, i.e. the spins in the edge states are perpendicular to the 2D plane. Nevertheless, it is not a necessary condition in general 2D topological insulators. In general the spins of edge states are not always perpendicular to the plane. To consider the spin

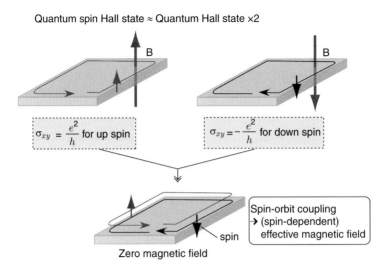

FIG. 17.3. Construction of a quantum spin Hall system (topological insulator) as a superposition of two subsystems, one is for up-spin and the other for down-spin.

directions of edge states, we note that in topological insulators time-reversal symmetry is assumed. Because of time-reversal symmetry, the edge state going clockwise and that going counter-clockwise are transformed to each other by the time-reversal operation, and therefore they have opposite spins (which may not be necessarily perpendicular to the plane). They are called "helical." They are degenerate due to Kramers' theorem, and are called Kramers pairs. They have a special property that any time-reversal-symmetric perturbation (such as nonmagnetic impurities or electron–electron interaction) cannot open a gap [119, 121]. These gapless edge states are topologically protected. This protection is due to the topological number, originating from the topological characterization of the band structure. Therefore, a topological insulator is insulating only in its bulk, and it is distinct from an ordinary insulator because of the topologically protected gapless states at the boundaries of the system.

As we have seen, the quantum Hall states are characterized by the Chern number N. This is an integer, and cannot be changed continuously. In a similar way, topological insulators are characterized by Z_2 topological numbers, taking two values, 0 and 1 (which are also called "even" and "odd"). When the Z_2 topological number is odd, it is a topological insulator and if it is even it is an ordinary insulator. The edge/surface states of topological insulators are characterized by the topological numbers calculated from the bulk wavefunctions [31, 74].

The two necessary conditions for topological insulators are (i) time-reversal symmetry, and (ii) strong spin–orbit coupling [7, 43]. Condition (i), time-reversal symmetry, is equivalent to nonmagnetic materials without an external magnetic field. As for (ii), systems without spin–orbit coupling are not topological insulators but ordinary insulators. Nevertheless, not all materials with strong spin–orbit coupling are topological insulators; one has to calculate the topological numbers in order to distinguish between topological and ordinary insulators.

17.1.2 *Topology and topological insulators*

Here we intuitively explain what "topology" in topological insulators means, without going into rigorous mathematical definitions. One of the well-known examples for "topology" is as follows. If one is allowed to do only a continuous deformation of a three-dimensional object, without detaching or attaching any parts, one can deform a doughnut into a coffee cup, but not to a ball. With these kinds of continuous deformation within some restrictions (i.e. without detaching or attaching), we can classify objects; the theory for this classification is called topology. Topology is a way to classify objects by identifying those which are connected by continuous change, within some restriction. In the above example, the classification is done in terms of the number g of holes (genus) in the three-dimensional objects. The doughnut and the coffee cup have $g = 1$, and the ball has $g = 0$. Two objects with the same g can be continuously deformed into each other, whereas those with different g cannot. In other words, g is invariant under continuous deformation, and is called a topological number or topological invariant.

The meaning of topology in topological insulators is similar. The idea is to classify nonmagnetic insulators by identifying those which can be continuously deformed into each other. The restriction is that the time-reversal symmetry is retained throughout the continuous change, and that the band gap does not close. The conclusion is that in two dimensions the classification is done by the Z_2 topological number ν, taking two values, $\nu = 0$ and $\nu = 1$ [20, 43]. Here Z_2 is a set of integers modulo 2, i.e. the set $\{0, 1\}$, and 0 and 1 can be called "even" and "odd." If the system considered has $\nu = 0$, it is an ordinary insulator, and if $\nu = 1$, it is a topological insulator. Now several questions arise: (i) how can we classify systems as topological insulators or ordinary insulators, and (ii) what properties do these two insulators have?

Question (i), namely the calculation of the Z_2 topological numbers, has been discussed in general in [20, 43], and we briefly explain the results in the following. We define here time-reversal invariant momenta (TRIM) as the momenta that satisfy $\mathbf{k} \equiv -\mathbf{k} \pmod{\mathbf{G}}$, where $\mathbf{G}$ is a reciprocal lattice vector. In the two-dimensional Brillouin zone there are four TRIM: $\mathbf{k} = \frac{1}{2}(n_1\mathbf{b}_1 + n_2\mathbf{b}_2)$ $(n_1, n_2 = 0, 1)$, where $\mathbf{b}_1$ and $\mathbf{b}_2$ are primitive vectors of the reciprocal lattice (see Fig. 17.4a). Let $\mathbf{k} = \Gamma_i$ $(i = 1, 2, 3, 4)$ denote the four TRIM. When the system is inversion-symmetric, the formula for the topological number is very simple. In this case the Hamiltonian satisfies $[H, P] = 0$, where P is the

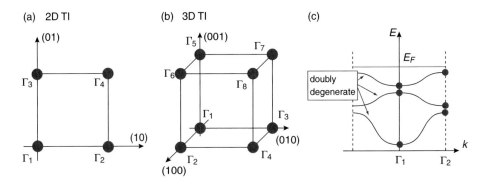

FIG. 17.4. (a) Four TRIM for 2D systems. (b) Eight TRIM for 3D systems. (c) Schematic of the bulk band structure of the system with time- and space-inversion symmetry.

inversion operation. If we convert this into Bloch form with Hamiltonian $H(\mathbf{k})$, we have $PH(\mathbf{k})P^{-1} = H(-\mathbf{k})$. At the TRIM $\boldsymbol{\Gamma}_i$, $\boldsymbol{\Gamma}_i \equiv -\boldsymbol{\Gamma}_i$ (mod$\mathbf{G}$) yields $PH(\boldsymbol{\Gamma}_i) = H(\boldsymbol{\Gamma}_i)P$. Therefore, the eigenstate at TRIM $\Psi(\boldsymbol{\Gamma}_i)$ is an eigenstate of P, and its eigenvalue (parity eigenvalue) can be either $\xi = 1$ (symmetric) or $\xi = -1$ (antisymmetric), because $P^2 = 1$. For the respective TRIM $\mathbf{k} = \boldsymbol{\Gamma}_i$, we consider the product of parity eigenvalues of Kramers pairs below the Fermi energy:

$$\delta_i = \prod_{m=1}^{N} \xi_{2m}(\boldsymbol{\Gamma}_i), \tag{17.1}$$

where $\xi_m(\boldsymbol{\Gamma}_i)$ denotes the parity eigenvalue of the mth eigenstate from the lowest energy states at $\boldsymbol{\Gamma}_i$. The $(2m-1)$th and $(2m)$th states are Kramers degenerate by time- and space-inversion (Fig. 17.4c), and they share the same parity eigenvalues: $\xi_{2m-1}(\boldsymbol{\Gamma}_i) = \xi_{2m}(\boldsymbol{\Gamma}_i)$. In Eq. (17.1) we used only the $(2m)$th eigenstates in order to avoid the double counting between the $(2m-1)$th and $(2m)$th states.

For 2D topological insulators the Z_2 topolological number ν is expressed in terms of these indices δ_i ($i = 1, 2, 3, 4$) as

$$(-1)^\nu = \prod_{i=1}^{4} \delta_i. \tag{17.2}$$

For 3D topological insulators, there are eight TRIM (see Fig. 17.4b), out of which four Z_2 topological numbers are defined. They are written in the form $\nu_0; (\nu_1\nu_2\nu_3)$, where the ν_k's are either 0 or 1, defined as a product of some of the indices δ_i:

$$(-1)^{\nu_0} = \prod_{i=1}^{8} \delta_i, \ (-1)^{\nu_k} = \prod_{i=(n_1 n_2 n_3), n_k=1} \delta_i \quad (k = 1, 2, 3). \tag{17.3}$$

Hence there are $2^4 = 16$ different phases. ν_0 is the most important among the four topological numbers, and we call the topological insulators with $\nu_0 = 1$ and those with $\nu_0 = 0$ strong topological insulators (STI) and weak topological insulators (WTI), respectively. On the other hand, ν_1, ν_2, and ν_3 depend on the crystallographic axes, and will become ill defined when the crystallographic translational symmetry is violated by (nonmagnetic) impurities, whereas ν_0 remains well-defined even in the presence of nonmagnetic disorder.

17.1.3 *Topological numbers*

We discuss how the edge states of 2D topological insulators are different from those of 2D ordinary insulators. Let us consider for example a semi-infinite plane of a 2D system in order to discuss edge states. We assume here translational symmetry along the boundary of the semi-infinite plane (edge), and the Bloch wavenumber k along this direction is a good quantum number. The band structure of the system with this geometry may have edge states. For ordinary insulators, even if there are edge states, they are as shown in Fig. 17.5(a); namely, the edge states come out of the conduction band (or valence band) and are absorbed back into the same conduction (or valence) band. In the presence of the spin–orbit coupling, these states are spin-split, as shown in Fig. 17.5(a). This is called Rashba splitting in the case of surface states. Because of the time-reversal symmetry, the states which are symmetric with respect to $k = 0$ are Kramers degenerate, and have opposite spins.

On the other hand, the band structure of the semi-infinite plane of the 2D topological insulator is schematically shown in Fig. 17.5(b). We note that the dispersion of the edge states traverses across the bulk gap, and connects the bulk valence and conduction bands. We can see that Figs. 17.5(a) and (b) cannot be transformed into each other by continuous deformation of the band structure, without closing the bulk gap. This holds as long as the time-reversal symmetry is preserved, which guarantees Kramers' theorem.

Figures 17.5 (a) and (b) are schematic diagrams showing only the simplest cases, and we can consider various types of edge states other than these. We can then ask ourselves how we can distinguish between ordinary insulators and topological insulators. This can be understood by introducing the topological number. We explain here with some examples how the "Z_2-ness" appears here in an intuitive way. For detailed explanations, readers are referred to Refs. [20, 21]. When the system is time-reversal symmetric, Kramers' theorem says that every state at $k = 0$ and $k = \pi$ is doubly degenerate. Therefore, in Fig. 17.5(b), the degeneracy (band crossing) in edge states at $k = 0$ will not be lifted unless the time-reversal symmetry is broken. On the other hand, when there are two Kramers pairs of edge states, the crossings at $k \neq 0$ are not protected, and they

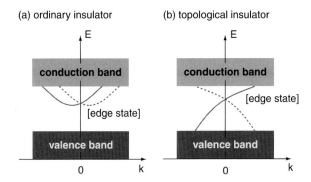

FIG. 17.5. Schematic diagram of edge states for (a) a 2D ordinary insulator and (b) a topological insulator. Solid and broken lines denote the edge states with opposite spin directions, because of the time-reversal symmetry.

will open a gap. When we draw this kind of band structure for several cases with various numbers of Kramers pairs of edge states, we can see that when the number of Kramers pairs is odd, the edge states are robust against time-reversal-invariant perturbations.

These arguments are based on the cases with wavenumber k being a good quantum number. The edge states remain gapless even when we include time-reversal-invariant perturbations which break translational symmetries, such as nonmagnetic impurities. Even in such cases Z_2 topological numbers remain well defined and such classifications are meaningful [43, 119, 121]. We note that similar arguments hold for surface states of 3D topological insulators when we replace k by $\mathbf{k}_\parallel$, the wavenumber along the surface. A detailed classification will be given in Section 17.3.

17.2   Two-dimensional (2D) topological insulators

17.2.1   *Edge states of 2D topological insulators*

In two-dimensional topological insulators, the Z_2 topological number and the edge states are related in the following way. When the Z_2 topological number is $\nu = 1$, the number of Kramers pairs of edge states at the Fermi energy is odd, whereas it is even, when $\nu = 0$ [20]. The reason why we cannot know the exact number but only know whether it is even or odd is that (time-reversal-invariant) perturbations can change the number of Kramers pairs by multiples of 2. Therefore the even-ness or odd-ness is unchanged under perturbations, while the number itself can change. In other words, whether there are even or odd numbers of Kramers pairs of edge states is a topological property. This topological number ν never changes under continuous change of some parameters, unless the bulk gap closes. When the bulk gap closes at some point, the topological number may change. The closing event of the bulk gap by changing parameters can be classified, and one can see whether the topological number changes or not [62, 65].

These edge states are not degenerate, and have fixed spin directions, which usually depend on the wavenumber k. Because the velocity $v = \dfrac{1}{\hbar} \dfrac{\partial E}{\partial k}$ is the slope of the dispersion, the two edge states in Fig. 17.5(b) are propagating in opposite directions, and they have opposite spins because of the time-reversal symmetry. As a result, these edge states carry pure spin current [64, 100].

As we mentioned earlier, these edge states are not backscattered by nonmagnetic impurities (see Fig. 17.6a). This can be explained intuitively in the following way. Scattering by nonmagnetic impurities is elastic, and the electron energy is invariant at the scattering. From Fig. 17.5(b), it follows that the two edge states with equal energy have opposite spins and thus the scattering between them is necessarily accompanied by spin flip. Therefore, nonmagnetic impurities cannot cause backscattering. A more general proof is in Refs. [119, 121]. Hence the edge-state transport becomes perfectly conducting. In a HgTe quantum well this perfect conduction of edge states has been experimentally observed [44, 45, 83].

Because the gapless edge states are protected by time-reversal symmetry, if one attaches a ferromagnet on the edge, the edge states become gapped and the system eventually becomes insulating when the Fermi energy is within the gap. In addition, when two ferromagnets are attached on the edge with some separation, some amount of charge is accumulated, and the amount of charge

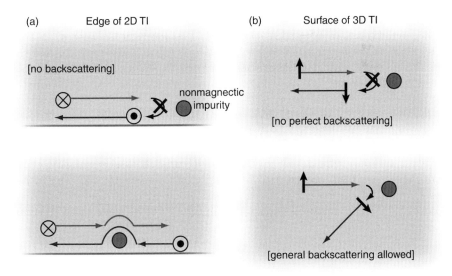

FIG. 17.6. Comparison of transport properties for (a) edge states of a 2D topological insulator and (b) surface states of a 3D topological insulator.

determined by the direction of the magnetizations of two ferromagnets becomes fractional, which can be less than the electronic charge [75].

Even if an interaction is introduced, a gap does not open in the helical edge states unless the time-reversal symmetry is spontaneously broken [48, 119, 121]. The helical edge states show perfect conduction, whereas the transport properties across junctions or constrictions will be affected by interactions [36, 101, 104, 109]. As a related subject, interactions may help systems to become topological insulators. In the presence of on-site interactions in the Hubbard model on a honeycomb lattice, a topological Mott insulator phase is theoretically predicted, which is due to a spontaneously generated spin–orbit coupling [79].

17.2.2 *Experiments on edge states of 2D topological insulators*

The CdTe/HgTe/CdTe quantum well was theoretically proposed to be a 2D toplogical insulator [8]. Compared with the usual cubic semiconductors such as GaAs and InSb, HgTe has an inverted band structure due to its strong spin–orbit coupling. In the usual cubic semiconductors, the conduction band and the valence band belong to Γ_6 and Γ_8 irreducible representations of the cubic group, respectively. The strong spin–orbit coupling pushes down the energy of the Γ_6 band below the Γ_8 band, and the resulting band structure has a zero gap within the Γ_8 band, with the degeneracy residing at $\mathbf{k} = 0$. This is called a zero-gap semiconductor. On the other hand, CdTe belongs to the class of usual cubic semiconductors. When HgTe is incorporated into a quantum well by sandwiching it between CdTe, the subband structure arises. When the thickness d of the HgTe layer is thinner than $d_c = 60$ Å, the subband structure is like that of CdTe; when d is thicker than $d_c = 60$ Å, the subband structure is like that of HgTe. It then follows that the quantum well with $d < d_c$ is an ordinary insulator and that with $d > d_c$ a 2D toplogical insulator. These can be distinguished by transport measurements. In the ordinary 2D insulator phase, there are no conducting channels when the Fermi energy is in the band gap and the charge conductance is zero. On the other hand, in a 2D topological insulator, there are two channels on the two sides of the system, and the system shows the two-channel conductance $G = 2e^2/h$. These have been confirmed in experiments [44, 45].

When we apply a magnetic field and break time-reversal symmetry, charge conductance has been observed to be rapidly suppressed [44, 45]. This is consistent with the theoretical proposal. These kinds of edge channels also lead to novel behavior of nonlocal transport properties. For example, multi-terminal conductance becomes $\frac{e^2}{h}$ times a simple fraction determined from the geometry of the terminals [83], which agrees with calculation by the Landauer–Büttiker formula. These experiments are good evidence for the existence of gapless helical edge states.

A similar theoretical proposal has been made for Type-II semiconductor quantum wells in InAs/GaSb/AlSb as well [55]. There is also a theoretical proposal for a bismuth ultrathin film [61], though it awaits experimental observation.

In the bismuth ultrathin film the edge states consist of three Kramers pairs, and two-terminal conductance is predicted to be three times $G = \dfrac{2e^2}{h}$ [115]. Such perfectly conducting channels affect not only charge transport but also thermoelectric transport such as the Seebeck coefficient [103].

17.3 Three-dimensional (3D) topological insulators

17.3.1 *Surface states of three-dimensional topological insulators*

In three dimensions, the four Z_2 topological numbers convey information on the surface states for arbitrary directions of surfaces. The surface states obey the following rule [26, 60, 85]. In order to know the surface states of a certain crystallographic surface, we first have to project the TRIM indices δ_i onto this surface. In this process, the TRIM in the original 3D crystal are mapped to the TRIM of the surface Brillouin zone. We then multiply the two indices δ_i for the two TRIM, which are mapped onto the identical point on the surface TRIM, and we associate each surface TRIM with the corresponding product $(= \pm 1)$. The resulting products $(= \pm 1)$ give information on the surface Fermi surfaces in the following way. We compare the products of the indices between two surface TRIM: (i) if they have opposite signs, there are odd numbers of surface Fermi surfaces intercepting between the two surface TRIM, and (ii) if they have the same signs, then there are even numbers of surface Fermi surfaces between them [26, 110]. Under a continuous change of parameters, these topological numbers will not change unless the bulk gap closes and unless the translational symmetry is not broken. When the bulk gap closes at some points in the Brillouin zone as parameters are changed, some of the Z_2 topological numbers can change [62, 63]. By classifying various cases of such gap closings, universal phase diagrams between topological and ordinary insulators have been found [63].

17.3.2 *Properties of surface states of 3D topological insulators*

One of the typical forms of surface states of a topological insulator is the single Dirac cone. As we will see later, in Bi_2Se_3 and Bi_2Te_3 the topological numbers are $1; (000)$, and the simplest possibility for the surface Fermi surface on the (111) surface is a single Fermi surface encircling the Γ-point. This is the case realized in these materials. The dispersion is linear in the wavenumber $\mathbf{k}_\parallel$, schematically shown in Fig. 17.7(a). This is called a Dirac cone. The typical form of the surface Hamiltonian is the Dirac type:

$$H_\parallel = \lambda(\sigma \times \mathbf{k}_\parallel)_z \tag{17.4}$$

where σ_i are the Pauli matrices representing spins, $\mathbf{k}_\parallel$ is the wavevector along the surface, and z is the axis normal to the surface. The eigenvalues are $E = s\lambda k_\parallel$ $(s = \pm 1)$ with eigenstates given by

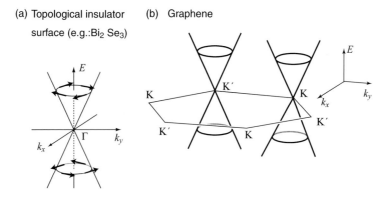

FIG. 17.7. Comparison between (a) the Dirac cone on the topological insulator surface and (b) the Dirac cones in graphene. The thick arrows in (a) represent the directions of spins.

$$|\psi_s\rangle = \frac{1}{\sqrt{2}} \begin{pmatrix} ise^{-i\phi} \\ 1 \end{pmatrix}, \tag{17.5}$$

where $e^{i\phi} = (k_x + ik_y)/k_\parallel$. Thus the eigenstates have fixed directions of spins in the direction $s\mathbf{k}/k \times \hat{z}$.

The Dirac cones can be found in other systems such as graphene. As compared with purely two-dimensional systems such as graphene, the surface Dirac cone of the 3D topological insulator has a unique property, that the number of Dirac cones in the Brillouin zone is odd. For example, on the (111) surface of Bi_2Se_3 there is a single Dirac cone at the Γ-point (see Fig. 17.7a). In contrast, in graphene (Fig. 17.7b), the Dirac cones are located at points K and K', and they are spin-degenerate. Thus the total number of Dirac cones within the Brillouin zone is four. In fact the Nielsen–Ninomiya theorem says that the number of Dirac cones in the two-dimensional system is always even. In this sense, the odd number of Dirac cones is unique to the surface of the topological insulators.

This odd number of Dirac cones exactly corresponds to the case of Z_2 topological number $\nu_0 = 1$. As in the 2D topological insulator, perturbations preserving time-reversal symmetry do not open a gap in the Dirac cone. It has been shown theoretically that the electrons in this single Dirac cone will not localize even if we increase nonmagnetic disorder [5, 70, 71, 86]. This is attributed to the π Berry phase when the electron wavefunction goes around the Dirac point. Another interesting effect of disorder is to induce a topological insulator from an ordinary insulator by disorder [28, 29, 49, 98].

On the other hand, a gap opens at the Dirac point, if we include perturbations which break time-reversal symmetry, for example by attaching a magnetic film onto the surface. For example, in the lowest order in $\mathbf{k}_\parallel$, the Hamiltonian under the Zeeman field is given by

$$H(\mathbf{k}) = \lambda(\sigma \times \mathbf{k}_{\parallel})_z - \mathbf{B} \cdot \sigma \qquad (17.6)$$

where $\mathbf{B}$ represents the Zeeman splitting due to the magnetic film. We can easily see that the eigenenergies are $E(\mathbf{k}) = \pm\sqrt{(\lambda k_y - B_x)^2 + (\lambda k_x + B_y)^2 + B_z^2}$. Hence the gap between the valence and the conduction bands is $2|B_z|$. Thus the Zeeman coupling in the z-direction opens a gap, while that in the xy-direction does not. This Dirac cone with a gap shows a Hall effect with half the quantum Hall conductivity, $\sigma_{xy} \sim \operatorname{sgn}(B_z)\dfrac{1}{2}\dfrac{e^2}{h}$.

This half-quantum Hall effect gives rise to the magnetoelectric effect [17, 76, 77]. Following Ref. [76], we consider a ferromagnetic film wrapping around the cylindrical 3D topological insulator, with the magnetization pointing radially outward (see Fig. 17.8). Then the surface states on the surface facing the ferromagnet will open a gap, showing the half-quantum Hall effect $\sigma_{xy} = \dfrac{1}{2}\dfrac{e^2}{h}$. Suppose we apply an electric field along the cylindrical axis of the topological insulator. It then induces a charge current around the cylinder with the current density $j = \sigma_{xy}E$, and it will eventually induce a magnetic field $H = j = \sigma_{xy}E$

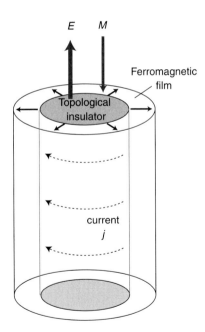

FIG. 17.8. A ferromagnetic film wrapped around a topological insulator is radially magnetized. When the electric field is applied in the axial direction, a magnetization is induced as a result of the Hall effect. This is the magnetoelectric effect.

$= \pm \frac{1}{2} \frac{e^2}{h} E$. Thus the whole system gives rise to the magnetoelectric effect with coefficient $\pm \frac{1}{2} \frac{e^2}{h} (= \alpha \varepsilon_0 c$ where is the fine structure constant, c is the speed of light, and ε_0 is the vacuum permittivity).

As a similar effect to this magnetoelectric effect, if one attaches a ferromagnetic film on the surface of the topological insulator, an external charge close to the surface will induce a magnetic field [17, 77]. In analogy to a mirror charge induced by an external charge close to the surface of a dielectric, this case corresponds to a mirror magnetic monopole. This effect awaits experimental verification.

There are other types of topological phases, such as topological superconductors. These topological phases are realizable only in systems with bulk gaps. The quantum Hall systems and topological insulators are among these topological phases. In addition, for example, the A phase in superfluid ^{3}He and $p + ip$ type superconductors belongs to these topological phases. These systems have a bulk gap, and they necessarily have boundary states characterized by topological numbers. Classification of such topological phases is possible, based on the symmetries of the system [95, 96]. By these methods, the topological surface states can be understood in a systematic way.

In some 3D topological insulators, dislocations in the crystal are accompanied with helical gapless states in some cases [80]. When the Burgers vector $\mathbf{B}$ for the dislocation satisfies the condition $\mathbf{B} \cdot \mathbf{M}_\nu = \pi$ (mod 2π), where $\mathbf{M}_\nu = \frac{1}{2}(\nu_1 \mathbf{b}_1 + \nu_2 \mathbf{b}_2 + \nu_3 \mathbf{b}_3)$, there necessarily appear helical gapless states along the dislocation. Hence, whether helical states appear on the dislocation or not depends only on the "weak" indices ν_1, ν_2, and ν_3, but not on ν_0. Among the strong topological insulators, Bi_2Se_3 and Bi_2Te_3 have $\mathbf{M}_\nu = 0$, and dislocations do not accompany helical states. On the other hand, $Bi_{1-x}Sb_x$ ($0.07 < x < 0.22$) has $\mathbf{M}_\nu = (\mathbf{b}_1 + \mathbf{b}_2 + \mathbf{b}_3)/2$ and some dislocations such as $\mathbf{B} = \mathbf{a}_1$ are accompanied by helical states. Other types of topological objects such as π-flux (flux with half the flux quantum) threading through the topological insulator are accompanied by bound states with spins and charges separated [73, 81].

17.3.3 *Materials for 3D topological insulators*

17.3.3.1 *Experiments of 3D topological insulators* We explain some materials for 3D topological insulators. The first material which is experimentally observed to be a 3D topological insulator is $Bi_{1-x}Sb_x$ ($0.07 < x < 0.22$) [37, 69]. The host material bismuth (Bi) is a semimetal, having small electron and hole pockets, and not an insulator. In order to create a topological insulator out of Bi, one should make it insulating. There are two ways to do this. One is to make it very thin, and it is proposed that it becomes a 2D topological insulator [61, 115], but this awaits experimental verification. The other way is to dope with antimony (Sb). The carrier pocket disappears by doping by some amount of Sb, and $Bi_{1-x}Sb_x$ with $0.07 < x < 0.22$ is expected to be a topological insulator [21].

In terms of the topological number, the host material Bi is trivial, and has the topological number $0; (000)$. We note that although Bi is a semimetal, there is a direct gap everywhere in the Brillouin zone, and Z_2 topological numbers are well defined (whereas they are not directly related to physical properties such as the robustness of the surface states). As a function of doping x, at $x = 0.04$ there occurs a band inversion at the L-points in the Brillouin zone, which involves the change of the parities of the occupied bands, giving rise to a change of topological number from $0; (000)$ to $1; (111)$. Then at $x = 0.07$ an indirect gap opens and the system becomes a 3D topological insulator. The system remains a 3D topological insulator up to $x = 0.22$ where the indirect gap closes again. In the topological insulator phase $(0.07 < x < 0.22)$, the indices for the TRIM are $+1$ for three L-points, and -1 otherwise. Because Bi and $Bi_{1-x}Sb_x$ cleaves at the (111) surface, we consider the (111) surface Brillouin zone, where the products of the indices turn out to be $+1$ for the Γ-point and -1 for the three M-points. Thus we expect that there are odd numbers of surface Fermi surfaces between the Γ- and M-points. In experiments the number of observed Fermi surfaces between the Γ- and M-points is five [37] or three [69], which completely agrees with the above calculation from topological numbers.

Experimental observation of the Fermi surface has been done by angle-resolved photoemission spectroscopy (ARPES) [37] and spin-resolved ARPES [38, 69]. The absence of backscattering has also been observed directly in experiments. Scanning tunnel spectroscopy (STS) can be used to detect the local density of states (LDOS) on the surface. When we take the Fourier transform of the observed STS image, we can identify the electronic waves of the surface scattered by disorder. This Fourier transform has a complicated interference pattern, which reflects various scattering processes between the wavenumbers on the Fermi surface. One can compare the Fourier transform of the STS image with the Fermi surface, which has been known either experimentally or theoretically. In $Bi_{0.9}Sb_{0.1}$, the Fourier transform of the STS image agree with the data from the Fermi surface quite well, *only when* we consider the spin directions of the states, and suppress the backscattering which involves spin flip. This means that the states on the surface Fermi surface are indeed spin-filtered, i.e. having a fixed direction of spins, and the backscattering is indeed suppressed [84].

Transport measurements have been done for $Bi_{1-x}Sb_x$ under a strong magnetic field [106, 107]. It is concluded that the magnetic oscillation is partially due to the surface carriers. There are also signals from bulk carriers, which are residual carriers in $Bi_{1-x}Sb_x$ due to inhomogeneity. These bulk carriers also contribute to transport. On the other hand, when we make it into a thin film and vary the film thickness, we can in principle separate the bulk and surface transport as a function of the film thickness [32]. Nevertheless, to observe the surface transport, we need a sample of very good quality, so that the bulk transport is much suppressed. This kind of transport measurement is highly desired.

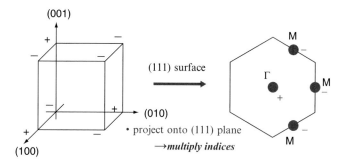

FIG. 17.9. Calculation of surface Fermi surface for Bi_2Se_3, out of the 3D Z_2 topological numbers.

In Bi_2Se_3 and Bi_2Te_3 [39, 40, 120, 124], the TRIM on the surface Brillouin zone on the (111) surface consist of one Γ-point and three M-points. Therefore, the product of projected indices is equal to -1 at the Γ-point and $+1$ at the other M-points (see Fig. 17.9). The surface states observed in experiments form a single Fermi surface around the Γ-point, which is the simplest possibility from the above consideration using the topological numbers. These surface states form a single Dirac cone (Fig. 17.7a).

The linear dispersion of surface Dirac cones in Bi_2Se_3 has been observed by ARPES [120]. The spin states have also been measured [40], and they have fixed directions nearly perpendicular to the surface wavenumber $\mathbf{k}_\parallel$ described in Fig. 17.7(a). As shown in Fig. 17.6(b), in this Dirac cone, perfect backscattering by nonmagnetic impurities is prohibited because the state with $\mathbf{k}$ and that with $-\mathbf{k}$ have opposite spins, while partial backscattering is suppressed but not prohibited. On the other hand, in Bi_2Te_3, the dispersion is linear only within close vicinity of the Γ-point. For larger wavenumber $\mathbf{k}$, on the other hand, the higher order terms (k^3) become larger, and the surface Fermi surface becomes no longer a circle, but like a snowflake shape with six vertices [12]. This kind of deformation of the Fermi surface is called warping [18]. These materials have a rather large band gap (~ 0.3eV for Bi_2Se_3). The surface states are topologically protected, and only spin-conserving scattering is allowed, as observed by STS measurements [125]. Under a strong magnetic field, the surface Dirac cones form Landau levels. Similarly to the Dirac cones in graphene, the energy levels of the n-th Landau levels are roughly proportional to $\sqrt{n}$, as is experimentally confirmed [14, 30]. As for transport magnetoresistance [11, 16, 106, 107] and cyclotron resonance [4, 102], measurements have been done. Some of them may involve bulk carriers, which complicates the theoretical analysis of the experimental data.

The relationship between 2D topological insulators and 3D topological insulator is usually complicated. When a 3D topological insulator is made into thin film, there occurs hybridization between the surface states of the top and the bottom surface, and the surface states become gapped. In some cases there remain edge states along the periphery of the thin film, which evolve into helical edge states in the thin limit. It typically occurs that as a function of film thickness, the 2D Z_2 topological number oscillates between 1 and 0 [52, 56, 57]. In Bi_2Se_3 such oscillating behavior has been observed [87, 126].

The search for ordered states by doping into Bi_2Se_3 is another interesting subject. Bi_2Se_3 doped with Cu becomes superconducting at low temperature [34], and Bi_2Se_3 doped with Mn becomes a ferromagnet [13, 35]. These doped systems are promising for making a junction with Bi_2Se_3 to open a gap in the surface states of Bi_2Se_3.

The search for materials for topological insulators is interesting and promising, because there might be a number of candidate materials. Recent experiments have revealed that $TlBiSe_2$ is also a 3D topological insulator. Its (111) surface has a Dirac cone, similar to Bi_2Se_3, with the bulk gap of 0.35 eV [46, 50, 93]. In addition, some Heusler alloys such as LuPtSb are theoretically proposed to be zero-gap semiconductors, similar to HgTe. If a gap is opened in these materials by making a quantum well or applying pressure, 2D topological insulators are expected to appear [10, 51].

17.3.3.2 *Towards new materials for topological insulators*   In topological insulators, the spin–orbit coupling should be large, and should exceed the gap without spin–orbit coupling. In the first paper by Kane and Mele [43], graphene is shown to be a 2D topological insulator. In graphene, the spin-orbit coupling is very weak and is usually neglected, resulting in the Dirac cones at the K and K′ points with a vanishing gap. When the spin–orbit coupling is included, a gap opens at the K and K′ points, which is theoretically estimated to be 10 mK. Although 10 mK is very tiny, it is still larger than zero, namely the size of the gap of graphene without spin–orbit coupling. Therefore, a 2D topological insulator is realized. In reality, however, because 10 mK is very small, it can be easily masked by other extrinsic effects such as disorder.

As another example, the usual cubic semiconductors such as Si or GaAs are ordinary (not topological) insulators. This is because the gap in these semiconductors is primarily by covalent bonds, not by the spin–orbit coupling; therefore, if we take a fictitious limit to reduce the spin–orbit coupling to zero in these materials, the gap remains open. The original gap (usually larger than 1 eV) is larger than the spin–orbit coupling. On the other hand, when the spin–orbit coupling is increased, it may exceed the original gap size; this may cause the band gap to close and then open again. This corresponds to the band inversion mentioned earlier, and causes a phase transition from ordinary to topological insulators. This occurs when the HgTe well thickness is varied through $d = d_c$. In general this band inversions may occur simultaneously at more than one point in the Brillouin zone. If the number of band inversions in the Brillouin zone is odd, this generally corresponds to a phase transition, whereas if it is even, it is not accompanied by a phase transition [62, 63].

Therefore, a necessary condition for a topological insulator is to choose materials with spin–orbit coupling larger than the original gap size, i.e. the gap size where the spin–orbit coupling is neglected. This implies that for judiciously chosen materials such that the original gap size is small, the spin–orbit coupling is not necessarily very large, in order to be a topological insulator. On the other hand, for experiments and potential applications, a larger band gap is better, because robustness as a topological insulator is determined by the size of the band gap. In this sense, a stronger spin–orbit coupling is important.

So far we have neglected interactions in topological insulators. Topological insulators with electron correlations have been discussed [48, 119, 121]. There are theoretical proposals for topological insulators spontaneously induced by electron–electron interactions [79, 127], and their applications to Ir oxides [72, 99].

It is an interesting coincidence that there is a close overlap between topological insulator materials and good thermoelectric materials. From this viewpoint, thermoelectric transport has been calculated for edge states of 2D topological insulators [103], surface states of 3D topological insulators [27], and helical states on the dislocations of 3D topological insulators [111]. For the edge states of 2D topological insulators [103] or the dislocation states of 3D topological insulators [111], the electrons do not undergo elastic backscattering, which is good for thermoelectric transport. Nevertheless, at finite temperatures, inelastic scattering causes decoherence of these 1D helical states, and the otherwise good thermoelectric transport will be reduced. It is therefore predicted that at lower temperature, such helical 1D states become gradually dominant over transport by bulk carriers, and the thermoelectric figure of merit increases.

17.3.4   *3D topological insulators and Majorana fermions*

Interface phenomena on the surface also feature topological insulators. For example, suppose one attaches two ferromagnets on the surface of a topological insulator. When their magnetizations are along the z and $-z$ directions, i.e. perpendicular to the surface, chiral edge channels are predicted to appear at the interface between the two ferromagnets [67]. On the other hand, when a superconductor is attached onto the topological insulator, the surface state hosts superconductivity as a result of the proximity effect. The cases of particular interest are Majorana fermions at an interface. They are predicted to emerge at the vortex core, at the interface of the superconductors, or at the interface between the ferromagnet and superconductor [22, 88] on the surface of topological insulator (see Fig. 17.10). Majorana fermions are characterized by their self-Hermitian nature: the second-quantized field operator γ satisfies $\gamma = \gamma^\dagger$. To satisfy this equality, particle–hole symmetry and the absence of the spin degree of freedom are required [117]. As mentioned, a topological insulator has a single Fermi surface near the Dirac point. In contrast, in conventional superconductors, there are two Fermi surfaces corresponding to the two directions of spins when

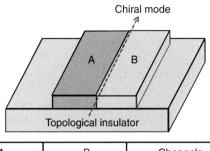

Chiral mode

A	B	Channels
Superconductor	Ferromagnet	Chiral Majorana fermion
Ferromagnet ↑	Ferromagnet ↓	Chiral fermion

FIG. 17.10. Chiral modes generated at the interface on the surface of a topological insulator.

time-reversal symmetry is respected. There are then two possibilities to form Cooper pairs in conventional superconductors: pairing between an electron with momentum $\mathbf{k}$ and spin ↑ and that with momentum $-\mathbf{k}$ and spin ↓, and also pairing between an electron with momentum $\mathbf{k}$ and spin ↓ and that with opposite momentum and spin. On the other hand, on the surface of a topological insulator, there is only a single choice of pairing on the Fermi surface, since momentum and spin are related to each other. *This reduction of the degrees of freedom— down to half—makes it possible to create Majorana fermions with the help of superconductivity*, since particle–hole symmetry is respected at zero energy in superconductors. These Majorana excitations obey non-abelian statistics, which has potential applications to fault-tolerant topological quantum computation and hence has received great attention [3, 66]. Majorana fermions are also predicted to appear in other superconducting systems, such as edge or vortex cores in chiral p-wave superconductors [15, 41], fractional quantum Hall systems [82], s-wave [92] and p-wave [58, 112] superfluids, films of a semiconductor with s-wave superconductivity and Zeeman splitting induced by the proximity effect [2, 94], nodal superconductors with spin–orbit interactions [91], and superconductors with pointlike topological defects [108].

When Majorana fermions are present, how are they reflected in physical quantities? In fact, some interesting behavior of physical quantities has been predicted in superconducting junctions formed on the surface of a topological insulator, regarding, for instance, crossed Andreev reflection [68], tunneling conductance [47, 59, 105], and the Josephson effect [23, 105]. Majorana bound states facilitate charge transport and hence lead to a zero-bias conductance peak of the tunneling conductance in ferromagnet/superconductor junctions on the surface of a topological insulator [105], similar to chiral p-wave superconductor junctions [33, 122]. An interferometer composed of a ferromagnet and a superconductor on the topological insulator enables us to break one complex fermion into two Majorana fermions and then to combine them again into a complex fermion; this process can be verified as a change in conductance [1, 24]. In addition, in various types of junctions, theoretical proposals have been given to observe the behavior of the Majorana fermions [19, 23, 123], and a junction with non-s-wave superconductors [24].

In addition to s-wave superconductivity treated in the above work, unconventional superconductivity on the surface of a topological insulator has been investigated [25, 53, 54, 89]. It has recently been shown [53, 54] how the interplay between unconventional superconductivity and ferromagnetism on the surface of a topological insulator gives rise to a number of effects with no counterpart in conventional metallic systems. In particular, zero-energy states on the surface of a d_{xy}-wave superconductor are demonstrated to be Majorana fermions, in contrast to the topologically trivial high-T_c cuprates. The dispersion of these states is highly sensitive to the orientation of the applied magnetic field [53, 54]. Besides topological insulators, the topological structure of superconductors also provides us with intriguing topics [6, 77, 90, 97, 113, 114, 116].

At present, the experimental study of topological insulators is in the early stages. However, in view of the recent experimental realization of superconductivity in the topological insulator Bi_2Se_3 by copper intercalation [34, 118] and also the magnetism of Bi_2Se_3 by magnetic dopants [13, 35], exciting experimental results on Majorana fermions are expected in the near future.

17.4 Summary

The field of topological insulators has been in rapid progress for several years, thanks to the interaction between theory and experiment. It has been a surprise that these kinds of topological phases are possible in ambient situations among known materials such as Bi_2Te_3. The well-known example of topological phases is the quantum Hall systems. As compared with quantum Hall states, the topological insulators are found not in extreme situations such as low temperature or high pressure. The number of candidate materials for topological insulators might be large, and the search for new materials will be a promising and important subject in the coming years. It is desired to search for topological insulators with a larger gap, those in which the carrier control is easier, or those which can become magnetic or superconducting by doping.

The edge states in 2D topological insulators or the surface states in 3D topological insulators are unique states, which cannot be easily realized in other systems. These states offer unique stages for novel phenomena, and are interesting for theory and experiment.

Acknowledgments

This research is supported in part by Grant-in-Aid for Scientific Research (No. 21000004, 22540327, 23103505 and 23740236) from the Ministry of Education, Culture, Sports, Science and Technology (MEXT).

References

[1] Akhmerov, A. R., Nilsson, J., and Beenakker, C. W. J. (2009). *Phys. Rev. Lett.*, **102**, 216404.

[2] Alicea, J. (2010). *Phys. Rev. B*, **81**, 125318.

[3] Alicea, J., Oreg, Y., Refael, G., Oppen, F., and Fisher, M. P. A. arXiv:1006.4395, to appear in *Nature Physics* (2011).

[4] Ayala-Valenzuela, O. E., *et al. arXiv*, 1004.2311.

[5] Bardarson, J. H. Tworzydło, J., Brouwer, P. W., and Beenakker, C. W. J. (2007). *Phys. Rev. Lett.*, **99**, 106801.

[6] Béri, B. (2010). *Phys. Rev. B*, **81**, 134515.

[7] Bernevig, B. A. and Zhang, S.-C. (2006). *Phys. Rev. Lett.*, **96**, 106802.

[8] Bernevig, B. A., Hughes, T. L. and Zhang, S.-C. (2006). *Science*, **314**, 1757.

[9] Bychkov, Y. A. and Rashba, E. I. (1984). *J. Phys. C: Solid State Phys.*, **17**, 6039.

[10] Chadov, S., *et al.* (2010). *Nature Mat.*, **9**, 541.

[11] Checkelsky, J. G., *et al.* (2009). *Phys. Rev. Lett.*, **103**, 246601.

[12] Chen, Y. L., Analytis, J. G., Chu, J.-H., Liu, Z. K., Mo, S.-K., Qi, X. L., Zhang, H. J., Lu, D. H., Dai, X., Fang, Z., Zhang, S. C., Fisher, I. R., Hussain, Z., and Shen, Z.-X. (2010). *Science*, **325**, 178.

[13] Chen, Y. L., Chu, J.-H., Analytis, J. G., Liu, Z. K., Igarashi, K., Kuo, H.-H., Qi, X. L., Mo, S. K., Moore, R. G., Lu, D. H., Hashimoto, M., Sasagawa, T., Zhang, S. C., Fisher, I. R., Hussain, Z., and Shen, Z. X. (2010). *Science*, **329**, 659.

[14] Cheng, P., *et al.* (2010). *Phys. Rev. Lett.*, **105**, 076801.

[15] Das Sarma, S., Nayak, C., and Tewari, S. (2006). *Phys. Rev. B*, **73**, 220502(R).

[16] Eto, K., *et al.* (2010). *Phys. Rev. B*, **81**, 195309.

[17] Essin, A. M., Moore, J. E., and Vanderbilt, D. (2009). *Phys. Rev. Lett.*, **102**, 146805.

[18] Fu, L. (2009). *Phys. Rev. Lett.*, **103**, 266801.

[19] Fu, L. (2010). *Phys. Rev. Lett.*, **104**, 056402.

[20] Fu, L. and Kane, C. L. (2006). *Phys. Rev. B*, **74**, 195312.

[21] Fu, L. and Kane, C. L. (2007). *Phys. Rev. B*, **76**, 045302.

[22] Fu, L. and Kane, C. L. (2008). *Phys. Rev. Lett.*, **100**, 096407.

[23] Fu, L. and Kane, C. L. (2009). *Phys. Rev. B*, **79**, 161408(R).

[24] Fu, L. and Kane, C. L. (2009). *Phys. Rev. Lett.*, **102**, 216403.

[25] Fu, L. and Berg, E. (2010). *Phys. Rev. Lett.*, **105**, 097001.

[26] Fu, L., Kane, C. L., and Mele, E. J. (2007). *Phys. Rev. Lett.*, **98**, 106803.

[27] Ghaemi, P., Mong, R. S. K., and Moore, J. E. (2010). *Phys. Rev. Lett.*, **105**, 166603.

[28] Groth, C. W., *et al.* (2009). *Phys. Rev. Lett.*, **103**, 196805.

[29] Guo, H.-M., Rosenberg, G., Refael, G., and Franz, M. (2010). *Phys. Rev. Lett.*, **105**, 216601.

[30] Hanaguri, T., *et al.* (2010). *Phys. Rev. B*, **82**, 081305.

[31] Hasan, M. Z. and Kane, C. L., *arXiv*, 1002.3895.

[32] Hirahara, T., *et al.* (2010). *Phys. Rev. B*, **81**, 165422.

[33] Honerkamp, C. and Sigrist, M. (1998). *J. Low Temp. Phys.*, **111**, 895.

[34] Hor, Y. S., Williams, A. J., Checkelsky, J. G., Roushan, P., Seo, J., Xu, Q., Zandbergen, H. W., Yazdani, A., Ong, N. P., and Cava, R. J. (2010). *Phys. Rev. Lett.*, **104**, 057001.

[35] Hor, Y. S., *et al.* (2010). *Phys. Rev. B*, **81**, 195203.

[36] Hou, C.-Y., Kim, E., and Chamon, C. (2009). *Phys. Rev. Lett.*, **102**, 076602.

[37] Hsieh, D., Qian, D., Wray L., Xia, Y., Hor, Y. S., Cava, R. J., and Hasan, M. Z. (2008). *Nature*, **452**, 970.

[38] Hsieh, D., *et al.* (2009). *Science*, **323**, 919.

[39] Hsieh, D., Xia, Y., Qian, D., Wray, L., Meier, F., Dil, J. H., Osterwalder, J., Patthey, L., Fedorov, A. V., Lin, H., Bansil, Grauer, A., D., Hor, Y. S., Cava, R. J., and Hasan, M. Z. (2009). *Phys. Rev. Lett.*, **103**, 146401.

[40] Hsieh, D., Xia, Y., Qian, D., Wray, L., Dil, J. H., Meier, F., Osterwalder, J., Patthey, L., Checkelsky, J. G., Ong, N. P., Fedorov, A. V., Lin, H., Bansil, A., Grauer, D., Hor, Y. S., Cava, R. J., and Hasan, M. Z. (2009). *Nature*, **460**, 1101.

[41] Ivanov, D. A. (2001). *Phys. Rev. Lett.*, **86**, 268.

[42] Kane, C. L. and Mele, E. J. (2005). *Phys. Rev. Lett.*, **95**, 226801.

[43] Kane, C. L. and Mele, E. J. (2005). *Phys. Rev. Lett.*, **95**, 146802.

[44] K'onig, M., Buhmann, H., Molenkamp, L. W., Hughes, T., Liu, C. X., Qi, X. L., and Zhang, S. C. (2008). *J. Phys. Soc. Jpn.*, **77**, 031007.

[45] K'onig, M., Wiedmann, S., Brune, C., Roth, A., Buhmann, H., Molenkamp, L. W., Qi, X. L., and Zhang, S. C. (2007). *Science*, **318**, 766.

[46] Kuroda, K., Ye, M., Kimura, A., Eremeev, S. V., Krasovskii, E. E., Chulkov, E. V., Ueda, Y., Miyamoto, K., Okuda, T., Shimada, K., Namatame, H., and Taniguchi, M. (2010). *Phys. Rev. Lett.*, **105**, 146801.

[47] Law, K. T., Lee, P. A., and Ng, T. K. (2009). *Phys. Rev. Lett.*, **103**, 237001.

[48] Lee, S. S. and Ryu, S. (2008). *Phys. Rev. Lett.*, **100**, 186807.

[49] Li, J., *et al.* (2009). *Phys. Rev. Lett.*, **102**, 136806.

[50] Lin, H., Markiewicz, R. S., Wray, L. A., Fu, L., Hasan, M. Z., and Bansil, A. (2010). *Phys. Rev. Lett.*, **105**, 036404.

[51] Lin, H., *et al.* (2010). *Nature Mat.*, **9**, 546.

[52] Linder, J., Yokoyama, T., and Sudbo, A. (2009). *Phys. Rev. B*, **80**, 205401.

[53] Linder, J., Tanaka, Y., Yokoyama, T., Sudbo, A., and Nagaosa, N. (2010). *Phys. Rev. Lett.*, **104**, 067001.

[54] Linder, J., Tanaka, Y., Yokoyama, T., Sudbo, A., and Nagaosa, N. (2010). *Phys. Rev. B*, **81**, 184525.

[55] Liu, C. X., *et al.* (2008). *Phys. Rev. Lett.*, **100** 236601.

[56] Liu, C.-X., *et al.* (2010). *Phys. Rev. B*, **81**, 041307.

[57] Lu, H.-Z., *et al.* (2010). *Phys. Rev. B*, **81**, 115407.

[58] Mizushima, T., Ichioka, M., and Machida, K. (2008). *Phys. Rev. Lett.*, **101**, 150409.

[59] Mondal, S., Sen, D., Sengupta, K., and Shankar, R. (2010). *Phys. Rev. B*, **82**, 045120.

[60] Moore, J. E. and Balents, L. (2007). *Phys. Rev. B*, **75**, 121306.

[61] Murakami, S. (2006). *Phys. Rev. Lett.*, **97**, 236805.

[62] Murakami, S. (2007). *New J. Phys.*, **9**, 356.

[63] Murakami, S. and Kuga, S. (2008). *Phys. Rev. B*, **78**, 165313.

[64] Murakami, S., Nagaosa, N., and Zhang, S.-C. (2003). *Science*, **301**, 1348.

[65] Murakami, S., *et al.* (2007). *Phys. Rev. B*, **76**, 205304.

[66] Nayak, C., Simon, S. H., Stern, A., Freedman, M., and Das Sarma, S. (2008). *Rev. Mod. Phys.*, **80**, 1083.

[67] Niemi, A. J. and Semenoff, G. W. (1986). *Phys. Rep.*, **135**, 99.

[68] Nilsson, J., Akhmerov, A. R., and Beenakker, C. W. J. (2008). *Phys. Rev. Lett.*, **101**, 120403.

[69] Nishide, A., Taskin, A. A., Takeichi, Y., Okuda, T., Kakizaki, A., Hirahara, T., Nakatsuji, K., Komori, F., Ando, Y., and Matsuda, I. (2010). *Phys. Rev. B*, **81**, 041309.

[70] Nomura, K., Koshino, M., and Ryu, S. (2007). *Phys. Rev. Lett.*, **99**, 146806.

[71] Ostrovsky, P. M., Gornyi, I. V., and Mirlin, A. D. (2007). *Phys. Rev. Lett.*, **98**, 256801.

[72] Pesin, D. and Balents, L. (2010). *Nature Phys.*, **6**, 376.

[73] Qi, X. L. and Zhang, S. C. (2008). *Phys. Rev. Lett.*, **101**, 086802.

[74] Qi, X.-L. and Zhang, S.-C. *arXiv*, 1008.2026.

[75] Qi, X. L., Hughes, T. L., and Zhang, S. C. (2008). *Nature Phys.*, **4**, 273.

[76] Qi, X. L., Hughes, T. L., and Zhang, S. C. (2008). *Phys. Rev. B*, **78**, 195424.

[77] Qi, X. L., Li, R. D., Zang, J. D., and Zhang, S. C. (2009). *Science*, **323**, 1184.

[78] Qi, X. L., Hughes, T. L., Raghu, S., and Zhang, S. C. (2009). *Phys. Rev. Lett.*, **102**, 187001.

[79] Raghu, S., *et al.* (2008). *Phys. Rev. Lett.*, **100**, 156401.

[80] Ran, Y., Zhang, Y., and Vishwanath, A. (2009). *Nature Phys.*, **5**, 298.

[81] Ran, Y., Vishwanath, A., and Lee, D. H. (2008). *Phys. Rev. Lett.*, **101**, 086801.

[82] Read, N. and Green, D. (2000). *Phys. Rev. B*, **61**, 10267.

[83] Roth, A., Brüne, C., Buhmann, H., Molenkamp, L. W., Maciejko J., Qi, X. L., and Zhang, S. C. (2009). *Science*, **325**, 294.

[84] Roushan, P., *et al.* (2009). *Nature*, **460**, 1106.

[85] Roy, R. (2009). *Phys. Rev. B*, **79**, 195322.

[86] Ryu, S., Mudry, C., Obuse, H., and Furusaki, A. (2007). *Phys. Rev. Lett.*, **99**, 116601.

[87] Sakamoto, Y., *et al.* (2010). *Phys. Rev. B*, **81**, 165432.

[88] Sato, M. (2003). *Phys. Lett. B*, **575**, 126.

[89] Sato, M. (2010). *Phys. Rev. B*, **81**, 220504(R).

[90] Sato, M. (2006). *Phys. Rev. B*, **73**, 214502.

[91] Sato, M. and Fujimoto, S. (2010). *Phys. Rev. Lett.*, **105**, 217001.

[92] Sato, M., Takahashi, Y., and Fujimoto, S. (2009). *Phys. Rev. Lett.*, **103**, 020401.

[93] Sato, T., Segawa, K., Guo, H., Sugawara, K., Souma, S., Takahashi, T., and Ando, Y. (2010). *Phys. Rev. Lett.*, **105**, 136802.

[94] Sau, J. D., Lutchyn, R. M., Tewari, S., and Das Sarma, S. (2010). *Phys. Rev. Lett.*, **104**, 040502.

[95] Schnyder, A. P., *et al.* (2008). *Phys. Rev. B*, **78**, 195125.

[96] Schnyder, A. P., Ryu, S., and Ludwig, A. W. W. (2009). *Phys. Rev. Lett.*, **102**, 196804.

[97] Schnyder, A. P., Ryu, S., Furusaki, A., and Ludwig, A. W. W. (2008). *Phys. Rev. B*, **78**, 195125.

[98] Shindou, R. and Murakami, S. (2009). *Phys. Rev. B*, **79**, 045321.

[99] Shitade, A., *et al.* (2009). *Phys. Rev. Lett.*, **102**, 256403.

[100] Sinova, J., Culcer, D., Niu, Q., Sinitsyn, N. A., Jungwirth, T., and MacDonald, A. H. (2004). *Phys. Rev. Lett.*, **92**, 126603.

[101] Strom, A. and Johannesson, H. (2009). *Phys. Rev. Lett.*, **102**, 096806.

[102] Sushkov, A. B., *et al. arXiv*, 1006.1008.

[103] Takahashi, R. and Murakami, S. (2010). *Phys. Rev. B*, **81**, 161302(R).

[104] Tanaka, Y. and Nagaosa, N. (2009). *Phys. Rev. Lett.*, **103**, 166403.

[105] Tanaka, Y., Yokoyama, T., and Nagaosa, N. (2009). *Phys. Rev. Lett.*, **103**, 107002.

[106] Taskin, A. A. and Ando, Y. (2009). *Phys. Rev. B*, **80**, 085303.

[107] Taskin, A. A., Segawa, K., and Ando, Y. (2010). *Phys. Rev. B*, **82**, 121302.

[108] Teo, J. C. Y. and Kane, C. L. (2010). *Phys. Rev. Lett.*, **104**, 046401.

[109] Teo, J. C. Y. and Kane, C. L. (2009). *Phys. Rev. B*, **79**, 235321.

[110] Teo, J. C. Y., Fu, L., and Kane, C. L. (2008). *Phys. Rev. B*, **78**, 045426.

[111] Tretiakov, O. A., Abanov, Ar., Murakami, S., and Sinova, J. (2010). *Appl. Phys. Lett.*, **97**, 073108.

[112] Tsutsumi, Y., *et al.* (2008). *Phys. Rev. Lett.*, **101**, 135302.

[113] Volovik, G. E. *The Universe in a Helium Droplet* (Oxford University Press, New York, 2003).

[114] Volovik, G. E. (2007). *Lect. Notes Phys.*, **718**, 31.

[115] Wada, M., Murakami, S., Freimuth, F., and Bihlmayer, G. *arXiv*, 1005.3912.

[116] Wen, X. G. and Zee, A. (2002). *Phys. Rev. B*, **66**, 235110.

[117] Wilczek, F. (2009). *Nature Phys.*, **5**, 614.

[118] Wray, L. A., Xu, S-Y., Xia, Y., Hor, Y. S., Qian, D., Fedorov, A. V., Lin, H., Bansil, A., Cava, R. J., and Hasan, M. Z. (2010). *Nature Phys.*, **6**, 855.

[119] Wu, C. J., Bernevig, B. A., and Zhang, S. C. (2006). *Phys. Rev. Lett.*, **96**, 106401.

[120] Xia, Y., Qian, D., Hsieh, D., Wray, L., Pal, A., Lin, H., Bansil, A., Grauer, D., Hor, Y. S., Cava, R. J., and Hasan, M. Z. (2009). *Nature Phys.*, **5**, 398.

[121] Xu, C. and Moore, J. E. (2006). *Phys. Rev. B*, **73**, 045322.

[122] Yamashiro, M., Tanaka, Y., and Kashiwaya, S. (1997). *Phys. Rev. B*, **56**, 7847.

[123] Yokoyama, T., Tanaka, Y., and Nagaosa, N. (2009). *Phys. Rev. Lett.*, **102**, 166801.

[124] Zhang, H. J., Liu, C. X., Qi X. L., Dai, X., Fang, Z., and Zhang, S. C. (2009). *Nature Phys.*, **5**, 438.

[125] Zhang, T., *et al.* (2009). *Phys. Rev. Lett.*, **103**, 266803.

[126] Zhang, Y., *et al.* (2010). *Nature Phys.*, **6**, 584.

[127] Zhang, Y., Ran, Y., and Vishwanath, A. (2009) *Phys. Rev. B*, **79**, 245331.

18 Spin Seebeck effect

E. Saitoh and K. Uchida

18.1 Introduction

The Seebeck effect, discovered by T. J. Seebeck in the 1820s, is the generation of an electric voltage as a result of a temperature gradient (see Fig. 18.1(a)) [1, 2]. In contrast, the spin Seebeck effect (SSE) stands for the generation of "spin voltage" as a result of a temperature gradient [3–15]. Spin voltage refers to the potential for electron spins, which drives a spin current [16–19], a flow of spin-angular momentum; when a conductor is attached to part of a magnet with

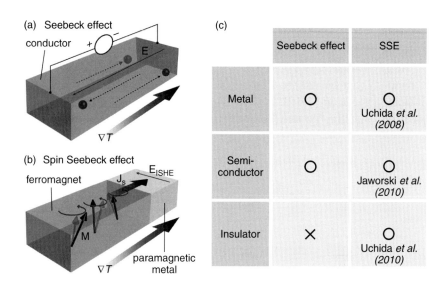

FIG. 18.1. (a) A schematic illustration of the conventional Seebeck effect. When a temperature gradient ∇T is applied to a conductor, an electric field $\mathbf{E}$ (electric voltage V) is generated along the ∇T direction. (b) A schematic illustration of the spin Seebeck effect (SSE). When ∇T is applied to a ferromagnet, a spin voltage is generated via magnetization ($\mathbf{M}$) dynamics, which pumps a spin current $\mathbf{J}_s$ into an attached paramagnetic metal. In the paramagnetic metal, this spin current is converted into an electric field $\mathbf{E}_{\mathrm{ISHE}}$ due to the inverse spin Hall effect (ISHE). (c) Difference between the Seebeck effect and the SSE. The SSE appears even in insulators.

a finite spin voltage, the spin voltage injects a spin current into the conductor (see Fig. 18.1(b)). The SSE is of crucial importance in spintronics [20–22] and spin caloritronics [23–28], since it enables simple and versatile generation of spin currents from heat. In 2008, Uchida *et al.* originally discovered the SSE in a ferromagnetic metal $Ni_{81}Fe_{19}$ film [3] by means of the spin-detection technique based on the inverse spin Hall effect [29–34] (ISHE) in a Pt film. In 2010, using the same experimental method, Jaworski *et al.* also observed this phenomenon in a ferromagnetic semiconductor GaMnAs in a low-temperature region [7, 13] and Uchida *et al.* revealed that the SSE appears even in magnetic insulators, such as $Y_3Fe_5O_{12}$ (YIG) [8], $LaY_2Fe_5O_{12}$ (La:YIG) [6], and $(Mn,Zn)Fe_2O_4$ [10]. The SSE was observed also in a half-metallic Heusler compound Co_2MnSi [14]. These observations indicate that the SSE is a universal phenomenon in magnetic materials.

The discovery of the SSE in magnetic insulators provides a crucial piece of information for understanding the physics of the SSE. The conventional Seebeck effect requires itinerant charge carriers, or conduction electrons, and therefore exists only in metals and semiconductors (see Fig. 18.1(c)). It appeared natural to assume that the same held for the SSE. In fact, originally the SSE was phenomenologically formulated in terms of thermal excitation of conduction electrons [3]. However, the observation of the SSE in insulators upsets this conventional interpretation; conduction electrons are not necessary for the SSE. This is direct evidence that the spin voltage generated by the SSE is associated with magnetic dynamics (see Fig. 18.1(b)). Based on this idea, various theoretical models have been proposed [5, 9, 11, 12].

18.2 Sample configuration and measurement mechanism

The observation of the SSE exploited the ISHE in two different device structures: one is a *longitudinal* configuration [8, 10], in which a spin current *parallel* to a temperature gradient is measured. The longitudinal configuration consists of the simplest and most straightforward structure, enabling the systematic and quantitative investigation of the SSE, although the longitudinal configuration is applicable only to insulators, as mentioned below. The other setup is a *transverse* configuration [3, 4, 6, 7, 13, 14], in which a spin current flowing *perpendicular* to a temperature gradient is measured. The transverse configuration has a more complicated device structure (hence is difficult to measure) than the longitudinal one, but it has been used for measuring the SSE in various materials ranging from ferromagnetic metals and semiconductors to insulators. In fact, the first observation of the SSE in $Ni_{81}Fe_{19}$ films was reported in the transverse configuration [3].

Figure 18.2(a) shows a schematic illustration of the longitudinal SSE device. The device structure is very simple, which consists of a ferromagnetic (F) slab or film covered with a paramagnetic metal (PM) film. When a temperature gradient ∇T is applied to the F layer in the direction perpendicular to the F/PM interface

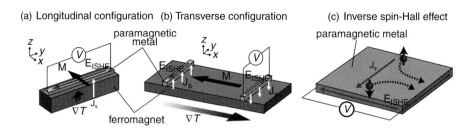

FIG. 18.2. (a), (b) Schematic illustrations of the longitudinal and transverse configurations for measuring the SSE. (c) A schematic illustration of the ISHE.

(z-direction), a spin voltage is thermally generated and injects a spin current with the spatial direction $\mathbf{J}_s$ and the spin-polarization vector $\boldsymbol{\sigma}$, parallel to the magnetization $\mathbf{M}$ of F, into the PM film along the ∇T direction (see Fig. 18.2(a)). In the PM wire, this injected spin current is converted into an electric field $\mathbf{E}_{\text{ISHE}}$ due to the ISHE. When $\mathbf{M}$ is along the x-direction, $\mathbf{E}_{\text{ISHE}}$ is generated in the PM along the y-direction because of the relation [29]

$$\mathbf{E}_{\text{ISHE}} = \frac{\theta_{\text{SH}}\rho}{A}\left(\frac{2e}{\hbar}\right)\mathbf{J}_s \times \boldsymbol{\sigma}, \qquad (18.1)$$

where θ_{SH}, ρ, and A are the spin Hall angle of PM, the electric resistivity of PM, and the contact area between F and PM, respectively (see Fig. 18.2(c)). In this equation, the dimensions of $\mathbf{J}_s$ are kg·m^2·s^{-2}; $\mathbf{J}_s$ is expressed as the flow of $\hbar/2$ per unit time. θ_{SH} is especially large in noble metals with strong spin–orbit interaction, such as Pt. Therefore, by measuring $\mathbf{E}_{\text{ISHE}}$ in the PM film, one can detect the longitudinal SSE electrically. Here we note that, if ferromagnetic metals were used as the F layer, the ISHE signal not only was suppressed by short-circuit currents [34] in the F layer due to the electric conduction of F but also was contaminated by the signal of the anomalous Nernst–Ettingshausen effect [35] in F. By using a spin Seebeck insulator such as YIG, these artifacts are eliminated.

Figure 18.2(b) shows a schematic illustration of the transverse SSE device, which consists of a rectangular-shaped F with one or several PM wires attached to the top surface of F. The typical length of the F layer along the x-direction is ~ 10 mm, much longer than the conventional spin-diffusion length [36]. In the transverse configuration, to generate the ISHE voltage induced by the SSE along the PM-wire direction, the F layer has to be magnetized along the ∇T direction (see Eq. (18.1) and Fig. 18.2(b) and note that $\mathbf{J}_s$ is parallel to the z-direction also in the transverse configuration). Therefore, the anomalous Nernst–Ettingshausen effect in the F layer vanishes due to the collinear orientation of ∇T and $\mathbf{M}$, enabling the pure detection of the transverse SSE in various magnetic materials.

18.3   Detection of spin Seebeck effect using inverse spin Hall effect

Now, we show some experimental results on the longitudinal SSE in ferrimagnetic insulator/Pt samples [8]. The sample consists of a single-crystalline YIG slab and a Pt film attached to a well-polished YIG (100) surface. The length, width, and thickness of the YIG slab (Pt film) are 6 mm (6 mm), 2 mm (0.5 mm), and 1 mm (15 nm), respectively. An external magnetic field $\mathbf{H}$ (with magnitude H) was applied in the x-y plane at an angle θ to the y-direction (see Fig. 18.3(a)). A temperature difference ΔT was applied between the top and bottom surfaces of the YIG/Pt sample. To detect $\mathbf{E}_{\text{ISHE}}$ generated by the longitudinal SSE, an electric voltage difference V between the ends of the Pt layer was measured.

Figure 18.3(b) shows V in the YIG/Pt sample as a function of ΔT at $H = 1\,\text{kOe}$. When $\mathbf{H}$ is applied along the x-direction ($\theta = 90°$), the magnitude of V is observed to be proportional to ΔT. The sign of the V signal at finite values of ΔT is clearly reversed by reversing the ∇T direction. In this setup, since YIG is an insulator, the complicating thermoelectric phenomena in itinerant magnets, such as the conventional Seebeck and Nernst–Ettingshausen effects, do not exist at all. As also shown in Fig. 18.3(b), the V signal disappears when $\mathbf{H}$ is along

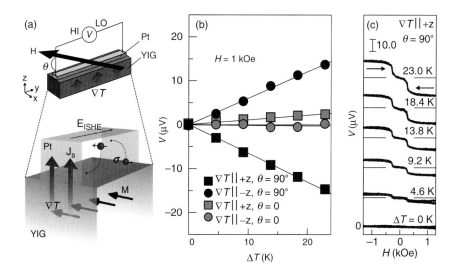

FIG. 18.3. (a) A schematic illustration of the longitudinal SSE and the ISHE in the YIG/Pt sample. (b) ΔT dependence of V in the YIG/Pt sample at $H = 1\,\text{kOe}$, measured when ∇T was applied along the $+z$ or $-z$ direction. The magnetic field $\mathbf{H}$ was applied along the x-direction ($\theta = 90°$) or the y-direction ($\theta = 0$). (c) H dependence of V in the YIG/Pt sample for various values of ΔT at $\theta = 90°$, measured when ∇T was along the $+z$ direction. All the data shown in this figure were measured at room temperature.

the y-direction ($\theta = 0$), a situation consistent with the symmetry of the ISHE induced by the longitudinal SSE (see Eq. (18.1) and Fig. 18.3(a)).

To confirm the origin of this signal, we measured the H dependence of V in the same YIG/Pt system. We found that the sign of V is reversed by reversing **H** when $\theta = 90°$ and $|H| > 500\,$Oe, indicating that the V signal is affected by the magnetization direction of the YIG layer (see Fig. 18.3(c)). We checked that the V signal disappears when the Pt layer is replaced with paramagnetic metal films with weak spin–orbit interaction, such as Cu. This behavior also supports the aforementioned prediction of the longitudinal SSE.

As shown in Ref. [10], the longitudinal SSE was observed even in sintered polycrystalline insulating magnets, such as $(Mn,Zn)Fe_2O_4$. This robustness of the longitudinal configuration enables the investigation of the SSE in a wide range of materials.

Next, we show some data on the transverse SSE in ferrimagnetic insulator/Pt samples [6]. The sample consists of a La-doped YIG (La:YIG) with Pt wires attached to the surface (see Fig. 18.4(a)). The single-crystalline La:YIG (111) film with thickness 3.9 μm was grown on a $Gd_3Ga_5O_{12}$ (GGG) (111) substrate by liquid phase epitaxy. Here, the surface of the La:YIG layer has an 8×4 mm^2 rectangular shape. Two 15-nm-thick Pt wires were then sputtered in an Ar atmosphere on the top of the La:YIG film. The length and width of the Pt wires are 4 mm and 0.1 mm, respectively. **H** and ∇T were applied along the x-direction (see Fig. 18.4(a)).

Figures 18.4(b) shows V between the ends of the Pt wires placed near the lower- and higher-temperature ends of the La:YIG layer as a function of ΔT at $H = 100\,$Oe. The magnitude of V is proportional to ΔT in both the Pt wires. Notably, the sign of V for finite values of ΔT is clearly reversed between the lower- and higher-temperature ends of the sample. This sign reversal of V is characteristic behavior of the ISHE voltage induced by the transverse SSE [3, 4, 6, 7, 13, 14].

As shown in Fig. 18.4(c), the sign of V at each end of the sample is reversed by reversing H. We also confirmed that the V signal vanishes when **H** is applied along the y-direction, a situation consistent with Eq. (18.1). This V signal disappears when the Pt wires are replaced by Cu wires with weak spin–orbit interaction. We checked that the signal also disappears in a $La:YIG/SiO_2/Pt$ system, where the La:YIG and Pt layers are separated by a thin (10 nm) film of insulating SiO_2, as well as in a GGG/Pt system, where the La:YIG layer is replaced by GGG, indicating that the direct contact between La:YIG and Pt is essential for the generation of the V signal. An extrinsic proximity effect or induced ferromagnetism in the Pt layers is also irrelevant because of the sign change of V between the ends of the La:YIG/Pt sample. All the results shown above confirm that the V signal observed here is due entirely to the transverse SSE in the La:YIG/Pt samples. Similar ISHE signals induced by the transverse SSE were observed also in ferromagnetic metal $Ni_{81}Fe_{19}/Pt$ systems [3, 4] and in semiconducting GaMnAs/Pt systems [7, 13].

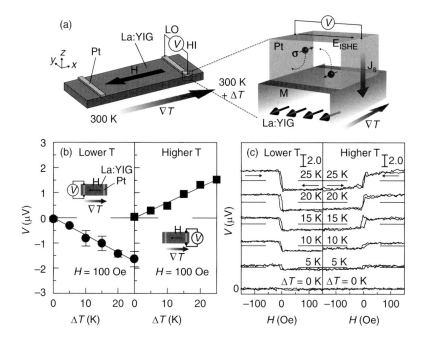

FIG. 18.4. (a) A schematic illustration of the transverse SSE and the ISHE in the La:YIG/Pt sample. (b) ΔT dependence of V in the La:YIG/Pt sample at $H = 100$ Oe, measured when the Pt wires were attached near the lower-temperature (300 K) and higher-temperature (300 K$+\Delta T$) ends of the La:YIG layer. (c) H dependence of V in the La:YIG/Pt sample for various values of ΔT.

18.4 Theoretical concept of spin Seebeck effect

We now present a qualitative picture of the physics underlying the SSE. Since the SSE appears even in magnetic insulators, it cannot fully be expressed in terms of thermal excitation of conduction electrons. The SSE in insulators cannot also be explained by equilibrium spin pumping [19, 29, 33, 34, 37–39], since the average spin-pumping current from thermally fluctuating magnetic moments is exactly canceled by the thermal (Johnson-Nyquist) spin-current noise. Therefore, the observed spin voltage requires us to introduce an unconventional nonequilibrium state between magnetic moments in F and electrons in an attached PM. The microscopic mechanism of the SSE was proposed both by Xiao *et al.* [5] and by Adachi *et al.* [9, 11, 15] by means of scattering and linear-response theories, respectively. In this section, we review the basic concepts of their calculations.

The thermally excited state in the SSEs at an F/PM interface can be described in terms of the effective magnon temperature T_m^* in F and the electron temperature T_e^* in PM, which can be different from the local thermodynamic temperature at a position. The ISHE signal derived from the net spin current is proportional to $T_{\mathrm{m}}^* - T_{\mathrm{e}}^*$, as shown below. The effective temperatures are related to thermal fluctuations through the fluctuation-dissipation theorem. The fluctuations of the magnetization $\mathbf{m}$ at the F/PM interface are excited by a random magnetic field $\mathbf{h} = \Sigma_j \mathbf{h}^{(j)}$ ($j = 0, 1, \ldots$), which represents the thermal disturbance from various sources (such as lattice, contacts, etc.). If we respectively denote $T^{(j)}$ and $\alpha^{(j)}$ as the temperature and the damping of dissipative source j, due to the fluctuation–dissipation theorem, the random field $\mathbf{h}$ satisfies the following equal-position time-correlation function:

$$\left\langle h_i^{(j)}(t)\, h_{i'}^{(j')}(t') \right\rangle = \left(\frac{2k_{\mathrm{B}} T^{(j)} \alpha^{(j)}}{\gamma M_{\mathrm{s}} V_{\mathrm{a}}} \right) \delta_{jj'} \delta_{ii'} \delta(t - t'), \qquad (18.2)$$

where k_{B} is the Boltzmann constant, γ is the gyromagnetic ratio, M_{s} is the saturation magnetization, $\alpha = \alpha^{(0)} + \alpha^{(1)} + \cdots$ is the effective damping parameter, $T_{\mathrm{m}}^* = [\alpha^{(0)} T^{(0)} + \alpha^{(1)} T^{(1)} + \cdots]/\alpha$ is the effective magnon temperature, and V_{a} is the magnetic coherence volume in F, which depends on the magnon temperature and the spin-wave stiffness constant D. Assuming the dissipative sources 0 and 1 are the F lattice and the PM contact respectively, then $T^{(0)}$, $\alpha^{(0)}$, $T^{(1)}$, and $\alpha^{(1)} = \gamma \hbar g_{\mathrm{r}} / 4\pi M_{\mathrm{s}} V_{\mathrm{a}}$ represent the bulk lattice temperature, the bulk Gilbert damping parameter, the electron temperature in the PM contact ($T^{(1)} = T_{\mathrm{e}}^*$), and the damping enhancement due to spin pumping with g_{r} being the real part of the mixing conductance for the F/PM interface. The net thermal spin current across the F/PM interface is given by the sum of a fluctuating thermal spin-pumping current $\mathbf{J}_{\mathrm{sp}}$ from F to PM proportional to T_{m}^* and a Johnson-Nyquist spin-current noise $\mathbf{J}_{\mathrm{fl}}$ from PM to F proportional to T_{e}^* [5, 40, 41]:

$$\mathbf{J}_{\mathrm{s}} = \mathbf{J}_{\mathrm{sp}} + \mathbf{J}_{\mathrm{fl}} = \frac{M_{\mathrm{s}} V_{\mathrm{a}}}{\gamma} \left[\alpha^{(1)} \mathbf{m} \times \dot{\mathbf{m}} + \gamma \mathbf{m} \times \mathbf{h}^{(1)} \right]. \qquad (18.3)$$

The dc component along the magnetization equilibrium direction (x-direction) reduces to

$$J_{\mathrm{s}} \equiv \langle \mathbf{J}_{\mathrm{s}} \rangle_x = 2\alpha^{(1)} k_{\mathrm{B}} (T_{\mathrm{m}}^* - T_{\mathrm{e}}^*). \qquad (18.4)$$

Therefore, when $T_{\mathrm{m}}^* > T_{\mathrm{e}}^*$ ($T_{\mathrm{m}}^* < T_{\mathrm{e}}^*$), spin currents are injected from F (PM) into PM (F) (see Figs. 18.5(b) and 18.5(c)). At an equilibrium state ($T_{\mathrm{m}}^* = T_{\mathrm{e}}^*$), no spin current is generated at the F/PM interface since the spin-pumping current, $\mathbf{J}_{\mathrm{sp}}$, is canceled out by the spin-current noise, $\mathbf{J}_{\mathrm{fl}}$ (see Fig. 18.5(a)). Equations (18.1) and (18.4) indicate that the magnitude and sign of the ISHE voltage induced by the SSE are determined by those of $T_{\mathrm{m}}^* - T_{\mathrm{e}}^*$ in the F/PM system under a temperature gradient. As demonstrated by the previous studies, this effective temperature difference $T_{\mathrm{m}}^* - T_{\mathrm{e}}^*$ is induced by the magnon-mediated [5, 6, 11, 12] and the phonon-mediated [9, 13, 15] processes; when nonequilibrium magnons and phonons travel around F, feeling a temperature gradient, T_{m}^* in F and/or T_{e}^* in PM are modulated by interacting with the traveling magnons and

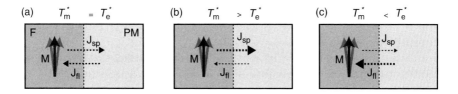

F$_{IG}$. 18.5. Mechanism of spin-current generation by the SSE at the F/PM interface.

phonons, respectively. Due to these processes, magnons in F and/or electrons in PM in a lower-temperature (higher-temperature) region feel the temperature information in a higher-temperature (lower-temperature) region. Therefore, the resultant T_m^* and/or T_e^* in the lower-temperature (higher-temperature) region increases (decreases).

An important clue to calculating the effective temperature distribution was provided by Sanders and Walton in 1977 [42]. They discussed the effective magnon-temperature (T_m^*) and phonon-temperature (T_p^*) distributions in a magnetic insulator, especially YIG, placed in a uniform temperature gradient and solved a simple heat-rate equation of the coupled magnon–phonon system under a situation similar to the transverse SSE configuration. The solution of the heat-rate equation yields a hyperbolic-sine $T_m^* - T_p^*$ profile with a decay length λ_m. In a magnetic insulator with weak magnetic damping, such as YIG and La:YIG, rapid equilibration of magnons is prevented and the resultant λ_m was shown to reach several milimeters. Making use of these results and assuming that T_p^* in F is equal to T_e^* in an attached PM, Xiao et al. formulated the magnon-mediated SSE and their calculation quantitatively reproduces the magnitude and spatial distribution of the SSE-induced ISHE signal observed in the transverse La:YIG/Pt system [5, 6]. The magnon-mediated SSE was formulated also by means of a many-body technique using nonequilibrium Green's functions by Adachi et al. [11] and numerical calculation based on a stochastic Landau–Lifshitz–Gilbert equation by Ohe et al. [12].

The temperature-dependent behavior of the SSE [9, 13] and the spatial distribution of the thermally generated spin voltage in ferromagnetic metals [3] may be explained by the phonon-based mechanism [9, 15], where the effective temperature difference $T_m^* - T_e^*$ is generated by nonequilibrium phonons. Phonon-mediated thermal spin pumping was formulated by Adachi et al. using a linear-response theory.

18.5 Thermal spin injection by spin-dependent Seebeck effect

Up to now, we have discussed the SSE: the generation of spin voltage from a temperature gradient in a macroscopic scale due to spin dynamics. Recently, spin currents were found to be able to be generated thermally also via conduction-electron' spin accumulation [26, 28]. Since the spin accumulation disappears within a spin-diffusion-length scale [36] from boundaries, this effect occurs only in nanoscale devices.

In ferromagnetic metals (FMs) and near FM/PM interfaces, the conduction-electron transport in a nonequilibrium state can be argued in terms of a spin-dependent conductivity $\sigma_{\uparrow,\downarrow}$ and a spin-dependent Seebeck coefficient $S_{\uparrow,\downarrow}$. A spin-dependent electric current density is described as

$$\mathbf{j}_{\uparrow,\downarrow} = -\sigma_{\uparrow,\downarrow} \left(\frac{1}{e} \nabla \mu_{\uparrow,\downarrow} + S_{\uparrow,\downarrow} \nabla T \right), \tag{18.5}$$

where e is the electron charge and $\mu_{\uparrow,\downarrow}$ is the spin-dependent electrochemical potential. Therefore, when a temperature difference is applied between FM and an attached PM, the conduction-electron spin current $\mathbf{j}_{\uparrow} - \mathbf{j}_{\downarrow}$ with accompanying spin accumulation $\mu_{\uparrow} - \mu_{\downarrow}$ is generated near the FM/PM interface.

To demonstrate thermal spin injection by the spin-dependent Seebeck coefficient $S_{\uparrow,\downarrow}$, Slachter $et~al.$ proposed a sample structure schematically illustrated in Fig. 18.6(a) [26]. The sample has a lateral spin-valve structure [17] comprising two ferromagnetic metals (FM_A and FM_B) and a PM wire. A temperature difference is generated between FM_A and PM via Joule heating by applying

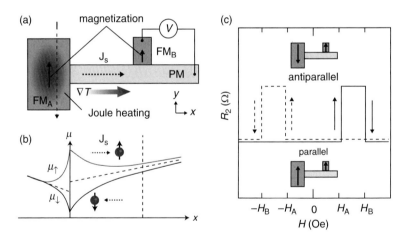

FIG. 18.6. (a) A schematic illustration of the lateral spin-valve structure for measuring the thermal spin injection by the spin-dependent Seebeck effect [26]. (b) Distribution of the thermally induced spin accumulation in the lateral spin-valve structure. (c) A schematic illustration of the magnetic-field H dependence of R_2, where R_2 is defined as $V = R_1 I + R_2 I^2 + \cdots$ with I and V being the charge current applied to FM_A and the electric voltage difference between FM_B and PM (see (a)). $H_A(H_B)$ denotes the coercive force of $FM_A(FM_B)$.

a charge current to FM_A (see Fig. 18.6(a)). When the interval between FM_A and FM_B is shorter than the spin-diffusion length of PM, the spin accumulation $\mu_\uparrow - \mu_\downarrow$ thermally generated at the FM_A/PM interface subsists at the FM_B/PM interface (see Fig. 18.6(b)). In this condition, an electric voltage V between FM_B and PM was shown to be dependent on whether the alignment of magnetizations of FM_A and FM_B is parallel or antiparallel, as depicted in Fig. 18.6(c), due to the thermal spin current injected into FM_B. This sample structure enables the pure detection of the thermally induced spin signals since no charge current flows in PM (see Fig. 18.6(a)) [26].

Thermal spin-current injection was also demonstrated in a ferromagnetic metal/oxide/silicon tunnel junctions by Le Breton *et al.* [28]. Here, spin currents injected into a silicon through an insulating tunnel barrier were detected by means of a spin-detection technique based on the Hanle effects. The observed phenomenon allows spin-current injection into silicon simply by applying a temperature gradient between the silicon and the attached ferromagnetic metal. Therefore, it may form a bridge between spin caloritronics and silicon-based semiconductor electronics.

18.6 Conclusion

In Sections 18.1–18.4, we briefly reviewed the experimental results and the theoretical models of the spin Seebeck effect (SSE). As examples, we showed the observations of the SSE in magnetic insulator/Pt samples, which were realized by the inverse spin Hall effect in Pt films. The experiments on magnetic insulators indicate that the SSE is conceptually and practically different from other thermo-spin effects based on conduction-electron spin transport (shown in Section 18.5); the SSE is attributed to thermal nonequilibrium between magnons in a ferromagnet and electrons in an attached paramagnetic metal, which can be described by *effective* magnon and electron temperatures. Since thermo-spin effects including the SSE enable simple and versatile spin-current generation from heat, it will be useful in basic spintronics research and spin-based device application.

References

[1] T. J. Seebeck, Repts. Prussian Acad. Sci. (1823).

[2] N. W. Ashcroft and N. D. Mermin, *Solid State Physics* (Saunders College, Philadelphia, 1976).

[3] K. Uchida, S. Takahashi, K. Harii, J. Ieda, W. Koshibae, K. Ando, S. Maekawa, and E. Saitoh, Nature **455**, 778 (2008).

[4] K. Uchida, T. Ota, K. Harii, S. Takahashi, S. Maekawa, Y. Fujikawa, and E. Saitoh, Solid State Commun. **150**, 524 (2010).

[5] J. Xiao, G. E. W. Bauer, K. Uchida, E. Saitoh, and S. Maekawa, Phys. Rev. B **81**, 214418 (2010).

[6] K. Uchida, J. Xiao, H. Adachi, J. Ohe, S. Takahashi, J. Ieda, T. Ota, Y. Kajiwara, H. Umezawa, H. Kawai, G. E. W. Bauer, S. Maekawa, and E. Saitoh, Nature Mater. **9**, 894 (2010).

[7] C. M. Jaworski, J. Yang, S. Mack, D. D. Awschalom, J. P. Heremans, and R. C. Myers, Nature Mater. **9**, 898 (2010).

[8] K. Uchida, H. Adachi, T. Ota, H. Nakayama, S. Maekawa, and E. Saitoh, Appl. Phys. Lett. **97**, 172505 (2010).

[9] H. Adachi, K. Uchida, E. Saitoh, J. Ohe, S. Takahashi, and S. Maekawa, Appl. Phys. Lett. **97**, 252506 (2010).

[10] K. Uchida, T. Nonaka, T. Ota, and E. Saitoh, Appl. Phys. Lett. **97**, 262504 (2010).

[11] H. Adachi, J. Ohe, S. Takahashi, and S. Maekawa, Phys. Rev. B **83**, 094410 (2011).

[12] J. Ohe, H. Adachi, S. Takahashi, and S. Maekawa, Phys. Rev. B **83**, 115118 (2011).

[13] C. M. Jaworski, J. Yang, S. Mack, D. D. Awschalom, R. C. Myers, and J. P. Heremans, Phys. Rev. Lett. **106**, 186601 (2011).

[14] S. Bosu, Y. Sakuraba, K. Uchida, K. Saito, T. Ota, E. Saitoh, and K. Takanashi, Phys. Rev. B **83**, 224401 (2011).

[15] K. Uchida, H. Adachi, T. An, T. Ota, M. Toda, B. Hillebrands, S. Maekawa, and E. Saitoh, Nature Mater. **10**, 737 (2011).

[16] J. C. Slonczewski, Phys. Rev. B **39**, 6995 (1989).

[17] F. J. Jedema, A. T. Filip, and B. J. van Wees, Nature **410**, 345 (2001).

[18] S. Takahashi and S. Maekawa, J. Phys. Soc. Jpn. **77**, 031009 (2008).

[19] Y. Kajiwara, K. Harii, S. Takahashi, J. Ohe, K. Uchida, M. Mizuguchi, H. Umezawa, H. Kawai, K. Ando, K. Takanashi, S. Maekawa, and E. Saitoh, Nature **464**, 262 (2010).

[20] S. A. Wolf, D. D. Awschalom, R. A. Buhrman, J. M. Daughton, S. von Molnar, M. L. Roukes, A. Y. Chtchelkanova, and D. M. Treger, Science **294**, 1488 (2001).

[21] I. Zutic, J. Fabian, and S. Das Sarma, Rev. Mod. Phys. **76**, 323 (2004).

[22] S. Maekawa, ed., *Concepts in Spin Electronics*, (Oxford University Press, Oxford, 2006).

[23] M. Johnson and R. H. Silsbee, Phys. Rev. B **35**, 4959 (1987).

[24] M. Hatami, G. E. W. Bauer, Q.-F. Zhang, and P. J. Kelly, Phys. Rev. Lett. **99**, 066603 (2007).

[25] G. E. W. Bauer, A. H. MacDonald, and S. Maekawa, eds., *Spin Caloritronics, Special Issue of Solid State Communications*, (Elsevier, Amsterdam, 2010).

[26] A. Slachter, F. L. Bakker, J.-P. Adam, and B. J. van Wees, Nature Phys. **6**, 879 (2010).

[27] F. L. Bakker, A. Slachter, J.-P. Adam, and B. J. van Wees, Phys. Rev. Lett. **105**, 136601 (2010).

[28] J.-C. Le Breton, S. Sharma, H. Saito, S. Yuasa, and R. Jansen, Nature **475**, 82 (2011).

[29] E. Saitoh, M. Ueda, H. Miyajima, and G. Tatara, Appl. Phys. Lett. **88**, 182509 (2006).

[30] S. O. Valenzuela and M. Tinkham, Nature **442**, 176 (2006).

[31] T. Kimura, Y. Otani, T. Sato, S. Takahashi, and S. Maekawa, Phys. Rev. Lett. **98**, 156601 (2007).

[32] T. Seki, Y. Hasegawa, S. Mitani, S. Takahashi, H. Imamura, S. Maekawa, J. Nitta, and K. Takanashi, Nature Mater. **7**, 125 (2008).

[33] O. Mosendz, J. E. Pearson, F. Y. Fradin, G. E. W. Bauer, S. D. Bader, and A. Hoffmann, Phys. Rev. Lett. **104**, 046601 (2010).

[34] K. Ando, S. Takahashi, J. Ieda, Y. Kajiwara, H. Nakayama, T. Yoshino, K. Harii, Y. Fujikawa, M. Matsuo, S. Maekawa, and E. Saitoh, J. Appl. Phys. **109**, 103913 (2011).

[35] H. B. Callen, Phys. Rev. **73**, 1349 (1948).

[36] J. Bass and W. P. Pratt Jr, J. Phys. Condens. Matter **19**, 183201 (2007).

[37] R. H. Silsbee, A. Janossy, and P. Monod, Phys. Rev. B **19**, 4382 (1979).

[38] Y. Tserkovnyak, A. Brataas, and G. E. W. Bauer, Phys. Rev. Lett. **88**, 117601 (2002).

[39] S. Mizukami, Y. Ando, and T. Miyazaki, Phys. Rev. B **66**, 104413 (2002).

[40] J. Foros, A. Brataas, Y. Tserkovnyak, and G. E. W. Bauer, Phys. Rev. Lett. **95**, 016601 (2005).

[41] J. Xiao, G. E. W. Bauer, S. Maekawa, and A. Brataas, Phys. Rev. B **79**, 174415 (2009).

[42] D. J. Sanders and D. Walton, Phys. Rev. B **15**, 1489 (1977).

Part III Spin-transfer torque

19 Introduction

T. Kimura

When a spin current enters a ferromagnet, a transfer of the spin angular momentum between the conduction electrons and the magnetization of the ferromagnet occurs because of the conservation of the spin angular momentum. This is known as a spin-transfer effect. The concept of the spin-transfer effect was introduced by L. Berger in 1984 [1]. Berger considered the exchange interaction between the conduction electron and the localized magnetic moment (*s-d* exchange interaction) and predicted that a magnetic domain wall can be moved by flowing the spin current. Unfortunately, at that time, since the microfabrication techniques required for the preparation of the ideal sample had not been developed, it was very difficult to distinguish such an effect from other spurious effects which are induced by the Oersted field and the Lorenz force. However, the spin-transfer effect was brought into the limelight by the progress in microfabrication techniques and the discovery of the giant magnetoresistance (GMR) effect in magnetic multilayers [2, 3]. In particular, after Slonczewski predicted that the magnetization can be reversed by the spin-transfer effect in a magnetic multilayered system [4], this phenomenon has attracted much attention for novel manipulation techniques of the magnetization. At the same time, Berger separately studied the spin-transfer torque in a similar system and predicted spontaneous magnetization precession [5]. After their proposals, the first experimental demonstration of spinwave excitation due to the spin-transfer torque was achieved by measuring a point contact magnetoresistance in a magnetic multilayer in 1998 [6], and in 1999 came the first experimental report of magnetization reversal due to spin-transfer torque achieved by using a Co/Cu/Co sandwich structure [8]. Thus, this innovation opens up a new paradigm for magneto-electronic device application such as magnetic random access memory (MRAM), fast programmable logic circuits, high-density recording media, and high-frequency devices for telecommunications. In particular, magnetization switching due to the spin-transfer torque is an attractive alternative to conventional field-induced switching in nanomagnetic devices since the electrical power consumption required for the switching decreases with decreasing size of the nanomagnetic elements. This provides architectural innovations for low-power writing of information with spintronic devices and also novel spintronic devices. In this chapter, we introduce theoretical and experimental studies on spin torques in nanostructured ferromagnetic systems.

19.1 Theoretical description of spin-transfer torque

We consider the spin-transfer effect in a ferromagnetic (F1) / nonmagnetic (N) / ferromagnetic (F2) trilayer structure, which is a typical structure for vertical spin devices, shown in Fig. 19.1. Here, the thickness, d_1, of the F1 is very thin, typically a few nm, while that of the F2, d_2, is very thick, over few tens of nm. We assume that the magnetization for F1, $\boldsymbol{M_1}$ tilts at an angle from the magnetization for F2, $\boldsymbol{M_2}$ at the initial state. When a positive voltage is applied in the sample, the electrons are injected from F2 to F1. Then, the electron spins whose directions tilts at an angle from $\boldsymbol{M_1}$ are injected into F1. The spin directions for the injected electrons are aligned with the direction of $\boldsymbol{M_1}$ because of the s-d exchange interaction. It should be noted that the electrons also extert a torque on $\boldsymbol{M_1}$ because of the action–reaction law (Newton's third law of motion). Therefore, the rotation of the magnetization $\boldsymbol{M_1}$ is induced by the torque due to the spin-current injection. This is known as the spin-transfer torque. According to Slonczewski's model [4], this spin-transfer torque is caused by transferring the transverse spin angular momentum from the electron into the magnetization. The transfer process acts like a mechanical torque on the magnetization, where the direction of the torque is given by $\boldsymbol{M_1} \times (\boldsymbol{M_1} \times \boldsymbol{M_2})$ [9, 10]. This enables us to switch the magnetization only by changing the polarity of the current.

When the injected spin current is not sufficiently large, the electron spins are aligned with the direction of $\boldsymbol{M_1}$, and $\boldsymbol{M_1}$ does not change its direction. When a large number of spins are injected into F1, the torque from the electron spins overcomes the torque from the magnetization (the damping torque). As a result, $\boldsymbol{M_1}$ becomes parallel to the direction of the electron spin by the spin-transfer

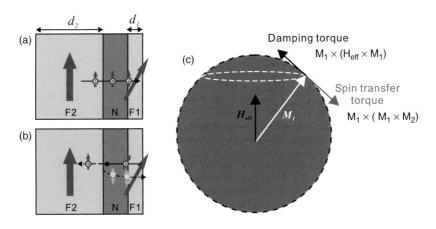

FIG. 19.1. Spin currents in a ferromagnetic (F1) /nonmagnetic (N) /ferromagnetic (F2) trilayered structure for (a) positive current and (b) negative current. (c) Schematic illustration of the spin-transfer torque and the damping torque.

torque. This means that $\boldsymbol{M_1}$ becomes parallel to $\boldsymbol{M_2}$. Oppositely, when a negative voltage is applied (Fig. 19.1(b)), the electrons should be injected from $\boldsymbol{M_1}$ into $\boldsymbol{M_2}$. In the ballistic model, electrons that are antiparallel to $\boldsymbol{M_2}$ are reflected at the N/F2 interface, then return back to F1. Such electrons should exert a torque, which rotates $\boldsymbol{M_1}$ antiparallel to $\boldsymbol{M_2}$. Thus, the direction of the torque can be switched by reversing the polarity of the flowing current.

In order, we analyze the influence of the spin-transfer torque more quantitatively. As described above, an additional torque, whose direction is given by $\boldsymbol{M_1} \times (\boldsymbol{M_1} \times \boldsymbol{M_2})$, is exerted on the magnetization $\boldsymbol{M_1}$. Since an electron has a spin angular momentum of $\hbar/2$[11], the spin transfer torque $\boldsymbol{T_s}$ is given by

$$T_s = \frac{\hbar}{2e} I_S \boldsymbol{S_1} \times (\boldsymbol{S_1} \times \boldsymbol{S_2}), \tag{19.1}$$

where I_s is the spin current injected into F1, and $\boldsymbol{S_1}$ and $\boldsymbol{S_2}$ are unit vectors for $\boldsymbol{M_1}$ and $\boldsymbol{M_2}$, respectively. So, $(\hbar/2e)I_S$ corresponds to the spin-angular momentum deposition per unit time.

According to Slonczewski's model based on ballistic transport [4, 9, 10], I_S can be expressed as

$$I_S = gI_e = \frac{I_e}{-4 + (1+P)^3(3 + \boldsymbol{S_1} \cdot \boldsymbol{S_2})/4P^{\frac{3}{2}}}. \tag{19.2}$$

Here, g is the spin-transfer efficiency. Then, we analyze the magnetization dynamics under the spin-transfer effect. By assuming a single-domain (macrospin) approximation and taking into account the contribution of the spin-transfer torque and the uniform torque in the entire film, the Landau–Lifschitz–Gilbert (LLG) equation for the free layer $\boldsymbol{M_1}$ is modified as

$$\frac{d\boldsymbol{M_1}}{dt} = \gamma \boldsymbol{M_1} \times \boldsymbol{H_{\mathrm{eff}}} - \alpha \boldsymbol{S_1} \times \frac{d\boldsymbol{M_1}}{dt} - \frac{\hbar}{2e} I_S \boldsymbol{S_1} \times (\boldsymbol{S_1} \times \boldsymbol{S_2}), \tag{19.3}$$

where γ and α are, respectively, the gyromagnetic ratio and the Gilbert damping parameter. $\boldsymbol{H_{\mathrm{eff}}}$ is the effective magnetic field on M_1, which corresponds to the sum of the applied, anisotropy, demagnetizing, and exchange fields.

In order to roughly understand the dynamical motion of $\boldsymbol{M_1}$ under the spin-transfer torque, we consider the case where $\alpha \ll 1$ and H_{eff} is antiparallel to $\boldsymbol{M_2}$. In this case, Eq. (19.3) can be simplified as follows.

$$\frac{d\boldsymbol{M_1}}{dt} = \gamma \boldsymbol{M_1} \times \boldsymbol{H_{\mathrm{eff}}} - \tilde{\alpha}\gamma \boldsymbol{S_1} \times (\boldsymbol{M_1} \times \boldsymbol{H_{\mathrm{eff}}}). \tag{19.4}$$

Here, $\tilde{\alpha}$ is the effective dumping parameter, which is given by

$$\tilde{\alpha} = \left(\alpha - \frac{\hbar}{2e} I_S \frac{1}{\gamma M_1 H_{\mathrm{eff}}}\right). \tag{19.5}$$

Thus, the damping parameter is effectively reduced by the spin-transfer torque. The positive damping torque dissipates the magnetization, which suppresses the precession motion. The negative damping torque amplifies the precessional motion. Therefore, the spin torque amplifies the precession for positive I_e, but attenuates the precession for negative I_e. Using the relation $\tilde{\alpha} = 0$, the threshold current I_{th} can be calculated as [11]

$$I_{\text{th}} = \frac{2e}{\hbar} \frac{\alpha \gamma M_1 H_{\text{eff}}}{g}. \tag{19.6}$$

One can understand the magnetization dynamics under the spin-transfer torque more clearly by solving Eq. (19.3) numerically. In the calculation, the uniaxial anisotropy along x and the demagenetizing field from the z-direction are considered. Figure 19.2(b) show the trajectory of the magnetization M_1 during the reversal process by the spin-current injection. Here, the initial magnetization is aligned with the $-x$ direction and no external magnetic field is applied. First, the magnetization starts to precess at a small cone angle, and promotes the precession while gradually increasing the opening angle of the precession, giving rise to the amplification of the spin precession. Once the magnetization during the

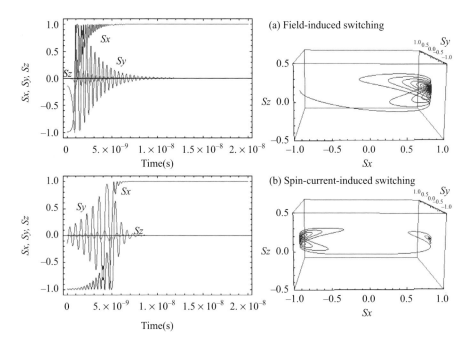

FIG. 19.2. Time-dependent vector components of the magnetization and trajectories during the magnetization reversal induced by (a) an external magnetic field and (b) spin-current injection.

precession reaches the angle $\phi = \pi/2$, which is the in-plane hard axis, the direction of the damping torque is reversed. After the reversal of the precession axis, the magnetization immediately aligns with the x-direction. This is the magnetization reversal process by the spin-transfer torque and therefore the damping torque is an important factor for the determination of the threshold of the switching current.

For comparison, the trajectory of the field-induced magnetization reversal is shown in Fig. 19.2(a). It should be noted that the magnetic potential energy is modified by the external magnetic field. In this case, the external field directly modifies the precession axis of magnetization. In field-induced reversal, when the magnetic field is larger than the other effective fields, the magnetization can be reversed. At first, the magnetization reaches the angle $\phi = \pi/2$, and then the magnetization rotates to the field direction attenuating the precession angle. Thus, large and different features are seen in the trajectories between the spin-transfer-induced and the field-induced switchings.

In spin-current-induced switching, the threshold current I_{th} is proportional to H_{eff} and the damping parameter as described in Eq. (19.6). [11] Since the largest term for H_{eff} is the demagnetizing field in conventional ferromagnetic thin films, the threshold current does not depend on the magnitude of the anisotropy so much. On the other hand, in field-induced switching, the switching field is directly related to the uni-axial anisotropy and strongly depends on the magnitude. Therefore, the switching field is not directly related to the switching current. Another important feature is the switching time τ_{sw} of the magnetization reversal due to the spin-current injection. τ_{sw} is defined by the time that the open angle of the magnetization precession reaches $\phi/2$. By solving the equation, τ_{sw} is roughly estimated as

$$\tau_{\mathrm{sw}} \approx \frac{2e}{\hbar} \frac{M_S}{g} \frac{1}{I - I_{\mathrm{th}}} \ln \frac{\pi/2}{\phi_0}. \tag{19.7}$$

Here, ϕ_0 is the initial angular deviation of the magnetic moment from its easy axis. The switching time reduces with increasing the injecting spin current. Therefore, injecting a large spin current is key for developing the ultrafast operation of spintronic devices. However, in general, there is a limitation of the magnitude of the spin current injecting into the ferromagnet because the spin current includes the charge current, which induces extra electro-migration under high bias current. Using the pure spin current, which does not include the charge current, may allow us to inject spin current larger than the above limitation [13, 14].

19.2 Perpendicular spin torque

The quantitative analysis of the spin-transfer torque is still a controversial issue. Equation (19.3) comes from the Slonzewski model, where he developed a theory that combines a density-matrix description of the spacer layer with a circuit theory. Stiles provided a clearer picture by solving the simplified Boltzmann equation with noncollinear spin configuration [9, 10]. Although the system is restricted only in the ohmic junction, this model allows us to treat the situations where the interface resistance does not necessarily dominate the transport and also where the layer thicknesses are less than the relevant mean-free paths. The model was found to well explain the experimental results obtained by the Cornell group [8, 15].

Heide pointed out an another important effect due to spin-current injection into a ferromagnetic material [16, 17]. According to his theory, a nonequilibrium spin current induces an exchange interaction between the magnetic films which is quite different from Ruderman–Kittel–Kasuya–Yosida (RKKY) exchange coupling. The effective torque due to the spin-current-induced exchange interaction is proportional to $Is(\boldsymbol{m} \times \boldsymbol{s})$. This interaction causes the torque, which is perpendicular to the Slonzewski spin-transfer torque and can be understood as the effective magnetic field along the $\boldsymbol{s}$, direction. Therefore, this term is called the "perpendicular spin torque" or a "field-like term" (Fig. 19.3).

Zhang *et al.* varied the above two terms to get a clearer physical picture by using a quasi-classical spin diffusion model [18]. They have shown that the key point is the spin accumulation associated with spin-dependent transmission/ reflection at the interfaces. More importantly, they have shown that it is the transverse component of the spin accumulation that contributes to the torque although the longitudinal part of spin accumulation does not induce any torque on the magnetization. They have also shown that the transverse component of the spin accumulation relaxes much faster than the longitudinal one because of the exchange interaction between the conduction electrons and local moments. According to their theory, the magnetization dynamics under the nonequilibrium spin current can be described by using the following equation

$$\frac{d\boldsymbol{m}}{dt} = \gamma \boldsymbol{m} \times (\boldsymbol{H}_{\text{eff}} + J\boldsymbol{m}_{\perp}) + \alpha \boldsymbol{m} \times \frac{d\boldsymbol{m}}{dt}, \tag{19.8}$$

where J is the magnitude of the spin current and $\boldsymbol{m}_{\perp}$ is the transverse component for the unit vector of the spin injector. The influence of the spin current is

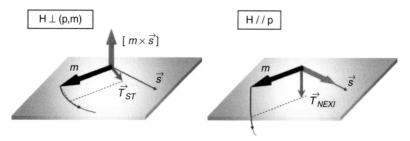

FIG. 19.3. Vector direction of spin–tranfer torque $\boldsymbol{T}_{\text{ST}}$ and perpendicular spin torque $\boldsymbol{T}_{\text{NEXI}}$.

expressed by the additional effective field $J\boldsymbol{m}_\perp$. By solving the spin diffusion equation, they found that $J\boldsymbol{m}_\perp = a\boldsymbol{s} \times \boldsymbol{m} + b(\boldsymbol{m} \times \boldsymbol{s}) \times \boldsymbol{m}$. Then, the equation becomes

$$\frac{d\boldsymbol{m}}{dt} = \gamma\boldsymbol{m} \times (\boldsymbol{H_e} + b\boldsymbol{s}) - \gamma a\boldsymbol{m} \times (\boldsymbol{s} \times \boldsymbol{m}) + \alpha M_s \boldsymbol{m} \times \frac{d\boldsymbol{m}}{dt}. \tag{19.9}$$

Thus, the transverse spin accumulation produces two effects. $\gamma a\boldsymbol{m} \times (\boldsymbol{s} \times \boldsymbol{m})$ is the in-plane torque, which corresponds to the spin-transfer torque introduce by Slonzewski. $\gamma b\boldsymbol{m} \times \boldsymbol{s}$ is the pependicular spin torque or field-like term due to the spin current, which is introduced by Heide. However, Zhang's theory states that the field-like term originates from the spin relaxation. Several mechanisms have been proposed as the origin of the perpendicular term. Momentum transfer, interlayer exchange coupling under a finite bias voltage, and a current-induced Oersted field could be the possible origins for this effect.

In metallic spin-valve structures, it has been shown that the field-like term is known to be very small [19]. However, in magnetic tunnel junctions, the field-like term becomes important especially under a high bias voltage. Kubota *et al.* [20] and Sankey *et al.* [21] have reported that the field-like term is proportional to the square of the voltage and it reaches 10–30% of the spin-transfer torque for a voltage bias of about 0.3 V. Furthermore, the sign of the field-like term is independent of the polarity of the bias. The obtained quadratic dependence is in agreement with the first-principles calculations by Heileger and Stiles [22]. Deac *et al.* demonstrated that the field-like term is also quadratic in the applied voltage their thermally excited FMR experiment [23]. However, Petit *et al.* observed that the field-like term changes sign when the voltage bias reverses [24]. Thus, the bias dependence of the field-like term is still a controversial issue.

19.3   Diffusive picture for injecting spin current

The spin current injected into $F2$ can be evaluated also by using the spin diffusion model when $\boldsymbol{M_1}$ and $\boldsymbol{M_2}$ are in the collinear configuration (parallel or antiparallel). The spatial distribution of the spin current is simply calculated in a F1/N1/F2/N2 structure using the spin resistance model. Here, the thicknesses of F1, F2, and N1 are d_{F1}, d_{F2}, and d_{N1} respectively. d_{N1} and d_{F2} are thinner than their spin diffusion lengths, while d_{F1} and d_{N2} are much longer than the spin diffusion length. In the structure, there are three F/N interfaces. Since each F/N interface is a source of the nonequilibrium spin-current, the system includes three diffusive spin-current sources. In order to simplify the situation, the system is divided into the three simple circuits shown in Fig. 19.4. Here, each circuit includes one spin-current source. From the superposition principle, the distribution of the spin current can be obtained by the sum of the distributions

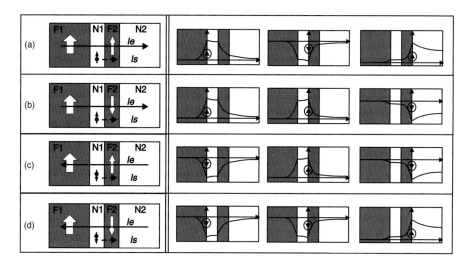

FIG. 19.4. Diffusive spin currents induced by nonequilibrium spin accumulation
in a F/N/F/N structure with the collinear magnetic configurations. In (a)
and (b) (c and d), the current flows from the right(left)-hand side to the
left(right)-hand side.

of the nonequilibrium spin current in the individual circuit. There are four
situations consisting of the parallel and antiparallel alignments with positive
and negative currents. The top figure at each configuration corresponds to a spin
injection from F1 and the middle two figures correspond to the backflows of the
spin current due to the N1/F2 and F2/N2 interfaces. Thus, the nonequilibrium
spin current injected into F2 consists not only of the spin injection from F2 but
also of the backflows from F1 itself. However, since the backflows of the spin
current from both interfaces induce the opposite spin torque in each other, the
influence of the backflow on M_2 can be neglected. Therefore, the spin current
injected into F2 is mainly dominated by the spin current generated from the
F1/N1 interface. When the current flows from the right-hand side to the left-
hand side, the injected spin current into F2 is parallel to M_1. This means
that the stable magnetization configuration is the parallel state. On the other
hand, when the current flows from the left-hand side to the right-hand side, the
injected spin current is antiparallel to M_1. Therefore, the antiparallel alignment
becomes stable when the current direction is reversed. Thus, Figs. 19.4(b) and
(c) are unstable situations for F2, and F2 is reversed by increasing the injecting
current.

Although the magnitudes of the injected spin current in Figs. 19.4(b) and (c)
are the same in the above explanation, the influence of the backflow of the spin
current should be considered for a more quantitative explanation. In Fig. 19.4(b),
the spin current injected from F1 and the spin current due to the N1/F2 interface

(Second figure from the top) are in the same direction. On the other hand, in Fig. 19.4(c), these are in the opposite direction, effectively resulting in a reduction of the magnitude of the spin current. Therefore, the absolute value of the switching current from antiparallel to parallel becomes larger than that from parallel to anti parallel. This tendency is consistent with the experimental results in metallic vertical devices. When d_{F2} is thicker than the spin diffusion length for F2, the sign of the injecting spin current depends on the position. This makes it difficult to reverse the magnetization by the spin-transfer torque.

19.4 Experimental study of magnetization reversal due to spin torque

In order to realize magnetization reversal due to spin-current injection, one has to flow the current into a small confined area. This restriction arises from the following two reasons. One is that the magnitude of the critical current density required for exciting the magnetization is quite large. In order to reduce the total current, the current-flowing area should be small. The other one is that the effect of the Oersted field induced by the flowing current should be reduced compared to the spin-transfer effect. The current flowing vertically in the film produces the circular Oersted field, which can create a different stable magnetization condition [26, 27]. The coercive fields for such domain structures reduce with increasing lateral dimension of the film. On the other hand, the critical current for the magnetization switching due to the spin-transfer torque decreases with decreasing the volume of the film. From theoretical and experimental studies, the lateral dimension for the magnetization switching due to the spin-transfer torque is around 100 nm. However, it should be noted that the Oersted-field effects were nonnegligible even in such small structures since they play an important role for increasing the initial spin-transfer torque [25].

To realize such conditions, the following three types of magnetic multilayered structures are used in most of the experiments for spin-transfer effects. The first one is a mechanical point contact, where a magnetic thin film or multilayer is contacted by a metallic sharp tip, shown in Fig. 19.5(a). This produces a nano-sized point contact whose diameter is typically a few tens of nanometers. The second one is a multilayered magnetic thin film with a lithography-defined point contact, where the metallic film is contacted to the magnetic multilayer through a nano-hole in the insulator with a diameter of a few tens of nanometers, as shown in Fig. 19.5(b). These two techniques locally produce an area with high current density in the magnetic multilayer, resulting in the excitation of the magnetization. However, since the excited area is magnetically connected to the large ferromagnet via the strong exchange interaction, the current required for exciting the magnetization becomes quite large. The third one is a nanopillar structure, which is fabricated by lithography and Ar ion milling (Fig. 19.5c). The typical lateral dimension of the nanopillar is 100 nm. Since the objective nanomagnet is isolated from the other ferromagnetic layer, the nanopillar is the

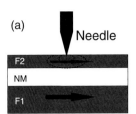
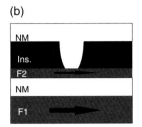
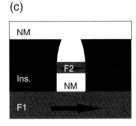

FIG. 19.5. Cross-sections for three representative device structures: (a) point contact, (b) lithographically defined point contact, and (c) nanopillar structures.

most suitable structure. In fact, this structure has been utilized in most of the experiments. However, it may be difficult to achieve a lateral dimension down to 50 nm in nanopillar structures.

The first clear experimental demonstration of spin-current-induced magnetization switching was done by the Cornell group at 4.2 K, where a lithographically defined point contact (Fig. 19.5b) was applied to a Co/Cu/Co trilayered structure [8]. Following this pioneering experiment, a second demonstration using a nanopillar structure (Fig. 19.5c), which provides better manipulation of the magnetization, was performed by the same group at room temperature [15]. They showed that the experimental results quantitatively agree with the theoretically calculated values based on the Slonzewski model. After the several experimental reports from the Cornnel group, similar experiments have been reported from Orsay [28]. There have been numerous experimental studies of these systems in the last decade [29, 30].

A typical experimental result of magnetization reversal due to spin-current injection is shown here. In order to observe spin-transfer-induced magnetization reversal, as mentioned above, a nanopillar structure consisting of a magnetic multilayer has been utilized, in general [31, 32]. Figure 19.6 shows a nanopillar structure consisting of Cu(100 nm)/Co(40 nm)/Cu(6 nm)/Co(2 nm)/Au(20 nm). Here, the shape of the nanopillar is an ellipse with dimensions of 120 nm × 390 nm. This structure enables us to flow a large amount of current above 20 mA perpendicularly to the magnetic layer because the Joule heating of the ferromagnetic layer can be suppressed by the top and bottom Cu electrodes with large dimensions. More importantly, magnetization switching (the magnetization direction) can be detected by measuring the voltage between the current probe because of the giant magnetoresistance effect. The resistance of the nanopillar and the differential resistance under the dc current is measured by the standard four-terminal method using a lock-in amplifier. Here, a dc current flowing from the bottom to the top is defined as positive "+". The magnetoresistance loop consists of the low and high resistance states with 8% resistance change, indicating that the parallel and antiparallel magnetic configurations are well

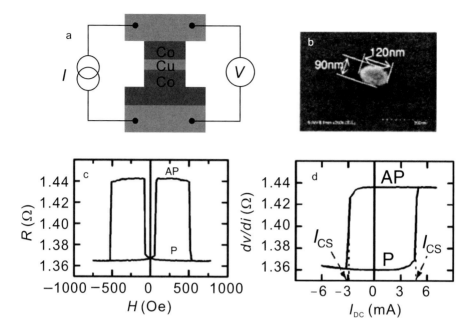

FIG. 19.6. Typical experimental result of spin-current-induced magnetization reversal at room temperature. (a) Schematic illustration of the nanopillar device together with the measured probe configuration; (b) scanning electron microscope image of the top view for the fabricated nanopillar; (c) four-terminal resistance as a function of the external magnetic field, and (d) differential resistance as a function of the injection dc current [31, 32].

stabilized. The differential resistance loop also shows sharp magnetic switching between the parallel and antiparallel states. As expected from the spin-torque model, the positive and negative currents stabilize the antiparallel and parallel configurations, respectively.

Spin-injection experiments were mainly performed in metallic magnetic multilayered structures in 2004. However, in metallic layered structures, the two-terminal resistance is typically several ohms and the magnetoresistance ratio is less than 10%. This results in a small voltage change ΔV of less than 1 mV. These poor voltage changes are a serious obstacle to integration with semiconductor technology. By using magnetic tunnel junctions instead of a metallic one, a large voltage change over hundreds of mV can be obtained. However, magnetic tunnel junctions were considered to produce large power consumption for spin-current-induced magnetization switching because of the large resistances of the tunnel junctions and the inelastic scattering of the tunneling process at high bias. The first experimental demonstration of spin-current induced magnetization switching was performed by Huai *et al.* using a

$CoFe/Al_2O_3$ barrier [33]. They found that the critical current density was smaller than those in metallic junctions. This is because the injection efficiency in tunnel juncitons is larger than that in metallic junctions. Shortly afterwords, the Cornel group also demonstrated spin-current-induced magnetization switching in the MTJ [34]. In magnetic tunnel devices, two-terminal resistances are typically in the kilo-ohm range and the magnetoresistance changes are much larger than 10%. These characteristics produce much larger voltage changes over hundreds of mV, which are suitable for high-speed read operation with semiconductor integrated devices. In particular, recent developments in MgO tunnel junctions provide a giant magnetoresistance change of over 500% even at room temperature [35, 36]. Moreover, the two-terminal resistance and the magnetoresistance ratio can be adjusted by the MgO thickness [35]. These technological jumps open up new possibilities for high-performance spin-transfer-torque random access memory.

It should also be noted that the present experimental results open up another interest in the spin-transfer physics since the electron transport in magnetic tunnel junctions at finite bias involves various electronic state energies both above and below the Fermi level. These situations are quite different from metallic systems, where the transport is mainly determined by the Fermi surface. In fact, as described in the previous section, the field-like term becomes large in a magnetic tunnel junction at finite bias. Essentially all the theoretical models proposed so far to obtain the spin-transfer torque from electronic transport calculations based on either quantum-mechanical or semiclassical descriptions have been derived in the limit of the weak nonequilibrium situation. Establishing a proper theoretical description of the spin-transfer torque in magnetic tunnel junctions at finite biases is a challenging and important issue from both the fundamental and technological viewpoints.

Apart from such vertical structures, laterally configured ferromagnteic/ nonmagnetic structures also produce spin currents [12]. Interestingly, one can generate a pure spin current, which does not include any charge current, by using a nonlocal injection scheme. This prevents the influence of the Oersted field [26] and electromigration-induced failures, leading to the injection of a large amount of spin current. Experimental demonstrations of magnetization switching due to pure spin-current injection have been carried out by using Py/Cu lateral structures [13, 14].

19.5   Magnetization dynamics due to spin-current injection

In the vertical magnetic device structures shown in Fig. 19.5, when a strong external magnetic field is applied to the device, the magnetizations of the two ferromagnetic layers align in parallel. In principle, the spin torque is zero because the injected spin direction is parallel to the magnetization. However, the thermal fluctuation of the magnetization and the self-induced Oersted field induce a small deviation of the magnetization from the completely parallel state. By increasing the injection spin current, the deviation is enhanced. In this situation, when the current flows from the bottom to the top electrode, the magnetization starts to precess with the axis parallel to the external magnetic field by the spin-transfer torque. When the magnitude of the spin-polarized current is not too large, the magnetization precession attenuates because of the Gilbert damping effect. However, the solution of Eq. (19.3) contains the steady state of the magnetization precession. So, when a large amount of spin current is injected into the free magnetic layer, the precession of the magnetization is stabilized by balancing the spin transfer torque and the Gilbert damping torque. An experimental demonstration of the steady-state precession mode was done by Tsoi *et al.* as a spin-wave excitation [6]. Interestingly, this experiment was performed earlier than the magnetization reversal due to the spin-transfer torque. They used a point contact in a Co/Cu multilayer. The spin-wave excitation was confirmed as the peak of the differential resistance in the static measurements. It should be noted that similar peaks of the differential resistance have been observed in experiments of spin-current-induced magnetization switching [8]. As an interesting experiment, Ji *et al.* demonstrated that a similar resistance peak can be observed in the absence of the second ferromagnetic layer. [37] This can be understood by assuming the formation of a nonuniform magnetic domain structure due to the strong nonuniform current injection in the vicinity of the ferromagnetic/nonmagnetic interface.

Kiselev *et al.* reported the first observation of the steady state precession of the magnetization in the microwave frequency range driven by a dc electrical current using a pillar structure of magnetic multilayers as shown in Fig. 19.7 [38]. Following this measurement, Rippard *et al.* observed similar high-frequency magnetization resonance with large quality factor by using a point contact spin injection technique [39]. They showed that the oscillation frequency can be tuned from 5 to above 40 GHz by adjusting the external magnetic field. Important progress since these two reports of the previous experiments are the dynamic measurements of the current-induced magnetization excitation in the frequency domain using the spectrum analyzer. The magnetization precession in the GHz range was clearly observed from the voltage signal originating from the GMR effects.

According to Ref. [38], the magnetization dynamics in a nanopillar under spin torque can be summarized by the phase diagram shown in Fig. 19.7(c). Here, the "S" state corresponds to the precession state with a small open angle. This state can be obtained both theoretically and experimentally. "L" corresponds to the precession state with a large open angle, which can also be obtained both theoretically and experimentally. The "W" state corresponds to a weakly microwave emitting mode. Numerical simulations based on the macro-spin model well explain the experimental results (Fig. 19.8). Here, two types in-plane precession (S-mode and L-mode) have been reproduced. Interestingly, the "W" state obtained in the experiment cannot be reproduced by the simulation. Instead of this, an out-of-plane precession mode has been obtained under

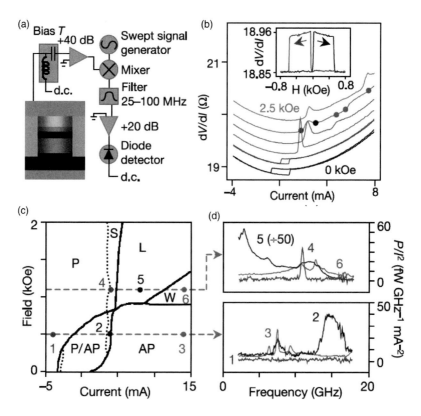

FIG. 19.7. (a) Schematic illustration of the sample structure together with the heterodyne mixer circuit. (b) Differential resistance versus current for several external magnetic fields: 0 (bottom), 0.5, 1.0, 1.5, 2.0 and 2.5 kOe (top). (c) Phase diagram of the dynamical magnetization state obtained from resistance and microwave data measured at room temperature. (d) Microwave spectra at each state in (c). Reproduced partially from [38].

high bias current in a numerical simulation with a single domain approximation (Fig. 19.8). Therefore, the "W" state is probably caused by the excitation of the nonuniform mode. The out-of plane precession mode has been observed in the magnetic multilayer using a perpendicular spin polarizer combined with an in-plane magnetized free layer [40].

It should be noted that the steady-state precession driven by the spin-transfer torque has rather different properties from the conventional precession driven by the external magnetic field. The most interesting feature in the spin-transfer-induced precession is that the frequency of the precessional motion depends on the damping constant. Since the magnetization precession is in the GHz range, this property can be utilized for a submicron macro-wave generator called a spin-torque nano-oscillator (STNO). STNO is one of the promising candidates in future nano-spintronic devices and magnetic recording technology

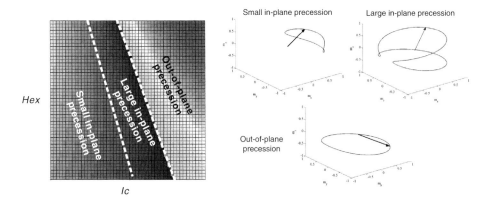

FIG. 19.8. Phase diagram obtained from the numerical simulation with the macro-spin approximation and trajectories for three representative precession modes (small angle in-plane, large angle in-plane, and out-of-plane precessions).

because the oscillation frequency and emitted microwave power can be tuned by changing the bias current and the external magnetic fields. Moreover, the STNO has great potential for novel active microwave devices such as broadband high-quality STNO. However, the output power is limited because of the small resistance and resistance change in the conventional metallic junction. STNOs based on magnetic tunnel junctions have been demonstrated with relatively large microwave power. Although the emission power is around one nW or less in the case of a STNO based on the metallic junction, it can reach 1 μW using the tunnel magnetoresistance effect [23]. The synchronization of an array of STNOs is a possible approach for increasing the emission power. Mutual phase locking between STNOs mediated by spin waves has been demonstrated by Kaka *et al.* [41] and simultaneously by Mancoff *et al.* [42]. Also a theoretical study predicted that an array of oscillators could be synchronized using electrical rather than magnetic coupling. More especially, the output power in the N oscillators becomes N^2 times as large as that for a single oscillator. In addition, the frequency linewidth is reduced by a factor of $1/N^2$. Therefore, the combination the STNOs with magnetic tunnel junctions and the phase-locking technique is a powerful tool for enhancing the output power of the oscillator.

Time-resolved studies of dynamics excited by spin-transfer torques has also been reported by Krivorotov *et al.* [43]. These measurements allow a direct view of the process of spin-transfer-driven magnetic reversal, and they determine the possible operating speeds for practical spin-transfer devices. The results provide rigorous tests of theoretical models for spin transfer and strongly support the spin-torque model over competing theories that invoke magnetic heating.

The ferromagnetic resonance can be excited by injecting the rf current, and can be electrically detected by observing the rf power absorption using a network analyzer. The precession of the free-layer magnetization due to the resonance yields a large oscillation of the resistance of the device. Therefore, when the resistance asymmetrically changes with respect to the magnetization angle, the voltage excited by the rf current injection produces a rectified effect analogous to homodyne detection. This is known as the spin-torque diode effect [44, 45].

A schematic illustration of the spin-torque diode effect is shown Fig. 19.9. The magnetization of the free layer is perpendicular to that of the fixed layer at the equilibrium condition. When a negative current is applied, the magnetization of the free layer rotates parallel to that of the fixed layer because of the spin-transfer torque. As a result, the resistance of the junction becomes small and the junction induces a small negative voltage for a given current. On the other hand, when a positive current is applied, the magnetization of the free layer favorably

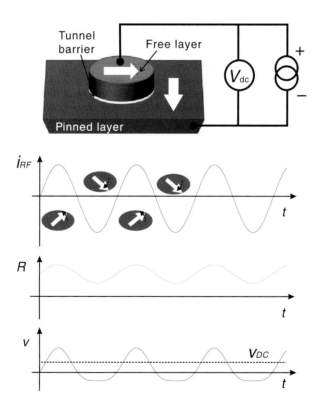

FIG. 19.9. Conceptual image of the principle of the spin torque diode effect. RF current injection induces the modulation of the device resistance. As a result, the induced voltage is rectified.

rotates to the antiparallel state. Therefore, the junction resistance becomes high and a large positive voltage is induced across the junction for a given current. By alternating the current direction at high speed, a positive voltage is induced in the junction as an average. This is a kind of homodyne detection. While the spin-transfer torque induces the in-plane rotation of the magnetization, the field-like torque induces a rotation perpendicular to the plane. As a result, only the resonance excited by the spin-transfer torque can contribute to rectify the rf current at the resonance frequency.

19.6 Domain wall displacement due to spin-current injection

As described in the above section, the magnetization in small ferromagnets can be switched by the spin-transfer torque. In such systems, nonequilibrium spin currents are produced by the sudden change of the spin-dependent conductivity at an F/N interface. As shown in Fig. 19.10 since the periodic domain structure in a magnetic thin film resembles a magnetic multilayered structure, a similar spin-transfer torque is thus expected to take place in the domain structure. As described in the previous section, L. Berger predicted that a magnetic domain wall can be moved by flowing a spin current through the domain wall before the prediction of magnetization reversal of a small nanomagnet [1]. Moreover, his group experimentally investigated such effects in permalloy films by using a pulsed current and succeeded in observing the domain wall displacement due to the current pulse. Although their experimental studies were really pioneering work, the detailed domain structures were not examined because of the limitations of the nanofabrication and domain-imaging techniques. Moreover, the current required to produce the domain wall displacement is extremely large because of the large sample dimension. Therefore, it was difficult to distinguish it from other spurious effects such as the Oersted field. Recent nanofabrication techniques made it possible to control the magnetic domain structure precisely. We are now able to control the magnetic configuration in nanostructured mag-

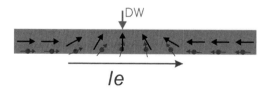

FIG. 19.10. Schematic illustration of the spin-current-induced domain wall motion.

nets using the geometrically induced magnetostatic interaction. This allows precise manipulation of a magnetic domain wall. More especially, lithographically defined ferromagnetic nanowires offer greater control of magnetic domain walls. Since spin-current-induced domain wall motion is expected to offer high potential applications such as race-track memory [46] and logic devices [47], this has been intensively studied both theoretically and experimentally in the last decade.

Gan *et al.* reported current-induced domain wall motion in a patterned ferromagnetic structure by means of MFM for the first time [48]. They found that the magnetic domain structure changes by the application of a strong current pulse with a magnitude of 2.5. It should be noted that the direction of the domain wall motion is always opposite to the flowing direction of the current. This seems to be consistent with the picture of the spin-transfer torque. Grollier *et al.* reported the current-induced domain wall motion in a 1 μm-wide magnetic multilayerd wire [49]. They found that the magnetic domain wall trapped around a notch was depinned by the current. In this structure, however, part of the current flows through the highly conductive Cu layer, whereby the current induced Oersted field may have given an additional contribution to the domain wall displacement, as suggested in their article. Tsoi *et al.* reported similar current-induced domain wall motion in a single ferromagnetic wire with several notches [50]. Kläui *et al.* performed an experimental study of the current-induced domain wall displacement in ferromagnetic ring structures [51]. After these experiments, numerous experimental studies were performed by various groups and similar results have been reported [52, 53]. However, in most of the experiments, the magnetic domain wall did not move only by the current injection. In order to induce the domain wall motion, an external magnetic field, which is much smaller than the value without current injection, was required. These are because the domain wall in the nanowire was trapped in the strong pinning potential. So, the weak pinning potential should be used for the current-induced domain wall motion.

The most elegant experiment on current-induced domain wall (DW) motion was demonstrated by Yamaguchi *et al.* [54]. They used a specially developed L-shaped magnetic wire with a round corner as schematically illustrated in Fig. 19.11. Here, one end of the L-shaped magnetic wire is connected to a diamond-shaped pad which facilitates DW nucleation, while the other end is a needle shape to prevent the nucleation of DW from this end. By the application of the magnetic field at an angle after initializing the domain structure, a magnetic domain wall is introduced in the vicinity of the corner. This creates the domain wall in the magnetic wire without any artificial pinning site. After the formation of the domain wall, they applied the current pulse. They showed that the domain wall moved along the flowing direction of the electrons by real-space observation of the domain wall position using MFM. In this situation, the domain wall can easily move without the application of the magnetic field.

Another elegant experiment has been done by Saitoh *et al.* [55]. By using a semi-circular-shaped wire, they showed that the potential profile the domain

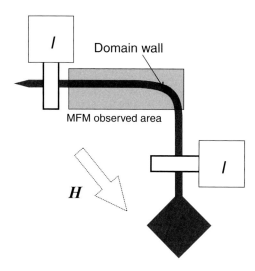

FIG. 19.11. Device structure for the real space observation of spin-current-induced domain wall motion by means of magnetic force microscopy.

wall can be tuned by the magnitude of the external magnetic field. They demonstrated that the domain wall motion can be resonantly excited by applying a high-frequency ac current whose current density is much smaller than the dc current density required for exciting the domain wall motion. This innovative demonstration that spin can be effectively excited by using the resonance was also really important for developing spin-torque applications.

19.7   Theoretical description of the spin-current-induced domain wall displacement

The spin-current-induced domain wall motion is simply described as follows. The flowing spin currents carried by moving the electrons adiabatically, follow the magnetization direction because the magnetization exerts a torque on the spin currents. There is a reaction torque on the magnetization that is proportional to the spin current. If the spin currents flow uniformly, this torque density simply translates the domain wall in the direction of the electron flow with a speed that is proportional to the spin current. However, in realistic cases, the complicated motions of the domain wall, which cannot be understood by the above description, have been observed in the experiments.

The discrepancies between the simple description and the realistic situation are mainly caused by undesired factors such as the current-induced

Oersted field, including damping, nonadiabatic torques, and extrinsic effects like pinning. The most important and outstanding issue concerns the nonadiabatic spin-torque component. Although the magnitude is much smaller than the adiabatic term, this plays an important role in wall dynamics. Zhang and Li have developed the LLG equation with the spin-transfer torque in an inhomogeneous domain structure [56]. They derive the LLG equation with spin-transfer torque with adiabatic and non-adiabatic terms.

$$\frac{\partial \boldsymbol{M}}{\partial t} = -\gamma \boldsymbol{M} \times \boldsymbol{H}_{\mathrm{eff}} + \frac{\alpha}{M_s} \boldsymbol{M} \times \frac{\partial \boldsymbol{M}}{\partial t} - \frac{b_{\mathrm{J}}}{M_s^2} \boldsymbol{M} \times \left(\boldsymbol{M} \times \frac{\partial \boldsymbol{M}}{\partial x} \right) - \frac{c_{\mathrm{J}}}{M_s} \boldsymbol{M} \times \frac{\partial \boldsymbol{M}}{\partial x}.$$

(19.10)

Here, $b_J = \mu_B/eM_s(1+\xi)$ and $c_J = \mu_B \xi/eM_s(1+\xi)$. ξ is defined by $\hbar/(SJ_{ex}\tau_{sf})$, where τ_{sf} is the spin-flip relaxation time. The third and fourth terms are, respectively, the adiabatic and nonadiabatic spin-transfer torques. The nonadiabatic term arises from the same mechanism as the field-like term discussed in the previous section. Microscopic descriptions of the spin-transfer torque for a magnetic domain wall have been carried out by several authors [57–59]. Tatara *et al.* derive a generation formula of the spin torque including adiabatic and nonadiabatic torque from the *s-d* exchange inteaction with the spin-conservation law. They also provide an elegant theoretical review of the current-driven domain wall motion [60]. However, at the moment, a theoretical description of the nonadiabatic term has not reached a consensus. Since the nonadiabatic term is not intrinsic to the material, it is highly sensitive to the micromagnetic structure of the wall [61].

19.8 Dynamics of magnetic domain wall under spin-current injection

In the early 2000s, most experimental studies covered the current-driven motion of domain walls under quasi-static conditions. However, since the domain walls must be moved on much shorter time-scales, the investigation of the subnanosecond dynamics of the domain wall is important. It is also important to minimize the influence of the thermal activation due to Joule heating on the domain wall motion. By using a short-pulsed current, one can suppress the heating effect. Luc Thomas *et al.* studied domain wall motion due to the application of a nanosecond pulse changing the pulse duration [63]. They showed that the probability of dislodging a domain wall, confined to a pinning site in a permalloy nanowire, oscillates with the length of the current pulse, as shown in Fig. 19.12. The oscillation period corresponds to the precessional period of the domain wall, which is determined by the wall's mass and the slope of the confining potential. These results are direct evidence of the precessional nature of the domain wall dynamics.

Estimation of the critical current of the domain wall motion has been an important issue for experimental and theoretical physicists in the past few years. This is strongly related to the adiabatic and non-adiabatic torques. In the field-driven domain wall motion, the existence of this propagation field is caused

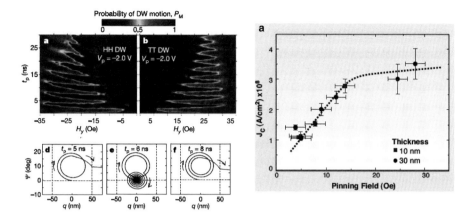

FIG. 19.12. (a), (b) Oscillatory dependence of current-driven magnetic domain wall motion on current pulse length. Contour plots of the probability of current-driven DW motion as a function of the current pulse length and the magnetic field. (c)–(f) Domain wall trajectory in phase space for different pulse lengths calculated by a one-dimensional model with a parabolic potential. (g) Critical current for depinning a vortex domain wall near zero field versus the depinning field, measured on 10 and 30 nm thick and 100 to 300 nm wide Py nanowires using current pulses 20–100 ns long. Reproduced partly from [45, 62].

by extrinsic effects such as defects or roughness. On the other hand, the original theories of the spin torque, in which only the adiabatic term is considered, suggested the existence of an intrinsic critical current density, even for an ideal ferromagnetic nanowire [57]. However, the critical current is reduced to zero by considering the nonadiabatic term. In such cases, the domain wall dynamics become similar to the field-induced dynamics. The IBM group investigated the relationship between the critical current and the strength of the pinning potential [46]. As shown in Fig. 19.12, when the pinning potential is weak, the critical current can be controlled by the magnitude of the pinning potential. When the pinning potential becomes strong, the critical current seems to saturate, implying the existence of an intrinsic threshold. However, it is difficult to explain the phenomena because other spurious effects such as pinning and the Oersted field are superimposed under such high current densities.

The velocity of the domain wall driven by the spin current is also an important parameter, especially for applications such as racetrack memory. Zhang and Li show that the domain wall is accelerated by the adiabatic spin torque and the nonadiabatic term causes the wall to continually move [56]. Thiaville *et al.* numerically investigated the influence of the nonadiabatic term on the domain wall velocity driven by the current [61]. According to their micromagnetic calculation, when the damping torque is equal to the nonadiabatic spin torque, the domain wall velocity linearly increases as a function of the spin current. When the nonadiabatic spin torque is larger than the damping torque, the current dependence of the domain wall velocity becomes similar to the field dependence of the velocity, where the velocity has a peak at a certain magnetic field (Walker break down). Hayashi *et al.* reported that the domain wall velocity driven by the current exceeds 100 m/s [66]. They also showed that the velocity shows the breakdown behavior at a certain current density.

Yamanouchi *et al.* demonstrated for the first time spin-current-induced domain wall motion in dilute magnetic semiconduoctor systems using GaMnAs film [68]. They showed that a domain wall in GaMnAs can be drive by a current with a density of 10^8 A/m^2, which is two or three orders of magnitude smaller than those for metallic wires. The value of the domain wall velocity obtained from the pulse current experiment was quite consistent with the adiabatic spin-transfer model. Moreover, intrinsic pinning appears. However, an important difference between metallic ferromagnetic wires such as Py wire and GaMnAs is the *p-d* exchange interaction and *s-d* exchange interaction. The theoretical model does not consider the strong spin–orbit interaction in GaMnAs. Thus, the detailed mechanism is still unsolved.

Very recently, Koyama *et al.* performed a beautiful experiment about the intrinsically induced threshold current by using a Co/Ni wire with perpendicular magnetic anisotropy. The width dependence and the external magnetic field dependence of the threshold current are quite consistent with the prediction of the adiabatic spin transfer model. Since the threshold current in magnetic wires with perpendicular magnetic anisotropy becomes much lower than in magnetic wires with in-plane magnetic anisotropy, a highly spin-polarized magnetic wire with perpendicular magnetic anisotropy may provide high controllability of the domain wall by a spin current.

19.9   Vortex motion due to spin-current injection

The magnetic vortex structure, which is stabilized in a ferromagnetic circular disk with a diameter less than a micron as shown in Fig. 19.13, has potential as a unit cell of high-density magnetic storage because of the negligible magnetostatic interaction and high thermal stability. A magnetic vortex with a single vortex core can be described by two topological quantities. One is the polarity, which corresponds to the magnetization direction of the vortex core. The polarity strongly correlates to the dynamical gyration motion of the vortex core and the displacement of the vortex core induced by the spin current. The other one is the chirality, which is the rotational direction of the magnetic moment whirling either clockwise (CW) or counterclockwise (CCW). The chirality determines the direction of the vortex shift induced by the in-plane magnetic field. Since such a magnetic vortex is a kind of confined vortex like a wall, vortex motion due to the

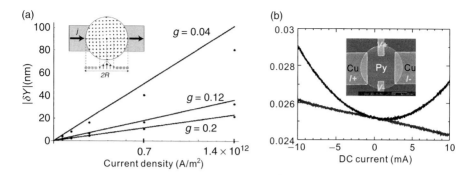

FIG. 19.13. (a) Theoretically and numerically calculated vortex displacement due to the spin-transfer torque. g is the geometrical factor defined by t/d, where t is the thickness and d is the diameter of the disk. Reproduced partly from [71]. (b) Experimentally observed differential planar Hall resistance in the absence of the magnetic field as a function of the dc current (black line) and that except for the parabolic component (gray line). The linear resistance change is caused by the vortex displacement due to spin-current injection. The parabolic component is caused by Joule heating. The inset shows a SEM image of the fabricated device together with the probe configuration. Reproduced partly from [72].

spin torque can be expected. Moreover, a large spin torque will be exerted on the vortex core with an exchange length of the order of a few nanometers. The vortex motion due to the adiabatic spin torque has been investigated theoretically [71]. According to the theory based on the adiabatic spin torque, the spin torque induces a force normal to the applied current, expressed as $\boldsymbol{G} \times \boldsymbol{v_s}$, where $\boldsymbol{G}$ and $\boldsymbol{v_s}$ are, respectively, the gyrovector, defined as the product of the vortex polarity and vorticity and the drift velocity of the electron spins. The vortex displacement is expected to be proportional to the spin current density, as shown in Fig. 19.13. The important difference from the current-induced domain wall in a magnetic wire is that there is no threshold current to induce the vortex displacement in a magnetic disk. This is due to the symmetric spin structure of the magnetic vortex.

Ishida *et al.* experimentally investigated the steady-state displacement of a vortex in a permalloy circular disk driven by a dc current [71]. They analyzed the small vortex motion due to dc current injection from the differential planar Hall resistance measurement combined with micro-Kerr measurements. They showed that the vortex moves linearly while increasing the dc current without the threshold current (Fig. 19.13). It was also shown that the annihilation field of the vortex can be tuned by dc current injection.

Kasai *et al.* observed the magnetic vortex resonance excited by rf current injection [74]. The electrical measurement based on the anisotropic magnetoresistance with the rf current successfully detects the resonant gyration motion of the vortex in a permalloy circular disk. They also studied the real-space observation of the vortex resonance by using time resolved X-ray microscopy with high spatial resolution less than 25 nm [75]. They estimated the spin polarization of the permalloy from the oscillation amplitude of the vortex core. Interestingly, they showed that the polarity of the vortex can be reversed by the application of a large rf current with a resonant frequency [77]. Since an extremely large quasi-static out-of-plane magnetic field was traditionally required for switching of the vortex core's polarity, this new manipulation method opens up a novel method for efficiently writing information in a memory device based the magnetic vortex. Their analysis was based on the adiabatic spin on transfer model. Bolte *et al.* also performed a similar experiment using a Py square dot [78]. They carefully analyzed the phase of the oscillation and found that the current-induced Oersted field also plays an important role for exciting the vortex core in the system with direct rf current injection. The magnitude of the driving force due to the Oersted field was found to be 30% for the total driving force. Therefore, in rf current excitation, optimization of the current-induced Oersted field, together with the spin-transfer torque, may provide more efficient and reliable manipulation methods for dynamical vortex motion.

19.10   Other new phenomena

Recent theoretical work suggests that the Rashba effect or the Dresselhaus effect provides a radically new mechanism for manipulating the magnetization in ferromagnetic systems. These current-induced spin–orbit effects are caused by the effective magnetic field induced by spin–orbit coupling between the spin of the electron and its momentum in the structural inversion asymmetry system [80, 81]. Electrons moving in an electric field experience a relativistic magnetic field in the electron's rest frame, as shown in Fig. 19.14. The direction of the spin can be manipulated by the electric field or unpolarized currents. Recently, evidences for the Rashba effect has been reported in magnetic metallic systems. Mirron *et al.* studied the Rashba effect in Co/Pt wires with perpendicular magnetic anisotropy. They studied the magnetization process of the Co/Pt wire under a transverse magnetic field, together with the Rashba field by means of the micro-magneto-optical Kerr effect [82]. They clearly observed the reduction (enhancement) of the coercive field at room temperature when the Rashba field was parallel (anti-parallel) to the transverse magnetic field (Fig. 19.14). Pi *et al.* observed a change of the magnetization direction due to the Rashba effect by using a homodyne detection technique. The important thing is that the current density is less than 1.0 A/cm^2 which is much smaller than the critical current density for the spin-transfer torque. The method, based on intrinsic spin–orbit

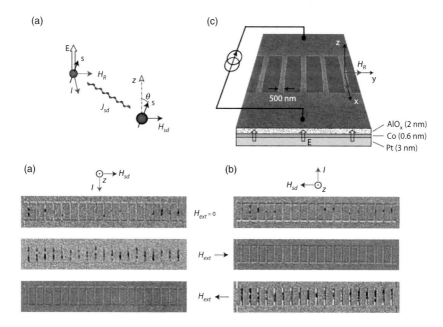

FIG. 19.14. (a) Conceptual image of the spin torque induced by the Rashba effect. (b) Scanning electron micrograph detail of the patterned Pt/Co/AlOx wire array and schematic vertical cross-section of the layer. (c) Differential Kerr microscopy images recorded after current pulse injection. Reproduced partly from [82].

interaction, may become an alternative way to induce a spin torque using an electric current.

Fernandez-Rossier *et al.* investigated the influence of the spin-transfer effect on the spin-wave excitation and predicted that the spin-transfer torque induces frequency shifts of the spin wave. They call this effect the "spin-wave Doppler shift." The spin-wave Doppler shift is simply understood by the spin current induced by the periodically modulated domain structures. As shown in Fig. 19.15, when the electrons flow from left to right, the precession is accelerated by the spin-transfer torque, resulting in enhancement of the spin-wave frequency. Oppositely, the spin-wave frequency decreases by negative flow of the electrons. Vlaminck and Bailleul have for the first time observed the spin-wave Doppler effect [84]. The relation between the frequency shift and the spin current is simply given by the following equation.

$$\Delta f = -\frac{P\mu_B}{eM_s} \mathbf{J} \cdot \mathbf{k}. \tag{19.11}$$

Here, $\mathbf{k}$ is the wavevector of the spin wave. Thus, the spin polarization can be determined from the frequency shift.

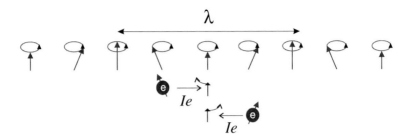

Fig. 19.15. Conceptual image of the spin-current-induced spin-wave Doppler effect.

Acknowledgment

The author would like to thank Dr. Nakata, Dr. S. Yakata, Dr. Yang and Prof. Y. Otani for their valuable discussion. This work is partially supported by JST CREST and NEDO.

References

[1] L. Berger, Exchange interaction between ferromagnetic domain wall and electric current in very thin metallic films, J. Appl. Phys. **55**, 1954 (1984).

[2] M. N. Baibich, J. M. Broto, A. Fert *et al.*, Giant magnetoresistance of (001)Fe/(001)Cr magnetic superlattices, Phys. Rev. Lett. **61**, 2472–2475 (1988).

[3] G. Binasch, P. Grunberg, F. Saurenbach, and W. Zinn, Enhanced magnetoresistance in layered magnetic structures with antiferromagnetic interlayer exchange, Phys. Rev. B **39**, 4828–4830 (1989).

[4] J. C. Slonczewski, Current-driven excitation of magnetic multilayers, J. Magn. Magn. Mater. **159**, L1–L7 (1996).

[5] L. Berger, Emission of spin waves by a magnetic multilayer traversed by a current, Phys. Rev. B **54**, 9353–9358 (1996).

[6] M. Tsoi, A. G. M. Jansen, J. Bass *et al.*, Excitation of a magnetic multilayer by an electric current, Phys. Rev. Lett. **80**, 4281–4284 (1998).

[7] M. Tsoi, A. G. M. Jansen, J. Bass, W.-C. Chiang, V. Tsoi, and P. Wyder, Generation and detection of phase-coherent current-driven magnons in magnetic multilayers, Nature **406**, 46–48 (2000).

[8] E. B. Myers, D. C. Ralph, J. A. Katine, R. N. Louie, and R. A. Buhrman, Current-induced switching of domains in magnetic multilayer devices, Science **285**, 867–880 (1999).

[9] M. D. Stiles and A. Zangwill, Anatomy of spin-transfer torque, Phys. Rev. B **66**, 014407 (2002).

[10] M. D. Stiles and J. Miltat, Spin dynamics in confined magnetic structures, in *Topics in Applied Physics*, vol. 101, B. Hillebrands and A. Thiavilles (Eds.), Springer, Heidelberg, 2006, pp. 225–308.

[11] J. Z. Sun, Current-driven magnetic switching in manganite trilayer junctions, J. Magn. Magn. Mater. **202**, 157–162 (1999).

[12] F. J. Jedema, A. T. Filip, and B. J. van Wees, Electrical spin injection and accumulation at room temperture in an allmetal mesoscopic spin valve, Nature **410**, 345–348 (2001).

[13] T. Kimura, Y. Otani, and J. Hamrle, Switching magnetization of a nanoscale ferromagnetic particle using nonlocal spin injection, Phys. Rev. Lett. **96**, 037201 (2006).

[14] T. Yang, T. Kimura, and Y. Otani, Giant spin-accumulation signal and pure spin-current-induced reversible magnetization switching, Nat. Phys. **4**, 851–854 (2008).

[15] J. A. Katine, F. J. Albert, R. A. Buhrman, E. B. Myers, and D. C. Ralph, Current-driven magnetization reversal and spinwave excitations in Co/Cu/Co Pillars, Phys. Rev. Lett. **84**, 3149–3152 (2000).

[16] C. Heide, Spin currents in magnetic films, Phys. Rev. Lett. **87**, 197201 (2001).

[17] C. Heide, P. E. Zilberman, and R. J. Elliott, Current-driven switching of magnetic layers, Phys. Rev. B **63**, 064424 (2001).

[18] S. Zhang, P. M. Levy and A. Fert, Mechanisms of spin-polarized current-driven magnetization switching, Phys. Rev. Lett. **88**, 236601 (2002).

[19] S. Urazhdin, N. O. Birge, W. P. Pratt Jr., and J. Bass, Current-driven magnetic excitations in permalloy-based multilayer nanopillars, Phys. Rev. Lett. **91**, 146803 (2003).

[20] H. Kubota, A. Fukushima, K. Yakushiji, T. Nagahama, S. Yuasa, K. Ando, H. Maehara, Y. Nagamine, K. Tsunekawa, and D. D. Djayaprawira, Quantitative measurement of voltage dependence of spin-transfer torque in MgO-based magnetic tunnel junctions, Nature Phys. **4**, 37 (2008).

[21] J. C. Sankey, Y. Cui, J. Z. Sun, J. C. Slonczewski, R. A. Buhrman, and D. C. Ralph, Measurement of the spin-transfer-torque vector in magnetic tunnel junctions, Nature Phys. **4**, 67 (2008).

[22] C. Heiliger and M. D. Stiles, *Ab Initio* Studies of the spin-transfer torque in tunnel junctions, Phys. Rev. Letts. **100**, 186805 (2008).

[23] A. M. Deac, A. Fukushima, H. Kubota, H. Maehara, Y. Suzuki, S. Yuasa, Y. Nagamine, K. Tsunekawa, D. D. Djayaprawira and N. Watanabe, Bias-driven high-power microwave emission from MgO-based tunnel magnetoresistance devices, Nature Phys. **4**. 803–809 (2008).

[24] S. Petit, C. Baraduc, C. Thirion, U. Ebels, Y. Liu, M. Li, P. Wang, and B. Dieny, Spin-torque influence on the high-frequency magnetization fluctuations in magnetic tunnel junctions, Phys. Rev. Lett. **98**, 077203 (2007).

[25] Y. Acremann, J. P. Strachan, V. Chembrolu *et al.*, Time-resolved imaging of spin transfer switching: Beyond the macrospin concept, Phys. Rev. Lett. **96**, 217202 (2006).

[26] J. A. Katine, F. J. Albert, and R. A. Buhrmana, Current-induced realignment of magnetic domains in nanostructured Cu/Co multilayer pillars, Appl. Phys. Lett. **76**, 17, 2000.

[27] K. Bussmann, G. A. Prinz, R. Bass, and J.-G. Zhu, Current-driven reversal in annular vertical giant magnetoresistive devices. Appl. Phys. Lett. **78**, 2029 (2001).

[28] J. Grollier, V. Cros, A. Hamzic *et al.*, Spin-polarized current induced switching in Co/Cu/Co pillars, Appl. Phys. Lett. **78**, 3663–3665 (2001).

[29] J. Z. Sun, D. J. Monsma, D. W. Abraham, M. J. Rooks, and R. H. Koch, Batch-fabricated spin-injection magnetic switches, Appl. Phys. Lett. **81**, 2202–2204 (2002).

[30] B. Ozyilmaz, A. D. Kent, D. Monsma, J. Z. Sun, M. J. Rooks, and R. H. Koch, Current-induced magnetization reversal in high magnetic fields in Co/Cu/Co nanopillars, Phys. Rev. Lett. **91**, 067203 (2003).

[31] T. Yang, T. Kimura, and Y. Otani, Spin-injection-induced intermediate state in a Co nanopillar, J. Appl. Phys. **97**, 064304 (2005).

[32] T. Yang, A. Hirohata, T. Kimura, and Y. Otani, Influence of capping layer on the current-induced magnetization switching in magnetic nanopillars, J. Appl. Phys. **99**, 073708 (2006).

[33] Y. Huai, F. Albert, P. Nguyen, M. Pakala, and T. Valet, Observation of spin-transfer switching in deep submicronsized and low-resistance magnetic tunnel junctions, Appl. Phys. Lett. **84**, 3118–3120 (2004).

[34] G. D. Fuchs, N. C. Emley, I. N. Krivorotov *et al.*, Spin-transfer effects in nanoscale magnetic tunnel junctions, Appl. Phys. Lett. **85**, 1205–1207 (2004).

[35] S. Yuasa, T. Nagahama, A. Fukushima,Y. Suzuki, and K. Ando, Giant room temperture magnetoresistance in single-crystal Fe/MgO/Fe magnetic tunnel junctions, Nature Mater. **3**, 868–871 (2004).

[36] S. S. P. Parkin, C. Kaiser, A. Panchula *et al.*, Giant tunnelling magnetoresistance at room temperture with MgO (100) tunnel barriers, Nature Mater. **3**, 862–867 (2004).

[37] Y. Ji, C. L. Chien, and M. D. Stiles, Current-induced spinwave excitations in a single ferromagnetic layer, Phys Rev. Lett. **90**, 106601 (2003).

[38] S. I. Kiselev, J. C. Sankey, I. N. Krivorotov *et al.*, Microwave oscillations of a nanomagnet driven by a spin-polarized current, Nature **425**, 380–383 (2003).

[39] W. H. Rippard, M. R. Pufall, S. Kaka, S. E. Russek, and T. J. Silva, Direct-current induced dynamics in Co_90Fe_10/Ni_80Fe_20 point contacts, Phys. Rev. Lett. **92**, 027201 (2004).

[40] D. Houssameddine, U. Ebels, B. Delat, B. Rodmacq, I. Firastrau, F. Ponthenier, M. Brunet, C. Thirion, J.-P. Michel, L. Prejbeanu-Buda, M.-C. Cyrille, O. Redon, and B. Dieny, Spin-torque oscillator using a perpendicular polarizer and a planar free layer, Nature Mater. **6**, 447–453 (2007).

[41] S. Kaka, M. R. Pufall, W. H. Rippard, T. J. Silva, S. E. Russek, and J. A. Katine, Mutual phase-locking of microwave spin torque nano-oscillators, Nature **437**, 389–392 (2005).

[42] F. B. Mancoff, N. D. Rizzo, B. N. Engel, and S. Tehrani, Phase-locking in double-point-contact spin-transfer devices, Nature **437**, 393–395 (2005).

[43] I. N. Krivorotov, N. C. Emley, J. C. Sankey, S. I. Kiselev, D. C. Ralph, and R. A. Buhrman, Time-domain measurements of nanomagnet dynamics driven by spin-transfer torques, Science **307**, 228–231 (2005).

[44] A. A. Tulapurkar, Y. Suzuki, A. Fukushima *et al.*, Spin-torque diode effect in magnetic tunnel junctions, Nature **438**, 339–342 (2005).

[45] J. C. Sankey, P. M. Braganca, A. G. F. Garcia, I. N. Krivorotov, R. A. Buhrman, and D. C. Ralph, Spin-transfer-driven ferromagnetic resonance of individual nanomagnets, Phys. Rev. Lett. **96**, 227601 (2006).

[46] S. S. P. Parkin, M. Hayashi, and L. Thomas, Magnetic domain-wall recetrack memory, Science **320**, 190–194 (2008).

[47] D. A. Allwood, G. Xiong, C. C. Faulkner, D. Atkinson, D. Petit, and R. P. Cowburn, Magnetic domain-wall Logic, Science, **309** 1688–92. (2005).

[48] L. Gan, S. H. Chung, K. H. Aschenbach, M. Dreyer, and R. D. Gomez, Pulsed-current-induced domain wall propagation in permalloy patterns observed using magnetic force microscope, IEEE Trans. Magn. **36**, 3047–3049 (2000).

[49] J. Grollier, D. Lacour, V. Cros *et al.*, Switching the magnetic configuration of a spin valve by current-induced domain wall motion, J. Appl. Phys. **92**, 4825–4827 (2002).

[50] M. Kläui, C. A. F. Vaz, J. A. C. Blanda *et al.*, Domain wall motion induced by spin polarized currents in ferromagnetic ring structures, Appl. Phys. Lett. **83**, 105–107 (2003).

[51] M. Tsoi, R. E. Fontana, and S. S. P. Parkin, Magnetic domain wall motion triggered by an electric current, Appl. Phys. Lett. **83**, 2617–2619 (2003).

[52] T. Kimura, Y. Otani, I. Yagi, K. Tsukagoshi, and Y. Aoyagi, Suppressed pinning field of a trapped domain wall due to direct current injection, J. Appl. Phys. **94**, 7226–7229 (2003).

[53] N. Vernier, D. A. Allwood, D. Atkinson, M. D. Cooke, and R. P. Cowburn, Domain wall propagation in magnetic nanowires by spin-polarized current injection, Europhys. Lett. **65**, 526–532 (2004).

[54] A. Yamaguchi, T. Ono, S. Nasu, K. Miyake, K. Mibu, and T. Shinjo, Real-space observation of current-driven domain wall motion in submicron magnetic wires, Phys. Rev. Lett. **92**, 077205 (2004).

[55] E. Saitoh, H. Miyajima, T. Yamaoka, and G. Tatara, Current-induced resonance and mass determination of a single magnetic domain wall, Nature **432**, 203–206 (2004).

[56] S. Zhang and Z. Li, Roles of nonequilibrium conduction electrons on the magnetization dynamics of ferromagnets, Phys. Rev. Lett. **93**, 127204 (2004).

[57] G. Tatara and H. Kohno, Theory of current-driven domain wall motion: Spin transfer versus momentum transfer, Phys. Rev. Lett. **92**, 086601 (2004).

[58] S. E. Barnes and S. Maekawa, Current-spin coupling for ferromagnetic domain walls in fine wires, Phys. Rev. Lett. **95**, 107204 (2005).

[59] S. E. Barnes and S. Maekawa, Generalization of Faraday's law to include nonconservative spin forces, Phys. Rev. Lett. **98**, 246601 (2007).

[60] G. Tatara, H. Kohno and J. Shibata, Microscopic approach to current-driven domain wall dynamics, Phys. Rep. **468**, 213–301 (2008).

[61] A. Thiaville, Y. Nakatani, J. Miltat, and Y. Suzuki, Micromagnetic understanding of current-driven domain wall motion in patterned nanowires, Europhys. Lett. **69**, 990–996 (2005).

[62] G. S. D. Beach, C. Nistor, C. Knutson, M. Tsoi, and J. L. Erskine, Dynamics of field-driven domain-wall propagation in ferromagnetic nanowires, Nature Mater. **4**, 741–744 (2005).

[63] L. Thomas, M. Hayashi, X. Jiang, R. Moriya, C. Rettner, and S. S. P. Parkin, Oscillatory dependence of current-driven magnetic domain wall motion on current pulse length, Nature **443**, 197–200 (2006).

[64] G. S. D. Beach, C. Knutson, C. Nistor, M. Tsoi, and J. L. Erskine, Nonlinear domain-wall velocity enhancement by spin-polarized electric current, Phys. Rev. Lett. **97**, 057203 (2006).

[65] L. Thomas, M. Hayashi, X. Jiang, R. Moriya, C. Rettner, and S. S. P. Parkin, Resonant amplification of magnetic domain-wall motion by a train of current pulses, Science **315**, 1553–1556 (2007).

[66] M. Hayashi, L. Thomas, C. Rettner, R. Moriya, and S. S. P. Parkin, Direct observation of the coherent precession of magnetic domain walls propagating along permalloy nanowires. Nature Phys. **3**, 21–25. (2007).

[67] D. Chiba, Y. Sato, T. Kita, F. Matsukura, and H. Ohno, Semiconductor (Ga;Mn)As/GaAs/(Ga;Mn)As tunnel junction, Phys. Rev. Lett. **93**, 216602 (2004).

[68] M. Yamanouchi, D. Chiba, F. Matsukura, and H. Ohno, Current-induced domain-wall switching in a ferromagnetic semiconductor structure, Nature **428**, 539–542 (2004).

[69] M. Yamanouchi, D. Chiba, F. Matsukura, T. Dietl, and H. Ohno, Velocity of domain-wall motion induced by electrical current in the ferromagnetic semiconductor (Ga,Mn)As, Phys. Rev. Lett. **96**, 096601 (2006).

[70] G. Meier, M. Bolte, R. Eiselt, B. Kruger, D.-H. Kim, and P. Fischer, Direct imaging of stochastic domain-wall motion driven by nanosecond current pulses, Phys. Rev. Lett. **98**, 187202 (2007).

[71] J. Shibata, Y. Nakatani, G. Tatara, H. Kohno, and Y. Otani, Current-induced magnetic vortex motion by spin-transfer torque, Phys. Rev. B **73**, 020403(R) (2006).

[72] T. Ishida, T. Kimura, and Y. Otani, Current-induced vortex displacement and annihilation in a single permalloy disk, Phys. Rev. B **74**, 014424–1 (2006).

[73] Q. Mistral, M. van Kampen, G. Hrkac *et al.*, Current-driven vortex oscillations in metallic nanocontacts, Phys. Rev. Lett. **100**, 257201 (2008).

[74] S. Kasai, Y. Nakatani, K. Kobayashi, H. Kohno, and T. Ono, Current-driven resonant excitation of magnetic vortices, Phys. Rev. Lett. **97**, 107204 (2006).

[75] S. Kasai, Y. Nakatani, K. Kobayashi, H. Kohno, and T. Ono, Current-driven resonant excitation of magnetic vortices, Phys. Rev. Lett. **97**, 107204 (2006).

[76] J. Stohr, Y. Wu, B. D. Hermsmeier *et al.*, Element-specific magnetic microscopy with circularly polarized X-rays, Science **259**, 658–661 (1993).

[77] K. Yamada, S. Kasai, Y. Nakatani *et al.*, Electrical switching of the vortex core in a magnetic disk, Nature Mater. **6**, 269-273 (2007).

[78] M. Bolte, G. Meier, B. Kruger *et al.*, Time-resolved X-ray microscopy of spin-torque-induced magnetic vortex gyration, Phys. Rev. Lett. **100**, 176601 (2008).

[79] V. S. Pribiag, I. N. Krivorotov, G. D. Fuchs *et al.*, Magnetic vortex oscillator driven by d.c. spin-polarized current, Nature Phys. **3**, 498–503 (2007).

[80] A. Manchon, and S. Zhang, Theory of nonequilibrium intrinsic spin torque in a single nanomagnet, Phys. Rev. B **78**, 212405 (2008).

[81] A. Manchon, and S. Zhang, Theory of spin torque due to spin–orbit coupling, Phys. Rev. B **79**, 094422 (2009).

[82] I. M. Miron, G. Gaudin, S. Auffret, B. Rodmacq, A. Schuhl, S. Pizzni, J. Vogel and P. Gambardella, Current-driven spin torque induced by the Rashba effect in a ferromagnetic metal layer, Nature Mater. **9**, 230–234 (2010).

[83] J. Fernandez-Rossier, M. Braun, A. S. Nunez, and A. H. MacDonald, Influence of a uniform current on collective magnetization dynamics in a ferromagnetic metal, Phys. Rev. B **69**, 174412 (2004).

[84] V. Vlaminck and M. Bailleul, Current-induced spin-wave Doppler shift, Science **322**, 410–413, 2008.

[85] J. C. Sankey, Y.-T. Cui, J. Z. Sun, J. C. Slonczewski, R. A. Buhrman, and D. C. Ralph, Measurement of the spin-transfer-torque vector in magnetic tunnel junctions, Nature Phys. **4**, 67–71 (2008).

[86] H. Kubota, A. Fukushima, K. Yakushiji *et al.*, Quantitative measurement of voltage dependence of spin-transfer torque in MgO-based magnetic tunnel junctions, Nature Phys. **4**, 37–41 (2008).

[87] M. D. Stiles and J. Miltat, Spin dynamics in confined magnetic structures in *Topics in Applied Physics*, vol. 101, B. Hillebrands and A. Thiavilles (eds.), Springer, Heidelberg, 2006, pp. 225–308.

[88] B. Georges, J. Grollier, V. Cros, and A. Fert, Impact of the electrical connection of spin transfer nano-oscillators on their synchronization: An analytical study, Appl. Phys. Lett. **92**, 232504 (2008).

20 Spin-transfer torque in uniform magnetization

Y. Suzuki

20.1 Torque and torquance in magnetic junctions

In this chapter, we discuss the effects of a spin current injected into a uniformly magnetized ferromagnetic cell. In Fig. 20.1, a schematic of a typical magnetoresistive junction with in-plane magnetization is shown. The junction consists of two ferromagnetic layers (e.g. Co and Fe) separated by a nonmagnetic metal interlayer (e.g. Cu, Cr, etc.) or insulating barrier layer (e.g. AlO and MgO). With

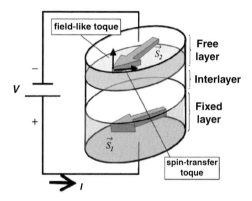

FIG. 20.1. Typical structures of magnetic nanopillars designed for spin injection magnetization switching (SIMS) experiments. The upper magnetic layer in the bird's-eye-view images acts as a magnetically free layer, whereas the lower magnetic layer is thicker than the free layer and acts as a spin polarizer. The spins in the lower layers are usually pinned to an antiferromagnetic material, which is placed below the pinned layer (not shown), as a result of exchange interaction at the bottom interface. Therefore, the spin polarizer layer is often called the "pinned layer," "fixed layer," or "reference layer." The interlayer between the two ferromagnetic layers is made of insulators such as MgO or nonmagnetic metals such as Cu. Here, the diameter of the pillar is around 100 nm. The free layer is typically a few nanometers thick. The large arrows indicate the direction of the total spin moment in each layer, and the two small arrows indicate two different spin torques.

a nonmagnetic metal interlayer, the junction is called a giant magnetoresistive (GMR) nanopillar, and with an insulating barrier layer a magnetic tunnel junction (MTJ).

The lateral shape of the pillar with in-plane magnetization is an ellipse or a rectangle with dimensions of about 200 nm × 100 nm or less. The angular momentum in the fixed layer, $\vec{S}_1$, which is opposite to the magnetization, is fixed along the long axis of the ellipse through an exchange interaction with an antiferromagnetic layer (e.g. PtMn). Without current injection, the angular momentum in the free layer, $\vec{S}_2$, also lies along the long axis of the ellipse because of magnetostatic shape anisotropy and is either parallel (P) or antiparallel (AP) with respect to $\vec{S}_1$. To induce asymmetry between the two magnetic layers, the thickness of the free layer is often less than that of the fixed layer.

When charge current is passed through this device, the electrons are first spin polarized by the fixed layer and then spin-polarized current is injected into the free layer through the nonmagnetic interlayer. This spin current interacts with the spins in the host material by an exchange interaction and exerts a torque. If the exerted torque is large enough, magnetization in the free layer is reversed or continuous precession is excited.

Such an electric current-induced spin torque in magnetic multilayers was first predicted theoretically [1, 2] and subsequently observed experimentally in metallic nanojunctions by excitation of spin waves [3] and spin-injection magnetization switching [4, 5]. The effects of the spin torque was also claimed to be observed in a perovskite system [6]. Further, spin-injection magnetization switching was also observed in magnetic tunnel junctions (MTJs) with an AlO barrier [7], and a MgO barrier [8, 9], and in magnetic semiconductor systems [10].

To simplify the problem, let us imagine an electron system in which the conduction electrons (s-electrons) and the electrons that hold local magnetic moments (d-electrons) interact with each other through exchange interactions (Fig. 20.2a). The exchange interaction (s-d exchange interaction) conserves the total spin angular momentum. Therefore, a decrease in the subtotal angular momentum of the conduction electrons equals increase in the subtotal angular momentum of the d-electron system. In the magnetic pillar, if the spin angular momentum of a conduction electron changes because of the s-d interaction during transport through the free layer, this amount of angular momentum should be transferred to the d electrons in the free layer. Therefore,

$$\frac{d\vec{S}_2}{dt} = \vec{I}_1^S - \vec{I}_2^S,$$ (20.1)

where $\vec{S}_2$ is the total angular momentum in the free layer. The spin currents $\vec{I}_1^S$ and $\vec{I}_2^S$ are obtained by integrating the spin current density flowing in the nonmagnetic interlayer or insulating barrier layer and nonmagnetic capping layer, respectively, over the cross-sectional area of the pillar. Since the free layer is

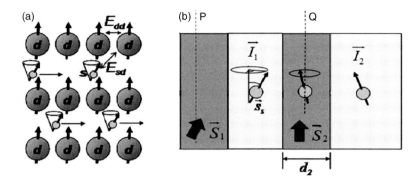

FIG. 20.2. (a) A simple *s-d* model to describe the spin-transfer effect. s-electrons flow among the localized *d*-electrons and contribute to a charge and spin current, while *d*-electrons create a single large local magnetic moment because of strong *d-d* exchange interaction. The *s-d* exchange interaction causes a precession of *s*- and *d*-electrons. Since *d*-electrons create a single large spin moment, the precession angle of the *d*-electron system is considerably smaller than that of *s*-electrons. (b) The injected spin shows a precession in a ferromagnetic layer as a consequence of the exchange interaction with *d*-local moments.

very thin, we neglected the spin–orbit interaction in it. Equation (20.1) indicates that a torque can be exerted on the local angular momentum as a result of spin transfer from the conduction electrons. This type of torque, which appears in Eq. (20.1), is called the "spin-transfer torque."

J. C. Slonczewski showed an intuitive way to evaluate the spin-transfer torque in MTJs [11, 12] by evaluating the spin currents inside ferromagnetic layers. We assume that the fixed layer is sufficiently thick; therefore, at cross-section P in the fixed layer (see Fig. 20.1), the conducting spins are relaxed and aligned parallel to $\vec{S}_1$. Those spin-polarized electrons are injected into the free layer. The injected spins are subjected to an exchange field made by the local magnetization and show precession motion. Here, we also assume that at cross-section Q inside the free layer, the spins of the conducting electrons have already lost their transverse spin component on average because of the decoherence of the precessions, and the spins have aligned parallel to $\vec{S}_2$ on average. Therefore, the spin currents at P and Q, $\vec{I}'^S_1$ and $\vec{I}'^S_2$, are parallel to $\vec{S}_1$ and $\vec{S}_2$, respectively. Since the spins of the conduction electrons at P and Q are either the majority or minority spins of the host material, the total charge current in the MTJ can be expressed as a sum of the following four components of the charge current as shown in Fig. 20.3:

$$I^C = I^C_{++} + I^C_{+-} + I^C_{-+} + I^C_{--} \tag{20.2}$$

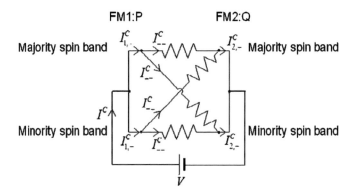

FIG. 20.3. Circuit model of a magnetic tunnel junction (after Ref. [12]).

Here, the suffixes $+$ and $-$ indicate the majority and minority spin channels, respectively. For example, I^C_{+-} represents a charge current flow from the fixed layer minority spin band into the free layer majority spin band. These charge currents are expressed using the conductance for each spin subchannel, $G_{\pm\pm}$.

$$\begin{cases} I^C_{\pm\pm} = V\, G_{\pm\pm} \cos^2 \frac{\theta}{2} \\ I^C_{\mp\pm} = V\, G_{\mp\pm} \sin^2 \frac{\theta}{2}. \end{cases} \tag{20.3}$$

Here, V is the applied voltage. The angle dependence of the conduction can be derived from the fact that the spin functions in the free layer are $|\text{maj.}\rangle = \cos(\theta/2)\,|{\uparrow}\rangle + \sin(\theta/2)\,|{\downarrow}\rangle$ for the majority spins and $|\text{min.}\rangle = \sin(\theta/2)\,|{\uparrow}\rangle - \cos(\theta/2)\,|{\downarrow}\rangle$ for the minority spins, and those in the fixed layer are $|{\uparrow}\rangle$ and $|{\downarrow}\rangle$, respectively. Since the spin quantization axes at P and Q are parallel to $\vec{S}_1$ and $\vec{S}_2$, respectively, the spin currents at P and Q are obtained easily as follows:

$$\begin{cases} \vec{I'}^S_1 = \frac{\hbar}{2} \frac{1}{-e} \left(I^C_{++} + I^C_{+-} - I^C_{-+} - I^C_{--} \right) \vec{e}_1 \\ \vec{I'}^S_2 = \frac{\hbar}{2} \frac{1}{-e} \left(I^C_{++} - I^C_{+-} + I^C_{-+} - I^C_{--} \right) \vec{e}_2, \end{cases} \tag{20.4}$$

where the unit vectors $\vec{e}_1$ and $\vec{e}_2$ are parallel to the majority spins in the fixed layer and the free layer, respectively. $-e$ and $\hbar/2 = h/(4\pi)$ are the charge and angular momentum of a single electron, respectively. Now, we apply the conservation of total angular momentum between the planes P and Q, i.e.

$$\left(\frac{d}{dt} \left(\vec{S}_1 + \vec{S}_2 \right) \right)_{ST} = \vec{I'}^S_1 - \vec{I'}^S_2. \tag{20.1'}$$

Here, we include angular momentum outside the planes since $\vec{S}_1$ and $\vec{S}_2$ move as macro-spins. Then, after a straightforward calculation, the total current and spin-transfer torque are obtained as follows:

$$
\begin{cases}
I^C = \left(\dfrac{G_{++}+G_{--}+G_{+-}+G_{-+}}{2} + \dfrac{(G_{++}+G_{--})-(G_{+-}+G_{-+})}{2} \vec{\mathbf{e}}_2 \cdot \vec{\mathbf{e}}_1 \right) V \\[2mm]
\quad = \left(\dfrac{G_P+G_{AP}}{2} + \dfrac{G_P-G_{AP}}{2} \cos\theta \right) V \\[2mm]
\left(\dfrac{d\vec{S}_2}{dt} \right)_{ST} = \dfrac{\hbar}{2} \dfrac{1}{-e} \left(\dfrac{G_{++}-G_{--}}{2} + \dfrac{G_{+-}-G_{-+}}{2} \right) (\vec{\mathbf{e}}_2 \times (\vec{\mathbf{e}}_1 \times \vec{\mathbf{e}}_2)) V \\[2mm]
\quad = T_{ST} (\vec{\mathbf{e}}_2 \times (\vec{\mathbf{e}}_1 \times \vec{\mathbf{e}}_2)) V.
\end{cases}
\tag{20.5}
$$

The first equation in Eq. (20.5) shows the $\cos\theta$ dependence of the tunnel conductance. The second equation shows the $\sin\theta$ dependence of the spin torque (note that $|\vec{\mathbf{e}}_2 \times (\vec{\mathbf{e}}_1 \times \vec{\mathbf{e}}_2)| = \sin\theta$). J. C. Slonczewski called T_{ST} the "torquance," which is an analogue of "conductance." In particular, in MTJs, the spin torques should be bias-voltage dependent because G_{++} is bias-voltage dependent. The direction of the spin-transfer torque is shown in Fig. 20.1(a) for $T_{ST} < 0$. This direction is the same as that in CPP GMR junctions (current perpendicular to the plane of the GMR junction). For a GMR junction, we should replace the second equation in Eq. (20.5) by the following equation,

$$
\begin{cases}
\left(\dfrac{d\vec{S}_2}{dt} \right)_{ST} = g\left(\theta\right) \dfrac{I^C}{-e} \dfrac{\hbar}{2} \vec{\mathbf{e}}_2 \times (\vec{\mathbf{e}}_1 \times \vec{\mathbf{e}}_2) \\[2mm]
g\left(\theta\right) = \left[-4 + \left(P^{-\frac{1}{2}} + P^{\frac{1}{2}} \right)^3 (3+\cos\theta)/4 \right]^{-1},
\end{cases}
\tag{20.5$'$}
$$

where $I^C/(-e)$ is the number of electrons flowing per unit time (I^C is the charge current). $g\left(\theta\right)$ expresses the efficiency of spin transfer obtained for the free-electron case [1] and is dependent on the spin polarization, P, of the conduction electron in the ferromagnetic layers and the relative angle between $\vec{S}_1$ and $\vec{S}_2$, i.e. θ.

20.2   Voltage dependence and field-like torque

One of the important features in MTJs is that the torque has a bias-voltage dependence because G_{++} has bias-voltage dependence. In Fig. 20.4(a), the theoretically predicted spin-transfer torque is plotted as a function of the bias voltage by fine lines [13]. As shown in the figure, the bias dependence of the spin-transfer torque is neither monotonic nor symmetric. The torque will be much higher at large negative bias even if the magnetoresistance is smaller at such a high bias. This slightly complicated behavior can be explained as follows from the second line in Eq. (20.5). Assume that the FM1 and FM2 are made of the same material. Now if we apply a voltage, due to the symmetric conditions for tunneling, the conductances G_{++} and G_{--} do not depend on the sign of the voltage. Thus the contribution to the torque from the $(G_{++} - G_{--}) \times V$ term

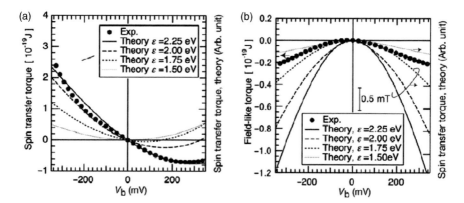

FIG. 20.4. Bias dependence of the spin torques. (a) Bias dependence of the spin-transfer torque. (b) Bias dependence of the field-like torque. Fine curves were obtained from the theoretical model calculations (after Ref. [13]) with different spin splitting parameters, ε. The points show the experimental results obtained for a CoFeB/MgO/CoFeB MTJ by exploiting the spin-torque diode effect (after Ref. [15]).

is odd with respect to the voltage. The conductances G_{+-} and G_{-+} are equal at zero bias. For positive voltage, G_{+-} decreases, whereas G_{-+} increases, thus giving a positive contribution to the spin torque. For negative voltage, due to the symmetry, we have the opposite situation and $(G_{+-} - G_{-+})$ changes sign. But as the voltage is also negative, the net contribution is again positive. Therefore, the $(G_{+-} - G_{+-}) \times V$ term is an even function of the voltage. This combination of odd and even terms gives an asymmetry in the spin torque as a function of voltage.

J. C. Sankey *et al.* [14] and H. Kubota *et al.* [15] experimentally observed the bias dependence of the torques. In Fig. 20.4(a), the experimentally obtained bias voltage dependence of the spin-transfer torque in a CoFeB/MgO/CoFeB magnetic tunnel junction is plotted with large circles [15]. The torque was measured by using the "spin-torque diode effect," which will be explained later in this chapter. The experimental observations [14, 15] essentially agree with the model calculation. Recent measurements also provide high-accuracy torque values avoiding possible considerable error in the high bias region [16].

The direction of $\vec{S}_2$ may change in two different ways. One is along the direction parallel to the spin-transfer torque, $(\vec{e}_2 \times (\vec{e}_1 \times \vec{e}_2))$. The other is in the direction parallel to $(\vec{e}_1 \times \vec{e}_2)$ (see Fig. 20.1a). If the torque is parallel to $(\vec{e}_1 \times \vec{e}_2)$, it has the same symmetry as the torque exerted by an external field. Therefore, the latter torque is called a field-like torque. It can also be called an "accumulation torque" based on one of its possible origins, or a "perpendicular torque" based on its direction with respect to the plane that includes both $\vec{e}_1$ and $\vec{e}_2$.

It has been pointed out that one of the important origins of the field-like torque in MTJs is the change in the interlayer exchange coupling through the barrier layer at a finite biasing voltage [13, 17]. In Figure 20.4(b), the theoretically obtained strength of the field-like torque is plotted as a function of the bias voltage [13]. As theoretically predicted for symmetrical MTJs, the field-like torque is an even function of the bias voltage. Its strength itself is less than 1/5 that of the spin-transfer torque. The experimental results obtained so far [15] seem to agree with this prediction.

The field-like torque could originate from other mechanisms that are similar to those responsible for the "β-term" in magnetic nanowires. Several mechanisms, such as spin relaxation [18, 19], Gilbert damping itself [20], momentum transfer [21], or a current-induced ampere field have been proposed for the origin of the β-term.

20.3   Landau–Lifshitz–Gilbert (LLG) equation in Hamiltonian form

To treat the dynamic property of the free-layer spin angular momentum, here we introduce the Landau–Lifshitz–Gilbert (LLG) equation including the spin-transfer torque and field-like torque as follows [1, 12, 22]:

$$\frac{d\vec{S}_2}{dt} = \gamma \vec{S}_2 \times \vec{H}_{eff} - \alpha \vec{e}_2 \times \frac{d\vec{S}_2}{dt}$$
$$+ T_{ST}(V) \, V \, \vec{e}_2 \times (\vec{e}_1 \times \vec{e}_2) + T_{FT}(V) \, V \, (\vec{e}_2 \times \vec{e}_1), \qquad (20.6)$$

The first term is the effective field torque; the second, the Gilbert damping; the third, the spin-transfer torque; and the fourth, the field-like spin torque. $\vec{S}_2 = S_2 \vec{e}_2$ is the total spin angular momentum of the free layer and is opposite to its magnetic moment, $\vec{M}_2$. If we may neglect the contribution from the orbital moment, $\mu_0 \vec{M}_2 = \gamma \vec{S}_2$, where $\mu_0 = 4\pi \times 10^{-7}$ [H/m] is the magnetic susceptability of the vacuum and γ is the gyromagnetic ratio, where $\gamma < 0$ for electrons ($\gamma = -2.21 \times 10^5$ [m/A · sec] for a free electron). $\vec{e}_2$ ($\vec{e}_1$) is a unit vector that expresses the direction of the spin angular momentum of the free layer (fixed layer). For simplicity, we neglect the distribution of the local spin angular momentum inside the free layer and assume that the local spins within each magnetic cell are aligned in parallel and form a coherent "macro-spin" [23, 24]. This assumption is not strictly valid since the demagnetization field and current-induced Oersted field inside the cells are not uniform. Such nonuniformities introduce incoherent precessions of the local spins and cause domain and/or vortex formation in the cell [24–26]. Despite the predicted limitations, the macro-spin model is still useful, because of both its transparency and its validity for small excitations. The effective field, $\vec{H}_{eff}$, is the sum of the external field, demagnetization field, and anisotropy field. It should be noted that the demagnetization field and the anisotropy field depend on $\vec{e}_2$. $\vec{H}_{eff}$ is derived from the magnetic energy, E_{mag}, and the total magnetic moment, M_2, of the free layer:

$$\vec{H}_{eff} = \frac{1}{\mu_0 M_2} \frac{\partial E_{mag}}{\partial \vec{e}_2}. \tag{20.7}$$

The first term in Eq. (20.6) determines the precession motion of $\vec{S}_2$. In the second term, α is the Gilbert damping factor ($\alpha > 0$, $\alpha \approx 0.007$ for Fe for example). V is the applied voltage, $T_{ST}(V) = \frac{\hbar}{2} \frac{1}{-e} \frac{1}{2} (G_{++} - G_{--} + G_{+-} - G_{-+})$ is the "torquance" that was defined in the second line in Eq. (20.5), while $T_{FT}(V)$ is an unknown coefficient that expresses the size of the "field-like torque."

The directions of the torques are illustrated in Fig. 20.5. The effective field torque promotes a precession motion of $\vec{S}_2$ around $-\vec{H}_{eff}$, while the damping torque tends to reduce the opening angle of the precession. By the effective field and damping torques, $\vec{S}_2$ exhibits a spiral trajectory and finally aligns antiparallel to the effective field if the junction current, I^C, is absent (Fig. 20.5a). It must be noted that the direction of $\vec{S}_2$ is opposite to that of its magnetic moment. The direction of the spin-transfer torque is also illustrated in Fig. 20.5(b) for the case where both $g(\theta)$ and I^C are positive. If the current, I^C, is sufficiently large, the spin-transfer torque overcomes the damping torque, resulting in negative effective damping. This negative damping results in an increase in the opening angle of the precession motion, i.e. an amplification of the precession takes place. Depending on the angular dependence of the effective damping, the amplification

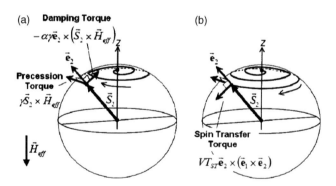

FIG. 20.5. Illustration of the direction of each torque and trajectory of the free-layer spin momentum for a nanopillar with perpendicular remnant magnetization. (a) In the absence of an electric current, the precession of the free layer spin is damped. (b) Under an electric current, if the spin-transfer torque overcomes the damping torque, the precession of the free-layer spin is amplified.

of the precession motion leads to a limit cycle (spin-transfer oscillation (STO)) or to total magnetization reversal (spin-injection magnetization switching (SIMS)).

To analyze the dynamics of the macro-spin system, we develop Hamilton's equation of motion of the system. Using spherical coordinate, i.e. (ϕ, θ), the Lagrangian and Rayleigh's dissipation function of the LLG equation (without spin torques) are expressed as follows [27];

$$\begin{cases} L\left(\phi, \dot{\phi}, \theta, \dot{\theta}\right) = S_2 \dot{\phi} \left(\cos\theta - 1\right) - E_{mag}\left(\phi, \theta\right) \\ W\left(\dot{\phi}, \dot{\theta}\right) = \frac{\alpha}{2} S_2 \left(\dot{\theta}^2 + \dot{\phi}^2 \sin^2\theta\right). \end{cases} \qquad (20.8)$$

The kinetic energy term in the Lagrangian is known as the spin Berry phase term. In classical mechanics, this term also results in the equation of motion of angular momentum. From the above Lagrangian, we may find the following Hermitian conjugate valuables:

$$\begin{cases} x^1 \equiv \phi \\ x^2 \equiv S_2 \left(\cos\theta - 1\right). \end{cases} \qquad (20.9)$$

Using this new coordinate system, Eq. (20.6) can be rewritten as

$$\dot{x}^i = \sum_{j=1}^{2} \varepsilon^{ij} \left(\partial_j E_{mag} - T_{FT} V \partial_j \left(\vec{e}_2 \cdot \vec{e}_1\right) + \alpha S_2 \dot{x}_j\right)$$

$$+ S_2^{-1} T_{ST} V \partial^i \left(\vec{e}_2 \cdot \vec{e}_1\right), \quad (i = 1, 2), \qquad (20.10)$$

where,

$$\begin{pmatrix} \partial_i \equiv \frac{\partial}{\partial x^i}, \qquad \partial^i \equiv \sum_{j=1}^{2} g^{ij} \partial_j \\ \left(\varepsilon^{ij}\right) \equiv \begin{pmatrix} 0 & 1 \\ -1 & 0 \end{pmatrix} \\ \left(g_{ij}\right) = \begin{pmatrix} \sin^2\theta & 0 \\ 0 & \frac{1}{S^2 \sin^2\theta} \end{pmatrix} = \left(g^{ij}\right)^{-1}. \end{pmatrix} \qquad (20.11)$$

Here, ε is the Levi-Civita symbol and g is the metric tensor. An explict form of Eq. (20.10) with respect to $\dot{x}^i$ can be obtained easily:

$$\dot{x}^i = F^i,$$

where

$$\begin{cases} F^i \cong \sum_{j=1}^{2} \varepsilon^{ij} \partial_j E_{mag+FT} - \alpha S^{-1} \partial^i E_{mag+ST} \\ E_{mag+FT} \equiv E_{mag} - T_{FT} V \left(\vec{e}_2 \cdot \vec{e}_1\right) \\ E_{mag+ST} \equiv E_{mag} - \alpha^{-1} T_{ST} V \left(\vec{e}_2 \cdot \vec{e}_1\right). \end{cases} \qquad (20.12)$$

Here, terms with α^2, αT_{ST}, and αT_{FT} are neglected. We also assumed $\partial_j T_{FT} = \partial_j T_{ST} = 0$. In Eq. (20.12), we clearly see that the spin-transfer term, which is a consequence of the spin current, directly affects the damping term. This Hamilton-type equation of motion in orthogonal curvilinear coordinates is useful to obtain an analytic understanding of the dynamics of spin-transfer torque.

20.4 Small-amplitude dynamics and anti-damping

20.4.1 Linearized LLG equation

Equations (6), (10), and (12) are all equivalent and describe the nonlinear response of a macro-spin in the junction under an applied magnetic field and a voltage. Before we discuss nonlinear behavior like switching, we derive a linearized equation of motion and discuss infinitesimal excitations [22].

The equilibrium point of the macro-spin, (x_0^1, x_0^2), under a static external field and dc bias voltage, (H_0^{ext}, V_0), can be obtained by solving

$$\left. \begin{pmatrix} \partial_2 E_{mag} - T_{FT} V_0 \partial_2 (\vec{e}_2 \cdot \vec{e}_1) + S^{-1} T_{ST} V_0 \partial^1 (\vec{e}_2 \cdot \vec{e}_1) \\ -\partial_1 E_{mag} + T_{FT} V_0 \partial_1 (\vec{e}_2 \cdot \vec{e}_1) + S^{-1} T_{ST} V_0 \partial^2 (\vec{e}_2 \cdot \vec{e}_1) \end{pmatrix} \right|_{(x_0^1, x_0^2)} = 0. \quad (20.13)$$

The linearized equation of motion is obtained by taking the deviation from the equilibrium point as new coordinates, i.e. $\left(x^1(t), x^2(t)\right) = \left(x_0^1 + \delta x^1(t), x_0^2 + \delta x^2(t)\right)$:

$$\delta \dot{x}^i(t) = \sum_{j=1}^{2} \delta x^j(t) \left. \partial_j F^i \right|_{(x_0^1, x_0^2)} + \left. \frac{\partial F^i}{\partial V} \delta V(t) \right|_{(x_0^1, x_0^2)}, \quad (20.14)$$

where $\delta V(t)$ is the time-dependent part of the bias voltage. The solution of the linearized LLG equation (20.14) is forced oscillatory motion around the equilibrium point, driven by a small rf voltage with frequency ω. Using the Fourier transformation, $\delta x(t) = \int d\omega \delta x(\omega) e^{-i\omega t}$, the above equation can be solved as follows:

$$\begin{pmatrix} \delta x^1(\omega) \\ \delta x^2(\omega) \end{pmatrix} \simeq \frac{1}{(\omega^2 - \omega_0^2 + i\omega \Delta \omega)} \begin{pmatrix} i\omega & -\Omega_{22} \\ \Omega_{11} & i\omega \end{pmatrix} \begin{pmatrix} -S_2^{-1} T'_{FT} \\ T'_{ST} \sin^2 \theta \end{pmatrix} \delta V(\omega), \quad (20.15)$$

where

$$\begin{cases} \hat{\Omega} \equiv (\Omega_{i,j}) \equiv \partial_j \partial_i E_{mag} \\ T'_{FT} \equiv \frac{\partial}{\partial V} T_{FT} V \\ T'_{ST} \equiv \frac{\partial}{\partial V} T_{ST} V. \end{cases}$$

Here, for a simplicity, we assumed orthogonal symmetry ($\Omega_{mag,12} = \Omega_{mag,21} = 0$) of the system and took the north pole of the spherical coordinates parallel to the spin direction of the fixed layer, i.e. $\vec{e}_1 = \vec{e}_z$ and $(\vec{e}_2 \cdot \vec{e}_1) = \cos \theta$. The resonance frequency, ω_0, and the full width at half maximum (FWHM) of the resonance, $\Delta\omega$, are given by the following equations:

$$\begin{cases} \omega_0^2 \cong \det\left[\hat{\Omega}\right] = \det\left[\partial_j \partial_i E_{mag}\right] \\ \Delta\omega \cong \alpha S_2^{-1} \vec{\nabla} \cdot \vec{\nabla} E_{mag+ST} = \alpha S_2^{-1} \sum_{i=1}^{2} \partial_i \partial^i E_{mag} + 2S_2^{-1} T_{ST} V_0 \cos\theta. \end{cases}$$

$$(20.16)$$

Equation (20.15) shows that both the spin-transfer torque and the field-like torque can excite a uniform mode (FMR mode) in the free layer. However, the phases of the FMR excitations differ by 90°. This difference in the precession phase is a result of the different directions of the respective torques (Fig. 20.1). In addition, the width of the resonance, $\Delta\omega$, is affected only by the spin-transfer torque exerted by the direct voltage, V_0. This is the (anti)damping effect of the spin-current injection. For a large dc bias, if $\Delta\omega$ becomes negative, the system is no longer stable and magnetization switching or auto-oscillation will take place. The field-like torque exerted by the direct voltage, V_0, changes the resonance frequency, ω_0, through a change in the equilibrium point.

20.4.2   Spin-torque diode effect

Both spin-transfer torque and field-like torque may excite a uniform mode of the magnetic free layer in magnetoresistive junctions. We may also observe it only by measuring the dc voltage across the junction as a function of the frequency of the applied rf voltage.

In Fig. 20.6, a mechanism for the rectification effect in magnetic tunnel junctions is schematically explained. To observe the spin-torque diode effect, we may apply an external field to set a specific relative angle between the free-layer and fixed-layer magnetizations. In Fig. 20.6(b), we show the case in which the free-layer and fixed-layer magnetizations are in-plane but perpendicular to each other. We then apply an alternative current to the junction. A negative

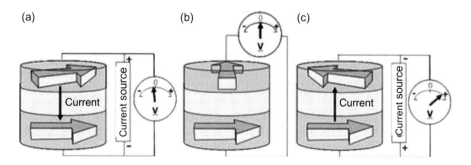

FIG. 20.6. Schematic explanation of the spin-torque diode effect: (a) negative current; (b) null current; (c) positive current (after Ref. [28]).

current induces a preferential parallel configuration of the spins. Thus, the resistance of the junction becomes smaller and we observe only a small negative voltage across the junction for a given current (Fig. 20.6a). A positive current of the same amplitude induces a preferential antiparallel configuration and the resistance becomes higher. We observe a larger positive voltage appearing across the junction (Fig. 20.6c). As a result, we observe a positive voltage on average. This is the spin-torque diode effect. This effect can be large if the frequency of the applied current matches the FMR frequency of the free layer. In other words, this effect provides a sensitive FMR measurement technique for the nanopillar moment excited by the spin torque and provides a quantitative measure of the spin torques.

From Eq. (20.5), an applied rf voltage across the MTJ of $\delta V \cos \omega t$ may cause a precession in the free-layer magnetization with the same frequency, $\delta\theta(t) = \delta\theta_0 \cos(\omega t + \phi)$. This also results in an oscillation of the junction resistance at the same frequency $\delta R(t) = \delta\theta(t)\partial_\theta R(\theta)$. The voltage across the junction induced by an application of the rf voltage is obtained by multiplying the oscillating part of the resistance and the current, i.e.

$$\delta\theta_0 \cos(\omega t + \varphi)(\partial_\theta R)\frac{\delta V \cos \omega t}{R} = \delta\theta_0 (\partial_\theta R)\frac{\delta V}{2R}(\cos\varphi + \cos(2\omega t + \varphi)).$$
$$(20.17)$$

Here, the frequencies of the induced voltages are zero (dc) and 2ω. This implies that, under spin-torque FMR excitation, the MTJs may possess a rectification function and a mixing function. Because of these new functions, A. A. Tulapulkar et al. referred to these MTJs as spin-torque diodes and to these effects as spin-torque diode effects [28]. These are nonlinear effects that result from two linear responses, i.e. the spin-torque FMR and Ohm's law.

When the MTJ is placed at the end of a waveguide, the explicit expression for the rectified dc voltage under a small bias voltage is given as follows

$$V_{dc,out} \cong \eta\frac{\partial_\theta \log(R(\theta))}{2S_2 \sin\theta}Re\left[\frac{-i\omega T'_{ST}\sin^2\theta + S_2^{-1}T'_{FT}\Omega_{11}}{\omega^2 - \omega_0^2 + i\omega\Delta\omega}\right]\delta V^2 \quad (20.18)$$

where δV is the rf voltage amplitude applied to the emission line and η is the coefficient used to correct the impedance matching between the MTJ and the waveguide with a characteristic impedance of Z_0, where

$$\eta = \left(\frac{2R(\theta)}{R(\theta) + Z_0}\right)^2. \quad (20.19)$$

If the emission line and the MTJ include some parasitic impedances (capacitance in most cases), we should employ an appropriate value of η to correct this effect [15].

This is a type of homodyne detection and is, thus, phase-sensitive. The motion of the spin, illustrated in Fig. 20.6, corresponds to that excited by the spin-transfer torque at the resonance frequency. However, the motion of the spin excited by the field-like torque shows a 90° difference in phase. As a consequence, only the resonance excited by the spin-transfer torque can rectify the rf current at the resonance frequency. In Fig. 20.7, the dc voltage spectra predicted for the spin-transfer torque excitation and for the field-like torque are both shown. The spectrum excited by the spin-transfer torque exhibits a single bell-shaped peak (dashed line), whereas that excited by the field-like torque is of dispersion type (dotted line). This very clear difference provides us with an elegant method to distinguish a spin-transfer torque from a field-like torque [28].

Figure 20.8 shows a schematic illustration of the measurement setup for spin-torque diode effect measurements with a cross-sectional view of the MTJ employed in ref. [28]. The rf voltage was applied through a bias-T from a high-frequency oscillator and the dc voltages across the MTJ was detected using a dc nanovoltmeter.

In Fig. 20.7, an example of the diode spectrum (closed dots) is shown together with a fitting curve based on the theoretical expression (Eq. (20.18)). The data were taken at room temperature (RT) without applying dc bias voltage. The observed spectrum has an asymmetrical shape and was well fitted by Eq. (20.18). By this fitting, the spectrum was decomposed into a contribution from the spin-transfer term and from the field-like term. The intensity and even the sign of the field-like term contribution at zero bias varied from sample to sample, while those of the spin-transfer term were reproducible. Therefore, it is thought that the contribution from the field-like term at zero bias voltage is very sensitive to small

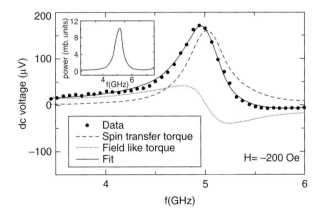

FIG. 20.7. Spin-torque diode spectra for a CoFeB/MgO/CoFeB MTJ. Data (closed dots) are well fitted by a theoretical curve that includes contributions from both the spin-transfer torque and the field-like torque. The inset shows an rf noise spectrum obtained for the same MTJ (after Ref. [28]).

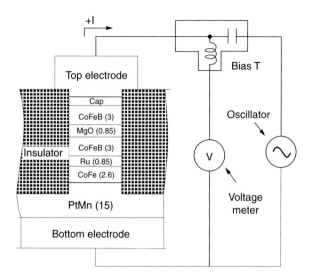

FIG. 20.8. Schematic diagram of the setup for measuring the spin-torque diode
    effect.

defects in the magnetic cell. By taking a sample that does not show a contribution
from the field-like term at zero bias, H. Kubota *et al.* have investigated the dc bias
voltage dependence of the spectra [15]. The results, which were already shown in
Figure 20.4, were well explained by band theory, in which bias-dependent spin-
subchannel conductivities and bias-dependent interlayer magnetic coupling were
taken into account [13].

   An expression for the rectified dc voltage at the peak of the spectrum for
$\theta = \pi/2$ is shown below together with that for $p - n$ junction semiconductor
diodes:

$$
(V_{dc,out})_{peak} = \begin{cases} \frac{1}{4} \frac{G_P - G_{AP}}{G_P + G_{AP}} \frac{\delta V^2}{V_c} & \text{(spin-torque diode)} \\ \frac{1}{4} \frac{\delta V^2}{k_B T/e} & \text{(p-n junction semiconductor diode)} \end{cases}, \quad (20.20)
$$

where $k_B T/e$ is the thermal voltage (25 mV at RT). For both cases, the rectified
voltage is a quadratic function of the applied rf voltage. Therefore, these detectors
are referred to as quadratic detectors. The output voltage is scaled by the critical
switching voltage, V_c, for the spin-torque diode and by $k_B T/e$ for the $p - n$
junction semiconductor diode. A typical critical switching voltage for MTJs was
about 300 mV and was about 10 times larger than $k_B T/e$ for the experiments
in ref. [8, 28]. Therefore, the output of the spin-torque diode was smaller than
that of the semiconductor diode. A reduction of the critical switching voltage
by using perpendicular magnetic anisotropy and magnetic field may enhance the
performance of the spin-torque diode [29].

20.5 Spin-transfer magnetization switching

In Fig. 20.9, a typical fabrication process for nanopillars from a magnetic tunnel junction (for research purposes) is shown. First, (a) a magnetic multilayer including a magnetic tunnel junction is sputter deposited. The multilayer consists of a bottom electrode layer, an antiferromagnetic exchange bias layer (MnPt, for example), a synthetic antiferromagnetic pinned layer (CoFeB/Ru/CoFe, for example), an MgO barrier layer, a magnetic free layer (CoFeB, for example), and a capping layer. The multilayer is then covered by a resist layer using a spin coater and transferred to an electron beam lithography machine. (b) After exposure and development, the sample with micro-patterned resist is transferred to an ion beam milling machine to remove parts of the multilayers and form magnetic pillars. (c) The outer side of the pillar is filled by a SiO_2 insulating layer. (d) The SiO_2 layer on the junction is lifted off with the resist by using

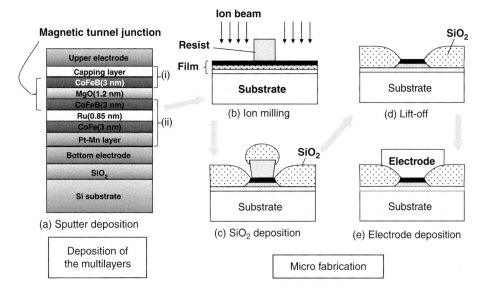

FIG. 20.9. An example of the sample fabrication process for a SIMS (spin-injection magnetization switching) experiment. (a) A magnetic multilayer including the tunneling barrier is first deposited under vacuum by a sputtering method. For memory applications, the film is deposited onto a C-MOS and wiring complex after a chemical-mechanical planarization process. (b) After resist coating is applied using a spin coater, the resist is patterned by electron beam lithography. Using the patterned resist, part of the film is etched by ion beam bombardment. (c) Interlayer insulator (SiO_2) deposition using a self-alignment technique. (d) The lift-off process to open a contact hole. (e) Deposition of the upper electrode.

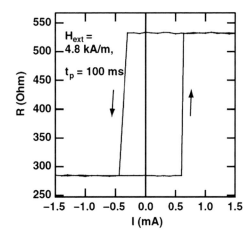

FIG. 20.10. A typical SIMS (spin-injection magnetization switching) hysteresis loop obtained for a CoFeB/MgO/CoFeB MTJ (magnetic tunnel junction) (after Kubota *et al.* [8]). The junction area, free-layer thickness, resistance area product, and MR ratio are 100 nm x 200 nm, 3 nm, 3 $\Omega\mu$ m^2, and about 100%, respectively. Measurements were performed at room temperature using electric current pulses of 100 msec duration. The resistance of the junction was measured after each pulse to avoid the effect of heating on the sample resistance.

a chemical solvent and ultrasonic scrubbing. Finally, (e) the top electrodes are deposited onto the junction under the vacuum.

A hysteresis loop obtained for a magnetic nanopillar comprising a CoFeB/MgO/CoFeB tunneling junction is shown in Fig. 20.10 [8]. The pillar has in-plane magnetization and elliptical cross-section with the dimensions 100 nm x 200 nm. A current was applied as a series of 100 msec wide pulses. In between the pulses, the sample resistance was measured to check the magnetization configuration while the pulse height was swept between −1.5 mA and +1.5 mA. By this method, the effect of the temperature increase during the application of the current on the resistance measurement could be eliminated. For the data shown in Fig. 20.16 below, the hysteresis measurement started at zero pulse height for the P state (285 Ω). An increase in the pulse height caused a jump from the P state to the AP state (560 Ω) at +0.6 mA. Further increase in the pulse height followed by a reduction to zero current did not affect the state. Subsequently, negative pulses were applied to the sample. At −0.35 mA, the sample switched its magnetization from the AP state to the P state. The average switching current density was about 6 × 10^6 A/cm^2. Intermediate resistance states between the P and the AP states were not observed during either of the two switching events: the switching events were always abrupt and complete. The slope of the hysteresis loop at the switching point is only due to discrete measurement points that were

not regularly placed because of the large change in the resistance. P to AP
and AP to P switching events occurred at different current levels because of
the dipole and the so-called orange peel coupling field from the pinned layer.
In the experiment, an external field of -4.8 kA/m was applied to cancel these
coupling fields. After the cancellation of the coupling fields, the hysteresis still
exhibited a certain shift because of the following intrinsic mechanisms. For the
MTJ nanopillars, the asymmetrical voltage dependence of the torque, which was
discussed in the previous section, causes a horizontal shift in the hysteresis curve.
For the GMR nanopillars, in contrast, the angle dependence of the spin-transfer
efficiency results in a significant shift in the hysteresis loop.

From Eq. (20.16), the critical current required to make the parallel (P) or
antiparallel configuration (AP) unstable can be obtained putting $\Delta\omega = 0$:

$$I_{c0} = \frac{V_{c0}}{R} = \frac{S_2 \Delta\omega_0}{2RT_{ST}}, \tag{20.21}$$

where $\Delta\omega_0$ is the linewidth for zero bias voltage. Above this voltage, the P or AP
configuration becomes unstable and magnetization switching or auto-oscillation
takes place.

Many efforts have been made to reduce the threshold current of the switching.
The first attempt is to reduce the total spin angular momentum, S_2, in the free
layer. SIMS requires effective injection of spin angular momentum that is equal to
that in the free layer. Therefore, reduction in S_2 results in a reduction in I_{c0}. F. J.
Albert showed that the threshold current of the SIMS is proportional to the free-
layer thickness [30]. Reduction in the free-layer thickness reduces S_2 and I_{c0}. S_2
can also be reduced by reducing the magnetization of the ferromagnetic material.
More especially, in a nanopillar with in-plane magnetization, since the magnetization also affects the size of the anisotropy field, I_{c0} is a quadratic function of
the magnetization. K. Yagami et al. reduced I_{c0} considerably by changing the
material of the free layer from CoFe (1.9×10^6 A/m) to CoFeB (0.75×10^6 A/m)
and obtained 1.7×10^7 A/cm^2 [31]. A second attempt is to use a double spin filter
structure. This method was originally proposed by L. Berger [32]. By using this
structure, Y. Huai et al. observed a substantial reduction of the threshold current
to 2.2×10^6 A/cm^2 [33]. The third attempt is to use perpendicular magnetic
anisotropy, which can reduce the size of the anisotropy field [34]. S. Yakata et al.
pointed that a free layer with a Fe-rich composition in FeCoB/MgO/FeCoB
stacking possesses a perpendicular crystalline anisotropy and results in reduction
of the switching current [35]. The perpendicular crystalline anisotropy partially
cancels the demagnetization field and reduces the size of the anisotropy field. By
this method, T. Nagase et al. obtained a significant reduction in the threshold
current under the required thermal stability factor for the MTJ nanopillars with
a CoFeB/[Pd/Co]x2/Pd free layer and a FePt/CoFeB pinned layer [36].

In Fig. 20.11, a trajectory for SIMS (a) is compared with a trajectory
for magnetic field-induced magnetization switching (b) in a nanopillar with
in-plane magnetization. The figure also illustrates the magnetic potential shapes

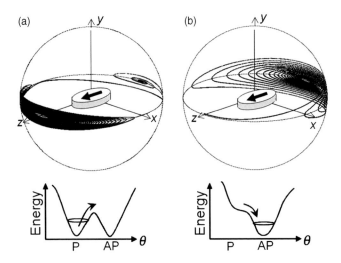

FIG. 20.11. Comparison of the magnetization processes driven by (a) spin-transfer torque and (b) an external magnetic field for in-plane magnetized nanopillars.

during switchings. In the absence of a current and an external magnetic field, the potential shows a double minimum for parallel (P) and antiparallel (AP) configurations of the local spin. For a particular case of SIMS, the spin-transfer torque does not affect the shape of the magnetic potential but amplifies the precession thereby providing energy to the local spin system. Once the orbital crosses the equator, it converges rapidly to the opposite direction since the spin-transfer torque extracts energy from the local spin system. In other words, the spin-transfer torque amplifies the precession in the front hemisphere, while enhancing the damping in the back hemisphere. In contrast to this process, the external magnetic field deforms the magnetic potential and the minimum on the P side disappears. Therefore, the local spin turns toward the AP side. The local spin system, however, keeps excess energy in the back hemisphere. As a result, it cannot stop at once and shows precessional motion (ringing) in the back hemisphere.

The interesting aspects of the spin-injection magnetization switching (SIMS) phenomenon are the small energy consumption and very high precession speed. To investigate the high-speed properties of the SIMS, high-speed pulse and high-speed time domain observations have been performed [35–41]. The first direct observation of precession switching was performed by Krivorotov et $al.$ for a $Ni_{80}Fe_{20}$ 4 nm/Cu 8 nm/$Ni_{80}Fe_{20}$ 4nm GMR nanopillar at 40 K [38]. A free layer was microfabricated in an elliptical shape with dimensions of 130×60 nm^2. To obtain reproducible trajectories for adiabatic switching, they maintained the initial angle between the fixed-layer spin and the free layer spin at about $30°$ by using an antiferromagnetic under layer to pin the spins in the fixed layer. Since the GMR nanopillars provide a very small output voltage, the authors averaged more than 10 000 traces using a sampling oscilloscope with a 12 GHz bandwidth. After a background subtraction process, they obtained a transient signal that corresponds to the adiabatic switching of the free-layer spin, as shown in Fig. 20.12 [38]. The precession of the free-layer spin was clearly observed. The amplitude of the precession was amplified in the early stage of the switching and was then damped before the transition from the P state to the AP state at around 2 nsec. The observed behavior was slightly different from that predicted by the simple macro-spin theory according to which continuous amplification of the precession should be observed until the transition. Krivorotov *et al.* explained this deviation by a dephasings among the traces. If the precession contains phase noise, the averaging process carried out by the sampling oscilloscope decreases the observed precession amplitude. The authors stated that the spectrum linewidth of about 10 MHz obtained from the dephasing rate agreed with that obtained from precession noise spectrum measurement. This fact implies that in their sample the phase noise dominated the spectrum linewidth of the precession. Krivorotov *et al.* also clearly showed [36] that, for large applied current, the switching time becomes multiples of the precession period (200 psec, for example) as was already pointed by Devolder *et al.* from their high-speed pulse measurements [43]. This also means that only one extreme point out of two in the orbital (see Fig. 20.11(a)) was responsible for

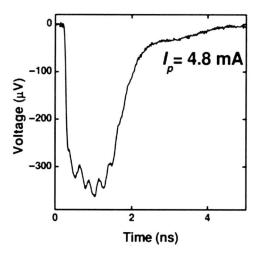

FIG. 20.12. Time-resolved measurement of spin-injection magnetization switching (SIMS) at 40 K. The observation was performed using a sampling oscilloscope. More than 10 000 traces were averaged for a NiFe/Cu/NiFe nanopillar with elliptical cross-sectional dimensions of 130 x 60 nm². The initial angle between the free-layer spin and the fixed-layer spin was about 30° (after Krivorotov *et al.* [38]).

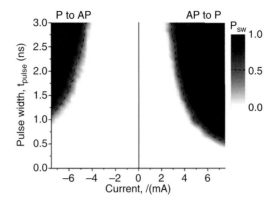

FIG. 20.13. 2D mapping of switching probability as a function of bias current
and applied external magnetic field.

the switching. They explained this fact from their asymmetrical configuration of
the magnetization. Tomita *et al.* showed high-speed switching for a nanopillar
with perpendicular magnetization (Fig. 20.13) down to 500 psec [44]. This is very
high-speed switching as for perpendicular magnetization. They also found about
150 psec initial delay in the switching and attributed it to a possible domain
formation/annihilation process.

20.6 Large-amplitude dynamics and auto-oscillation

From the beginning, it was thought that the electric current inside a ferromagnetic material may interact with the collective modes of spins and excite spin
waves [1, 2, 45]. Actually, before confirmation of spin-injection magnetization
reversal (SIMS), spin dynamics in magnetic nanopillars resulting from spin
injection were observed as anomalies in the derivative conductance spectra [3,
4, 46]. The first and complete observation of microwave emission from magnetic
nanopillars with in-plane magnetization was performed by Kiselev *et al.* in 2003
[47]. They employed a Co/Cu/Co GMR nanopillar with a 130 x 70 x 2 nm^3 free
layer and applied a direct current (greater than I_{c0}) and an external magnetic
field (greater than H_c) at the same time. The external field preferred parallel (P)
configuration of the spins, while the direct current preferred antiparallel (AP)
configuration. Under this situation, the P state is unstable and the switching from
the P to the AP state is prevented by the external field. As a result, the free-
layer spin is driven into a cyclic trajectory (limit cycle) with frequency typically
in the GHz range. Because of the GMR effect, the resistance of the pillar also
oscillates with the continuous precession of the free-layer spins. The oscillation
of the resistance under a direct current bias results in an rf (radio frequency)
voltage that can be detected by a spectrum analyzer or rf diode. For currents
up to 2.4 mA, the spectrum intensity normalized by the square of the current

is almost unchanged. The peak frequency matches the FMR (ferromagnetic resonance) frequency of the free layer and does not shift significantly under this magnitude of current. A further increase in the applied current, however, results in a strong increase in the peak height and significant lowering of the peak frequency (red shift). Such behavior was understood as the spontaneous excitation of the precession motion of the macro-spin. The maximum rf power obtained was several tens of pW. This is the spin-transfer oscillation (STO), a manifestation of the spin current.

Clearer evidence of the onset of auto-oscillation in an MTJ is shown in Fig. 20.14 [48]. When the injection current is less than $I_{c,0}$, an increase in the injection current results in a linear reduction in the peak width, as it was explained previously from Eq. (20.16) (Fig. 20.14b). The threshold current $(I_{c,0})$, which is indicated by an arrow in Fig. 20.14(b), corresponds to the current at which

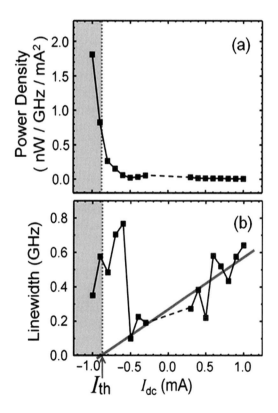

FIG. 20.14. Rf auto-oscillation properties observed in Fe/MgO/Fe single crystalline MTJs (AP state). The junction size is 220×420 nm^2. (a) Peak power density as a function of the biasing current. The power is normalized by I^2. (b) FWHM as a function of the biasing current (after Ref. [48]).

the peak width reduces to zero, if a leaner reduction holds until $I_{c,0}$. In practice, when the injection current is around the threshold current, there is a sudden increase in the peak width. The peak width has its maximum value slightly below $I_{c,0}$. Further increase in the injection current reduces the peak width and results in a sudden increase in the output power. These observations provide clear evidence of the threshold properties, which are in good agreement with the theory developed by J. V. Kim *et al.* [49]. The widths of the spectral lines are, however, very wide when compared to the widths of the spectral lines for the CPP-GMR nanopillars and nanocontacts. Often MTJs provide much larger output power but also much larger linewidth compared with those in GMR nanopillars and point contacts [50]. Application of an external field along the hard axis in the magnetic cell, and fabrication of the MTJ with high current density contribute to obtaining narrower lines, keeping the output power large [51–56] (see Fig. 20.15).

The condition for a limit cycle can be understood using the LLG equation involving the spin-transfer torque (Eq. (20.6)) and the magnetic energy of the macro-spin. The change in the magnetic energy during one cycle of the isoenergy trajectory of the free-layer spin is estimated as follows [24]:

$$\Delta E_{mag}(E) = -\gamma S_2 \oint_{E_{mag}-E} \left\{ -\alpha(-\gamma) \left| \vec{H}_{eff} \times \vec{e}_2 \right|^2 \right.$$
$$\left. + \frac{T_{ST}V}{S_2} (\vec{e}_1 \times \vec{e}_2) \cdot \left(\vec{H}_{eff} \times \vec{e}_2 \right) \right\} dt, \qquad (20.22)$$

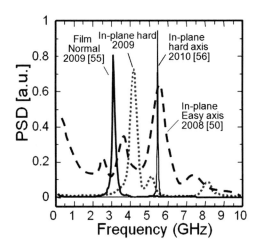

FIG. 20.15. Comparison of several oscillation output spectra taken for magnetic tunnel junctions.

where the integral should be evaluated for one cycle of an isoenergy trajectory with energy E by taking time as a parameter. The first term in the integral is always negative and expresses energy consumption through the Gilbert damping. The second term in the integral can be positive depending on the sign of the current and expresses the energy supply from the current source through the spin-transfer torque. The condition for a stable limit cycle at energy E is

$$\begin{cases} \Delta E_{mag}\left(E\right) = 0 \\ \frac{d\Delta E_{mag}(E)}{dE} < 0. \end{cases} \qquad (20.23)$$

As can be seen in the above equations, the condition is sensitive the dependence of E_{mag} and the spin-transfer torque on the angle θ. More especially, in the nanopillar with perpendicular magnetization, a higher order crystalline anisotropy can also play a role. The conditions describing the threshold current of SIMS (Eq. (20.21)) and the condition to obtain STO (Eq. (20.23)) separate the possible dynamic phases appearing in magnetic nanopillars.

The phase diagram of a nanopillar with in-plane magnetization under an external field and current injection obtained by Kiselev et $al.$ [47] is illustrated in Fig. 20.16. For a zero external field and zero current the system is in the bistable state (P/AP in the figure). The application of a positive (negative) current causes the SIMS to undergo a transition from the P (AP) to the AP (P) state and stabilizes the AP (P) state (dotted line (i)). The system shows

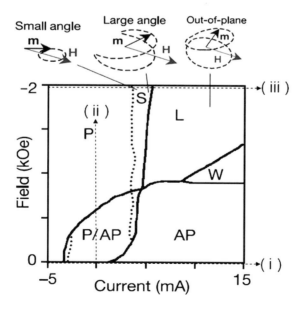

FIG. 20.16. Phase diagram observed for the GMR nanopillar with in-plane magnetization (after Kiselef et $al.$ [47]).

hysteresis along the line (i). For zero current, if we apply a negative (positive, not shown) external field, the system switches to the P (AP) state (dotted line (ii)). The system again shows hysteresis along the line (ii). Now, we apply a large negative external field, −2 kOe for example. At zero applied current, the system is in the P state with small precession of the spin caused by thermal excitation. Under such a large field, even if we supply a positive current larger than the threshold current of the SIMS, switching does not occur. Alternatively, the precession starts to be enhanced significantly and spontaneous oscillation starts. A further increase in the current changes the orbital form from small-angle oscillation to large-angle oscillation and then out-of-plane oscillation (dotted line (iii)). Corresponding to the change in the orbital form, the oscillation frequency first shows a significant red shift and then a blue shift. Along the line (iii), the system does not show hysteresis.

In Fig. 20.17, isoenergy contours of a rectangular shaped magnetic nanopillar with in-plane magnetization are shown. The magnetic energy is lowest at magnetization points P and P′. These points are stable equilibrium points. The small-angle oscillation trajectory corresponds to an isoenergy contour around P and P′. A larger bias current may sustain a higher energy orbital and it may approach the R point (large-angle orbital). The R point is at a saddle point in the energy landscape. This is an unstable equilibrium point. The trajectory that includes the R point separates the region that includes small-angle oscillation orbits and the region that includes out-of-plane orbits. Therefore, it is known as a separatorix. Since infinite time is needed to approach a saddle point, the period in the separatorix is also infinite. As a result, a red shift occurs when the trajectory approaches the separatorix. A large enough bias voltage may excite an out-of-plane orbit. The out-of-plane orbit shows a blue shift when it estranges from the separatorix. Both north and south poles are also unstable equilibrium points.

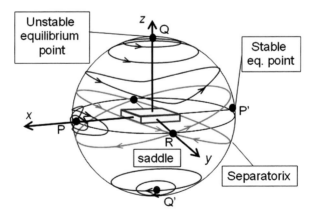

FIG. 20.17. Isoenergy contours of a magnetic nanopillar with in-plane magnetization. Several equilibrium points are also listed.

When an intermediate negative field and a large positive current were applied, a new phase "W" shown in Fig. 20.16 appeared. The very wide spectra observed in region W were attributed to chaotic motion of the spins. The overall phase diagram was well explained by the micromagnetic simulation including the region "W". It was shown that vortex generation and annihilation were the main origin of the chaotic behavior in "W" [26]. Deac extended the phase diagram to the positive field case using a nanopillar with a pinned layer and showed that a combination of a positive field and a negative current also produces STO [57]. The phase diagram of a nanopillar with perpendicular magnetization was obtained by S. Mangin *et al.* [34] and was quite different from the in-plane case. The STO from a nanopillar consisting of a free layer with in-plane magnetization, a perpendicularly magnetized polarizer, and a reference layer with in-plane magnetization was observed by Houssameddine *et al.* [58].

Apart from magnetic nanopillars, the STO have also been observed in the case of magnetic nanocontacts. In 2004, Rippard *et al.* demonstrated that the linewidth of the rf emission spectrum emitted by a magnetic nanocontact can be as narrow as 1.89 MHz by applying a perpendicular magnetic field [59]. The obtained linewidth corresponds to a very large Q-factor of about 18 000. Here Q is defined as $Q =$ (peak frequency)/(linewidth). After this report the linewidth of the STO was investigated both experimentally [60–63, 54] and theoretically [51, 64–66]. Kim *et al.* employed a general model of a nonlinear oscillator and showed that the special point in the STO compared to the other oscillators is a strong amplitude dependence of the oscillation frequency. This nonlinear coupling and thermal fluctuations produce a significant phase noise and dominate the linewidth. Therefore, the linewidth is proportional to the absolute temperature and depends on the size of the nonlinear coupling between amplitude and frequency. By finding a configuration with small amplitude–frequency coupling one may achieve in principle a very small linewidth.

Acknowledgments

The author would like to express his gratitude to the members of the Nanospintronics Research Center in AIST Tsukuba for fruitful collaboration and discussions. The author also thanks the group members in Osaka University.

References

[1] J. C. Slonczewski, *J. Magn. Magn. Mat.* **159**, L1–L7 (1996).

[2] L. Berger, *Phys. Rev. B* **54**, 9353–9358 (1996).

[3] M. Tsoi, A. G. M. Jansen, J. Bass, W. C. Chiang, M. Seck, V. Tsoi, and P. Wyder, *Phys. Rev. Lett.* **80,** 4281 (1998); **81**, 492 (1998) (Erratum).

[4] E. B. Myers, D. C. Ralph, J. A. Katine, R. N. Louie, and R. A. Buhrman, *Science* **285**, 867–870 (1999).

[5] J. A. Katine, F. J. Albert, R. A. Buhrman, E. B. Myers, and D. C. Ralph, *Phys. Rev. Lett.* **84**, 3149–3152 (2000).

[6] J. Z. Sun, *J. Magn. Magn. Mat.* **202**, 157–162 (1999).

[7] Y. Huai, F. Albert, P. Nguyen, M. Pakala, and T. Valet, *Appl. Phys. Lett.* **84**, 3118–3120 (2004).

[8] H. Kubota, A. Fukushima, Y. Ootani, S. Yuasa, K. Ando, H. Maehara, K. Tsunekawa, D. D. Djayaprawira, N. Watanabe, and Y. Suzuki, *Jpn. J. Appl. Phys.* **44**, L1237-L1240 (2005).

[9] Z. Diao, D. Apalkov, M. Pakala, Y. Ding, A. Panchula, and Y. Huai, *Appl. Phys. Lett.* **87**, 232502 (2005).

[10] D. Chiba, Y. Sato, T. Kita, F. Matsukura, and H. Ohno, *Phys. Rev. Lett.* **93**, 216602 (2004).

[11] J. C. Slonczewski, *Phys. Rev.* B **39**, 6995–7002 (1989).

[12] J. C. Slonczewski, *Phys. Rev.* B **71**, 024411 (2005).

[13] I. Theodonis, N. Kioussis, A. Kalitsov, M. Chshiev, and W. H. Butler, *Phys. Rev. Lett.* **97**, 237205 (2006).

[14] J. C. Sankey, Y.-T. Cui, J. Z. Sun, J. C. Slonczewski, R. A. Buhrman, and D. C. Ralph, *Nature Phys.* **4**, 67–71 (2008).

[15] H. Kubota, A. Fukushima, K. Yakushiji, T. Nagahama, S. Yuasa, K. Ando, H. Maehara, Y. Nagamine, K. Tsunekawa, D. D. Djayaprawira, N. Watanabe, and Y. Suzuki, *Nature Phys.* **4**, 37–41 (2008).

[16] C. Wang, Y. T. Cui, J. Z. Sun, J. A. Katine, R. A. Buhrman, and D. C. Ralph, *Phys. Rev.* B **79**, 10 (2009); C. Wang, Y.-T. Cui, J. A. Katine, R. A. Buhrman, and D. C. Ralph, *Nature Phys.* **7**, 496 (2011).

[17] D. M. Edwards, F. Federici, J. Mathon, and A. Umerski, *Phys. Rev.* B **71**, 054407 (2005).

[18] S. Zhang and Z. Li, *Phys. Rev. Lett.* **93**, 127204 (2004).

[19] A. Thiaville, Y. Nakatani, J. Miltat, and Y. Suzuki, *Europhys. Lett.* **69**, 990 (2005).

[20] S. E. Barnes and S. Maekawa, *Phys. Rev. Lett.* **95**, 107204 (2005).

[21] G. Tatara and H. Kohno, *Phys. Rev. Lett.* **92**, 086601 (2004).

[22] Y. Suzuki and H. Kubota, *J. Phys. Soc. Jpn.* **77**, 031002 (2008).

[23] J. Z. Sun, *Phys. Rev.* B **62**, 570 (2000).

[24] M. D. Stiles and J. Miltat, Spin dynamics in confined magnetic structures III, in *Topics in Applied Physics* **101**, B. Hillebrands, A. Thiaville, eds., Springer, 2006, pp. 225–308.

[25] J. Miltat, G. Albuquerque, A. Thiaville, and C. Vouille, *J. Appl. Phys.* **89**, 6982–6984 (2001).

[26] K.-J. Lee, A. Deac, O. Redon, J.-P. Nozières, and B. Dieny, *Nature Mater.* **3**, 877 (2004).

[27] H. Kohno and G. Tatara, Theoretical aspects of current-driven magnetization dynamics, in *Nanomagnetism and Spintronics*, T. Shinjo, eds., Elsevier, Amsterdam 2009.

[28] A. A. Tulapurkar, Y. Suzuki, A. Fukushima, H. Kubota, H. Maehara, K. Tsunekawa, D. D. Djayaprawira, N. Watanabe, and S. Yuasa, *Nature* **438,** 339 (2005).

[29] S. Ishibashi, T. Seki, T. Nozaki, H. Kubota, S.Yakata, A. Fukushima, S. Yuasa, H. Maehara, K. Tsunekawa, D. D. Djayaprawira, and Y. Suzuki, *Appl. Phys. Express* **3,** 073001 (2010).

[30] F. J. Albert, N. C. Emley, E. B. Myers, D. C. Ralph, and R. A. Buhrman, *Phys. Rev. Lett.* **89,** 226802 (2002).

[31] K. Yagami, A. A. Tulapurkar, A. Fukushima, and Y. Suzuki, *Appl. Phys. Lett.* **85,** 5634–5636 (2004).

[32] L. Berger, *J. Appl. Phys.* **93,** 7693 (2003).

[33] Y. Huai, M. Pakala, Z. Diao, and Y. Ding, *Appl. Phys. Lett.* **87,** 222510 (2005).

[34] S. Mangin, D. Ravelosona, J. A. Katine, M. J. Carey, B. D. Terris, and E. E. Fullerton, *Nature Mater.* **5,** 210–215 (2006).

[35] S. Yakata, H. Kubota, Y. Suzuki, K. Yakushiji, A. Fukushima, S. Yuasa, and K. Ando, *J. Appl. Phys.* **105,** 07D131 (2009).

[36] T. Nagase, K. Nishiyama, M. Nakayama, N. Shimomura, M. Amano, T. Kishi, H. Yoda, C1.00331, American Physics Society March Meeting, New Orleans (2008).

[37] A. A. Tulapurkar, T. Devolder, K. Yagami, P. Crozat, C. Chappert, A. Fukushima, and Y. Suzuki, *Appl. Phys. Lett.* **85,** 5358 (2004).

[38] I. N. Krivorotov, N. C. Emley, R. A. Buhrman, and D. C. Ralph, *Phys. Rev. B* **77,** 054440 (2008).

[39] T. Aoki, Y. Ando, D. Watanabe, M. Oogane, and T. Miyazaki, *J. Appl. Phys.* **103,** 103911 (2008).

[40] I. N. Krivorotov, N. C. Emley, J. C. Sankey, S. I. Kiselev, D. C. Ralph, and R. A. Buhrman, *Science* **307,** 228 (2005).

[41] T. Devolder, J. Hayakawa, K. Ito, H. Takahashi, S. Ikeda, P. Crozat, N. Zerounian, J.-V. Kim, C. Chappert, and H. Ohno, *Phys. Rev. Lett.* **100,** 057206 (2008).

[42] H. Tomita, K. Konishi, T. Nozaki, H. Kubota, A. Fukushima, K. Yakushiji, S. Yuasa, Y. Nakatani, T. Shinjo, M. Shiraishi, and Y. Suzuki, *Appl. Phys. Express* **1,** 061303 (2008).

[43] T. Devolder, C. Chappert, J. A. Katine, M. J. Carey, and K. Ito, *Phys. Rev. B* **75,** 064402 (2007).

[44] H. Tomita, T. Nozaki, T. Seki, T. Nagase, K. Nishiyama, E. Kitagawa, M. Yoshikawa, T. Daibou, M. Nagamine, T. Kishi, S. Ikegawa, N. Shimomura, H. Yoda, and Y. Suzuki, *IEEE Trans Magn.* **47,** 1599 (2011).

[45] J. C. Slonczewski, *J. Magn. Magn. Mat.* **195,** L261–L267 (1999).

[46] M. Tsoi, A. G. M. Jansen, J. Bass, W.-C. Chiang, V. Tsoi, and P. Wyder, *Nature,* **406,** 46 (2000).

[47] S. I. Kiselev, J. C. Sankey, I. N. Krivorotov, N. C. Emley, R. J. Schoelkopf, R. A. Buhrman, and D. C. Ralph, *Nature* **425**, 380–383 (2003).

[48] R. Matsumoto, A. Fukushima, K. Yakushiji, S. Yakata, T. Nagahama, H. Kubota, T. Katayama, Y. Suzuki, K. Ando, S. Yuasa, B. Georges, V. Cros, J. Grollier, and A. Fert, *Phys. Rev.* B **80**, 174405 (2009).

[49] J.-V. Kim, Q. Mistral, C. Chappert, V. S. Tiberkevich, and A. N. Slavin, *Phys. Rev. Lett.* **100**, 167201 (2008).

[50] A. M. Deac, A. Fukushima, H. Kubota, H. Maehara, Y. Suzuki, S. Yuasa, Y. Nagamine, K. Tsunekawa, D. D. Djayaprawira, and N. Watanabe, *Nature Phys.* **4**, 803 (2008).

[51] A. V. Nazarov, H. M. Olson, H. Cho, K. Nikolaev, Z. Gao, S. Stokes, and B. B. Pant, *Appl. Phys. Lett.* **88**, 162504 (2006).

[52] D. Houssameddine, S. H. Florez, J. A. Katine, J.-P. Michel, U. Ebels, D. Mauri, O. Ozatay, B. Delaet, B. Viala, L. Folks, B. D. Terris, and M.-C. Cyrille, *Appl. Phys. Lett.* **93**, 022505 (2008).

[53] T. Devolder, L. Bianchini, J.-V. Kim, P. Crozat, C. Chappert, S. Cornelissen, M. Op de Beeck, and L. Lagae, *J. Appl. Phys.* **106**, 103921 (2009).

[54] S. Cornelissen, L. Bianchini, G. Hrkac, M. Op de Beeck, L. Lagae, J.-V. Kim, T. Devolder, P. Crozat, C. Chappert, and T. Schrefl, *Euro. Phys. Lett.* **87**, 57001 (2009).

[55] T. Wada, T. Yamane, T. Seki, T. Nozaki, Y. Suzuki, H. Kubota, A. Fukushima, S. Yuasa, H. Maehara, Y. Nagamine, K. Tsunekawa, D. D. Djayaprawira, and N. Watanabe, *Phys. Rev.* B **81**, 104410 (2010).

[56] H. Maehara, H. Kubota, T. Seki, K. Nishimura, H. Tomita, Y. Nagamine, K. Tsunekawa, D. D. Djayaprawira, A. Fukushima, S. Yuasa, K. Ando, and Y. Suzuki, 55th Magnetism and Magnetic Materials Conference, Atlanta (2010).

[57] A. Deac, Y. Liu, O. Redon, S. Petit, M. Li, P. Wang, J.-P. Nozières, and B. Dieny, *J. Phys.: Condens. Matter* **19,** 165208 (2007).

[58] D. Houssameddine, U. Ebels, B. Delaet, B. Rodmacq, I. Firastrau, F. Ponthenier, M. Brunet, C. Thirion, J.-P. Michel, L. Prejbeanu-Buda, M.-C. Cyrille, O. Redon, and B. Dieny, *Nature Mater.* **6**, 447 (2007).

[59] W. H. Rippard, M. R. Pufall, S. Kaka, T. J. Silva, and S. E. Russek, *Phys. Rev.* B **70**, 100406R (2004).

[60] J. C. Sankey, I. N. Krivorotov, S. I. Kiselev, P. M. Braganca, N. C. Emley, R. A. Buhrman, and D. C. Ralph, *Phys. Rev.* B **72**, 224427 (2005).

[61] Q. Mistral, J.-V. Kim, T. Devolder, P. Crozat, C. Chappert, J. A. Katine, M. J. Carey, and K. Ito, *Appl. Phys. Lett.* **88,** 192507 (2006).

[62] S. Petit, C. Baraduc, C. Thirion, U. Ebels, Y. Liu, M. Li, P. Wang, and B. Dieny, *Phys. Rev. Lett.* **98**, 077203 (2007).

[63] K. V. Thadani, G. Finocchio, Z.-P. Li, O. Ozatay, J. C. Sankey, I. N. Krivorotov, Y.-T. Cui, R. A. Buhrman, and D. C. Ralph, *Phys. Rev. B* **78**, 024409 (2008).

[64] J.-V. Kim, *Phys. Rev. B* **73**, 174412 (2006).

[65] J.-V. Kim, V. Tiberkevich, and A. N. Slavin, *Phys. Rev. Lett.* **100**, 017207 (2008).

[66] V. S. Tiberkevich, A. N. Slavin, and J.-V. Kim, *Phys. Rev. B* **78**, 092401 (2008).

21 Magnetization switching due to nonlocal spin injection

T. Kimura and Y. Otani

21.1 Generation and absorption of pure spin current

A laterally configured ferromagnetic (F)/nonmagnetic (N) hybrid structure combined with nonlocal spin injection allows us to create a flow of spins without accompanying a flow of electrical charges, i.e. a pure spin current [1–8]. Figure 21.1 shows a schematic illustration of nonlocal spin injection. A bias voltage for the spin injection is applied between the ferromagnet and the left-hand-side nonmagnet. In this case, the spin-polarized electrons are injected from the ferromagnet and are extracted from the left-hand side of the nonmagnet. This results in the accumulation of nonequilibrium spins in the vicinity of the F/N junctions. Since the electrochemical potential on the left-hand side is lower than that underneath the F/N junction, the electron flows by the electric field. On the right-hand side, although there is no electric field, the diffusion process from the nonequilibrium into the equilibrium state induces the motion of the electrons. Since the excess up-spin electrons exist underneath the F/N junction, the up-spin electrons diffuse into the right-hand side. On the other hand, the deficiency of the down-spin electrons induces the incoming flow of the down-spin electrons opposite to the motion of the up-spin electron. Thus, a pure spin current, which carries the spin angular momentum without electric charges, can be induced by the nonlocal spin injection.

The induced pure spin current can be detected by using another ferromagnetic voltage probe. When the pure spin current is injected into the ferromagnet, a shift of the electrostatic potential of the ferromagnet is induced because of the spin-dependent conductivity. The sign of the potential shift depends on the relative angle between the spin direction of the injecting spin current and the magnetization direction. When the direction of the injecting spin is parallel to the majority (minority) spin for the spin detector, the electrostatic potential of the spin detector shifts positively (negatively). Therefore, when the voltage between the ferromagnet and the right-hand side of the nonmagnet is measured by sweeping the magnetic field, a clear voltage change is observed. The voltage normalized by the exciting current is known as a spin signal [1].

As described above, the driving force of the spin current in the N is the diffusion of the nonequilibrium electrons into the equilibrium state. Since the spatial distribution of the nonequilibrium spin accumulation in the N is modified

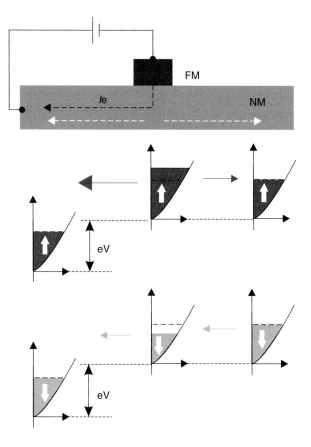

FIG. 21.1. Schematic illustration of the nonlocal spin injection together with the density of states for the up-spin and down-spin electrons in a nonmagnetic metal for the left-hand, center, and right-hand sides.

by connecting the additional structure, the distribution of the spin current is also modified. For example, in a single F/N junction shown in Fig. 21.2(a), the spatial distribution of the spin accumulation symmetrically decays from the junction. Therefore, the spin current flows symmetrically also into both sides. On the other hand, when one connects an additional material on the right-hand side through a low resistive ohmic junction as in Fig. 21.2(b), the spatial distribution of the spin accumulation is strongly modified. Thus, one can selectively extract the spin current.

To demonstrate the above spin current absorption effect, two kinds of lateral spin valves (LSVs) have been prepared [9]. One is a conventional lateral spin valve consisting of the Py injector and the detector bridged by a Cu strip (device A). The other one is a lateral spin valve with a middle Py wire (device B). Here,

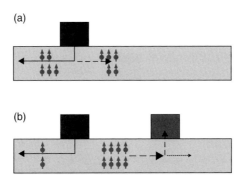

FIG. 21.2. Schematic illustrations of the flow of the spin current in (a) a single
    F/N junction and (b) an F/N junction with an F contact.

the center–center distance between the injector and the detector for device A
is 600 nm while that for the device B is 460 nm. The thickness and width of
the Py wires are 30 nm and 100 nm, respectively. The Cu strip is 80 nm in
thickness and 300 nm in width. Although the geometrical disorder due to the
additional ferromagnetic contact may also violate the spin coherence and the spin
accumulation, such an effect should be negligible because of the large difference
in thickness between Cu and Py.

Figure 21.3(a) shows the spin signal observed in device A, where a spin valve
signal with a magnitude of 0.2 mΩ is clearly observed. Since the center–center
distance between the injector and detector for device B is shorter than that for
device A, one may naively expect that a larger spin signal is expected to be
observed in device B. However, as in Fig. 21.3(b), a quite small spin signal, less
than 0.05 mΩ, is observed in device B. This is due to the influence of the spin
current absorption into the middle Py wire. Since the nonequilibrium spins want
to diffuse into the equilibrium states as fast as possible, the spins are preferably

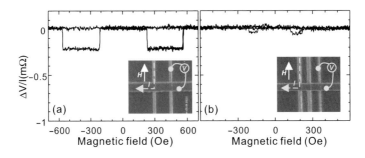

FIG. 21.3. (a) Nonlocal spin-valve signal for a conventional lateral spin valve
    and that for a lateral spin valve with a middle Py wire. The insets show SEM
    images of the measured device and the probe configurations for the nonlocal
    spin-valve measurements.

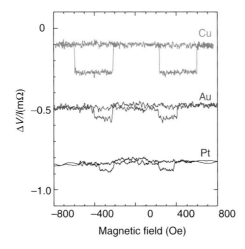

FIG. 21.4. Nonlocal spin-valve signals for Cu, Au, and Pt middle wires measured at room temperature.

absorbed into the middle Py wire which has much stronger spin relaxtion than that of the Cu wire. These results clearly suggest that the spin accumulation in the Cu is strongly suppressed by the middle Py2 wire connected to the Cu. It should be noted here that in the spin signal shown in Fig. 21.3(b) no change is observed at the switching field of the middle wire. This implies that the magnitude of the spin current absorption does not depend on the magnetization configuration of the middle wire.

It was also demonstrated that the spin accumulation in the Cu is suppressed by connecting the nonmagnetic wire with a strong spin relaxation [10]. Figure 21.4 shows the spin signals with various nonmagnetic middle wires. Here, the center-center distance between the injector and detector is fixed at 600 nm. For the middle Cu wire, the obtained spin signal is 0.18 mΩ, which is almost the same as that without the middle wire. Large reductions of the spin signals are obsereved in the Au and Pt middle wires. These indicate that the nonequilibrium spin currents are strongly relaxed by the Pt and Au wires while the Cu gives weak relaxation of the spin current. Thus, one can evaluate the magnitude of the spin relaxation of a material from the magnitude of the spin signal.

21.2 Efficient absorption of pure spin current

In the F/N junction shown in Fig. 21.1, the injection efficiency η_{I} of the spin current is given by the following equation [3, 10]

$$\eta_{\mathrm{I}} = \frac{2R_{\mathrm{SF}}}{2R_{\mathrm{SF}} + R_{\mathrm{SN}}} P_{\mathrm{F}}, \tag{21.1}$$

where R_{SF} and R_{SN} are the spin resistances for the F and N, respectively. The spin resistance is a measure of the difficulty of flow of the spin current and is defined by $2\rho\lambda/((1 - P^2)S)$ with resistivity ρ, spin diffusion length λ, spin polarization P, and effective cross-section for the spin current S. R_{SF} is, in general, much smaller than R_{SN} because of the extremely short spin diffusion length for the F. Therefore, η_I becomes quite small. When one connects an additional material on the right-hand side of the N, whose spin resistance is R_{SA}, the injection efficiency is modified as

$$\eta_I = \frac{R_{SF}}{R_{SF} + R_{S1}} P_F \qquad (21.2)$$

where R_{S1} is the effective spin resistance and is given by

$$R_{S1} = R_{SN} \frac{R_{SA} \cosh(d/\lambda_N) + (R_{SA} + R_{SN}) \sinh(d/\lambda_N)}{2R_{SA} + R_{SN}} e^{-\frac{d}{\lambda}}. \qquad (21.3)$$

So, in order to increase η_I, R_{S1} should be smaller than R_{SF}. As in Eq. (21.2), R_{S1} can be reduced by connecting an additional material. When the material has a small spin resistance, η_I drastically decreases.

Figure 21.5 shows η_I as a function of R_{SA} for various distances d in Py/Cu LSVs, assuming $R_{SPy}/R_{SCu} \approx 0.05$. η_I increases with decreasing R_{SA}, especially when d is much shorter than λ_{Cu}. Thus, the spin current induced in the N is effectively extracted by connecting the material with a small spin resistance nonlocally.

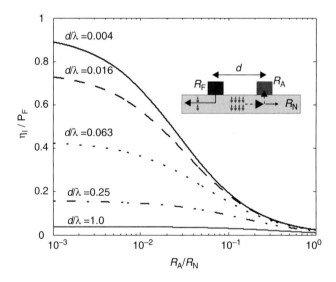

FIG. 21.5. Efficiency of spin current absorption as a function of the spin resistance for various separation distances.

21.3 Efficient injection of spin current

In the previous section, the injection efficiency η_I of the spin current is enhanced by reducing the effective spin resistance R_I of the injecting part. As in Eq. 21.1, η_I is determined by the balance between R_{SF} and R_{SI}. Therefore, η_I also increases by increasing R_{SF}. Although R_{SF} is mainly determined by the material, the effective cross-section A for the spin current can be geometrically controlled. Since the spin diffusion length for Fs is extremely short, A for Fs is given by the size of the F/N junction. In this section, we introduce the experimental demonstration that the size of the ohmic F/N junction is an important geometrical factor for obtaining large spin polarization in Ns and that both the spin polarization and the spin resistance of the F are enhanced by adjusting the junction size [11].

As mentioned above, the difference in the junction size between injector and detector gives rise to a significant difference in the spin resistance. Then, the spin signal ΔR in the present device can be given by

$$\Delta R \approx \frac{P_{Py}^2 R_{SPy}^P R_{SPy}^W}{R_{SCu} \sinh\left(\frac{d}{\lambda_{Cu}}\right)}, \tag{21.4}$$

where, R_{SPy}^P and R_{SPy}^W are, respectively, the spin resistances of the Py pad and the Py wire. Here, the spin current diffusions into the horizontal Cu arms are neglected because the vertical Cu arms are connected to the Py with the small spin resistances.

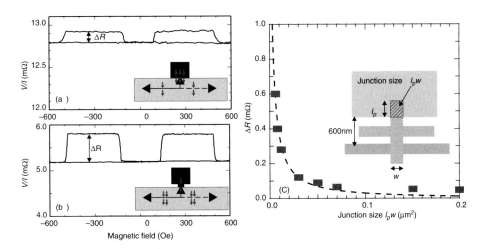

FIG. 21.6. Nonlocal spin-valve (NLSV) signals of (a) a large-junction device and (b) a small-junction device with the probe configurations. (c) Spin signal in the NLSV measurement as a function of the junction size $l_p w$. The dotted curve is the best fit to the data points using Eq. (21.4).

As mentioned above, reducing the size of the ohmic junction between the Py pad and the Cu wire increases the spin resistance of the Py pad. To change the junction size between the Py pad and the Cu wire, the length of the Cu wire on the Py pad is adjusted, as seen in the SEM images in Figs. 21.6(a) and 21.6(b). The junction size dependence of the spin signal has been investigated by changing the spin resistance of the Py pad while keeping the same electrode spacing of 600 nm. The obtained spin signal is plotted as a function of the junction size in Fig. 21.6(c). The spin signal increases by reducing the junction size and is well reproduced by Eq. 21.4, where the spin signal is inversely proportional to the junction size. From the fitting parameters, the spin diffusion length of the Cu wire, that of the Py wire, and the spin polarization are, respectively, found to be 1.5 μm, 0.25, and 3.5 nm at 77 K.

21.4   Magnetization switching due to injection of pure spin current

The switching mechanism due to spin torque is explained by a model proposed by Slonczewski in which the torque exerted on the magnetization is proportional to the injected spin current. This clearly indicates that the spin current is essential to realizing magnetization switching due to spin injection. Most of the present spin-transfer devices consist of vertical multilayered nanopillars in which typically two magnetic layers are separated by a nonmagnetic metal layer [12, 13]. In such vertical structures, the charge current always flows together with the spin current, during which undesirable Joule heat is generated. As mentioned above, by optimizing the junction, the pure spin current can be effectively injected into the nanomagnet because of the spin absorption [11]. Therefore, the magnetization of the nanomagnet can be switched nonlocally. To test this idea, a nanoscale ferromagnetic particle is configured for a lateral nonlocal spin injection device as in Figs. 21.7(a) and (b) [14].

The device for the present study consists of a large Permalloy (Py) pad 30 nm in thickness, a Cu cross 100 nm in width and 80 nm in thickness, and a Py nanoscale particle, 50 nm in width, 180 nm in length, and 6 nm in thickness. A gold wire 100 nm in width and 40 nm in thickness is connected to the Py particle to reduce the effective spin resistance, resulting in high spin current absorption into the Py particle. The magnetic field is applied along the easy axis of the Py particle. Here, the dimensions of the Py pad and Cu wires are chosen large so that a charge current of up to 15 mA can flow through them.

To confirm that the spin current from the Py injector is injected into the Py particle, nonlocal spin-valve measurements are performed. As in Fig. 21.7(c), the field dependence shows a clear spin signal with a magnitude of 0.18 mΩ, ensuring that the spin current reaches the Py particle. Then, the effect of the nonlocal spin injection into the Py particle was examined by using the same probe configuration. Before performing the nonlocal spin injection, the magnetization configuration is set in the antiparallel configuration by controlling the external magnetic field.

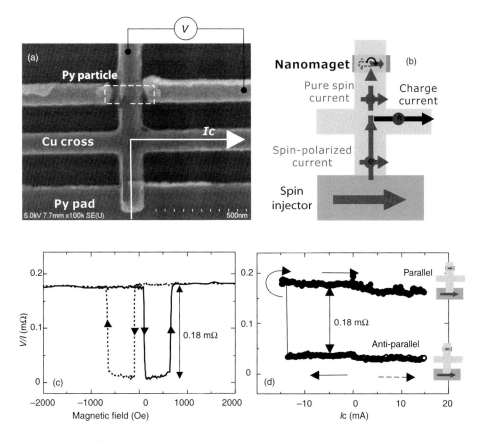

FIG. 21.7. (a) Scanning electron microscope image of the fabricated lateral spin valve. The device consists of a large 30 nm thick Py injector, a Cu cross 100 nm wide and 80 nm thick, and a Py nanoscale particle, 50 nm wide, 180 nm long, and 6 nm thick. (b) Schematic illustration of nonlocal spin injection using a lateral spin-valve geometry. Pure spin current is effectively absorbed into the nanomagnet. (c) Field dependence of the nonlocal spin signal. The changes in resistance at low and high fields correspond to the relative magnetic switching of the Py injector and particle, from parallel to antiparallel states and vice versa. (d) Nonlocal spin-valve signal after the pulsed current injection as a function of the current amplitude with corresponding magnetization configurations.

The nonlocal spin injection is performed by applying large pulsed currents up to 15 mA in the absence of the magnetic field. As shown in Fig. 21.7(d), when the magnitude of the pulsed current is increased positively in the antiparallel state, no signal change is observed up to 15 mA. On the other hand, for the negative scan, an abrupt signal change is observed at −14 mA. The change in resistance at

−14 mA is 0.18 mΩ, corresponding to that of the transition from antiparallel to parallel states. This means that the magnetization of the Py particle is switched only by the spin current induced by the nonlocal spin injection. The spin current for switching is estimated from the experiment to be about 200 μA, which is reasonable compared with the values obtained for conventional pillar structures. However, the switching from the parallel to antiparallel state has not been achieved in the present device. This is mainly due to the low spin-injection efficiency.

To improve the efficiency of the injecting spin current, a newly designed sample has been fabricated, as shown in Fig. 21.8(a) [15]. The new sample consists of two Py/Au nanopillars on a Cu wire. As shown in Fig. 21.8(a), the junction size between the Py/Cu in the new sample is effectively diminished, leading to the efficient generation of the pure spin current. Figure 21.8(b) shows the nonlocal spin-valve signal as a function of the external field. The obtained spin signal is around 4 mΩ, much larger than that of the previous device. Then, nonlocal spin injection with variable dc current between contacts 3 and 6 is applied to perform the magnetization switching. The sample is preset to a parallel state at which both magnetizations are aligned in the positive field direction. As can be seen in Fig. 21.8(b), when the current is increased, the nonlocal spin-valve signal sharply decreases at about 4.5 mA, indicating a clear magnetization reversal. According to the change in the nonlocal spin-valve signal, the parallel state is transformed into an antiparallel state (denoted B), which is switched back to the parallel state by a negative dc current of 5 mA. Thus, reversible magnetization switching between antiparallel and parallel states is realized by

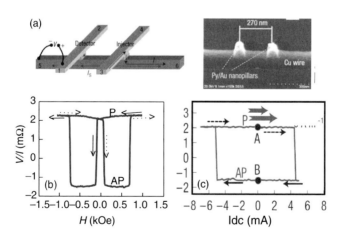

FIG. 21.8. (a) SEM image and schematic illustration of the improved nonlocal spin injection device. (b) Giant spin signal and (c) reversible magnetization switching by pure spin current injection observed in the improved device.

means of nonlocal spin injection with the specially developed device consisting of perpendicular nanopillars and lateral magnetic nanostructures.

Very recently, Zou and Ji have demonstrated nonlocal switching of a Py nanodot by using a specially developed LSV structure [16, 17]. They prepared a lateral spin valve with a 5 nm thick ferromagnetic Py detector. This structure enabled them to inject the pure spin current entirely into the Py detector. As a result, the magnetization of the Py detector is reversed by a sufficiently large spin torque. The interesting thing is that the structure includes the interface barriers both at the injecting and detecting junctions. According to the spin diffusion model, the interface resistance strongly suppresses the spin-current diffusion into the ferromagnet. To understand the result more quantitatively, other effects such as the magnetic interface anisotropy may have to be considered.

Acknowledgment

The author would like to thank Dr. Yang and Prof. Y. Otani for their valuable discussion. This work is partially supported by JST CREST and NEDO.

References

[1] F. J. Jedema, A. T. Filip, and B. J. van Wees, Nature (London) **410**, 345 (2001).

[2] F. J. Jedema, M. S. Nijboer, A. T. Filip, and B. J. van Wees, Phys. Rev. B **67**, 085319 (2003).

[3] S. Takahashi, H. Imamura, and S. Maekawa, in *Concepts in Spin Electronics*, edited by S. Maekawa (Oxford Univ Press, Oxford, 2006).

[4] M. Urech, V. Korenivski, N. Poli, and D. B. Haviland, Nano Lett. **6**, 871, (2006).

[5] T. Kimura and Y. Otani, J. Phys.: Condens. Matter, **19**, 165216 (2007).

[6] S. O. Valenzuela, Int. J. Mod. Phys. B **23**, 2413 (2009).

[7] R. Godfrey and M. Johnson, Phys. Rev. Lett. **96**, 136601 (2006).

[8] G. Mihajlovic *et al.*, Appl. Phys. Lett. **97**, 112502 (2010).

[9] T. Kimura, J. Hamrle, Y. Otani, K. Tsukagoshi, and Y. Aoyagi, Appl. Phys. Lett. **85**, 3501 (2004).

[10] T. Kimura, J. Hamrle, and Y. Otani, Phys. Rev. B **72**, 014461 (2005).

[11] T. Kimura, Y. Otani, and J. Hamrle, Phys. Rev. B **73**, 132405 (2006).

[12] M. Tsoi, A. G. M. Jansen, J. Bass, W.-C. Chiang, M. Seck, V. Tsoi, and P. Wyder, Phys. Rev. Lett. **80**, 4281 (1998).

[13] F. J. Albert, N. C. Emley, E. B. Myers, D. C. Ralph, and R. A. Buhrman, Phys. Rev. Lett. **89**, 226802 (2002).

[14] T. Kimura, Y. Otani, and J. Hamrle, Phys. Rev. Lett. **96**, 037201 (2006).

[15] T. Yang, T. Kimura, and Y. Otani, Nature Phys. **4**, 851 (2008).

[16] H. Zou and Y. Ji, J. Magn. Magn. Mater. **323**, 2448 (2011).

[17] H. Zou, S. Chen, and Y. Ji, Appl. Phys. Lett. **100**, 012404 (2012).

22 Magnetic domains and magnetic vortices

Y. Otani and R. Antos

One of the most remarkable manifestations of recent progress in magnetism is the establishment of microfabrication procedures employing modern magnetic materials. Electron or ion beam lithography combined with conventional thin film deposition techniques yield a variety of laterally patterned nanoscale structures such as arrays of magnetic nanodots or nanowires [1,2]. Among them, submicron ferromagnetic disks and wires have drawn particular interest due to their possible applications in high-density magnetic data storage [3], magnonic crystals [4], and logic operation devices [5]. In this way control of magnetic domains and domain wall structures is one of the most important issues from the viewpoint of both applied and basic research. Although there is a variety of nanoscale micromagnetic structures, we limit our discussion in this chapter to static and dynamic properties of magnetic vortex structures.

It has been revealed both theoretically and experimentally that for particular ranges of dimensions of cylindrical and other magnetic elements (Fig. 22.1) a curling in-plane spin configuration (vortex) is energetically favored, with a small region of the out-of-plane magnetization appearing at the core of the vortex [6–8]. Such a system, which is sometimes referred to as a magnetic soliton [9] and whose potentialities have already been discussed in a few recent review papers [10,11],

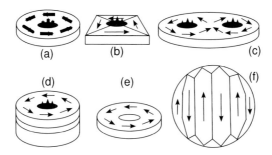

FIG. 22.1. Examples of vortices appearing in a cylindrical (a), rectangular (b), elliptic (c), multilayered (d), and ring-shaped (e) element. Each vortex's center contains an out-of-plane polarized core except for the ring. Classical multidomain structures appear in larger elements where the anisotropy energy is predominant (f).

is thus characterized by two binary properties ("topological charges"), a chirality (counterclockwise or clockwise direction of the in-plane rotating magnetization) and a polarity (the up or down direction of the vortex core's magnetization), each of which suggests an independent bit of information in future high-density nonvolatile recording media.

For this purpose various properties have been investigated, such as the appearance and stability of vortices when subjected to quasi-static or short-pulse magnetic fields and variations of these properties when the dots are densely arranged into arrays. The properties are identified with experimentally measured and theoretically calculated quantities called nucleation and annihilation fields, effective magnetic susceptibilities, etc.

Most recently, the time-resolved response to applied magnetic field pulses or spin-polarized electrical currents with subnanosecond resolution has been extensively studied, providing results on the time dependence of the location, size, shape, and polarity deviations of the vortex cores, eigenfrequencies and damping of time-harmonic trajectories of the cores, the switching processes, and the spin waves involved. In this chapter we will review recent achievements in this research area with particular interest in submicron cylindrical ferromagnetic disks with negligible magnetic anisotropy, for which permalloy (Py) has been chosen as the most typical material. We will demonstrate the theoretical background of the research topic according to the description by Hubert and Schäfer [6] (Section 22.1) and briefly describe the achievements in analytical approaches (Section 22.2) and experimental techniques (Section 22.3). Then we will review the research of various authors devoted to steady state excitations (Section 22.4), dynamic switching of vortex states (Section 22.5), and excitations of magnetostatically coupled vortices (Section 22.6).

Finally we will summarize the present state of research with respect to future prospects and possible applications (Section 22.7). We will accompany our description by our simulations using the Object-Oriented Micromagnetic Framework (OOMMF) [12], and in some cases by demonstrative examples provided by their original authors.

22.1 Micromagnetic equations

The unique spin distributions favored in ferromagnetic materials are governed by the exchange interaction between nearest neighbor spins $\mathbf{s}_i$, $\mathbf{s}_j$ described by the Heisenberg Hamiltonian

$$H = -\sum_{i,j} J_{i,j} \mathbf{s}_i \cdot \mathbf{s}_j, \qquad (22.1)$$

or by more general formulas if particular anisotropies are taken into account. For the sake of solving many-spin problems, the discrete spin distribution is replaced by the magnetization $\mathbf{M}(\mathbf{r}, t)$, a continuous function of space and time, or by the unit magnetization $\mathbf{m} = \mathbf{M}/M_\mathrm{s}$, where M_s is the saturation magnetization.

Accordingly, the total energy of a ferromagnet is determined by the integrated total sum, $E_{tot} = \int (E_{exch} + E_{an} + E_d + E_{ext} + \cdots)dV$, which demonstrates the competition among exchange, anisotropy, demagnetizing, external-field, and other forms of energy (such as magneto-elastic interaction or magnetostriction which we do not consider in our discussion), more precisely as

$$E_{tot} = A \int \left[(\nabla \mathbf{m})^2 + \frac{K}{A} \cdot f(\mathbf{m}) - \frac{M_S^2}{2\mu_0 A}\mathbf{m}(\mathbf{r}) \cdot \mathbf{h}_d + \frac{M_S H_{ext}}{A}\mathbf{m}(\mathbf{r}) \cdot \mathbf{h}_{ext} \right] dV$$

$$= A \int \left[(\nabla \mathbf{m})^2 + \frac{1}{\xi_K^2} \cdot f(\mathbf{m}) - \frac{1}{\xi_M^2}\mathbf{m}(\mathbf{r}) \cdot \mathbf{h}_d + \frac{1}{\xi_H^2}\mathbf{m}(\mathbf{r}) \cdot \mathbf{h}_{ext} \right] dV, \quad (22.2)$$

where A is the exchange stiffness constant, μ_0 the permeability of the vacuum, and H_d, and H_{ext} are the demagnetizing (stray) field, and external magnetic field, respectively. The function $f(\mathbf{m})$ describes the magnetic anisotropy energy landscape spacially varying with the unit magnetization $\mathbf{m}$.

Here one should notice that the coefficients for the energy terms K/A, $M_S^2/(2\mu_0 A)$ and $M_S H_{ext}/A$ have units of the inverse square of length. The inverse square root of these three coefficients define respectively the anisotropy characteristic length $\xi_K \left(= \sqrt{A/K}\right)$ given by the magnetic anisotropy energy the magnetostatic characteristic length $\xi_M \left(= \sqrt{(2\mu_0 A)/M_S^2}\right)$ given by the magnetostatic energy, and the external field characteristic length $\xi_H \left(= \sqrt{A/M_S H_{ext}}\right)$ given by the external magnetic field. The length ξ_K means the length required for twisting a unit angle (radian) in an exchange coupled chain of magnetic moments, and is for example about 4 nm for Fe. Therefore a ferromagnet gets homogeneously magnetized when the size of the ferromagnet is smaller than ξ_K, whereas an inhomogeneous magnetic distribution such as domain walls appears when the size is larger than ξ_K. The domain wall width and the wall energy can be respectively expressed as $\pi\xi_K$ and $2A/\xi_K + 2K\xi_K$ using ξ_K. The second characteristic length ξ_M represents the length determined as a result of competition between the exchange energy and magnetostatic energy, corresponding to the core radius (~ 4 nm) of the magnetic vortex confined in a Fe disk structure. The last characteristic length ξ_H indicates the apparent effect of the externally applied magnetic field in length-scale. Therefore the other two characteristic lengths ξ_K and ξ_M above are influenced by the application of the external field such that $1/\xi_K^2 \Rightarrow 1/\xi_{K0}^2 \pm 1/\xi_H^2$ and $1/\xi_M^2 \Rightarrow 1/\xi_{M0}^2 \pm 1/\xi_H^2$, where the subscript K0 or M0 represents the initial state.

The exchange interaction forces the nearest spins to align into a uniform distribution, and the demagnetizing field creates the opposite effect on the long-range scale. It can be evaluated via a potential d as

$$H_d = -\nabla\Phi_d, \qquad \nabla^2\Phi_d = -M_S\rho_d, \qquad (22.3)$$

whose sources are volume and surface "magnetic charges"

$$\rho_d = -\nabla \cdot \mathbf{m}, \qquad \sigma_d = \mathbf{m} \cdot \mathbf{n} \tag{22.4}$$

where $\mathbf{n}$ is a unit vector normal to the surface of the magnetized element. Hence, the magnetization tends to align parallel to the surface in order to minimize the surface charges, leading to the occurrence of vortex distributions as depicted in Figs. 22.1(a)–(e). Moreover, the singularity at the center of a vortex is replaced by an out-of-plane magnetized core in order to reduce the exchange energy. On the other hand, in large samples, where the anisotropy energy predominates over the surface effects of the disk edges, the magnetization forms conventional domain patterns with magnetization aligned along easy axes (Fig. 22.1f).

When we slowly apply an external magnetic field, the competition among all the energies breaks the symmetry of the vortex, shifting its core so that the area of magnetization parallel to the field enlarges, until the vortex annihilates (at the "annihilation field"), resulting in the saturated (uniform) state. Then, when we reduce the external field, the uniform magnetization changes into a curved "C-state," until the vortex nucleates again (at the "nucleation field"). Reducing the field further to negative values causes a symmetrically analogous process, as depicted in Fig. 22.2.

For the vortex dynamics the main area of interest is the range of states before the vortex annihilates, which is represented in Fig. 22.2 by the slightly curved line whose tangent $\chi = \delta M / \delta H$ is called the effective magnetic susceptibility

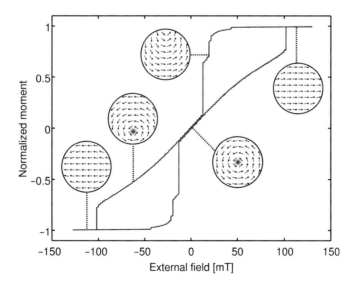

FIG. 22.2. Hysteresis loop representing the process of quasi-static switching of a cylindrical Py disk.

defined both statically and dynamically as a function of frequency $\chi(\omega)$. The dynamic response to fast changes of the external field is considerably different from that described by the hysteresis loop, and is in general governed by the Landau–Lifshitz–Gilbert (LLG) equation

$$\frac{\partial \mathbf{m}}{\partial t} = -\gamma \mathbf{m} \times \mathbf{H}_{eff} + \alpha \mathbf{m} \times \frac{\partial \mathbf{m}}{\partial t}, \tag{22.5}$$

where γ denotes the gyromagnetic ratio, α the Gilbert damping parameter, and t the time. Instead of applying an external field, the vortex distribution can be excited by an electrical current propagating through the ferromagnetic disk [14–16]. It has been revealed that this process, referred to as spin-transfer torque (STT), can be (in the adiabatic approximation) evaluated as an additional term on the right-hand side of the LLG equation

$$\mathbf{T}_{STT}^{(1)} = -(\mathbf{v}_S \cdot \Delta)\mathbf{m} \tag{22.6}$$

$$\mathbf{T}_{STT}^{(2)} = -g\frac{\hbar I_e}{2e}\mathbf{m}_2 \times (\mathbf{m}_2 \times \mathbf{m}_1) \tag{22.7}$$

where the first equation corresponds to the in-plane current with velocity $\mathbf{v}_S = \mathbf{j}_e P g \mu_B / 2e M_S$, whereas the second corresponds to the perpendicular current propagating through a multilayer depicted in Fig. 22.1(d) from the bottom to the top ferromagnetic layer (with magnetization distributions $\mathbf{m}_1$, $\mathbf{m}_2$, respectively), both of which are separated by a thin nonmagnetic interlayer (F/N/F). Here $\mathbf{j}_e$, P, g, μ_B, e, $\hbar$, and I_e denote the current density, spin polarization, g value of an electron, Bohr magneton, electronic charge, Planck constant, and total electrical current, respectively.

22.2 Analytical approaches

Although most authors have adopted OOMMF for its generality, simplicity, and accuracy, the development of analytical approaches is very useful for analyzing various fundamental aspects of dynamic processes. Among many attempts to reduce the number of parameters involved in vortex dynamics, perhaps the most frequently used is treating the vortex as a quasi-particle whose motion (or motion of its center $\mathbf{a} = [a_x, a_y]$ is described by an equation derived from the LLG equation by Thiele [7] for magnetic bubbles and adopted by Huber [18] to vortex systems,

$$\mathbf{G} \times \frac{\partial \mathbf{a}}{\partial t} = \frac{1}{R^2}\frac{\partial E_{tot}}{\partial \mathbf{a}} - \overset{\leftrightarrow}{\mathbf{D}} \cdot \frac{\partial \mathbf{a}}{\partial t}, \tag{22.8}$$

where $\mathbf{G} = -2\pi p L \mu_0 M_S \hat{\mathbf{z}}/\gamma$ is the gyrovector with $P = \pm 1$ denoting the vortex's polarity (the positive value stands for the up direction, parallel to the unit vector $\hat{z}$) and L denoting the disk's thickness, and where $\overset{\leftrightarrow}{\mathbf{D}} = -2\pi L \alpha \mu_0 M_S (\hat{\mathbf{x}}\hat{\mathbf{x}} + \hat{\mathbf{y}}\hat{\mathbf{y}})/\gamma$ is the second-order dissipation tensor. Thiele's equation of motion has thus found use as one of the most convenient approaches for dealing with vortex dynamics and has further been generalized to include an additional term of "mass times acceleration" [19,20] or to take into account STT [21].

To perform simulations with Thiele's equation, one needs to evaluate $E_{tot}(\mathbf{a})$ as a function of the vortex center's position. For this purpose, two approximations have been utilized; the "rigid vortex" model [22–25], assuming the static susceptibility, and the "side charges free" model [26], which assumes the magnetization on the disk edges to be constantly parallel to the surfaces. It has been revealed that the latter approximation applied to an isolated disk gives considerably better agreement with rigorous numerical simulations [27]. However, when applied to a pair or arrays of magnetostatically coupled disks, the rigid vortex model gives a reasonable tendency while the other model fails due to the absence of the side surface charges which are particularly responsible for the magnetostatic interaction between disks [28].

To study excitations of vortices more precisely, some authors have analytically solved the LLG equation by assuming small deviations of the static vortex distribution. They start with the description of the unit magnetization vector $\mathbf{m}(r, \chi)$ by angular parameters $\theta(r, \chi)$, $\varphi(r, \chi)$,

$$m_x + im_y = \sin\theta e^{i\varphi}, \quad m_z = \cos\theta \tag{22.9}$$

where r, χ are polar coordinates determining the lateral position within the disk. The small deviations of the static magnetization distribution (described as $\theta_{stat} = \theta_0(r)$, $\varphi_{stat} = \varphi_0(\chi) = q\chi$ where q denotes the vorticity of the system being $+1$ for a normal vortex or -1 for an antivortex) can be written as

$$\theta(r, \chi) = \theta_0(r) + \vartheta(r, \chi), \tag{22.10}$$

$$\varphi(r, \chi) = qr + [\sin\theta_0(r)]^{-1}\mu(r, \chi), \tag{22.11}$$

leading to the solution in the form [29]

$$\vartheta(r, \chi) = \sum_n \sum_{m=-\infty}^{+\infty} f_{nm}(r) \cos(m\chi + \omega_{nm}t + \delta_m) \tag{22.12}$$

$$\mu(r, \chi) = \sum_n \sum_{m=-\infty}^{+\infty} g_{nm}(r) \sin(m\chi + \omega_{nm}t + \delta_m) \tag{22.13}$$

where $[n, m]$ is a full set of numbers labeling magnon eigenstates and δm are arbitrary phases. This approach has been successfully applied to both antiferromagnets [30,31] and ferromagnets [32–42], and has revealed eigenfrequencies and eigenfunctions of spin-wave modes propagating in cylindrical disks and S-matrices of magnon–vortex scattering.

22.3 Experimental techniques

Experimental measurements of quasi-static properties of magnetic elements giving clear evidence of vortex structures, including the core's shapes and quasi-static switching processes, have been carried out by magnetic force microscopy (MFM) [7], spin-polarized scanning tunneling microscopy [8], magnetoresistance and Hall effect measurements [43–48], Lorentz transmission electron microscopy [49,50], magneto-optical Kerr effect (MOKE) measurements [51–59], photoelectron emission microscopy [60–62], scanning electron microscopy with spin-polarization analysis (SEMPA) [63,64], and others.

On the other hand, different techniques have to be employed for time-resolved dynamic measurement such as time-resolved Kerr microscopy (TRKM) in the scanning [65–72] or wide-field mode [73–75], photoemission electron microscopy combined with pulsed X-ray lasers [76], Brillouin light scattering (BLS) [77], time-resolved MFM [78], ferromagnetic resonance (FMR) techniques [79,80], vector network analyzers [81], superconducting quantum interference device magnetometry [82], and others. The most typical measurement technique, TRKM, often referred to as a "pump–probe" technique, combines a Kerr microscope of high space–time resolution achieved by an ultrashort-pulse laser light source and high-quality microscopic imaging (the "probe"), and a system for operating ultrafast pulse excitations achieved practically via various transmission line configurations as depicted in Fig. 22.3 (the "pump"). The source for the

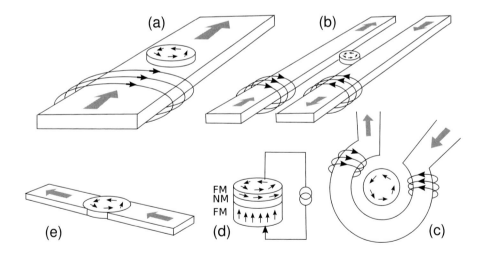

FIG. 22.3. Various types of ultrafast excitations: in-plane (a) or out-of-plane (b,c) magnetic field pulses are generated by electrical current pulses propagating through transmission lines with appropriate geometries; out-of-plane (d) and in-plane (e) currents induce excitations based on spin-transfer torque.

excitation current can be generated either by a pulse generator (triggered by the laser control device) or by a photoconductive switch (when laser pulses are split between probe and pump pulses). The wavelength of light is often halved by a second-harmonic generation device to increase the spatial resolution of measurement. The time dependence of magnetization evolution after excitation is determined by changing the delay time between the pump and the probe. To obtain an appropriate signal-to-noise ratio, the pump–probe measurement must be repeated many times with exactly the same initial condition, referred to as a stroboscopic method. An example of TRKM measurement [63,64], showing radial modes propagating from the edges of a Co cylindrical dot excited by an out-of-plane field, is displayed in Fig. 22.4.

22.4 Steady state motion phenomena

Various authors have studied dynamic excitations of vortices in cylindrical disks to observe a rich spectrum of modes [83]. Besides the existence of the radial modes

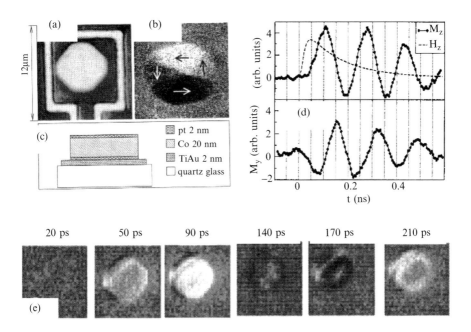

FIG. 22.4. Example of TRKM measurement performed by Acremann *et al.* [63,64] on a Co disk with an optical micrograph (a), SEMPA measurement of the static domain configuration (b), multilayer specification of the sample (c), time-resolved evolution of the M_z, M_y differences (d), and snapshots of M_z at particular times revealing the propagation of radial modes from the edges (which are excited by the out-of-plane field) towards the center.

excited by the out-of-plane field (Fig. 22.4) [63,64], it has also been revealed that low-energy modes (those near the ground state) excited by the in-plane field can be classified into two elementary types. The first type, referred to as the gyrotropic mode, is an oscillatory motion of the vortex core around its position in equilibrium, whose numerical simulation is displayed in Fig. 22.5 for a Py disk with a diameter of 100 nm and thickness of 20 nm. This type of motion has been predicted as the solution of Thiele's equation (22.8), as the analytical solution of

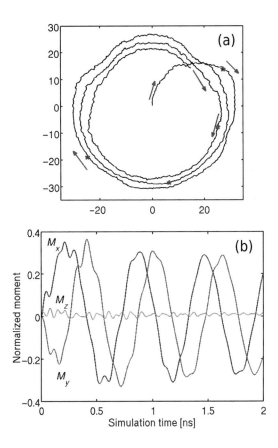

FIG. 22.5. Trajectory of the vortex core during and after an externally applied in-plane magnetic field pulse of strength $B_x = 60$ mT and duration 200 ps (a). After turning the field off, the vortex core exhibits a typical spiral motion around its equilibrium position at the disk's center ([0,0] nm). Asterisks denote the time points of 0.2, 0.5, 1, and 1.5 ns. The time evolution of the normalized total magnetic moment is displayed in (b). For the simulation, typical values for Py were used, $M_s = 860 \times 10^3$ A/m, A $= 1.3 \times 10^{-11}$ J/m, $\gamma = 2.2 \times 10^5$ m/As, and $\alpha = 0.01$.

the LLG equation in angular variables (Eqs. (22.12) and (22.3)), and has been extensively studied by micromagnetic simulations and various experiments. It has been revealed that the core's initial motion is parallel or antiparallel to the applied magnetic field pulse, depending on the vortex's "handedness" (the polarity relative to the chirality) [76]. However, the clockwise or counterclockwise sense of the core's spiral motion only depends on the vortex's polarity and is independent of the chirality. Owing to this rule, the vortex polarity can be magneto-optically measured via this dynamic motion, even though the small size of the vortex core makes the static magneto-optical measurement very difficult. In our example (Fig. 22.5) the motion's frequency is slightly above 1 GHz, and is decreased by reducing the disk's aspect ratio (thickness over diameter) [27].

The second type, referred to as magnetostatic modes, are high-frequency spin-wave excitations. It has been theoretically predicted [29] and experimentally observed [84] that there are azimuthal modes with degenerate frequency (frequency doublets), corresponding to the two values of the azimuthal magnon number $m = \pm |m|$ in Eqs. (22.12) and (22.13). In small disks (where the size of the out-of-plane polarized core becomes comparable to the entire size of the disk) this degeneracy is lifted (i.e. the frequency doublet becomes split), which has been explained via spin-wave–vortex (or magnon–soliton) interactions [85]. However, it has also been shown that removing the core (replacing the disk by a wide ring or introducing strong easy-plane anisotropy) retains the degeneracy of the doublet, so that no splitting occurs [84,86]. The whole process of the gyrotropic motion and a higher frequency doublet is displayed in Fig. 22.6.

The dynamic manipulation of vortex states by means of STT has become one of the most attractive subjects from both the fundamental and application viewpoints. Therefore, the current-induced motion of vortices has also been investigated to reveal phenomena analogous to those managed by the field excitation [87,88].

22.5 Dynamic switching

During the last few months, immensely intensive work has been carried out to study the process of dynamic switching of vortex polarities and chiralities, which is particularly important for the data storage application. Traditionally, to switch the vortex core's polarity, an extremely large quasi-static out-of-plane magnetic field was required. Moreover, to control chirality, the disk had to be fabricated with a geometric asymmetry, e.g. with a "D-shape" [49,89] or other shapes [61]. In contrast, dynamic processes have revealed considerable advantages.

Several authors have recently demonstrated [90–92] that a short pulse of in-plane magnetic field of a certain amplitude and duration excites the vortex so that a pair of a new vortex and an antivortex is created, the new vortex possessing the opposite polarity, and that the antivortex annihilates together with the old vortex core, as depicted in Fig. 22.7 for a cylindrical disk with a diameter of 200 nm and a thickness of 20 nm. This process is fully controllable

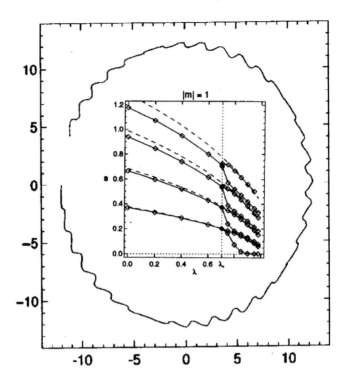

FIG. 22.6. Trajectory of the vortex core obtained by Ivanov *et al.* [29] using
the analytical approach as described by Eqs. 22.9 and 22.13 with damping
neglected, showing that the spiral motion is affected by high-frequency
oscillations. The inset in the middle shows the eigenfrequency spectrum of the
doublet $|m| = 1$ (denoting the "quantum number of the angular momentum")
as functions of the intrinsic easy-plane anisotropy λ. For $\lambda < \lambda_c$ (high
easy-plane anisotropy) the vortex only possess the in-plane components of
magnetization (no out-of-plane core appears); for $\lambda > \lambda_c$ (low anisotropy) the
vortex with an out-of-plane polarized core becomes responsible for significant
frequency splitting.

by applying an appropriate field pulse whose amplitude is considerably smaller
than those which are necessary for quasi-static switching. The process has also
been successfully observed by experimental measurements [93], and its variations
and further details are presently being researched [94,95].

For applications in spintronics, to control the switching process via an electrical current is of particular interest. In this respect various authors have recently
carried out theoretical [96–99] end experimental [78] studies to reveal that similar
dynamic switching processes are possible by applying STT excitations in both
configurations as described by Eqs. (22.6) and (22.7).

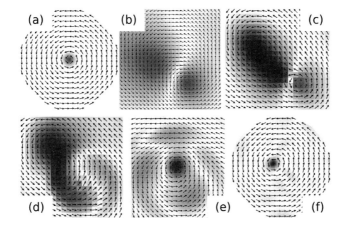

FIG. 22.7. Process of dynamic switching of the vortex polarity. An external field pulse of strength $B_x = 100$ mT and duration 30 ps is applied to the vortex at $t = 0$ (a). Shortly after turning the field off ($t = 33$ ps) the vortex distribution becomes slightly deviated (b), at $t = 57$ ps a pair of a new vortex and an antivortex is created (c), at $t = 67$ ps the antivortex annihilates together with the old vortex (d) which creates a point source of spin waves which are scattered by the new vortex at $t = 78$ ps (e). Finally, (f) shows a later distribution of the new vortex with the opposite polarity at $t/1$ ns. Each arrow in (b)–(e) represents the in-plane component of the magnetization vector in a grid of 2×2 nm, whereas the shades of gray correspond to the out-of-plane component.

Similarly to the quasi-static case, to change the vortex chirality requires introducing some geometric asymmetry into the process. For this purpose, Choi *et al.* [100] have applied a perpendicular current pulse to an F/N/F nanopillar, where the asymmetry is due to the magnetostatic interaction between the vortices in the two ferromagnetic layers. Recently the authors of this chapter have shown that introducing an asymmetry into the process of chirality switching is not necessary, because such asymmetry can be represented by the chirality itself [101]. However, the full and reproducible control of both binary parameters of a vortex is still a demanding task.

22.6 Magnetostatically coupled vortices

Many studies of quasi-static processes have been performed on pairs, chains, and two-dimensional arrays of magnetostatically coupled vortices, but little attention has been paid to their dynamic properties, although putting magnetic disks near each other is of high importance for improving the density of data storage and studying the propagation of micromagnetic excitations through such arrays.

Among dynamic studies, the pioneering experiments have been performed by means of Brillouin Right-scattering spectroscopy [102,103].

Recently the dynamics of magnetostatically coupled vortices was studied in pairs of disks placed near each other laterally [28], vertically (as an F/N/F nanopillar) [104], and as a pair of two vortices located inside a single elliptic dot [105,106]. It has been revealed, e.g. that the eigenfrequency of the synchronized steady-state motion of two vortices in the lateral arrangement is split into four distinct levels whose values depend on the lateral uniformity of the vortices' excitation and on the combination of their polarities (but are independent on chiralities) [28].

Large arrays of coupled vortices have also been investigated to reveal a close analogy with crystal vibrations (or phonon modes) in two-dimensional atomic lattices [107,108]. The dispersion relations and the corresponding densities of states of propagating waves of vortex excitations were found to vary with different ordering of vortex polarities regularly arranged within nanodisk arrays. Moreover, for arrays of disks with two different alternating diameters, a forbidden band gap has been observed (Fig. 22.8), pointing to the analogy with the band gap between the acoustic and optical phonon branches in atomic lattices [109,110]. Collective excitation modes have also been studied in small (3×3) arrays of nanodisks [111] and as analytical calculations using the Bloch theorem which enables us to deal with infinite arrays [41].

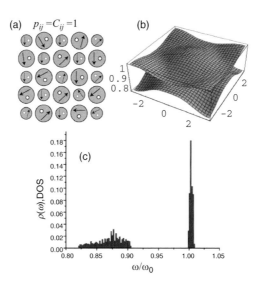

FIG. 22.8. Schematic diagram (a), dispersion relation in the first Brillouin zone (b), and density of states (c) of a vortex array within magnetostatically coupled disks with two different alternating radii (100 and 120 nm) and uniform polarities and chiralities ($p_{ij} = C_{ij} = 1$) throughout the lattice [108,109].

22.7   Conclusions and perspectives

We have reviewed the most fundamental achievements on the dynamic properties of magnetic vortices with a particular interest in soft cylindrical ferromagnetic disks. We have demonstrated the basic theoretical background widely used in analytical and numerical simulations, and briefly listed the utilized experimental approaches (including a description of TRKM as the most typical method of measurement). Then we have demonstrated the most significant results achieved by various authors. First, we have shown that the elementary excitations near the ground vortex state (steady state motions) are an important starting point to understand the whole principle of spin vortex dynamics. Then we have reported the recent results of ultrafast magnetic-field and STT-based switching of the vortex binary properties (polarity and chirality) which are of high importance for the possible future application in nonvolatile magnetic recording media. Finally we have briefly presented a few potentialities of vortices densely arranged into arrays or multilayers, which were found important, e.g. for designing novel artificial metamaterials which possess propagation modes based on magnetostatic interactions between nearest-neighbor elements.

Contemporary research continues to push the limits of the up-to-date theoretical and measurement capabilities and explors new directions of studying fundamental physical phenomena and utilizing them in new or higher-level applications. As regards the theoretical capabilities, reducing the element sizes to the true nanoscale requires the generalization of models to allow for the effect of surfaces and interfaces on the atomic scale [112], quantum and nonlinear effects (such as nonlinear optical excitations of spins usable for entirely optical switching [113,114]), etc. for which first-principles calculations will probably be employed. As regards experiment, the tendency of reducing sizes will require not only the improvement of the spatial resolution [115] (for which novel techniques are interesting, such as magnetic exchange force microscopy with atomic resolution [116]), but also further increase of the time resolution of dynamic measurements [11]. Moreover, since the stroboscopic measurement of the ultrafast dynamics requires repetition with always equal initial conditions, this method cannot be used to study possible stochastic processes, which offers a challenge to developing new experimental concepts. Another challenge from both the theoretical and the experimental viewpoints is the modulation of the dynamic properties of vortices by virtual fab [10] or artificial defects designed by tricky methods of deposition [117] or etching [118], leading to a considerable increase in the vortices' sensitivity to fields with particular strengths or frequencies. We can thus conclude that the results of numerous dynamics studies have pointed out various advantages and new potentialities in data storage, nanoscale probing of magnetic thin-film structures, and other types of sensing or controlling.

References

[1] R. Skomski: J. Phys.: Cond. Matter **15** (2003) R841.

[2] J. I. Martin, J. Nogues, K. Liu, J. L. Vicent, and I. K. Schuller: J. Magn. Magn. Mater. **256** (2003) 449.

[3] T. Thomson, G. Hu, and B. D. Terris: Phys. Rev. Lett. **96**, (2006), 257204; O. Hellwig, A. Berger, T. Thomson, E. Dobisz, H. Yang, Z. Bandic, D. Kercher, and E. Fullerton: Appl. Phys. Lett. **90** (2007) 162516.

[4] S. Neusser, and D. Grundler: Adv. Mater. **21** (2009) 2927; A. V. Chumak, V. S. Tiberkevich, A. D. Karenowska, A. A. Serga, J. F. Gregg, A. N. Slavin, and B. Hillebrands: Nat. Commun. **1** (2010) 141.

[5] D. A. Allwood, G. Xiong, C. C. Faulkner, D. Atkinson, D. Petit, and R. P. Cowburn: Science **309** (2005) 1688.

[6] A. Hubert and R. Schafer: *Magnetic Domains: The Analysis of Magnetic Microstructures*, Springer, Berlin, 1998.

[7] T. Shinjo, T. Okuno, R. Hassdorf, K. Shigeto, and T. Ono: Science **289** (2000) 930.

[8] A. Wachowiak, J. Wiebe, M. Bode, O. Pietzsch, M. Morgenstern, and R. Wiesendanger: Science **298** (2002) 577.

[9] A. M. Kosevich, B. A. Ivanov, and A. S. Kovalev: Phys. Rep. **194** (1990) 117.

[10] S. D. Bader: Rev. Mod. Phys. **78** (2006) 1.

[11] G. Srajer, L. H. Lewis, S. D. Bader, A. J. Epstein, C. S. Fadley, E. E. Fullerton, A. Hoffmann, J. B. Kortright, K. M. Krishnan, S. A. Majetich, T. S. Rahman, C. A. Ross, M. B. Salamon, I. K. Schuller, T. C. Schulthess, and J. Z. Sun: J. Magn. Magn. Mater. **307** (2006) 1.

[12] M. J. Donahue and D. G. Porter: OOMMF User's Guide, Version 1.0, Interagency Report NIST IR 6376, Gaithersburg, MD (1999).

[13] W. Heisenberg: Z. Physik **49** (1928) 619.

[14] J. C. Slonczewski: J. Magn. Magn. Mater. **159** (1996) L1.

[15] L. Berger: Phys. Rev. B **54** (1996) 9353.

[16] G. Tatara, H. Kohno, J. Shibata, Y. Lemaho, and K.-J. Lee: J. Phys. Soc. Jpn. **76** (2007) 054707.

[17] A. A. Thiele: Phys. Rev. Lett. **30** (1973) 230.

[18] D. L. Huber: Phys. Rev. B **26** (1982) 3758.

[19] G. M. Wysin: Phys. Rev. B **54** (1996) 15156.

[20] K. Y. Guslienko: Appl. Phys. Lett. **89** (2006) 022510.

[21] J. Shibata, Y. Nakatani, G. Tatara, H. Kohno, and Y. Otani: J. Magn. Magn. Mater. **310** (2007) 2041.

[22] K. Y. Guslienko and K. Metlov: Phys. Rev. B **63** (2001) 100403.

[23] K. Y. Guslienko, V. Novosad, Y. Otani, H. Shima, and K. Fukamichi: Appl. Phys. Lett. **78** (2001) 3848.

[24] K. Y. Guslienko, V. Novosad, Y. Otani, H. Shima, and K. Fukamichi: Phys. Rev. B **65** (2001) 024414.

[25] S. Savel'ev and F. Nori: Phys. Rev. B **70** (2004) 214415.

[26] K. L. Metlov and K. Y. Guslienko: J. Magn. Magn. Mater. **242–245** (2002) 1015.

[27] K. Y. Guslienko, B. A. Ivanov, V. Novosad, Y. Otani, H. Shima, and K. Fukamichi: J. Appl. Phys. **91** (2002) 8037.

[28] J. Shibata, K. Shigeto, and Y. Otani: Phys. Rev. B **67** (2003) 224404.

[29] B. A. Ivanov, H. J. Schnitzer, F. G. Mertens, and G. M. Wysin: Phys. Rev. B **58** (1998) 8464.

[30] B. A. Ivanov and D. D. Sheka: Phys. Rev. Lett. **72** (1994) 404.

[31] B. A. Ivanov, A. K. Kolezhuk, and G. M. Wysin: Phys. Rev. Lett. **76** (1996) 511.

[32] D. D. Sheka, B. A. Ivanov, and F. G. Mertens: Phys. Rev. B **64** (2001) 024432.

[33] B. A. Ivanov and G. M. Wysin: Phys. Rev. B **65** (2002) 134434.

[34] B. A. Ivanov and C. E. Zaspel: Appl. Phys. Lett. **81** (2002) 1261.

[35] D. D. Sheka, I. A. Yastremsky, B. A. Ivanov, G. M. Wysin, and F. G. Mertens: Phys. Rev. B **69** (2004) 054429.

[36] B. A. Ivanov and C. E. Zaspel: J. Appl. Phys. **95** (2004) 7444.

[37] A. Y. Galkin, B. A. Ivanov, and C. E. Zaspel: J. Magn. Magn. Mater. **286** (2005) 351.

[38] B. A. Ivanov and C. E. Zaspel: Phys. Rev. Lett. **94** (2005) 027205.

[39] C. E. Zaspel, B. A. Ivanov, J. P. Park, and P. A. Crowell: Phys. Rev. B **72** (2005) 024427.

[40] D. D. Sheka, C. Schuster, B. A. Ivanov, and F. G. Mertens: Eur. Phys. J. B **50** (2006) 393.

[41] A. Y. Galkin, B. A. Ivanov, and C. E. Zaspel: Phys. Rev. B **74** (2006) 144419.

[42] B. A. Ivanov, A. Y. Merkulov, V. A. Stephanovich, and C. E. Zaspel: Phys. Rev. B **74** (2006) 224422.

[43] T. Kimura, Y. Otani, and J. Hamrle: Appl. Phys. Lett. **87** (2005) 172506.

[44] P. Vavassori, M. Grimsditch, V. Metlushko, N. Zaluzec, and B. Ilic: Appl. Phys. Lett. **86** (2005) 072507.

[45] Y. S. Huang, C. C. Wang, and A. O. Adeyeye: J. Appl. Phys. **100** (2006) 013909.

[46] T. Ishida, T. Kimura, and Y. Otani: Phys. Rev. B **74** (2006) 014424.

[47] T. Ishida, T. Kimura, and Y. Otani: J. Magn. Magn. Mater. **310** (2007) 2431.

[48] M. Hara and Y. Otani: J. Appl. Phys. **101** (2007) 056107.

[49] M. Schneider, H. Hoffmann, and J. Zweck: Appl. Phys. Lett. **79** (2001) 3113.

[50] M. Schneider, H. Hoffmann, S. Otto, T. Haug, and J. Zweck: J. Appl. Phys. **92** (2002) 1466.

[51] R. P. Cowburn, D. K. Koltsov, A. O. Adeyeye, M. E. Welland, and D. M. Tricker: Phys. Rev. Lett. **83** (1999) 1042.

[52] N. Kikuchi, S. Okamoto, O. Kitakami, Y. Shimada, S. G. Kim, Y. Otani, and K. Fukamichi: J. Appl. Phys. **90** (2001) 6548.

[53] Y. Otani, H. Shima, K. Guslienko, V. Novosad, and K. Fukamichi: Phys. Stat. Sol. A **189** (2002) 521.

[54] V. Novosad, K. Y. Guslienko, H. Shima, Y. Otani, S. G. Kim, K. Fukamichi, N. Kikuchi, O. Kitakami, and Y. Shimada: Phys. Rev. B **65** (2002) 060402.

[55] H. Shima, K. Y. Guslienko, V. Novosad, Y. Otani, K. Fukamichi, N. Kikuchi, O. Kitakami, and Y. Shimada: J. Appl. Phys. **91** (2002) 6952.

[56] M. Grimsditch, P. Vavassori, V. Novosad, V. Metlushko, H. Shima, Y. Otani, and K. Fukamichi: Phys. Rev. B **65** (2002) 172419.

[57] M. Natali, I. L. Prejbeanu, A. Lebib, L. D. Buda, K. Ounadjela, and Y. Chen: Phys. Rev. Lett. **88** (2002) 157203.

[58] V. Novosad, M. Grimsditch, J. Darrouzet, J. Pearson, S. D. Bader, V. Metlushko, K. Guslienko, Y. Otani, H. Shima, and K. Fukamichi: Appl. Phys. Lett. **82** (2003) 3716.

[59] M. Natali, A. Popa, U. Ebels, Y. Chen, S. Li, and M. E. Welland: J. Appl. Phys. **96** (2004) 4334.

[60] K. S. Buchanan, K. Y. Guslienko, A. Doran, A. Scholl, S. D. Bader, and V. Novosad: Phys. Rev. B **72** (2005) 134415.

[61] T. Taniuchi, M. Oshima, H. Akinaga, and K. Ono: J. Appl. Phys. **97** (2005) 10J904.

[62] T. Taniuchi, M. Oshima, H. Akinaga, and K. Ono: J. Electron. Spectrosc. Relat. Phenom. **144–147** (2005) 741.

[63] Y. Acremann, C. H. Back, M. Buess, O. Portmann, A. Vaterlaus, D. Pescia, and H. Melchior: Science **290** (2000) 492.

[64] Y. Acremann, A. Kashuba, M. Buess, D. Pescia, and C. H. Back: J. Magn. Magn. Mater. **239** (2002) 346.

[65] T. J. Silva and A. B. Kos: J. Appl. Phys. **81** (1997) 5015.

[66] W. K. Hiebert, A. Stankiewicz, and M. R. Freeman: Phys. Rev. Lett. **79** (1997) 1134.

[67] M. R. Freeman, W. K. Hiebert, and A. Stankiewicz: J. Appl. Phys. **83** (1998) 6217.

[68] G. E. Ballentine, W. K. Hiebert, A. Stankiewicz, and M. R. Freeman: J. Appl. Phys. **87** (2000) 6830.

[69] M. R. Freemann and W. K. Hiebert: Topics Appl. Phys. **83** (2002) 93.

[70] J. P. Park, P. Eames, D. M. Engebretson, J. Berezovsky, and P. A. Crowell: Phys. Rev. B **67** (2003) 020403(R).

[71] R. J. Hicken, A. Barman, V. V. Kruglyak, and S. Ladak: J. Phys. D: Appl. Phys. **36** (2003) 2183.

[72] V. V. Kruglyak, A. Barman, R. J. Hicken, J. R. Childress, and J. A. Katine: J. Appl. Phys. **97** (2005) 10A706.

[73] A. Neudert, J. McCord, R. Schafer, and L. Schultz: J. Appl. Phys. **97** (2005) 10E701.

[74] A. Neudert, J. McCord, D. Chumakov, R. Schafer, and L. Schultz: Phys. Rev. B **71** (2005) 134405.

[75] A. Neudert, J. McCord, R. Schafer, R. Kaltofen, I. Monch, H. Vinzelberg, and L. Schultz: J. Appl. Phys. **99** (2006) 08F302.

[76] S.-B. Choe, Y. Acremann, A. Scholl, A. Bauer, A. Doran, J. Stohr, and H. A. Padmore: Science **304** (2004) 420.

[77] K. Perzlmaier, M. Buess, C. H. Back, V. E. Demidov, B. Hillebrands, and S. O. Demokritov: Phys. Rev. Lett. **94** (2005) 057202.

[78] K. Yamada, S. Kasai, Y. Nakatani, K. Kobayashi, H. Kohno, A. Thiaville, and T. Ono: Nat. Mater. **6** (2007) 269.

[79] G. N. Kakazei, P. E. Wigen, K. Y. Guslienko, R. W. Chantrell, N. A. Lesnik, V. Metlushko, H. Shima, K. Fukamichi, Y. Otani, and V. Novosad: J. Appl. Phys. **93** (2003) 8418.

[80] G. N. Kakazei, P. E. Wigen, K. Y. Guslienko, V. Novosad, A. N. Slavin, V. O. Golub, N. A. Lesnik, and Y. Otani: Appl. Phys. Lett. **85** (2004) 443.

[81] V. Novosad, F. Y. Fradin, P. E. Roy, K. S. Buchanan, K. Y. Guslienko, and S. D. Bader: Phys. Rev. B **72** (2005) 024455.

[82] H. Shima, V. Novosad, Y. Otani, K. Fukamichi, N. Kikuchi, O. Kitakamai, and Y. Shimada: J. Appl. Phys. **92** (2002) 1473.

[83] K. Y. Guslienko, W. Scholtz, R. W. Chantrell, and V. Novosad: Phys. Rev. B **71** (2005) 144407.

[84] X. Zhu, Z. Liu, V. Metlushko, P. Grutter, and M. R. Freeman: Phys. Rev. B **71** (2005) 180408.

[85] J. P. Park and P. A. Crowell: Phys. Rev. Lett. **95** (2005) 167201.

[86] F. Hoffmann, G. Woltersdorf, A. N. Slavin, V. S. Tiberkevich, A. Bischof, D. Weiss, and C. H. Back: Phys. Rev. B **76** (2007) 014416.

[87] J. Shibata, Y. Nakatani, G. Tatara, H. Kohno, and Y. Otani: Phys. Rev. B **73** (2006) 020403(R).

[88] S. Kasai, Y. Nakatani, K. Kobayashi, H. Kohno, and T. Ono: Phys. Rev. Lett. **97** (2006) 107204.

[89] T. Kimura, Y. Otani, H. Masaki, T. Ishida, R. Antos, and J. Shibata: Appl. Phys. Lett. **90** (2007) 132501.

[90] N. Massart and Y. Otani: "Collective motion in ferromagnetic dots arrays with vortex configuration—Preparation for a time-resolved Kerr-effect experiment," Scientific option traineeship report, Quantum Nano-Scale Magnetics Laboratory, RIKEN, Japan (2004).

[91] Q. F. Xiao, J. Rudge, B. C. Choi, Y. K. Hong, and G. Donohoe: Appl. Phys. Lett. **89** (2006) 262507.

[92] R. Hertel, S. Gliga, M. Fahnle, and C. M. Schneider: Phys. Rev. Lett. **98** (2007) 117201.

[93] B. Van Waeyenberge, A. Puzic, H. Stoll, K. W. Chou, T. Tyliszczak, R. Hertel, M. Fahnle, H. Bruckl, K. Rott, G. Reiss, I. Neudecker, D. Weiss, C. H. Back, and G. Schutz: Nature (London) **444** (2006) 461.

[94] V. P. Kravchuk, D. D. Sheka, Y. Gaididei, and F. G. Mertens: J. Appl. Phys. **102** (2007) 043908.

[95] S. Choi, K.-S. Lee, K. Y. Guslienko, and S.-K. Kim: Phys. Rev. Lett. **98** (2007) 087205.

[96] S.-K. Kim, Y.-S. Choi, K.-S. Lee, K. Y. Guslienko, and D.-E. Jeong: Appl. Phys. Lett. **91** (2007) 082506.

[97] Y. Liu, S. Gliga, R. Hertel, and C. M. Schneider: Appl. Phys. Lett. **91** (2007) 112501.

[98] D. D. Sheka, Y. Gaididei, and F. G. Mertens: Appl. Phys. Lett. **91** (2007) 082509.

[99] J.-G. Caputo, Y. Gaididei, F. G. Mertens, and D. D. Sheka: Phys. Rev. Lett. **98** (2007) 056604.

[100] B. C. Choi, J. Rudge, E. Girgis, J. Kolthammer, Y. K. Hong, and A. Lyle: Appl. Phys. Lett. **91** (2007) 022501.

[101] A. Roman and Y. Otani: Phys. Rev. B **80** (2009) 140404-1.

[102] K. Y. Guslienko, K. S. Buchanan, S. D. Bader, and V. Novosad: Appl. Phys. Lett. **86** (2005) 223112.

[103] K. S. Buchanan, P. E. Roy, M. Grimsditch, F. Y. Fradin, K. Y. Guslienko, S. D. Bader, and V. Novosad: Nat. Phys. **1** (2005) 172.

[104] K. S. Buchanan, P. E. Roy, F. Y. Fradin, K. Y. Guslienko, M. Grimsditch, S. D. Bader, and V. Novosad: J. Appl. Phys. **99** (2006) 08C707.

[105] C. Mathieu, C. Hartmann, M. Bauer, O. Buettner, S. Riedling, B. Roos, S. O. Demokritov, B. Hillebrands, B. Bartenlian and C. Chappert, D. Decanini, F. Rousseaux, E. Cambril, A. Muller, B. Hoffmann, and U. Hartmann: Appl. Phys. Lett. **70** (1997) 2912.

[106] J. Jorzick, S. O. Demokritov, B. Hillebrands, B. Bartenlian, C. Chappert, D. Decanini, F. Rousseaux, and E. Cambril: Appl. Phys. Lett. **75** (1999) 3859.

[107] J. Shibata and Y. Otani: Phys. Rev. B **70** (2004) 012404.

[108] J. Shibata, K. Shigeto, and Y. Otani: J. Magn. Magn. Mater. **272–276** (2004) 1688.

[109] J. Shibata, K. Shigeto, and Y. Otani: paper presented at the 58th Annual Meeting of the Physical Society of Japan, Tohoku University, Sendai, March 2003.

[110] R. Antos, J. Hamrle, H. Masaki, T. Kimura, J. Shibata, and Y. Otani: Proc. SPIE **6479** (2007) 647907.

[111] G. Gubbiotti, M. Madami, S. Tacchi, G. Carlotti, and T. Okuno: J. Appl. Phys. **99** (2006) 08C701.

[112] S. Rohart, V. Repain, A. Thiaville, and S. Rousset: Phys. Rev. B **76** (2007) 104401.

[113] C. D. Stanciu, A. V. Kimel, F. Hansteen, A. Tsukamoto, A. Itoh, A. Kirilyuk, and Th. Rasing: Phys. Rev. B **73** (2006) 220402(R).

[114] C. D. Stanciu, F. Hansteen, A. V. Kimel, A. Kirilyuk, A. Tsukamoto, A. Itoh, and Th. Rasing: Phys. Rev. Lett. **99** (2007) 047601.

[115] K. W. Chou, A. Puzic, H. Stoll, D. Dolgos, G. Schutz, B. Van Waeyenberge, A. Vansteenkiste, T. Tyliszczak, G. Woltersdorf, and C. H. Back: Appl. Phys. Lett. **90** (2007) 202505.

[116] U. Kaiser, A. Schwarz, and R. Wiesendanger: Nature (London) **446** (2007) 522.

[117] R. L. Compton and P. A. Crowell: Phys. Rev. Lett. **97** (2006) 137202.

[118] K. Kuepper, L. Bischoff, Ch. Akhmadaliev, J. Fassbender, H. Stoll, K. W. Chou, A. Puzic, K. Fauth, D. Dolgos, G. Schutz, B. Van Waeyenberge, T. Tyliszczak, I. Neudecker, G. Woltersdorf, and C. H. Back: Appl. Phys. Lett. **90** (2007) 062506.

23 Spin-transfer torque in nonuniform magnetic structures

T. Ono

23.1 Magnetic domain wall

Weiss pointed out in his paper on spontaneous magnetization in 1907 that ferromagnetic materials are not necessarily magnetized to saturation in the absence of an external magnetic field [1]. Instead, they have magnetic domains, within each of which magnetic moments align. The formation of the magnetic domains is energetically favorable because this structure can lower the magnetostatic energy originating from the dipole–dipole interaction. The directions of magnetization of neighboring domains are not parallel. As a result, between two neighboring domains, there is a region in which the direction of magnetic moments gradually changes. This transition region is called a magnetic domain wall (DW).

Recent developments in nanolithography techniques make it possible to prepare nanoscale magnets with simple magnetic domain structure which is suitable for basic studies on magnetization reversal and also for applications. For example, in a magnetic wire with submicron width, two important processes in magnetization reversal, nucleation and propagation of a magnetic DW, can be clearly seen. As shown in Fig. 23.1(a), in a very narrow ferromagnetic wire, the magnetization is restricted to being directed parallel to the wire axis due to the magnetic shape anisotropy. When an external magnetic field is applied against the magnetization, a magnetic DW nucleates at the end of the wire and magnetization reversal proceeds by the propagation of this DW through the wire (Fig. 23.1b, c). This textbook DW motion has been observed experimentally thanks to the developments of nanotechnology, and there are many interesting reports even on the magnetic field-driven DW motion [2–13].

23.1.1 *Magnetic vortex*

Another typical example of nonuniform magnetic structure is a magnetic vortex which is realized in a ferromagnetic disk. As mentioned above, ferromagnetic materials generally form domain structures to reduce their magnetostatic energy. In very small ferromagnetic systems, however, the formation of DWs is not energetically favored. Specifically, in a disk of ferromagnetic material of micrometer or submicrometer size, a curling spin configuration—that is, a magnetization vortex (Fig. 23.2)—has been proposed to occur in place of domains. When the

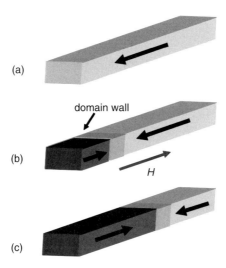

(a)

domain wall

(b)

H

(c)

FIG. 23.1. Schematic illustration of the magnetization reversal process in a magnetic wire.

dot thickness becomes much smaller than the dot diameter, usually all spins tend to align in-plane. In the curling configuration, the spin directions change gradually in-plane so as not to lose too much exchange energy, but to cancel the total dipole energy. In the vicinity of the dot center, the angle between adjacent spins then becomes increasingly larger when the spin directions remain confined in-plane. Therefore, at the center of the vortex structure, the magnetization within a small spot will turn out-of plane and parallel to the plane normal [14].

Figure 23.2 is the first proof of such a vortex structure with a nanometer-scale core where the magnetization rises out of the dot plane [15,16].The sample is an array of 3×3 dots of permalloy ($Ni_{81}Fe_{19}$) with 1 μm in diameter and 50 nm thickness. At the center of each dot, bright or dark contrast is observed, which corresponds to the positive or negative stray field from the vortex core. The direction of the magnetization at the center turns randomly, either up or down, as reflected by the different contrast of the center spots. This is reasonable since up and down-magnetizations are energetically equivalent without an external applied field and do not depend on the vortex orientation: clockwise or counterclockwise. MFM observations were performed also for an ensemble of permalloy dots with varying diameters, nominally from 0.3 to 1 μm (Fig. 23.3). The image in Fig. 23.3(a) was taken after applying an external field of 1.5 T along an in-plane direction. Again, the two types of vortex core with up and down magnetization are observed. In contrast, after applying an external field of 1.5 T normal to the substrate plane, the center spots exhibit the same contrast (Fig. 23.3b) indicating that all the vortex core magnetizations have been oriented into the field direction. The size of the core cannot be determined from the images

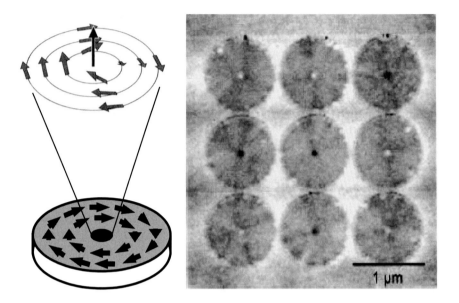

FIG. 23.2. MFM image of an array of permalloy dots 1 μm in diameter and 50 nm thick with the schematic spin structure (magnetic vortex and vortex core) in a disk [15].

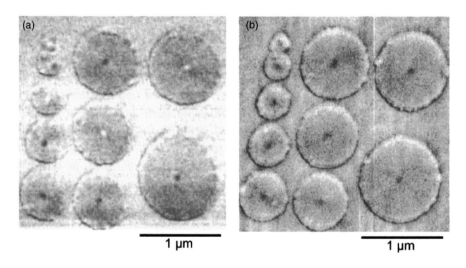

FIG. 23.3. MFM images of an ensemble of 50-nm-thick permalloy disks with diameters varying from 0.3 to 1 μm: (a) after applying an external field of 1.5 T along an in-plane direction and (b) parallel to the plane normal, respectively [15].

since the spatial resolution of MFM is much larger than the theoretical core size. The core size was determined to be 9 nm by using spin-polarized scanning tunneling microscopy which has an atomic-scale resolution [17].

The experimental confirmation of the existence of the vortex core by MFM studies [15, 16] stimulated the subsequent intensive studies on the dynamics of the vortex core. It has been clarified that a vortex core displaced from the stable position (dot center) exhibits a spiral precession around it during the relaxation process [18–20]. This motion has a characteristic frequency which is determined by the shape of the disk. Thus, the disk functions as a resonator for the vortex core motion. Excitation of a magnetic vortex by the spin-transfer torque will be discussed in Section 23.3.

23.2 Current-driven domain wall motion

23.2.1 *Basic idea of current-driven domain wall motion*

As a typical and instructive example of spin-transfer torque in a nonuniform magnetic structure, let us consider the interaction between an electric current and a DW. Figure 23.4(a) is an illustration of a DW between two domains in a

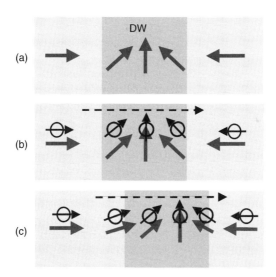

FIG. 23.4. Schematic illustration of current-driven DW motion. (a) A DW between two domains in a magnetic wire. The arrows show the direction of the magnetic moments. The magnetic DW is a transition region of the magnetic moments between domains, and the direction of the moments gradually changes in the DW. (b) The spin of a conduction electron follows the direction of the local magnetic moments because of the *s-d* interaction. (c) As a reaction, the local magnetic moments rotate in reverse, and in consequence, the electric current displaces the DW.

magnetic wire. Here, the arrows show the direction of local magnetic moments. The magnetic DW is a transition region of the magnetic moments between domains, and the direction of the moments gradually changes in the DW. What will happen if an electric current flows through a DW? Suppose a conduction electron passes though the DW from left to right. During this travel, the spin of the conduction electron follows the direction of local magnetic moments because of the *s-d* interaction (Fig. 23.4b). As a reaction, the local magnetic moments rotate reversely (Fig. 23.4c), and, in consequence, the electric current can displace the DW.

The current-driven DW motion described above was first discussed by Berger in a space-integrated form from the change of angular momentum of the conduction electrons after crossing a DW [21, 22]. Then, Bazaliy *et al.* proposed an expression for the local torque due to the spin transfer inside a DW [23]. For a wide DW, the conduction electrons' spin is expected to follow the direction of the local magnetic moment. In this adiabatic limit, the spin-transfer torque is obtained as the differential change of the angular momentum of conduction electrons, and can be expressed as

$$-\frac{Jg\mu_B P}{2eM_s}\frac{\partial \boldsymbol{m}}{\partial x} \equiv -(\boldsymbol{u}\cdot\nabla)\boldsymbol{x}, \qquad (23.1)$$

where J is the current density, g is the g-value, P is the spin polarization of current, e is the electron charge, M_s is the saturation magnetic moment, and $\boldsymbol{m}$ is a unit vector along the local magnetization. Therefore, the magnetization dynamics is governed by the following modified Landau–Lifshitz–Gilbert equation

$$\dot{\boldsymbol{m}} = \gamma H_{eff}\times\boldsymbol{m} + \alpha\boldsymbol{m}\times\dot{\boldsymbol{m}} - (\boldsymbol{u}\cdot\nabla)\boldsymbol{m}, \qquad (23.2)$$

where γ is the gyromagnetic ratio, H_{eff} is the effective magnetic field, α is the Gilbert damping constant, and an overdot is used to denote the time derivative. The adiabatic torque moves a DW by changing its structure periodically between the Bloch wall and Nl wall (Fig. 23.5). The energy barrier to be overcome to change the domain wall structure is called intrinsic pinning, which determines the threshold current density, J_c, for the DW motion [24].

Berger and his collaborators performed several experiments on magnetic films [25, 26]. It needed huge currents to move a DW in a magnetic film due to the large cross-section. Recent developments in nanolithography techniques make it possible to prepare nanoscale magnetic wires, resulting in a review of their pioneering work. The current-driven DW motion provides a new strategy to manipulate a magnetic configuration without any assistance from the magnetic field, and will improve drastically the performance and functions of recently proposed spintronic devices, whose operation is based on the motion of a magnetic DW [27–30]. Reports on this subject have been increasing in recent years from both the theoretical [31–35] and experimental [36–69] points of view because of

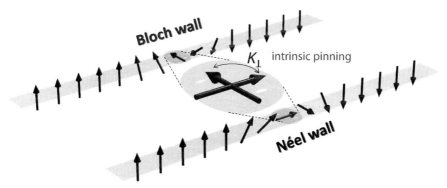

FIG. 23.5. Schematic of Bloch and Néel domain walls. A current flowing in the wire exerts a spin-transfer torque on the local spins inside the domain wall and rotates them in the sample plane, resulting in a periodic change of the domain wall structure between the Bloch and Néel walls. $K_\perp$ is the energy density of magnetic anisotropy associated with the rotation of the domain wall spins, which determines the intrinsic pinning for the current-driven domain wall motion.

its scientific and technological importance. However, most of the results cannot be reviewed here due to the limitation of space.

23.2.2 *Direct observation of current-driven domain wall motion by magnetic force microscopy*

The result of direct observation of the current-driven DW motion by means of magnetic force microscopy (MFM) is shown in Fig. 23.6 [40]. The sample is a magnetic wire of 10 nm thick $Ni_{81}Fe_{19}$ with the width of 240 nm. A single DW is imaged as a bright contrast, which corresponds to the stray field from a positive magnetic charge (Fig. 23.6a), an indication that a head-to-head DW is realized as illustrated schematically in Fig. 23.6(d). The position and the shape of the DW were unchanged after several MFM scans, indicating that the DW was pinned by a local structural defect, and that a stray field from the probe was too small to change the magnetic structure and the position of the DW. After the observation of Fig. 23.6(a), a pulsed current was applied through the wire in the absence of a magnetic field. The current density and the pulse duration were 7.0×10^{11} A/m^2 and 5 μs, respectively. Figure 23.6(b) shows an MFM image after the application of the pulsed current from left to right. The DW is displaced from right to left by the application of the pulsed current. Thus, the direction of the DW motion is opposite to the current direction. Furthermore, the direction of the DW motion can be reversed by switching the current polarity as shown in Fig. 23.6(c). These results are consistent with the spin-transfer mechanism that is described in Section 23.3.1.

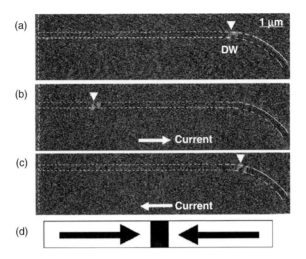

FIG. 23.6. (a) MFM image after the introduction of a head-to-head DW. The DW is imaged as a bright contrast, which corresponds to the stray field from positive magnetic charge. (b) MFM image after an application of a pulsed current from left to right. The current density and pulse duration are $7.0 \times 10^{11} A/m^2$ and 5 µs, respectively. The DW is displaced from right to left by the pulsed current. (c) MFM image after the application of a pulsed current from right to left. The current density and pulse duration are $7.0 \times 10^{11} A/m^2$ and 5 µs, respectively. The DW is displaced from left to right by the pulsed current. (d) Schematic illustration of a magnetic domain structure inferred from the MFM image. The DW has a head-to-head structure [40].

The same experiments for a DW with different polarities, a tail-to-tail DW, were performed to examine the effect of a magnetic field generated by the electric current (Oersted field). The introduced DW is imaged as a dark contrast in Fig. 23.7(a), which indicates that a tail-to-tail DW is formed as schematically illustrated in Fig. 23.7(d). Figures 23.7(a), (b), and (c) show that the direction of the tail-to-tail DW displacement is also opposite to the current direction. The fact that both head-to-head and tail-to-tail DWs are displaced opposite to the current direction indicates clearly that the DW motion is not caused by the Oersted field. The successive MFM images with one pulsed current applied between each consecutive image shown in Figs. 23.8(a)–(k) demonstrate that a DW can be displaced in any position in the nanowire by the current-driven DW motion.

23.2.3 *Beyond the adiabatic approximation: Non-adiabatic torque*

It was shown that the DW position in a wire can be controlled by tuning the intensity, the duration, and the polarity of the pulsed current, and thus the

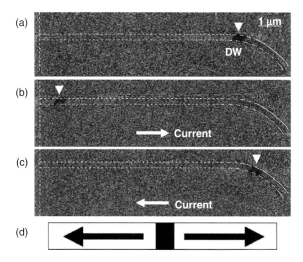

FIG. 23.7. (a) MFM image after the introduction of a tail-to-tail DW. The DW is imaged as a dark contrast, which corresponds to the stray field from negative magnetic charge. (b) MFM image after the application of a pulsed current from left to right. The current density and pulse duration are $7.0 \times 10^{11} \text{A/m}^2$ and 5 μs, respectively. The DW is displaced from right to left by the pulsed current. (c) MFM image after the application of a pulsed current from right to left. The current density and pulse duration are $7.0 \times 10^{11} \text{A/m}^2$ and 5 μs, respectively. The DW is displaced from left to right by the pulsed current. (d) Schematic illustration of a magnetic domain structure inferred from the MFM image. The DW has a tail-to-tail structure [40].

current-driven DW motion has the potentiality for spintronic device applications such as novel memory and storage devices [27–30]. However, there was a big discrepancy between the experimental results and the theoretical prediction. The experimentally obtained threshold current densities for NiFe wires are the order of 10^{11}–10^{12} A/m^2 [40, 45]. These values are more than an order of magnitude smaller than the theoretical value and also that obtained from the micromagnetic simulation [24, 34, 35]. To solve this discrepancy, a new term called the non-adiabatic spin transfer term or beta term was proposed to be included in the Landau–Lifshitz–Gilbert (LLG) equation. Because DWs are never infinitely wide, the adiabatic spin-transfer torque (Eq. 23.1) is a kind of approximation, and the deviation from adiabaticity should be taken into account. From the mathematical point of view, the form of the only possible other torque is $\boldsymbol{m} \times [(\boldsymbol{u} \cdot \triangledown)\boldsymbol{m}]$, because $\dot{\boldsymbol{m}}$ has to be orthogonal to $\boldsymbol{m}$. Therefore, the LLG equation is modified to

$$\dot{\boldsymbol{m}} = \gamma H_{eff} \times \boldsymbol{m} + \alpha \boldsymbol{m} \times \dot{\boldsymbol{m}} - (\boldsymbol{u} \cdot \triangledown)\boldsymbol{m} + \beta \boldsymbol{m} \times [(\boldsymbol{u} \cdot \triangledown)\boldsymbol{m}]. \qquad (23.3)$$

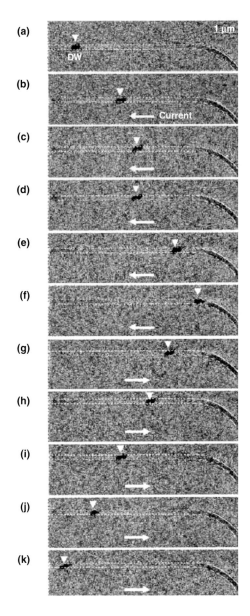

FIG. 23.8. Successive MFM images with one pulse applied between each consecutive image. The current density and the pulse duration were $7.0 \times 10^{11}\,\mathrm{A/m}^2$ and 5 µs, respectively. Note that a tail-to-tail DW is introduced, which is imaged as a dark contrast [40].

Because this equation can be rewritten as

$$\dot{\boldsymbol{m}} = [\gamma H_{eff} - \beta(\boldsymbol{u} \cdot \nabla)\boldsymbol{m}] \times \boldsymbol{m} + \alpha \boldsymbol{m} \times \dot{\boldsymbol{m}} - (\boldsymbol{u} \cdot \nabla)\boldsymbol{m}, \qquad (23.4)$$

the nonadiabatic torque works like an effective magnetic field. This is the reason that the nonadiabatic torque is often called the field-like torque. For a wire with DW pinning potentials due to defects, J_c is expected to increase linearly with the strength of the DW pinning potential, because the nonadiabatic torque works as the effective field, which has been experimentally observed in NiFe wires [61]. Another important consequence of the inclusion of the nonadiabatic torque is the theoretical prediction that the DW velocity is proportional to β/α Therefore, the value of β is crucial for J_c and the DW velocity, although there have been only a few reports of the estimation of β [51, 68]

23.2.4 Domain wall motion by adiabatic torque and intrinsic pinning

As discussed in the previous sections, both adiabatic and nonadiabatic torques should be taken into account in the current-driven DW motion. The question is which torque dominates the current-driven DW motion. The current-driven DW motion in NiFe wires is believed to be dominated by the nonadiabatic torque, because J_c is proportional to the strength of the DW pinning [61]. Recently, clear evidence has been reported that the intrinsic pinning determines the threshold, and thus that the adiabatic spin torque dominates the DW motion, in a perpendicularly magnetized Co/Ni nanowire as described below [69].

The dependence of J_c on the wire width w was investigated, and the result is summarized in Fig. 23.9(a) [69]. J_c reduces from 5×10^{11} A/m^2 to 2×10^{11} A/m^2 as w reduces, and then increases below $w = 70$ nm. Thus J_c has a clear minimum value around $w = 70$ nm. For the stationary DW in a perpendicularly magnetized film like a Co/Ni multilayer, generally the Bloch wall is stable. In the nanowire, however, the energy of the Bloch wall increases as w reduces, and finally the Néel wall is expected to be stable, because the demagnetizing field of the transverse direction of the wire increases. The energy difference between the Bloch and Néel walls is lowest at the boundary. As discussed in Section 23.3.1, the energy difference between the Bloch and Néel walls determines the intrinsic pinning, which determines J_c for the DW motion (Fig. 23.5). Therefore, the observed minimum of J_c in Fig. 23.9(a) suggests the existence of the intrinsic pinning of a DW.

To confirm the above scenario, it is necessary to identify the structure of the stationary DW in the wires. For this purpose, the resistance of a DW in the wire was measured. The resistance of the Néel wall is expected to be larger than that of the Bloch wall due to the anisotropic magnetoresistance effect, because the local spin inside the DW points in the parallel direction with respect to the measuring current, whereas it points in the perpendicular direction in the Bloch wall. Figures 23.9(b)–(e) show the histograms of the DW resistance measured for the wires with $w = 92$, 76, 59, and 40 nm, respectively. For $w = 92$ and 76

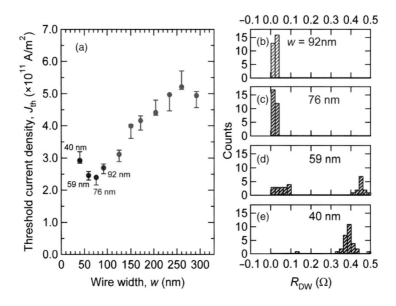

FIG. 23.9. (a) The threshold current density as a function of the width of the Co/Ni wire. (b)–(e) Histograms of the domain wall resistance in wires with $w = 92$, 76, 59, and 40 nm. The measurement of domain wall resistance was repeated 30 times in each wire.

nm, the DW resistance is distributed in the vicinity of zero. Then, it split up into two peaks for $w = 59$ nm. Finally a single peak at around 0.4 Ω appears when $w = 40$ nm. These results indicate that the DW structure changes from Bloch to Néel in the vicinity of $w = 59$ nm by reducing w. The minimum J_c is observed near this border. It was also confirmed that there is no systematic correlation between the w dependence of J_c and that of the DW pinning field. This is in clear contrast to the result for NiFe wires, that J_c is proportional to the DW pinning field. Therefore, these results offer strong evidence that J_c is dominated by the intrinsic pinning in the Co/Ni nanowires.

23.2.5   *Toward applications of current-driven domain wall motion*

There are several proposals for application of current-driven DW motion to spintronic devices such as novel memory and storage devises [27–30], and device operations have already been demonstrated [61, 67]. However, there are several issues to be overcome for practical applications: (1) low threshold current density; (2) high DW velocity; and (3) stability of DW position. These three conditions should be simultaneously satisfied to realize practical devices.

Although the high velocity of 100 m/s has been demonstrated for NiFe nanowires [59], it has also been shown that the threshold current density increases with the DW pinning field [61]. This could become a problem in applications, because the DW position should be stabilized with high pinning potential against the thermal agitation. Recently very attractive simulation results have been coming out, which suggest that it is possible to reduce the threshold current density while keeping the high thermal stability of the DW position for nanowires with perpendicular magnetic anisotropy [62, 63]. Promising experimental results, which support these simulations, have been reported [64–66, 69–72]. It was shown that a single nanosecond current pulse can control precisely the DW position from notch to notch in a Co/Ni wire with perpendicular magnetic anisotropy in spite of the large DW depinning field from the notch of 400 Oe [65]. It was also revealed that both J_c and the DW velocity are insensitive to the external magnetic field which certifies the robust operation of DW devices [69, 70]. A stable domain wall motion was observed up to the temperature at which perpendicular magnetic anisotropy vanishes [72]. Moreover, the current required for domain wall motion was independent of the device temperature [71, 72]. A relatively high DW velocity of 60 m/s was reported for the current density of 1.0×10^{12} A/m^2, which can be understood by the adiabatic spin-transfer model [70]. These observed characteristics make the Co/Ni system a promising candidate for practical applications.

23.3   Current-driven excitation of magnetic vortices

23.3.1   *Current-driven resonant excitation of magnetic vortices*

As is clear from Fig. 23.4, the underlying physics of the current-driven DW motion is that the electric currents can apply a torque on the magnetic moment when the spin direction of the conduction electrons has a relative angle to the local magnetic moment. This leads us to the hypothesis that any type of spin structure with spatial variation can be excited by a spin-polarized current in a ferromagnet.

The ideal example of such a noncollinear spin structure is a curling magnetic structure ("magnetic vortex") realized in a ferromagnetic circular nanodot described in Section 23.1.2. In this section, current-induced dynamics of a vortex core in a ferromagnetic dot will be discussed. It is shown that a magnetic vortex core can be resonantly excited by an ac current through the disk when the current frequency is tuned to the resonance frequency originating from the confinement of the vortex core in the disk [73]. The core is efficiently excited by the ac current due to the resonant nature and the resonance frequency is tunable by the disk shape. It is also demonstrated that the direction of a vortex core can be switched by utilizing the current-driven resonant dynamics of the vortex [76, 78].

Figure 23.10(a) shows the simulation results of the time evolution of the core position when an ac current ($f = f_0 = 380$ MHz and $J_0 = 3 \times 10^{11}$A/m^2) is applied to a disk with $r = 410$ nm and $h = 40$ nm [73]. Once the ac current is applied, the vortex core first moves in the direction of the electron flow or spin current. This motion originates from the spin-transfer effect. The off-centered core is then subjected to a restoring force toward the disk center.

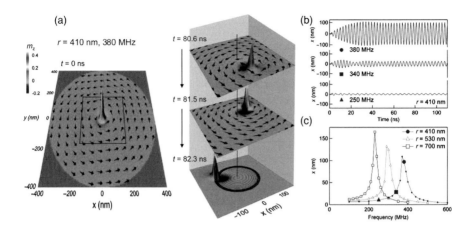

FIG. 23.10. (a) Time evolution of the magnetic vortex under application of an ac current. The magnetization direction $\mathbf{m} = (m_x, m_y, m_z)$ inside the disk on the xy plane was obtained by micromagnetic simulation. The 3D plots indicate m_z with the $m_x - m_y$ vector plots superimposed. The plot on the left represents the initial state of the vortex core situated at the center of the disk with $r = 410$ nm. The 3D plots on the right show the vortex on the steady orbital at $t = 80.6, 81.5$, and 82.3 ns after applying the ac current ($f_0 = 380$ MHz and $J_0 = 3 \times 10^{11} \mathrm{A/m}^2$). These plots are close-ups of the square region around the disk center indicated by the black square in the plot on the left. The time evolution of the core orbital from $t = 0$ to 100 ns is superimposed only on the $t = 82.3$ ns plot. (b) Time evolution of the vortex core displacement (x) for three excitation frequencies $f = 250, 340$, and 380 MHz ($r = 410$ nm and $J_0 = 3 \times 10^{11} \mathrm{A/m}^2$). (c) Radius of the steady orbital as a function of the frequency for the disks with $r = 410, 530$, and 700 nm [73].

However, because of the gyroscopic nature of the vortex (i.e. a vortex moves perpendicularly to the force), the core makes a circular precessional motion around the disk center [18]. The precession is amplified by the current to reach a steady orbital motion where the spin transfer from the current is balanced with the damping, as depicted in Fig. 23.10(a). The direction of the precession depends on the direction of the core magnetization as in the motion induced by the magnetic field [18, 20]. It should be noted that the radius of the steady orbital on resonance is larger by more than an order of magnitude as compared to the displacement of the vortex core induced by a dc current of the same amplitude [74]. Thus, the core is efficiently excited by the ac current due to resonance.

Figure 23.10(b) shows the time evolutions of the x-position of the vortex core for three different excitation frequencies $f = 250, 340$, and 380 MHz. The steady state appears after around 30 ns on resonance ($f = 380$ MHz). For $f = 340$ MHz slightly off resonance, the amplitude beats first, and then the steady state with smaller amplitude appears. The vortex core shows only a weak motion for $f = 250$ MHz, which is quite far from resonance. Figure 23.10(c) shows the radii of the steady orbitals as a function of the current frequency for the disks with $r = 410$, 530, and 700 nm. Each dot exhibits the resonance at the eigenfrequency of the vortex motion.

In order to experimentally detect the resonant excitation of a vortex core predicted by the micromagnetic simulation, the resistance of the disk was measured while an ac excitation current was passed through it at room temperature in the configuration shown in Fig. 23.11. A scanning electron microscope image of the sample is shown in Fig. 23.11. Two wide Au electrodes with 50 nm thickness, through which an ac excitation current is supplied, are also seen. The amplitude of the ac excitation current was 3×10^{11} A/m^2. Figure 23.12(a) shows the resistances as a function of the frequency of the ac excitation current for the disks with three different radii $r = 410$, 530, and 700 nm. A small but clear dip is observed for each disk, signifying the resonance. The radius dependence of the resonance frequency is well reproduced by the simulation, as shown in Fig. 23.12(b).

Following the indirect evidence of the excitation of the current-induced vortex core described above, the current-induced gyration motion of a vortex core was directly observed by using time-resolved soft X-ray transmission microscopy [79].

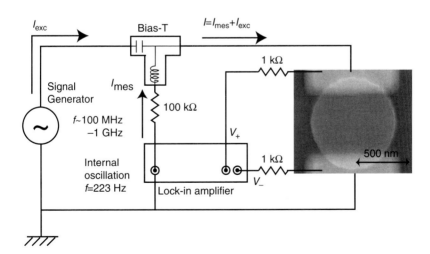

FIG. 23.11. Scanning electron microscope image of the sample along with a schematic configuration used for the measurements. The detection of the vortex excitation was performed by resistance measurements with a lock-in technique (223 Hz and current $I_{\mathrm{mes}} = 15$ μA) under the application of an ac excitation current $I_{\mathrm{exe}} = 3 \times 10^{11}$ A/m^2 [73].

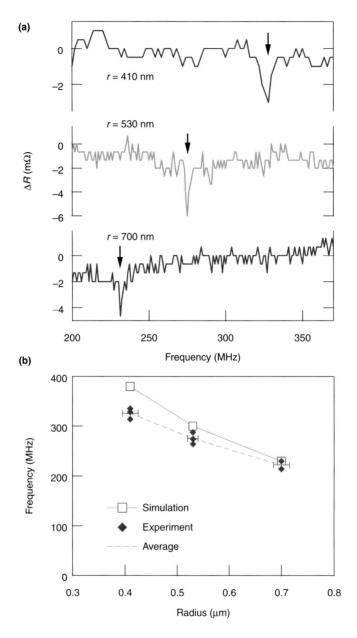

FIG. 23.12. (a) Experimental detection of the current-driven resonant excitation of a magnetic vortex core. The resistances are indicated as a function of the frequency of the AC excitation current for the disks with three different radii $r = 410, 530$, and 700 nm. (b) Radius dependence of the resonance frequency. The open rectangles and the filled diamonds indicate the simulation and the experimental results, respectively. The experimental results for eight samples are plotted. The dashed line is the averaged value of the experimental data [73].

By analyzing the radius of the vortex core gyration as a function of the excitation frequency, the spin polarization of the current in the disk was estimated to be 0.67.

23.3.2 *Switching a vortex core by electric current*

It was found that higher excitation currents induce even the switching of the core magnetization during the circular motion [76]. Figures 23.13(a)–(f) are successive snapshots of the calculated results for the magnetization distribution during the process of core motion and switching, showing that the reversal of the core magnetization takes place in the course of the circular motion without going out of the dot. Noteworthy is the development of an out-of-plane magnetization (dip) which is opposite to the core magnetization (Figs. 23.13a–d).

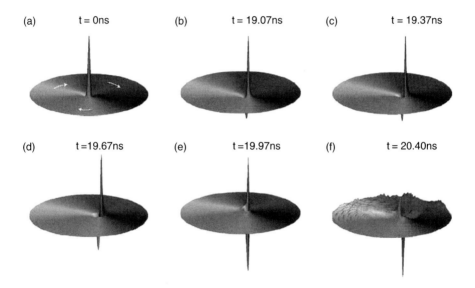

FIG. 23.13. Perspective view of the magnetization with a moving vortex structure. The height is proportional to the out-of-plane (z) magnetization component. (a) Initially, a vortex core magnetized upward is at rest at the disk center. (b) On application of the ac current, the core starts to make a circular orbital motion around the disk center. There appears a region with downward magnetization (called a "dip" here) on the inner side of the core. (c), (d), (e) The dip grows slowly as the core is accelerated. When the dip reaches the minimum, reversal of the initial core starts. (f) After the completion of the reversal, the stored exchange energy is released to a substantial amount of spin waves. A positive "hump" then starts to build up, which will trigger the next reversal. Calculation with $h = 50$ nm, $R = 500$ nm and $J_0 = 4 \times 10^{11} \mathrm{A/m}^2$ [76].

The predicted current-induced switching of the vortex core was confirmed by the magnetic force microscopy (MFM) observation as described below [76]. First, the direction of the core magnetization was determined by MFM observation. A dark spot at the center of the disk in Fig. 23.14(b) indicates that the core magnetization is directed upward with respect to the plane of the paper. The core direction was checked again after the application of an ac excitation current of frequency $f = 290$ MHz and amplitude $J_0 = 3.5 \times 10^{11}$ A/m^2 through the disk, with a duration of about 10 sec. As shown in Fig. 23.14(c), the dark spot

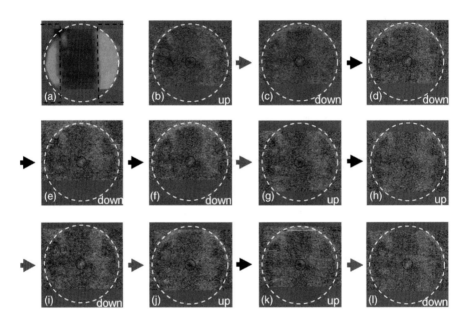

FIG. 23.14. MFM observation of electrical switching of a vortex core. (a) AFM image of the sample. A permalloy disk fills the white circle. The thickness of the disk is 50 nm, and the radius is 500 nm. Two wide Au electrodes with 50 nm thickness, through which an ac excitation current is supplied, are also seen. (b) MFM image before the application of the excitation current. The dark spot at the center of the disk (inside the small circle) indicates that the core magnetization is directed upward with respect to the plane of the paper. (c) MFM image after the application of the ac excitation current at a frequency $f = 290$ MHz and amplitude $J_0 = 3.5 \times 10^{11}$ A/m^2 through the disk with a duration of about 10 s. The dark spot at the center of the disk in (b) changed ia the bright spot, indicating the switching of the core magnetization from up to down. (b)–(l) Successive MFM images with excitation current applied similarly between consecutive images. The switching of the core magnetization occurs from (b) to (c), (f) to (g), (h) to (i), (i) to (j), and (k) to (l) [76].

at the center of the disk changed into a bright spot after the application of the excitation current, indicating that the core magnetization has been switched. Figures 23.14(b)–(l) are successive MFM images with an excitation current applied between each consecutive image. It was observed that the direction of the core magnetization after application of the excitation current was changed randomly. This indicates that the switching occurred frequently compared to the duration of the excitation current (about 10 s) and the core direction was determined at the last moment when the excitation current was turned off.

Figure 23.15 shows the core velocity as a function of excitation time which was obtained by the micromagnetic simulation. The sudden decreases of velocity correspond to the repeated core-switching events. Worth noting is that the core switches when its velocity reaches a certain value, $v_{switch} \approx 250$ m/s here, regardless of the value of the excitation current density. This is the crucial key to understanding the switching mechanism together with the existence of the dip structure which appears just before the core switching. The rotational motion of the core is accompanied by the magnetization dynamics in the vicinity of the core. This magnetization dynamics in the disk plane produces a so-called damping torque perpendicular to the plane according to the second term of Eq. (23.2), which generates the dip structure seen in Figs. 23.13(b)–(c). The higher core velocity leads to the stronger damping torque, and eventually the core switching occurs at the threshold core velocity. If the core switching is governed by the core velocity, the switching should occur regardless of how the

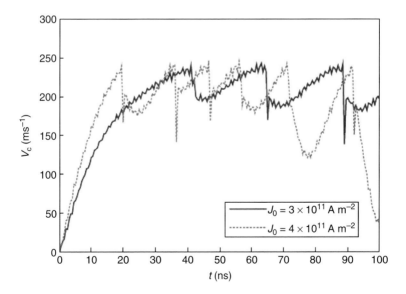

FIG. 23.15. Magnitude of the core velocity as a function of time, for two choices of the current density, showing that the maximum velocity is the same [76].

core achieves the threshold velocity. In fact, the core switching was also observed by resonant excitation with an ac magnetic field [77].

The current necessary for the switching is only several mA, while the core switching by an external magnetic field needs a large magnetic field of several kOe as described [16]. Although the repeated vortex core switching by a continuous ac current was presented here, it will be possible to control the core direction by a current with an appropriate waveform. In fact, it was shown that a nanosecond single current pulse can switch the core magnetization [78]. This method has advantages over core switching by using the resonance effect described above; it gives a short switching time as well as controllability of the core direction which is indispensable in applications. Current-induced vortex core switching can be used as an efficient data writing method for a memory device in which the data are stored in a nanometer size core. To realize such a vortex core memory, it is indispensable to develop techniques for electrically switching and detecting the direction of the core magnetization. While the current-induced vortex core switching corresponds to the electrical writing method, it was demonstrated recently that the core direction (information) can be detected by using a three-terminal device in which the tunneling magnetoresistance junction is integrated onto a ferromagnetic disk [80].

References

[1] P. Weiss, J. Phys., **6**, 661 (1907).

[2] T. Ono, H. Miyajima, K. Shigeto, and T. Shinjo, Appl. Phys. Lett. **72**, 1116 (1998).

[3] T. Ono, H. Miyajima, K. Shigeto, K. Mibu, N. Hosoito, and T. Shinjo, Science, **284** 468–470 (1999).

[4] K. Shigeto, T. Shinjo, and T. Ono, Appl. Phys. Lett. **75**, 2815 (1999).

[5] A. Himeno, T. Ono, S. Nasu, K. Shigeto, K. Mibu, and T. Shinjo, J. Appl. Phys. **93**, 8430 (2003).

[6] A. Himeno, T. Okuno, T. Ono, K. Mibu, S. Nasu, and T. Shinjo, J. Magn. Magn. Mater. **286**, 16 (2005).

[7] J. Grollier, P. Boulenc, V. Cros, A. Hamzic, A. Vaures, A. Fert, and G. Faini, Appl. Phys. Lett. **83**, 509 (2003).

[8] D. A. Allwood, G. Xiong, and R. P. Cowburn, Appl. Phys. Lett. **85**, 2848 (2005).

[9] A. Himeno, S. Kasai, and T. Ono, Appl. Phys. Lett. **87**, 243108 (2005).

[10] A. Himeno, T. Ono, S. Nasu, T. Okuno, K. Mibu, and T. Shinjo, J. Magn. Magn. Mater. **272–276**, 1577 (2004).

[11] D. Atkinson, D. A. Allwood, G. Xiong, M. D. Cooke, C. C. Faulkner, and R. P. Cowburn, Nature Mater., **2**, 85 (2003).

[12] G. S. D. Beach, C. Nistor, C. Knutson, M. Tsoi, and J. L. Erskine, Nature Mater., **4**, 741 (2005).

[13] Y. Nakatani, A. Thiaville, and J. Miltat, Nature Mater., **2**, 521 (2003).

[14] A. Hubert and H.Schafer, *Magnetic Domains* (Springer, Berlin, 1998).

[15] T. Shinjo, T. Okuno, R. Hassdorf, K. Shigeto, and T. Ono, Science **289**, 930 (2000).

[16] T. Okuno, K. Shigeto, T. Ono, K. Mibu, and T. Shinjo, J. Magn. Magn. Mater. **240**, 1 (2002).

[17] A. Wachowiak, J. Wiebe, M. Bode, O. Pietzsch, M. Morgenstern, and R. Wiesendanger, Science **298**, 577 (2002).

[18] K. Yu. Guslienko, B. A. Ivanov, V. Novosad, Y. Otani, H. Shima, and K. Fukamichi, J. Appl. Phys. **91**, 8037 (2002).

[19] J. P. Park, P. Eames, D. M. Engebretson, J. Berezovsky, and P. A. Crowell, Phys. Rev. B **67**, 020403 (2003).

[20] S.-B. Choe, Y. Acremann, A. Scholl, A. Bauer, A. Doran, J. Stöhr, and H. A. Padmore, Science **304**, 420 (2004).

[21] L. Berger, J. Appl. Phys. **55**, 1954 (1984).

[22] L. Berger, J. Appl. Phys. **71**, 2721 (1992).

[23] Y. B. Bazaliy, B. Jones, and S. C. Zhang, Phys. Rev. B **57** (1998) R3213.

[24] G. Tatara and H. Kohno, Phys. Rev. Lett. **92**, 086601 (2004).

[25] P. P. Freitas and L. Berger, J. Appl. Phys. **57**, 1266 (1985).

[26] C.-Y. Hung and L. Berger, J. Appl. Phys. **63**, 4276 (1988).

[27] D. A. Allwood, G. Xiong, M. D. Cooke, C. C. Faulkner, D. Atkinson, N. Vernier, and R. P. Cowburn, Science, **296**, 2003 (2002).

[28] J. J. Versluijs, M. A. Bari, and J. M. D. Coey, Phys. Rev. Lett. **87**, 026601 (2001).

[29] S. S. P. Parkin, U.S. Patent No. 6834005 2004.

[30] H. Numata, T. Suzuki, N. Ohshima, S. Fkami, K. Ishiwata, and N. Kasai, Symp. on VLSI Tech. Dig., pp. 232–233 (2007).

[31] Z. Li and S. Zhang, Phys. Rev. Lett. **92**, 207203 (2004).

[32] S. Zhang and Z. Li, Phys. Rev. Lett. **93**, 127204 (2004).

[33] X. Waintal and M. Viret, Europhys. Lett. **65**, 427 (2004).

[34] A. Thiaville, Y. Nakatani. J. Miltat and N. Vernie, J. Appl. Phys. **95**, 7049 (2004).

[35] A. Thiaville, Y. Nakatani. J. Miltat, and Y. Suzuki, Europhys. Lett. **69**, 990 (2005).

[36] H. Koo, C. Krafft, and R. D. Gomez, Appl. Phys. Lett. **81**, 862 (2002).

[37] M. Tsoi, R. E. Fontana, and S. S. P. Parkin, Appl. Phys. Lett. **83**, 2617 (2003).

[38] M. Klaui, C. A. F. Vaz, J. A. C. Bland, W. Wemsdorfer, G. Fani, E. Cambril, and L. J. Heyderman, Appl. Phys. Lett. **83**, 105 (2003).

[39] T. Kimura, Y. Otani, K. Tsukagoshi, and Y. Aoyagi, J. Appl. Phys. **94**, 07947 (2003).

[40] A. Yamaguchi, T. Ono, S. Nasu, K. Miyake, K. Mibu, and T. Shinjo, Phys. Rev. Lett. **92**, 077205 (2004); Phys. Rev. Lett. **96**, 179904(E) (2006).

[41] C. K. Lim, T. Devolder, C. Chappert, J. Grollier, V. Cros, A. Vaures, A. Fert, and G. Faini, Appl. Phys. Lett. **84**, 2820 (2004).

[42] N. Vernier, D. A. Allwood, D. Atkinson, M. D. Cooke, and R. P. Cowburn, Europhys. Lett. **65**, 526 (2004).

[43] M. Yamanouchi, D. Chiba, F. Matsukura, and H. Ohno, Nature **428**, 539 (2004).

[44] E. Saitoh, H. Miyajima, T. Yamaoka, and G. Tatara, Nature **432**, 203 (2004).

[45] M. Klaui, C. A. F. Vaz, J. A. C. Bland, W. Wernsdorfer, G. Faini, E. Cambril, L. J. Heyderman, F. Nolting, and U. Rdiger., Phys. Rev. Lett. **94**, 106601 (2005).

[46] M. Klaui, P.-O. Jubert, R. Allenspach, A. Bischof, J. A. C. Bland, G. Faini, U. Rdiger, C. A. F. Vaz, L. Vila, and C. Vouille, Phys. Rev. Lett. **95**, 026601 (2005).

[47] A. Yamaguchi, S. Nasu, H. Tanigawa, T. Ono, K. Miyake, K. Mibu, and T. Shinjo, Appl. Phys. Lctt. **86**, 012511 (2005).

[48] D. Ravelosona, D. Lacour, J. A. Katine, B. D. Terris, and C. Chapper, Phys. Rev. Lett. **95**, 117203 (2005).

[49] A. Yamaguchi, K. Yano, H. Tanigawa, S. Kasai, and T. Ono, Jpn. J. Appl. Phys. **45**, 3850 (2006).

[50] G. S. D. Beach, C. Knutson, C. Nistor, M. Tsoi, and J. L. Erskine, Phys.Rev. Lett. **97**, 057203 (2006).

[51] M. Hayashi, L. Thomas, Y. B. Bazaliy, C. Rettner, R. Moriya, X. Jiang, and S. S. P. Parkin, Phys. Rev. Lett. **96**, 197207 (2006).

[52] K. Fukumoto, W. Kuch, J. Vogel, F. Romanens, S. Pizzini, J. Camarero, M. Bonfim, and J. Kirschner, Phys. Rev. Lett. **96**, 097204 (2006).

[53] M. Laufenberg, W. Bhrer, D. Bedau, P.-E. Melchy, M. Klui, L. Vila, G. Faini, C. A. F. Vaz, J. A. C. Bland, and U. Rdiger, Phys. Rev. Lett. **97**, 046602 (2006).

[54] L. Thomas, M. Hayashi, X. Jiang, R. Moriya, C. Rettner, and S. S. P. Parkin, Nature, **443**, 197 (2006).

[55] M. Yamanouchi, D. Chiba, F. Matsukura, T. Dietl, and H. Ohno, Phys. Rev. Lett. **96**, 096601 (2006).

[56] M. Klaui, M. Laufenberg, L. Heyne, D. Backes, and U. Rdiger, Appl. Phys. Lett. **88**, 232507 (2006).

[57] Y. Togawa, T. Kimura, K. Harada, T. Akashi, T. Matsuda, A. Tonomura, and Y. Otani, Jpn. J. Appl. Phys. **45**, L683 (2006).

[58] Y. Togawa, T. Kimura, K. Harada, T. Akashi, T. Matsuda, A. Tonomura, and Y. Otani, Jpn. J. Appl. Phys. **45**, L 1322 (2006).

[59] M. Hayashi, L. Thomas, C. Rettner, R. Moriya, Y. B. Bazaliy, and S. S. P. Parkin, Phys. Rev. Lett. **98**, 037204 (2007).

[60] L. Thomas, M. Hayashi, X. Jiang, R. Moriya, C. Rettner, and S. S. P. Parkin, Science **315**, 1553 (2007).

[61] S.S.P. Parkin *et al.*, Science **320**, 190 (2008).

[62] S. Fukami *et al.*, J. Appl. Phys. **103**, 07E718 (2008).

[63] S.W. Jung, *et al.*, App. Phys. Lett. **92**, 202508 (2008).

[64] H. Tanigawa *et al.*, Appl. Phys. Express **1**, 011301 (2008).

[65] T. Koyama *et al.*, Appl. Phys. Express **1**, 101303 (2008).

[66] H. Tanigawa *et al.*, Appl. Phys. Express **2**, 053002 (2009).

[67] D. Chiba *et al.*, Appl. Phys. Express **3**, 073004 (2010).

[68] C. Burrowes *et al.*, Nature Phys. **6**, 17 (2010).

[69] T. Koyama *et al.*, Nature Mater. **10**, 194 (2011).

[70] T. Koyama *et al.*, Appl. Phys. Lett. **98**, 192509 (2011).

[71] K. Ueda *et al.*, Appl. Phys. Express **4**, 063006 (2011).

[72] H. Tanigawa *et al.*, Appl. Phys. Express **4**, 013007 (2011).

[73] S. Kasai, Y. Nakatani, K. Kobayashi, H. Kohno, and T. Ono, Phys. Rev. Lett. **97**, 107204 (2006).

[74] J. Shibata, Y. Nakatani, G. Tatara, H. Kohno, and Y. Otani, Phys. Rev. B **73**, 020403 (2006).

[75] T. Ishida, T. Kimura, and Y. Otani, Phys. Rev. B **74**, 014424 (2006).

[76] K. Yamada, S. Kasai, Y. Nakatani, K. Kobayashi, H. Kohno, A. Thiaville, and T. Ono, Nature Mater. **6**, 269 (2007).

[77] B.V. Waeyenberge, A. Puzic, H. Stoll, K. W. Chou, T. Tyliszczak, R. Hertel, M. Fähnle, H. Brückl, K. Rott, G. Reiss, I. Neudecker, D. Weiss, C. H. Back, and G. Schütz, Nature **444**, 461 (2006).

[78] K. Yamada *et al.*, Appl. Phys. Lett. **93**, 152502 (2008).

[79] S. Kasai *et al.*, Phys. Rev. Lett. **101**, 237203 (2008).

[80] K. Nakano *et al.*, Appl. Phys. Express, **3**, 053001 (2010).

24 Spin torques due to large Rashba fields

S. Zhang

24.1 Introduction

The Rashba field [1] is a special type of spin–orbit interaction; it requires broken spatial-inversion symmetry. In semiconductor quantum wells, a gate voltage or an intrinsic band mismatch between two contacting layers can efficiently break the inversion symmetry. In particular, for the two-dimensional electron gas in a quantum well, the Rashba interaction takes the following form

$$H_R = \alpha_r \mathbf{s} \cdot (\hat{\mathbf{z}} \times \mathbf{p}) \tag{24.1}$$

where $\hat{\mathbf{z}}$ is a unit vector representing the direction of the growth which confines the electrons within the layers, $\mathbf{p}$ is the momentum of the electron, and $\mathbf{s}$ is the spin of the conduction electron. The Rashba Hamiltonian is odd under the spatial inversion transformation, i.e. $H_R(\mathbf{r}) = -H_R(-\mathbf{r})$, while the conventional spin–orbit coupling $H_{\text{so}} \propto \mathbf{s} \bullet [\nabla V(r) \times \mathbf{p}]$ is even under the transformation if the potential is even. The Rashba coefficient α_r is usually treated as a phenomenological or empirical constant for a given material and it is related to the spin–orbit splitting Δ_{so} of the valence p bands and the gap E_g between the s conduction and p valence bands [2],

$$\alpha_r = c \langle \frac{dU_v}{dz} \rangle \left[\frac{1}{E_g^2} - \frac{1}{(E_g + \Delta_{so})^2} \right] \tag{24.2}$$

where c is a numerical constant, and U_v is the total potential acting on the holes of the valence band. Since U_v can be tuned by the gate voltage, the Rashba coupling can be controlled by experiments.

In metals, the Rashba spin–orbit coupling is much less familiar compared to the semiconductor heterostructure. Due to efficient electron screening in metals, the Rashba interaction is limited within the length-scale of the election screening from the location of the broken symmetry. At a metal surface or interface, the structure inversion symmetry is naturally broken. The symmetry consideration would suggest the presence of the Rashba Hamiltonian in the same form as Eq. (24.1) with a key modification: unlike semiconductors where the Rashba coupling is present for the entire quantum well, the Rashba interaction in metals is localized within the screening length from the interface. Since the screening length in metals is usually of the order of just $1\,\mathring{A}$, the effect of the Rashba Hamiltonian on the transport and magnetic properties would be profound only for ultrathin magnetic films. In this chapter, we discuss how a qualitative estimate of the Rashba coefficient α_r at metallic interfaces can be reasonably carried out. Then, we study the effect of the Rashba Hamiltonian on the manipulation of the current-driven magnetization.

24.2   Rashba coupling at metallic interfaces

While the rigorous spin–orbit Hamiltonian is encoded in the relativistic Dirac equation for a full electric potential, the reduced spin–orbit Hamiltonian such as the Rashba model is practically more useful since only a partial potential in the solids may be approximately known. Furthermore, a model Hamiltonian would make theoretical calculations analytically tractable. We start our estimation of the Rashba coefficient by introducing a symmetry-broken potential $V(\mathbf{r})$ in addition to periodic potentials $V_p(\mathbf{r})$ that determine the electronic bands. If we treat $V(\mathbf{r})$ as a perturbation to the Bloch electrons, the energy of the Bloch electron is then

$$\epsilon_{\mathbf{k}\sigma} = \epsilon_{\mathbf{k}\sigma}^0 - e\boldsymbol{\nabla}V(\mathbf{r})\cdot\mathbf{r}_{\mathbf{k}\sigma} \tag{24.3}$$

where $\epsilon_{\mathbf{k}\sigma}^0$ is the energy of the Bloch state in the absence of $V(\mathbf{r})$. We have assumed that the potential varies slowly on the scale of the lattice constant and thus it is treated as a parameter rather than a dynamic variable. One may also view $\nabla V(\mathbf{r})$ in Eq. (24.3) as a slowly varying effective electric field. Thus, the second term represents the dipolar interaction of the electric field on the Bloch electron. For simplicity, the band index n is discarded. The expectation value of the position is defined as $\mathbf{r}_{\mathbf{k}\sigma} = \int d^3 r \psi_{\mathbf{k}\sigma}^*(\mathbf{r})\mathbf{r}\psi_{\mathbf{k}\sigma}(\mathbf{r})$ where $\psi_{\mathbf{k}\sigma}(\mathbf{r})$ is the Bloch wavefunction. The Bloch states can be constructed from local Wannier orbitals $\psi_{\mathbf{k}}^0 = e^{i\mathbf{k}\cdot\mathbf{r}}\sum_{\mathbf{R}} e^{-i\mathbf{k}\cdot\mathbf{R}}W_{\mathbf{R}}(\mathbf{r})$ where the Wannier function $W_{\mathbf{R}}(\mathbf{r})$ is usually taken as a linear combination of the atomic orbital at lattice site $\mathbf{R}$. The atomic spin–orbit is conventionally modeled by $H_{so} = (\Delta_{so}/\hbar^2)(\mathbf{r}\times\mathbf{p})\cdot\mathbf{s}$ where $\Delta_{so} = \hbar\mu_B/mc^2\langle dU/rdr\rangle$ and U is the potential of the atomic orbital. Then, one can explicitly include the spin–orbit coupling H_{so} in the Bloch wavefunction by treating the spin–orbit coupling as a perturbation,

$$\psi_{\mathbf{k}}(\mathbf{r}\sigma) = \psi_{\mathbf{k}\sigma}^0(\mathbf{r}) + \sum_{\mathbf{k}'\sigma'}\frac{\langle\psi_{\mathbf{k}\sigma}^0|H_{so}|\psi_{\mathbf{k}'\sigma'}^0\rangle}{\epsilon_{\mathbf{k}\sigma}^0 - \epsilon_{\mathbf{k}'\sigma'}^0}\psi_{\mathbf{k}'\sigma'}^0(\mathbf{r}). \tag{24.4}$$

To first order in the spin–orbit coupling, we have

$$\mathbf{r}_{\mathbf{k}\sigma} = \frac{2\Delta_{so}}{\hbar^2}\sum_{\mathbf{k}'\sigma'}\frac{\langle\psi_{\mathbf{k}\sigma}^0|\mathbf{r}|\psi_{\mathbf{k}'\sigma'}^0\rangle\langle\psi_{\mathbf{k}'\sigma'}^0|\mathbf{r}\cdot(\mathbf{p}\times\mathbf{s})|\psi_{\mathbf{k}\sigma}^0\rangle}{\epsilon_{\mathbf{k}\sigma}^0 - \epsilon_{\mathbf{k}'\sigma'}^0} \tag{24.5}$$

where we have dropped the zero-order term $\langle\psi_{\mathbf{k}\sigma}^0|\mathbf{r}|\psi_{\mathbf{k}\sigma}^0\rangle$. Let us further assume that the Wannier function is nearly spherical about the site $\mathbf{R}$ so that the above equation is simplified to

$$\mathbf{r}_{\mathbf{k}\sigma} = \alpha_{so}(\hbar\mathbf{k} \times \mathbf{s}) \tag{24.6}$$

where

$$\alpha_{so} = \frac{2\Delta_{so}}{3\hbar^2} \sum_{\mathbf{k}'\sigma'} \frac{|\langle \psi^0_{\mathbf{k}\sigma}|\mathbf{r}|\psi^0_{\mathbf{k}'\sigma'}\rangle|^2}{\epsilon_{\mathbf{k}\sigma} - \epsilon_{\mathbf{k}'\sigma'}} \tag{24.7}$$

where we have used the isotropic assumption $|\langle\mathbf{k}|x|\mathbf{k}'\rangle|^2 \approx |\langle\mathbf{k}|y|\mathbf{k}'\rangle|^2 \approx |\langle\mathbf{k}|z|\mathbf{k}'\rangle|^2$. Equation (24.6) can be identified as a "side-jump" induced by the atomic spin-orbit coupling: a Bloch state with momentum $\mathbf{k}$ shifts the center of the wavepacket side ways with respect to the center of the atom: the transverse displacement is perpendicular to the momentum and the spin; this provides a mechanism for the anomolous Hall effect proposed by Berger [3]. If we take the spin–orbit energy Δ_{so} in transition metals to be about several eV, we would find that the side-jump is about 1 Å, two to three orders of magnitude larger than the Compton length $\lambda_c = \hbar/mc$. By substitutions, Eq. (24.6) into Eq. (24.3), one may express the second term of Eq. (24.3) as

$$H_R = e\alpha_{so}(\nabla V(\mathbf{r}) \times \mathbf{p}) \cdot \mathbf{s}. \tag{24.8}$$

It is noted that the above coupling, Eq. (24.8), is several orders of magnitude larger than the conventional Pauli spin–orbit coupling, $H_{pauli} = (\hbar\mu_B/2mc^2)(\nabla V(\mathbf{r}) \times \mathbf{p}) \cdot \mathbf{s}$; this is known as the enhancement factor due to the contribution of the strong atomic spin–orbit coupling [4]. The much weaker symmetry-breaking potential $V(\mathbf{r})$ does not directly provide a spin–orbit coupling, rather it couples to the dipole moment (or side jump) induced by the atomic spin–orbit coupling, Eq. (24.6), through the dipolar interaction. If we apply Eq. (24.8) to the interface potential $V(\mathbf{r}) = V(z)$, one arrives at the Rashba spin–orbit coupling at the metal interface

$$H_R = \frac{\alpha_r}{\hbar}(\hat{\mathbf{z}} \times \mathbf{p}) \cdot \mathbf{s} \tag{24.9}$$

where $\alpha_r = e\hbar\alpha_{so}V'(z)$. In metals, the interface potential $V(z)$ is screened and $V'(z)$ is nonzero only within the screening length from the interface.

The above argument qualitatively supports the existence of the Rashba coupling in materials with spin–orbital coupled bands. The quantitative determination of the Rashba coefficient is, however, not an easy task even if one can accurately calculate the band structures by including the spin–orbit coupling with a full Kohn–Sham potential. This is because the Rashba coupling is a model Hamiltonian which is not directly related to the specific term in the microscopic Hamiltonian. Recently, there have been several attempts [5, 6] to calculate the Rashba coefficient at interfaces via first principles methods. The essential scheme in these works is to identify certain portions of the band structure derived from ab-initio calculations that can be matched by the model calculation based on the Rashba Hamiltonian with fitting parameters.

24.3   Current-driven spin torque with Rashba coupling

For the conventional current-driven spin torque, noncollinear magnetization, e.g. a spin valve with the anti-parallel magnetic moments of the two layers or a magnetic wire with domain wall patterns, is required. In these systems, the spin torques comes from the change of the angular momentum carried by the spin-polarized current. The Rashba Hamiltonian works differently: it provides a momentum-dependent magnetic field to the conduction electron and gives rise to a net magnetic field as long as the summation of all electron momenta is nonzero as in the case of the current flowing in the system. To evaluate the effective field generated by the current in more detail, we consider the following simplified model for the conduction electrons in ferromagnetic metals,

$$ H = \frac{p^2}{2m} - \frac{J}{\hbar}\mathbf{s} \cdot \hat{\mathbf{M}} + \frac{\alpha_r}{\hbar}\mathbf{s} \cdot (\hat{\mathbf{z}} \times \mathbf{p}) + \sum_i V_i(\mathbf{r} - \mathbf{R}_i) \tag{24.10} $$

where $\hat{\mathbf{M}}$ represents the direction of the magnetization and V_i are the potentials of the random distributed impurities. In the limiting case where the ferromagnetic exchange interaction is much larger than the Rashba coefficient, i.e. $J \gg \alpha_r p_F$ where p_F stands for the Fermi momentum, one can approximately treat $\mathbf{s}$ as parallel to the magnetization and the effect of the Rashba term on the magnetization is equivalent to an effective magnetic field $\mathbf{H}_{eff} = -\alpha_r(\hat{\mathbf{z}} \times \langle \mathbf{p} \rangle) \propto -\alpha_r(\hat{\mathbf{z}} \times \mathbf{j}_e)$ where $\mathbf{j}_e$ is the electrical current. The Rashba field is thus perpendicular to both the directions of the current and of the normal of the layers. In a conventional $3d$ ferromagnetic metal, the exchange J is indeed several orders of magnitude larger than $\alpha_r p_F$ and hence the simple estimation of the effective field given above is rather robust: the effective field is independent of the impurity potential V_i in Eq. (24.10).

When the exchange and the Rashba couplings are comparable in strength, the spin of the conduction electron is no longer parallel to the magnetization direction. In this case, we should first determine the spin states from the first three terms of Eq. (24.10) and then calculate the nonequilibrium electron spin density from the impurity scattering. The calculation will involve detailed scattering potentials. If we limit our calculation to the two-dimensional interface states, the one-electron eigenenergy and wavefunction are [7]

$$ E_{\mathbf{k}}^{\pm} = \frac{\hbar^2 k^2}{2m} \pm \frac{1}{2}|\alpha_r(\hbar\mathbf{k} \times \hat{\mathbf{z}}) + J\hat{\mathbf{M}}| \tag{24.11} $$

and

$$ \Psi_{\mathbf{k}}^{\pm} = \frac{1}{\sqrt{2A}} \begin{pmatrix} \pm e^{i\gamma_k} \\ 1 \end{pmatrix} \exp(i\mathbf{k} \cdot \mathbf{r}) \tag{24.12} $$

where $\tan\gamma_k = (\alpha_R \hbar k_x + J\sin\theta)/(\alpha_R \hbar k_y - J\cos\theta)$ and A is the area of the film.

To calculate the nonequilibrium spin density and the spin torque in the presence of the current, one uses the Boltzmann equation for the two bands,

$$eE_x\left(-\frac{\partial f_0^\pm}{\partial k_x}\right) = c_i \int d^2k' W_{kk'}^\pm (f_k^\pm - f_{k'}^\pm) \qquad (24.13)$$

where c_i is the impurity concentration and E_x is the electric field in the x-direction, $f_0^\pm$ is the equilibrium distribution of the two bands ($\pm$), and the scattering probability is

$$W_{kk'}^\pm = |\langle \Psi_k^\pm |V_i| \Psi_{k'}^\pm \rangle|^2 \delta(E_k^\pm - E_{k'}^\pm) \qquad (24.14)$$

and V_i is the impurity scattering potential. We note that we have neglected the interband scattering, i.e. the impurity is spin-independent. Instead of assuming a particular form of scattering potential, one may simply take the isotropic scattering probability $W_{kk'}^\pm = W$; this is equivalent to a constant relaxation time approximation that is valid in the limiting case under consideration. In this case, the Boltzmann distribution in Eq. (24.13) has a simple solution,

$$f_k^\pm = f_0^\pm - eE_x\tau\left(-\frac{\partial f_0^\pm}{\partial k_x}\right) \qquad (24.15)$$

where τ is the relaxation time ($e > 0$). With the above distribution, the spin density $\delta \mathbf{m}$ and the spin current $\mathbf{j}_s$ can be readily evaluated, i.e.

$$\delta \mathbf{m} = \int d^2k (f_k^+ - f_k^-)(\hat{\mathbf{x}} \cos\gamma_k - \hat{\mathbf{y}} \sin\gamma_k) \qquad (24.16)$$

and

$$\mathbf{j}_s = e \int d^2k (v_x^+ f_k^+ - v_x^- f_k^-)(\hat{\mathbf{x}} \cos\gamma_k - \hat{\mathbf{y}} \sin\gamma_k) \qquad (24.17)$$

where $v_x^\pm = (1/\hbar)\partial E_k^\pm/\partial k_x$ is the velocity. Once $\delta \mathbf{m}$ is obtained, the Rashba field is $\mathbf{H}_{eff} = J\delta \mathbf{m}$.

In the limiting case of $J \gg \alpha_R p_F$, one can analytically integrate out the momentum in Eqs. (24.16) and (24.17). Up to first order in $\alpha_R p_F/J$,

$$\cos\gamma_k = -\cos\theta + \frac{\alpha_R p_F}{J}(k_y \sin^2\theta + k_x \sin\theta \cos\theta) \qquad (24.18)$$

and

$$\sin\gamma_k = \sin\theta + \frac{\alpha_R p_F}{J}(k_x \cos^2\theta + k_y \sin\theta \cos\theta). \qquad (24.19)$$

By substituting the above expressions into Eqs. (24.16) and (24.17) and by noticing that the terms linear in k_y are zero after integration, we have

$$\mathbf{H}_{eff} = \frac{\alpha_R m P}{\hbar e M_s}(\hat{\mathbf{z}} \times \mathbf{j}_e).\qquad(24.20)$$

This is essentially the same as the result from the simple estimation earlier. The result is strikingly robust: the current-induced field is independent of the impurity potential and the exchange coupling J, at least in the limit of large J.

For an arbitrary ratio of $\alpha_R p_F/J$, a numerical calculation can be readily carried out [7, 8].

24.4 Manipulating magnetization by the current

The Rashba field, Eq. (24.20), differs from the conventional current-induced spin torques in several significant ways. First, the Rashba field requires the current flowing in the direction parallel to the interface while the conventional spin-transfer torque in multilayers must have the current perpendicular to the layers. Second, the Rashba field competes with other magnetic fields such as the anisotropy and demagnetization field in determining the equilibrium states of the magnetization, while the conventional spin torques compete with the magnetic damping which is not a conservative field. Third, the Rashba field is a single surface or interface effect and thus it exists even for a single uniformly magnetized layer. The common feature of the Rashba field and the spin torque is that they both originate from the vicinity of the interface: the former has a length-scale of the screening length and the latter is associated with the ferromagnetic decoherent length.

A very convincing experimental verification of the Rashba field has been recently carried out [9]. An ultrathin Co layer of only 0.7 nm in thickness is asymmetrically sandwiched between Pt and Al_2O_3 layers. Note that the total Rashba field would be zero if the Co layer was sandwiched by the same material, e.g. Pt, because the Rashba field from two sides of the Co interfaces has an equal but opposite direction. The presence of the Pt layer moves the easy axis of the Co layer out of the layer plane. By carefully studying the domain wall nucleation in the presence of the current, it is found that the direction of the Rashba field is indeed in the plane and perpendicular to the current, consistent with Eq. (24.20). The Rashba field magnitude is about 1000 gauss for a current density of about $10^7 A/cm^2$.

For a single domain layer, the analysis of the Rashba field with respect to the magnetization is rather straightforward. Consider the in-plane current in the x-direction and the Rashba field in the y-direction. For a magnetic layer whose easy axis is the y-axis, the Rashba field can switch the direction of the magnetization. The critical switching current is determined by $H_r = H_K$ where H_K is the anisotropy field, i.e.

$$j_c = \frac{\hbar e H_K M_s}{\alpha_R m P}. \tag{24.21}$$

On the other hand, if the easy axis lies in the x- or z-directions, the Rashba field can rotate the magnetization away from the easy axis when a current is applied. The rotation angle θ is simply given by $\tan\theta = H_R/H_K$.

24.5 Discussions and conclusions

The Rashba spin–orbit coupling at a ferromagnetic interface is mainly controlled by the intrinsic atomic spin–orbit coupling where broken inversion symmetry is required. It is also possible to tune the Rashba coupling by external means. An electric field applied perpendicular to the interface can alter the Rashba field. For example, suppose that an electric potential is applied across a bilayer Co/MgO where the insulating MgO layer is thick enough to avoid electric breakdown. The electric field in the MgO layer is $E_a = V_a/t_{MgO}$ where t_{MgO} is the thickness; this field will be screened at the ferromagnetic–metal interface and thus result in a screening potential $V_s \approx \epsilon\lambda E_a$ [10]. Since it is experimentally viable to apply E_a of the order of 0.3 V/nm [11], one can achieve a very large V_s so that the Rashba field can be significantly tuned. Such tunability of the Rashba field could be very useful for devices based on the current-driven control of the magnetization.

Apart from the Rashba spin–orbit coupling, there are other forms of spin–orbit couplings. For a ferromagnetic material with a broken bulk inversion symmetry, the Dresselhaus spin–orbit coupling [12] is well known. Then a similar current-driven field can be found. In fact, the experimental verification of the current-driven effect has already been carried out [13]. We should point out that the spin–orbit couplings without broken inversion symmetry such as the Luttinger spin–orbit coupling Hamiltonian [14], give a much smaller contribution to the current-driven field because the effective magnetic field is zero in the first order of the spin–orbit coupling [15, 16].

We conclude this chapter by emphasizing two of the most challenging issues in the current-induced Rashba field: to develop a quantitative theory for the determination of Rashba coefficients of specific magnetic materials and to design innovative devices by using the Rashba effect for advanced technologies.

References

[1] Y. A. Bychkov and E. I. Rashba, J. Phys. C **17**, 6039 (1984).
[2] R. Winkler, in *Spin–Orbit Effects in Two-Dimensional Electron and Hole Systems*, Springer Tracts in Modern Physics (Springer, Berlin, 2003), Vol. 191.
[3] L. Berger, Phys. Rev. B **2**, 4559 (1970).
[4] R. C. Fivaz, Phys. Rev. **183**, 586 (1969).
[5] M. Nagano, A. Kodama, T. Shishidou, and T. Oguchi, J. Phys.: Condens. Matter **21**, 064239 (2009).

 [6] G. Bihlmayer, Yu. M. Koroteev, P. M. Echenique, E. V. Chulkov, and S. Blugel, Surf. Sci. **600**, 3888 (2006).

 [7] A. Manchon and S. Zhang, Phys. Rev. B **78**, 212405 (2008).

 [8] A. Matos-Abiague and R. L. Rodríguez-Suárez, Phys. Rev. B **80**, 094424 (2009).

 [9] I. M. Miron, G. Gaudin, S. Auffret, B. Rodmacq, A. Schuhl, S. Pizzini, J. Vogel, and P. Gambardella, Nature Mater. **9**, 230 (2010).

[10] S. Zhang, Phys. Rev. Lett. **83**, 640 (1999).

[11] Endo, M. *et al.*, Appl. Phys. Lett. **96**, 212503 (2010).

[12] G. Dresselhaus, Phys. Rev. **100**, 580 (1955).

[13] Chernyshov, A. *et al.*, Nature Phys. **5**, 656–659 (2009).

[14] M. Luttinger, Phys. Rev. **102**, 1030 (1956).

[15] A. Manchon and S. Zhang, Phys. Rev. B **79**, 094422 (2009).

[16] I. Garate and A. H. MacDonald, Phys. Rev. B **80**, 134403 (2009).

Index

The page numbers in **bold** refer to tables and figures.

adiabatic local-density approximation
  (ALDA, Stoner model) 108
Ag, spin–orbit coupling parameter 200,
  201
Aharonov–Casher spin interference 217–24
  experiment 220–2, **223**
  theory 217–20
Aharonov–Bohm (AB) effect 217
Al
  experimental spin Hall angles **237**
  spin–orbit coupling parameter 200, **201**
AlGaAs 52, 59, 61, 187
  quantum well (QW) 187
Al'tshuler–Aronov–Spivak (AAS) effect 220
angle-resolved photoemission spectroscopy
  (ARPES) 43, 181, 189–90, 285
  linear dispersion of surface Dirac
    cones 179, 286
angular momentum transfer 89
anisotropic magnetoresistance (AMR) 118
annihilation field, magnetic domain
  walls/vortices 385
anomalous Hall effect (AHE) 34, 152, 177,
  185, 194, 203–4, 252, **253**, 426
Au
  experimental spin Hall angles **237**
    spin–orbit coupling parameter 200, **201**
  thermal Hall effect 140

band bending in semiconductor **54, 209**
band gap engineered NiFe/AlGaAs
  barrier/GaAs structures 56
band index n 33
Barnett effect 143
Berger, anomalous Hall effect 426
Berger's vector 284
Berry's phase 65–7, 78–9, 83, 149, 152, 180,
  217, 219, 282, 351
  topological spin current 33
Bethe ansatz and form factor expansion 168
$Bi_{1-x}Sb_x$ 190, 284
  transport measurements 285
Bi_2Se_3
  copper intercalation 290

doped with Cu and Mn 287
linear dispersion of surface Dirac
  cones 181, 286
magnetic dopants 290
surface Fermi surface **286**
Bi_2Te_3, TRIM on surface Brillouin zone 286
Bloch electron 156, 425
  energy 425
  wavefunctions 156, 425
Bloch and Néel domain walls **407**
  resistance 411
Blonder–Tinkham–Klapwijk (BTK)
  theory 41
Bohr magneton 111, 150
Boltzmann equation 428
  semi-classical electrons 33
  steady state 196–7
Born approximation 196
breathing Fermi surface model 115
Brillouin light scattering (BLS) 388
Brillouin zone 36, 51, 189, 275, 281–2,
  285–7, **394**
bulk inversion asymmetry (BIA) 209

CdTe/HgTe/CdTe, quantum well (QW) 184,
  190, 280
charge conservation law 9
charge current, defined 9
Chern form 170
Chern number N 274
Chern–Simon term 152, 169
CMS *see* Co_2MnSi
Co, sharp rise in damping with temperature
  decrease 115
Co-based full-Heusler compounds 38
Co/Cu/Co trilayer structure,
  lithographically defined point
  contact 320
Co/Pt wire, Rashba effect 334
Co_2MnSi
  calculated element and spin-resolved
    density of states (DOS) **39**
  magnon excitation at the CMS/MgO
    interface 45

Co$_2$MnSi (*cont.*)
  MTJs, temperature dependence of the
      TMR ratio **44**
CoFeB layer, CMS/MgO interface 45
coherent potential approximation (CPA) 117
conformal field theory (CFT) 168
Cooper pairs, conventional
      superconductors 289
Coulomb interaction 141
Coulomb potential, induced by atomic
      core 209
Cu
  CuIr, experimental spin Hall angles **237**
  nonlocal spin-valve signal 227, **374**
  spin–orbit coupling parameter 200, **201**
cubic semiconductors 287
current-perpendicular-to-plane (CPP), giant
      magnetoresistive (GMR) devices 39,
      45
currents in interface plane (CIP) GMR
      devices 136

Datta–Das device concept, Rashba
      SO-coupled system 253, 259, 265, 267
density functional theory (DFT), local spin
      density approximation (LSDA) 117
density of states (DOS)
  quasi-particle 41
  up- (down-) spin channel 36
diffusion constant 19
diffusion current 18
  continuity equation 19
  density 19
diffusive spin-currents, **318**
dilute magnetic semiconductor (DMS) 58
Dirac cone 179, 281–2
Dirac equation and spin 5–6, 181
Dirac Hamiltonian 6, 282
Dirac relativistic electrons
  quantum electrodynamics (QED) 149
  see also Fermi–Dirac
domain wall (DW) motion 329–31, 405
  adiabatic spin-transfer torque 331–2,
      411–12
  critical current for current-induced domain
      wall motion 330–1
  current-driven domain wall
      motion 329–32, 405–13, **405**
  current-driven excitation of magnetic
      vortices 332–34, 413–19
  intrinsic pinning for domain wall
      motion 406, 411
  Oersted field 408
  quasi-one-dimensional ferromagnetic
      wire 112

donor–acceptor mixed stack charge transfer
      compounds, spin-Peierls systems 167
DOS *see* density of states (DOS)
Dresselhaus Hamiltonian 259
Dresselhaus spin–orbit (SO) field **264**
  coupling 430
Dresselhaus spin–orbit (SO)
      interactions 212–13
  bulk inversion asymmetry (BIA) 209
  manipulating magnetization 334
drift current density 19
Dyakonov–Perel (DP) mechanism 177, 213,
      214, 253
Dzyaloshinskii–Moriya (DM) spin–orbit
      interaction 152, 156

Einstein relation 20
Einstein–de Haas effects 143
electric polarization **160**
electro-spinon 167
electromagnons 82–3, 161–4
  exchange-striction effect 162
  multiferroics 161–4
  and phonons 82–3
spin motive force 83
  theoretical dispersions and spectra **164**
  theoretical optical absorption spectra
      163
electron beam lithography processing 230,
      320, 382, 406
  see also nanolithography
electron cyclotron resonance (ECR)
      dry-etching 220
electron diffusion in a one-dimensional
      system **18**
electron number, conservation 15
electron system, exchange interactions 344
electron wavefunction in semiconductors 210
electrostatic potential 7
ensemble Monte Carlo (EMC)
      calculations 262, **264**
equilibrium carrier density 21
exchange spin current 25–32
  Landau–Lifshitz–Gilbert equation 25–6, 27
  spin-wave spin current 28–31
exchange-striction effect 162
exciton, soliton and anti-soliton 168
experimental techniques for spin-polarization
      measurement
  current-perpendicular-to-plane
      magnetoresistive device
      (CPP-GMR) 39, 45
  nonlocal spin devices 229
  point-contact Andreev reflection
      (PCAR) 39–40

superconducting tunneling spectroscopy (STS) 41–2

Faraday's law 65–6, 68
fast Fourier transform (FFT) spectra, SdH oscillations **213**
Fe/GaAs(001) interface, bias dependence of a photoexcited current **56**
Fermi energy 21
  breathing Fermi surface model 115
Fermi velocity for up- (down-) spin channel 36
Fermi–Dirac distribution of electrons 105
  distribution function 16, **17**
  incoherent spin current 15–17
ferroelectricity
  electromagnons in multiferroics 161
  spin-current model 154–6
ferromagnetic metal (FM) junction 41
  superconducting tunneling spectroscopy (STS) 41
ferromagnetic resonance (FMR) 80–2
  'anti'- (a-FMR) 80
  field-driven FMR 80
  FMR with thermal gradients 81
  spin Seebeck effect 80
  spin-coupled interface resistance 23
  spin-motive forces 80–2
  spin-torque 80, 354
ferromagnetic structures
  spin Faraday's law 65–6
  spin-motive forces 65–84
  total energy 384
ferromagnetic/nonmagnetic (F/N) junction 21, 229
  diffusive spin-currents **318**
  electrical conductivity 22
  electrochemical potential 22
  scattering theory **104**
  spin diffusion length 22
  spin polarization 23
  spin-coupled interface resistance 23
ferromagnetic/nonmagnetic/ferromagnetic trilayer structure, spin currents **312**
ferromagnetism 36
  half-metallicity 37–8
  s-d model 108
Fick's law 19

GaAlAs/GaAs heterostructures 254
GaAs+CsOCs, photoemission spectrum of spin polarization **52**
GaAs
  photoemission spectrum **52**
  schematic band structure **51**

GaAs STM tip 58
GaMnAs 140, 141
  Nernst effect 140
  spin Seebeck effect (SSE) 297
  spin-current-induced domain wall motion 332
Gauss's theorem 9
giant magneto-heat resistance 141
giant magnetoresistance (GMR) 138
  nanopillar GMR structure 344
  magnetic multilayers 311
  Valet and Fert's model 36
Gilbert damping 26, 28, 88, 90, 93, 109, 119, 250, 350
  constant 88, 90, 109, 350
  cubic substitutional alloys 118
  dimensionless coefficient 163
  parameter **119**
  and resistivity for fcc Ni **119**
  zero-temperature and experimental room temperature values **119**
Ginzburg–Landau theory 165
Goldstone modes 161
graphene, topological insulator 287
gyromagnetic ratio 406

half-Heusler compounds 38
half-metallic ferromagnets 37–8
half-metals
  current-perpendicular-to-plane device 39, 45
  magnetic tunnel junctions 43
Hall detectors **257**
Hall effect see spin Hall effect; thermal Hall effect
Hamiltonian 6, 8
  Dirac 6, 282
  Dresselhaus 259
  Heisenberg's 25, 27, 384
  one-electron 195
  Rashba 182, 259, 424, 427
  in spin space 107
head-to-head domain wall **408**
heat conductance, tunnel junctions 141, **142**
heat engines and motors 143
Heisenberg, 1D antiferromagnetic Heisenberg model 167
Heisenberg equation of motion 26, 29
Heisenberg's Hamiltonian 25, 27, 384
helicity-dependent photocurrents 52–3
Hermitian conjugate valuables 351
Heusler compounds 37–8, 287
HgTe
  quantum spin Hall state 189
  quantum well (QW) 180, 190, 279

HgTe/CdTe, quantum well (QW) 184, 189
Hilbert space, electron-band manifolds 33
homodyne detection 355

incoherent spin current 15–24
    diffusion equation 18
    Fermi–Dirac distribution 15–17
    spin diffusion equation 19–23
InGaAs, quantum well (QW) 59
intrinsic pinning 406, 411
inverse spin Hall effect (iSHE) 177, 185,
    199, 252, 253, 299–300
    electronic measurement 230-5
    spin pumping 244–7

Johnson-Nyquist spin-current noise 301
Jordan–Wigner transformation 167
Julliere's formula 43

$k \cdot p$ perturbation theory 210
Kerr effect, magneto-optical (MOKE) 388
Kerr rotation detection **186**
Kerr rotation microscopy 216, 388
Klein–Gordon equation 5
Kondo effect 203
Kramer's pairs 274
    edge states at Fermi energy 278
    parity eigenvalues below Fermi energy 276

La:$Y_2Fe_5GaO_{12}$ (La:YIG) 297, 300
    transverse SSE and ISHE **301**
Lagrangian, spin Berry phase term 351
Lagrangian formalism 10–13
Landau–Lifshitz–Gilbert (LLG) equation 27,
    69, 90, 313, 386
    exchange spin current 25–6, 27
    Gilbert damping term 98
    gyromagnetic ratio 407
    in Hamiltonian form 349–51
    linearized 352
    nonadiabatic spin transfer (beta) term
        409
    Rayleigh's dissipation function 351
    Runge–Kutta method 163, 165
    spin-transfer torque 386
        adiabatic and non-adiabatic terms 330
Landauer–Buttiker formalism 113, 189
Landau–Lifshitz (LL) equation 90, 93
Landau–Lifshitz–Gilbert–Slonczewski
    (LLGS) equation 93, 101
Larmor theorem 98
LaSrMnO (LSMO) structure 43–4
lateral spin valves (LSVs) 227, 374
    see also spin valves, nonlocal spin injection
        device, nonlocal spin valve

lattice vibrations see phonons
Levi–Civita tensor 101, 351
local density of states (LDOS) 285
local spin density approximation (LSDA),
    density functional theory (DFT)
    117
Lorentz force in a magnetic field 177, **178**
Lorentz transmission electron
    microscopy 388
Luttinger spin–orbit coupling
    Hamiltonian 430

magnetic circular dichroism (MCD) 54
magnetic domain walls/vortices 382–95,
    402–5
    analytical approaches 386–8
    annihilation field 385
    current-driven excitation 413–19
    current-induced Oersted field 334
    dispersion relation in first Brillouin
        zone **394**
    dynamic switching 391–2, **393**
    examples **382**
    experimental techniques 388–9
    field-driven/static 78–9
    high-density magnetic storage 332, **333**
    low-energy modes
        gyrotropic mode of magnetic vortex
            390
        magnetostatic mode of magnetic
            vortex 391
    magnetostatically coupled vortices 393–4
    micromagnetic equations 383–5
    object-oriented micromagnetic framework
        (OOMMF) 383–7
    steady-state displacement 333
    steady-state motion phenomena 389–90
    vortex core switching by electric
        current 417–20
    see also domain wall
magnetic force microscopy (MFM) 58, 288,
    328, 388, 407–9, 418–19
    core velocity as a function of time **420**
magnetic insulators, spin (wave) Seebeck
    effect 136–7
magnetic junctions, torque and
    torquance 343–6
magnetic momentum, adiabatic
    electrons 66–7
magnetic order and exchange spin
    interaction 25
magnetic oxides 37
magnetic random access memory
    (MRAM) 311
magnetic soliton 382

magnetic tunnel junctions (MTJs) 39, 43–5, 69–72, 80, 322, 344
  bias-voltage dependence 347, **348**
  circuit model **346**
  CoFeB/MgO/CoFeB 348, **355**
  Fe/MgO/Fe single crystalline MTJs **363**
  spin caloritronics 140
magnetization dynamics, spin-current injection 322–4
magnetization switching
  nonlocal spin injection 372–81
  spin torque 319–22, 357–61
  spin current 372–81
magneto intersubband scattering (MIS) 212
magneto-electric (ME) effect 153
magneto-electronic circuit theory 91–2
magneto-heat resistance 141–2
magneto-optical Kerr effect (MOKE) 388
magneto-Peltier and Seebeck effects 138
magneto-Seebeck effect, lateral spin valves 139
magnetoelectric (ME) effect 170, 284
magnetoresistive devices
  half-metals, current-perpendicular-to-plane device 39, 45
  magnetic tunnel junctions 43
  Valet and Fert's model 36
magnetostatic shape anisotropy 344
magnets, spin polarization 36–46
  ferromagnets 36
  half-metallic ferromagnets 37–8
magnons *see* electromagnons
Majorana fermions 288–90
metal/oxide/semiconductor (MOS) junctions, spin-dependent tunneling 55
metallic structures, spin dependent thermoelectric phenomena 138
metric tensor 351
Mg, spin–orbit coupling parameter 200, **201**
micromagnetic equations 383–5
microscopic derivations, spin pumping and spin transfer 103–8
microscopy *see* magnetic force microscopy (MFM); spin-polarized scanning tunneling microscopy (spin STM); time-resolved Kerr microscopy (TRKM)
Mo, experimental spin Hall angles **237**
Moore's law, thermodynamic bottleneck 136
Mott-skew scattering 177, 182, **183**
multiferroics 149–70
  electromagnons 161–4
  generic consideration 152–3

quasi-one-dimensional quantum multiferroics 167
spin Hamiltonian for RMnO$_3$ 157–60
spin-current model of ferroelectricity 154–6
ultrafast switching of spin chirality by optical excitation 165–6

nanolithography 230, **320**, 382, 406
nanopillar structures **320**, **324**
  free ferromagnetic layer **350**
  high-speed switching **362**
  magnetization dynamics 323
  spin-torque nano-oscillator (STNO) 324–5
nanowire spin valves 140
Nb, experimental spin Hall angles **237**
Néel domain wall (DW) 74, 407
  in Py **119**, **121**
Nernst effect 139
  GaMnAs 140
Ettingshausen effect 139
Ni, damping 115
Ni$_{81}$Fe$_{19}$/Pt film **245**, 297
NiFe alloys 117
  binary alloys **118**
  permalloy (Ni$_{80}$Fe$_{20}$) 116, 117, 122
NiFe/AlGaAs barrier/GaAs structures 56
NiMnSb electrodes 38
Noether's theorem 150
non-adiabatic spin transfer (beta) term, LLG 409
non-uniform magnetic structures, spin-transfer torque (STT) 402–23
nonlocal spin Hall effect 201–3, 230-9
nonlocal spin injection device 227-30, 372
  schematic illustrations 228, 230, **373**
nonlocal spin-valve (NLSV) signal 227-30, 374
  conventional lateral spin valve 227, **374**
nonrelativistic approximation 7–8

object-oriented micromagnetic framework (OOMMF), magnetic domains/vortices 383–7
Oersted field 319, 349
  current-induced 334
  domain wall (DW) motion 408
1D antiferromagnetic Heisenberg model 167
Onsager reciprocal theorem 153
  spin pumping and spin transfer 99–102
Onsager relation, verification 235
Onsager–Kelvin identity of thermopower 143
Onsager–Kelvin relation 137

optical excitation, ultrafast switching of spin
       chirality 165–6
optically induced and detected spin
       current 49–62
   Hall bar **257**
   optical generation of spins 49
   optical spin detection 58–62
       ferromagnet/semiconductor Schottky
           diodes 60–1
       spin-polarized light-emitting diodes
           (spin LED) 58–9
   optical spin injection 55–8
       photoexcitation 55
       Schottky diodes 55–6
   photoexcitation model 50–4
   photon energy dependence 56
   spin injection into Si 62
   spin polarization in GaAs 49

p-n junctions 253, 255
paramagnetic metal (PM) film 297–8
Pauli exclusion principle 15
Pauli spin matrices 4, 8
Pd, experimental spin Hall angles **237**
permalloy ($Ni_{80}Fe_{20}$) 116, 378
   hysteresis loop **385**
   Py–Si tunneling junctions 139
perovskite system, spin torque 344
perpendicular spin-transfer torque 315, **316**
persistent spin helix (PSH) 253, 256,
       259–60, 262–4
   Rashba spin–orbit interactions 215
phonons
   phonon-drag effect, thermopower at lower
       temperatures 138
   spin-force Doppler shifts for phonons 77
photoelastic modulator (PEM) **53**
photoemission spectroscopy (PES) 43
photoexcitation 49, 55
   model 50–4
   schematic configuration **53**
photon energy dependence, optical
       magnetocurrent 56
point contacts **320**
point-contact Andreev reflection
       (PCAR) 39–40
Pt
   experimental spin Hall angles **237**
   spin–orbit coupling parameter 200, **201**
pump–probe technique *see* time-resolved
       Kerr microscopy (TRKM)
pure spin current 229, 378–9
   see also spin-current
Py *see* permalloy

quantization axis 20
quantum electrodynamics (QED), Dirac
       relativistic electrons 149
quantum spin Hall effect 179, **180**, 272–94
   2-D topological insulators 179, 189, **273**,
       278–80
   first signatures 189
   half-quantum Hall effect 283
   quantum spin Hall systems 272–7
quantum well (QW)
   AlGaAs 187
   band bending 254
   CdTe/HgTe/CdTe 184, 189, 280
   GaAs 59, 61, 182
   Heusler alloys 287
   HgTe 180, 279
   InGaAs 59
   persistent spin helix (PSH) condition 215
   semiconductor 424
   triangular 254–5
   two-dimensional electron gas (2DEG) 210,
       254–5
quasi-one-dimensional quantum
       multiferroics 167
quasi-particle tunneling from SC 41

Rashba 2DEG system 216
Rashba fields 183, **264**, 334–5, 424–9
   Co/Pt wire 334
   current-driven spin torque with Rashba
       coupling **427**
   induced spin torque **335**
Rashba Hamiltonian 182, 259, 424, 427
Rashba spin–orbit interactions
   Co/Pt wire 334
   coupling at metallic interfaces 425–6
   current-driven spin torque 427–8
   Datta–Das device concept 253, 259, 265,
       267
   electrical control **211**
   gate controlled Rashba SOI 212–13, 215
   manipulating magnetization 334, 429–30
   persistent spin helix (PSH) condition 215
   in semiconductors 212–13
   structural inversion asymmetry (SIA)
       209
reflection coefficient 106
$RMnO_3$
   exchange-striction effect 162
   spin Hamiltonian 157–60
   superexchange interactions **158**
   theoretical phase diagram **159**
Ruderman–Kittel–Kasuya–Yosida (RKKY)
       exchange coupling 316

scanning electron microscope, image of
  typical device for SHE
  measurements **236**
scanning tunnel spectroscopy (STS) 285
scattering theory 104, **104**
spin-flip scattering 106
scattering-matrix expressions 112
scattering-matrix formalism 107
Schottky diodes 55–6, 60–1
Schrödinger equation 5–6, 8, 108
Schrödinger–Poisson simulation, conduction
  band profile 258–9
Seebeck coefficient 138
semiconductors
  quantum wells 424
  Rashba spin–orbit interactions 212–13
  spin generation and manipulation based on
    SOI 209–24
  see also GaAs
Sharvin conductance 114
Sharvin resistances 106
Shubnikov–de Haas (SdH) oscillations 212,
  213
  fast Fourier transform (FFT) spectra **213**
sine-Gordon model, Bethe ansatz and form
  factor expansion 168
single-pole relaxation 10
single-step (shadow) or multiple-step
  electron-beam lithography
  processing 230
Skyrmion lattice structure 152, 169
Slater–Pauling rule 38
Slonczewski's spin-transfer torque 91, 93,
  312
  ballistic transport 313
  evaluation 345
  precession motion 345
  'torquance' 347
solids, semiclassical equation of motion of
  electrons 34
soliton, defined 382
soliton and anti-soliton 168
spectroscopy *see* angle-resolved
  photoemission spectroscopy
  (ARPES); spin-resolved
  photoemission spectroscopy
  (SP-PES); superconducting tunneling
  spectroscopy (STS)
spin angular momentum 3–4, 50, 89
spin battery and enhanced Gilbert
  damping 97
Spin caloritronics 136
  heat engines and motors 143
  magneto-heat resistance 141–2

spin Seebeck effect 144
spin-dependent thermoelectric phenomena
  in metallic structures 138–40
  magneto-Peltier and Seebeck effects 138
  thermal Hall effects 139
  thermal spin-transfer torques 140–2
spin channel, density of states (DOS) 36
spin chirality, optical excitation 165–6
spin current 3–14
  absorption, efficiency 375–7, **378**, 380
  carried by a spin wave 31
  exact definition 10
  generation and absorption 372–5
  magnetic moment of spin 4
  magnetization switching 372–81
  spin–wave spin current 28–31
  surface, topological insulators 35
spin-current injection 227
  diffusive picture 317–18
  domain wall displacement 327–30
  efficient injection 228, 377, **378**, 380
  magnetization dynamics 322–6
  vortex motion 332–4
spin diffusion equation 19–23
spin diffusion length 21, 201, 228
spin diffusion and spin precession in narrow
  2DEG 262-4
spin dynamics 244–50
spin Faraday's law 65–6
spin field effect transistor (FET) 62
spin filtering efficiency 54
spin-flip scattering 106
spin generation and manipulation based on
  spin–orbit interaction in
  semiconductors 209–24
  Aharonov–Casher spin interference 217–23
  gate controlled Rashba SOI 212–13
  origin of SOI 209–11
  spin Hall effect based on Rashba and
    Dresselhaus SOI 212–13
  spin relaxation and its suppression 214–16
spin Hall effect (SHE) 151, 177–90, 194–205,
  253, 261
  anomalous Hall effect 34, 152, 177, 185,
    194, 203
  based on Rashba and Dresselhaus
    SOI 178, 212–13
  experimental observation 185–9, 227–39,
    244–50, 254
    dynamics 244–50
    electronic nonlocal detection 227–39
    electronic spin Hall experiments 230–8
    induced modulation of magnetization
      dynamics 248–50

spin Hall effect (SHE) (*cont.*)
  family of spin Hall effects 184
  ferromagnets **139**
  historical background 177–80
  inverse spin Hall effect 24, 185, **199**, 229, 244–7, 299–300
  Kerr rotation detection **186**
  non-local spin Hall effect 201–3, 227-239
  non-local spin injection and detection 227–8
  schematics 230, 231, 235, **249**, **253**, **273**
  side jump and skew scattering 194
    diffusive metals 195–8
  spin and charge currents 199
  spin–orbit coupling 200
  spin–orbit interaction 181–3, 195
  see also spin-injection Hall effect
spin Hall effect (SHE) transistor 265–8, **266**
spin Hamiltonian, for $RMnO_3$ 157–60
spin-injection Hall effect (SiHE) 185, 252–70, **253**
  experiment 253–8
  nonequilibrium polarization dynamics along the $[1\bar{1}0]$ channel 259–60
  prospectives 269
  schematic **253**
  spin diffusion and spin precession in narrow 2DEG bars 262–4
  spin-dependent Hall effect 252
  theory discussion 258–64
spin injection into Si 62
spin injection magnetization switching (SIMS) experiments **343**, 357–61
spin-injection induced magnetization reversal 319–22, **321**, **343**, 357–61, **358**
spin low-energy diffraction (SPLEED) detector 43
spin-motive forces in ferromagnetic structures 65–84, 382–95, 402–5
  calculation of magnetic momentum for adiabatic electrons 66–7
  examples 74–9
  ferromagnetic resonance (FMR) 80–2
  field-driven/static magnetic vortices 78–9
  Landau–Lifshitz equations 69
  magnetic momentum, spin electric and magnetic fields 66
  magnons and phonons 82–3
  plain Néel domain wall 74
  spin electric field E_s 71
  spin Faraday's law 65, 68
  spin fields and Faraday's law 68
  spin-force Doppler shifts for phonons 77
  spin Seebeck effect 80

spin valves and MTJs 71–2
STT (spin transfer torque), spin valves and magnetic tunnel junctions (MTJ) 69–72
voltage steps and magnetoresistance 77
see also magnetic domain walls/vortices
spin-orbit interaction (SOI) in semiconductors
  gate controlled Rashba SOI 212–13
  origin of SOI 181, 209–11
  Rashba-type **183**
  spin Hall effect (SHE) 181–3
spin–orbit (SO) coupling 181, 200, **201**, 252
spin–orbit (SO) field, Rashba, Dresselhaus, combined **264**
spin-Peierls systems, donor–acceptor mixed stack charge transfer compounds 167
spin-polarization analysis (SEMPA) 388, 389
spin polarization in GaAs 49
spin polarization in magnets 36–46
  current-perpendicular-to-plane magnetoresistive device with half metals 39, 45
  experimental techniques for measurement 39–43
  ferromagnets 36
  half-metallic ferromagnets 37–8
  magnetic tunnel junctions with half-metals 43
  magnetoresistive devices with half-metals 43–5
  point-contact Andreev reflection (PCAR) 39–40
  spin-resolved photoemission spectroscopy (SP-PES) 43
  superconducting tunneling spectroscopy (STS) 41–2
spin-polarized electron tunneling from Ni STM tip into GaAs substrate 58
spin-polarized inverse photoemission **50**
spin-polarized light-emitting diodes (spin LED) **50**, 58–9
spin-polarized scanning tunneling microscopy (spin STM) **50**, 57
spin precession and decay **26**
spin pumping vs. spin transfer 87–125
  discrete vs. homogeneous 87
  first principles calculations 114–22
  microscopic derivations 103–13
    spin pumping 109–13
    spin transfer torque 89, 103–8
  normal metal–ferromagnet systems **94**
  phenomenology 89–102
    mechanics 89

Onsager reciprocity relations between
    STT and spin pumping 88, 99–102
  spin transfer torque (STT) 89–98
  technology pull and physics push 87
  theory vs. experiments 123
  see also spin-transfer torque
spin relaxation 10, 20, 21, 198
spin relaxation time 21, 198
spin-resolved photoemission spectroscopy
    (SP-PES) 43
spin-dependent Seebeck effect 303, **304**
  thermal spin injection 304
spin Seebeck effect (SSE) 144, 296–308
  detection using inverse spin Hall
      effect 299–300
  longitudinal and transverse
      configurations **298**
  and magneto-Peltier effects 138
  sample configuration and measurement
      mechanism 297–8
  schematic illustration **296**
  spin caloritronics 138, 144
  spin-current generation by the SSE at the
      F/PM interface **303**
  spin-motive forces 80
  theoretical concept 301–2
spin-transfer effect, *s-d* model **345**
spin-transfer oscillation (STO) 363, 367
spin-transfer torque (STT) 311–429
  adiabatic and non-adiabatic spin
      torques 330
  diffusive picture for injecting spin
      current 317–18
  spin diode effect **326**, 353–6, **353**
  domain wall displacement due to
      spin-current injection 327–8
  due to large Rashba fields 424–9
  experimental study of magnetization
      reversal 319–21
  field-like spin torque 348
  giant magnetoresistance (GMR) effect in
      magnetic multilayers 311
  introduction 311–18
  Landau–Lifshitz–Gilbert (LLG)
      equation 386
  large-amplitude dynamics and
      auto-oscillation 362–6
  layered normal metal–ferromagnet
      system **92**
  LLG equation 386
  LLG equation in Hamiltonian form
      349–51
  magnetization dynamics 322–6
  magnetization processes compared **360**
  magnetization switching 319–22, 357–61

  due to nonlocal spin injection 372–81
  non-uniform magnetic structures
      402–23
  perpendicular spin transfer torque
      315–16, 348
  phenomenology 89
  small-amplitude dynamics and
      anti-damping 352
  and spin pumping 89–98, 103–8
  spin accumulation torque 348
  theoretical description 312
  thermal 140–2
  uniform magnetization 343–70
  vector direction **317**
  vortex motion due to spin-current
      injection 332–4
spin valves
  half-metallic contacts **142**
  spin caloritronics 140
  see also lateral spin valves (LSVs);
      nonlocal spin valve
spin–torque nano-oscillator (STNO)
      324–5
spin–wave
  Doppler effect **336**
  of a one-atomic chain **29**
spin–wave spin current 28–31
  spin current carried by spin wave 31
  spin–wave formulation 28–30
spintronics 65, 209
steady state motion phenomena 389–90
Stern–Gerlach spin filter 216, **217**
Stirling's formula 16
Stoner–Wohlfarth model 232
structural inversion asymmetry (SIA) 209
superconducting tunneling spectroscopy
      (STS) 41–2
superconductor (SC) electrode **42**
superconductors
  Cooper pairs 289
  topological superconductors 284, 289
superexchange interactions, $RMnO_3$ **158**
surface spin current, topological
      insulators 35

Ta, experimental spin Hall angles **237**
tail-to-tail domain wall **409**
Taylor expansion 17
thermal Hall effect 139
  AuFePt structures 140
thermal spin-transfer torques 140–2
  magnetic tunnel junctions 140
  spin valves 140
  textures 140
Thiele's equation 387

Thomas term, Pauli equation 210
time-dependent vector components,
    magnetization reversal **314**
time-resolved Kerr microscopy
    (TRKM) 388, **389**
time-reversal invariant momenta
    (TRIM) 275, 281–3
  3D systems **276**
  2D systems **276**
Tomonaga–Luttinger behavior 167
topological insulators 75–6, 179, 189,
    272–94
  definition 75–6
  edge states **181**, 278–81
    comparison of transport properties **279**
    schematic diagram **278**
  first - $Bi_{1-x}Sb_x$ 190, 284
  graphene 287
  induced magnetization **283**
  materials, 190, 284, 287
  necessary conditions 274–5
  surface states **279**, 281
    chiral modes generated at the
      interface **289**
  three-dimensional 179, **273**, **279**, 281–90
    Majorana fermions 288–90
    materials for 284
  two-dimensional 179, **273**, 278–80
    relationship with 3-D 286
    see also quantum spin Hall effect
topological numbers 277
topological spin current 33–5
  Berry's phase 33
  bulk topological spin current 33–4
  surface topological spin current 35
topological superconductors 284, 289
topology, defined 275
torque
  non-adiabatic 408–10
  voltage dependence and field-like
    torque 347–9
  see also spin-transfer torque
torque correlation model (TCM) 115
torque and torquance, magnetic
    junctions 343–7
total vector spin chirality, time
    evolution **166**

transistor, spin Hall effect (SHE) 265–8,
    266
tunnel junctions, heat conductance 141,
    142
tunneling magnetoresistance (TMR) 37, 39,
    42–4
  temperature dependence **45**
tunneling spin polarization 36
  P_T 41
two-dimensional electron gas (2DEG) 210,
    252–66, 269
  2DEG channel 213
  quantum well (QW) 210
two-dimensional hole gas (2DHG) 252–4
two-dimensional mapping of switching
    probability **382**

Umklapp scattering 165

valence band (p-symmetry) 49
Valet–Fert model 36, 249
  giant-magnetoresistive devices 36
  'vector potential', in k-space 34
voltage dependence and field-like
    torque 347–9
vortex motion
  due to spin-current injection 332–4
  see also magnetic domain walls/vortices

Walker ansatz 113
Walker breakdown field 124
Wannier function WR(r) 425
'wavefunction matching' (WFM) scheme 117
Wiedemann–Franz Law 137–8
  breakdown 141

$Y_3Fe_4GaO_{12}$ (YIG)/Pt bilayer 247
  longitudinal SSE and ISHE **299**
$Y_3Fe_5GaO_{12}$ (YIG) 297

Zeeman coupling 283
Zeeman energy 29
Zeeman field 281–2
Zeeman interaction 8–9, 26
Zeeman splitting 41
  dilute magnetic semiconductor (DMS) 58
zinc-blende structure materials 37